Theory and Decision Library C

Game Theory, Social Choice, Decision Theory, and Optimization

Volume 46

More information about this series at http://www.springer.com/series/6618

Michel Grabisch

Set Functions, Games and Capacities in Decision Making

Springer

Michel Grabisch
Paris School of Economics
Université Paris I Panthéon-Sorbonne
Paris, France

ISSN 0924-6126 ISSN 2194-3044 (electronic)
Theory and Decision Library C
ISBN 978-3-319-30688-9 ISBN 978-3-319-30690-2 (eBook)
DOI 10.1007/978-3-319-30690-2

Library of Congress Control Number: 2016941938

Printed on acid-free paper

This Springer imprint is published by Springer Nature
The registered company is Springer International Publishing AG Switzerland

Foreword

The book by Michel Grabisch is about a fascinating mathematical object that has received different names and has been studied by different communities: set functions, capacities, pseudo-Boolean functions, and cooperative games, to mention just a few. Results on these objects were often proven several times, often under slightly different forms, in different communities. The book has two main parts.

The first one is devoted to a detailed presentation of this mathematical object and its main properties. In particular, Michel gives a detailed presentation of the notion of core and of the integrals based on nonadditive measures. In this first part, the learning curve is steep, but the reward is quite worth the effort. Many results scattered in the literature are arranged and proved here in a unified framework. I have no doubt that this first part will serve as a reference text for all persons working in the field.

The second part deals with applications of this mathematical object. Three of them are emphasized: decision-making under risk and uncertainty, multiple criteria decision-making, and belief and plausibility measures in the spirit of Dempster and Shafer. A reader interested in these areas of application can directly start reading the book with one of these chapters. The style of exposition is such that the reader is given many useful hints and precise references to the first part of the book. It will be most useful to anyone willing to use the tools and concepts in his/her own research.

The book is quite rich and very pleasant to read. The technical parts are well organized, and difficult points are always illustrated by figures and examples. The application parts are lucidly written and should be accessible to many readers.

This should not be surprising. Michel has been a major figure in the area since nearly 30 years. He has fully succeeded in transforming his deep knowledge of the field into a very rich and quite readable text.

CNRS and Université Paris Dauphine
Paris, France
March 2016

Denis Bouyssou

Preface

Set functions are mappings that assign to subsets of a universal set a real number and appear in many fields of mathematics (pure and applied) and computer sciences: combinatorics, measure and integration theory, combinatorial optimization, reliability, graph theory, cryptography, operations research in general, decision theory, game theory, etc. While additive nonnegative set functions (called measures) have been studied in depth and from a long time in measure theory, nonadditive set functions received less attention and only from, roughly speaking, the last 50 or 60 years. As the foregoing list shows, (nonadditive) set functions appeared in many different fields, under different names, and most often in an independent way. As far as possible, we have tried to give the historical origins of the concepts presented in this monograph. One of the most prominent seminal work is undoubtedly the one of Gustave Choquet, who proposed in 1953 the concept of *capacity* (monotone set function). Largely ignored during several decades, reinvented in 1974 by Michio Sugeno under the name *fuzzy measures*, capacities have become a central tool in all areas of decision-making, in particular, owing to the pioneering work of David Schmeidler in 1986. At the same time of Choquet's work on capacities, Lloyd Shapley studied another type of set functions, namely, *transferable utility games in characteristic form* (which we call here "game" for brevity), introduced by John von Neumann and Oskar Morgenstern, giving rise to what is known today as cooperative game theory. Submodular games happened to be of particular importance in combinatorial optimization through the work of Jack Edmonds, and many results concerning this class of games have been shown independently in both domains. Lastly, set functions, viewed as real-valued functions on the vertices of the unit hypercube, have been studied in the 1960s by Peter Hammer under the name *pseudo-Boolean functions* and constitute an important tool in operations research. This brief historical perspective, first, explains the title of this book, which is a compromise between the laconic "set functions" and the verbose "set functions, capacities, games, and pseudo-Boolean functions in decision-making, game theory, and operations research," and, second, gives an idea about the difficulty to have a clear view about what is known on set functions, from a mathematical point of view. As it is common in sciences and especially in our times where specialization reigns (a feature that will certainly worsen with the mania of evaluation and bibliometrics), scientific communities work independently and ignore that some of them share the same (mathematical) concerns. This book is an attempt to give a unified view of set functions and their avatars in the above-mentioned fields, mainly decision-making, cooperative game theory, and operations research, focusing on mathematical properties and presented in a way which is free of any particular applicative context. I mainly work with a finite universal set, first because most of the application fields concerned here consider finite sets (with the exception of decision under uncertainty and risk) and second because infinite sets require radically different mathematical tools, in the present case, close to those of measure theory (needless to say, a rigorous treatment of this would require a second volume, at least as thick as this one, a task which is probably beyond my capabilities!). The seven chapters are divided into three parts:

- Chapter 1 (introductive) establishes the notation and gathers the main mathematical ingredients which are necessary to understand the book.
- Chapters 2–4 (fundamental), which represent almost ⅔ of the book, form the mathematical core of the book. They give the mathematical properties of set functions, games, and capacities (Chap. 2), of the core of games, that is, the set of measures dominating a given game (Chap. 3), and of the various integrals defined w.r.t. games and capacities, mainly the Choquet and Sugeno integrals. At very few exceptions, all proofs are given.
- Chapters 5–7 (applicative) are devoted to applied domains: decision under risk and uncertainty (Chap. 5), decision with multiple criteria (Chap. 6), and Dempster-Shafer and possibility theory (Chap. 7). Clearly, each of these topics would have required a whole book, and at least for the two first ones, already many books are available on the topic. My philosophy was therefore different from the other chapters, and I tried to emphasize there the use of capacities. For these reasons, few proofs are given, but those given concern results which are either new or difficult to find in the literature. Chapter 7 is a bit in the spirit of the fundamental chapters, and therefore almost all proofs are provided. This is because, unlike the two chapters on decision-making, the topic is not so well known and still lacks comprehensive monographs.

The applicative chapters can be read independently from one another. It is also possible to read them without having studied in depth the fundamental chapters, because necessary concepts and results from these chapters are always clearly indicated and referenced.

The idea of writing this book germinated in my mind many years ago while teaching a course on capacities and Choquet integral applied to decision-making to second year master's students. I started the writing in 2012 and realized that it will take much time and go far beyond the initial project, when I saw that the first three pages of my handwritten lecture notes developed little by little into the hundred pages of Chap. 2. Anyway, the trip through the world of set functions was long, exhausting, but fascinating. Such a trip would have never existed if Prof. Michio Sugeno would not have permitted me to stay 1 year in his laboratory in 1989–1990, where I discovered his work on fuzzy measures and fuzzy integrals. I owe him to have introduced me to this beautiful world, which has become my main topic of research, and for this, I would like to express my most sincere gratitude to him. Many thanks are due also to his colleague of that time Toshiaki Murofushi, from whom I learned so much. My thoughts go also to the late Jean-Yves Jaffray and Ivan Kramosil, who were outstanding scientists in this domain and good friends.

I would like to thank many colleagues who have accepted to spend time in reading parts of this book. Needless to say, they greatly contributed to the quality of the book. In particular, many thanks are due to Alain Chateauneuf, Miguel Couceiro, Yves Crama, Denis Feyel, Peter Klement, Ehud Lehrer, Jean-Luc Marichal, Massimo Marinacci, Michel Maurin, Radko Mesiar, Pedro Miranda, Bernard Monjardet, Hans Peters, and Peter Sudhölter. Special thanks go to Ulrich Faigle for providing a proof of Theorem 3.24 and material on Walsh basis; Tomáš Kroupa for providing material on the Fourier transform and drawing my attention to the cone of supermodular games; Peter Wakker for invaluable comments on Chap. 5 (as well as on English!); Denis Bouyssou, Christophe Labreuche, Patrice Perny, and Marc Pirlot for in-depth discussion on Chap. 6; and finally to Thierry Denœux and Didier Dubois for fruitful discussion on Chap. 7 and drawing my attention to the possibilistic core, as well as to the ontic vs. epistemic view of sets.

This long task of writing would not have been possible without enough free time to do it and without the support of my institution. My sincere gratitude goes to Bernard Cornet, head of the research unit, and to Institut Universitaire de France, for having protected me against too many administrative and teaching tasks. Last but not least, countless thanks are due to my wife, Agnieszka Rusinowska, researcher in mathematical economics, for her unfailing support, understanding, and love, as well as for many comments on the last three chapters.

Despite all my efforts (and those of my colleagues), the book may contain typos, errors, gaps, and inaccuracies. Readers are encouraged to report them to me for future editions (if any), and all that remains for me now is to wish the readers a nice trip in the world of set functions.

Paris, France
January 2016

Michel Grabisch

Contents

Chapter 1
Introduction

We introduce in this chapter the notation used throughout the book (see also the list of symbols), as well as the necessary mathematical background to make the book self-contained. Section 1.2 gives useful results from combinatorics, some of them being quite standard while some others are specific, and most often proofs are provided. The rest of the chapter (Sect. 1.3) gives condensed summaries about binary relations, partially ordered sets, inequalities and polyhedra, linear programming, convexity, convex optimization, etc. Of course, no proof is provided there and readers who are unfamiliar with these topics and finding the exposition rather dry should consult the mentioned references.

1.1 Notation

(i) The set of positive integers is denoted by $\mathbb{N}$, while $\mathbb{N}_0$ denotes the set of nonnegative integers. As usual, $\mathbb{C}, \mathbb{Q}, \mathbb{R}, \mathbb{Z}$ denote the set of complex, rational, real numbers, and integers, respectively;

(ii) Closed intervals of numbers are denoted by $[x, y]$, while open intervals are denoted by $]x, y[$, and semi-open intervals by $]x, y]$ and $[x, y[$;

(iii) 2^X is the power set of X; i.e., $2^X = \{A \subseteq X\}$;

(iv) For two real-valued functions f, g on some space X, we write $f \geqslant g$ if $f(x) \geqslant g(x)$ for all $x \in X$. Similarly, $f \geqslant 0$ means that $f(x) \geqslant 0$ for all $x \in X$;

(v) The domain and range of a function f are denoted by $\mathbf{dom} f$ and $\mathbf{ran} f$. The *support* of a function f valued on $\mathbb{R}_+$ is the subset of its domain where the function is positive. It is denoted by $\mathbf{supp}(f)$;

(vi) The identity function is denoted by Id;

M. Grabisch, *Set Functions, Games and Capacities in Decision Making*, Theory and Decision Library C 46, DOI 10.1007/978-3-319-30690-2_1

(vii) The *signum function* is defined from $\mathbb{R}$ to $\{-1, 0, 1\}$ as

$$\operatorname{sign}(x) = \begin{cases} 1 & \text{if } x > 0 \\ 0 & \text{if } x = 0 \\ -1 & \text{if } x < 0; \end{cases}$$

(viii) We do not use any special notation for vectors and matrices, except for the zero vector, denoted by $\mathbf{0}$, and the vector whose all coordinates are equal to 1, denoted by $\mathbf{1}$. Whether x denotes a single variable or a vector should be clear from context. If x is a vector, it is meant that it is a column vector, and its components are denoted by $x_1, x_2, \ldots$. Transposition is indicated by $\cdot^\top$. Hence $x^\top = (x_1, \ldots, x_n)$ denotes a row vector. The inner product of two vectors x, y is denoted by $\langle x, y\rangle$ and is taken as $x^\top y$ unless otherwise indicated. For n-dimensional real vectors, we take $\langle x, y\rangle = \sum_{i=1}^n x_i y_i$;

(ix) As a consequence of (iv), for two vectors $x, y \in \mathbb{R}^n$, $x \leqslant y$ means that $x_i \leqslant y_i$ for $i = 1, \ldots, n$. We write $x < y$ if $x \leqslant y$ and $x_i < y_i$ for some $i \in \{1, \ldots, n\}$, and we write $x \ll y$ if $x_i < y_i$ for $i = 1, \ldots, n$;

(x) For a vector $x \in \mathbb{R}^X$ and $Y \subset X$, the vector $x_{|Y}$ is the restriction of x to Y. The same notation is used for functions: the restriction of $f : X \to \mathbb{R}$ to $Y \subset X$ is denoted by $f_{|Y}$, with $f_{|Y} : Y \to \mathbb{R}$, $y \in Y \mapsto f_{|Y}(y) = f(y)$;

(xi) For a vector $x \in \mathbb{R}^X$ and $Y \subset X$, we may write for simplicity $x = (x_Y, x_{-Y})$ instead of $(x_{|Y}, x_{|X\setminus Y})$. Similarly, for two vectors $x, z \in \mathbb{R}^X$, we write (x_Y, z_{-Y}) instead of $(x_{|Y}, z_{|X\setminus Y})$;

(xii) The set $\{1, \ldots, n\}$ is sometimes denoted by $[n]$;

(xiii) Finite families of elements $x_1, x_2, \ldots, x_n$ are denoted by $\{x_i\}_{i=1,\ldots,n}$, while countable families $x_1, \ldots, x_2, \ldots$ are denoted by $\{x_n\}$;

(xiv) For any finite set N, the set of permutations on N is denoted by $\mathfrak{S}(N)$, or simply $\mathfrak{S}(n)$ if $|N| = n$;

(xv) The characteristic function of a set A is denoted by 1_A;

(xvi) As far as possible, sets are denoted by capital letters, like $A, B, S, T, \ldots$, elements of sets by small letters, like $i, j, x, y, \ldots$, and collections of sets are denoted by calligraphic capital letters, like $\mathcal{A}, \mathcal{B}, \mathcal{F}, \ldots$. We sometimes omit braces for singletons, writing $A \cup i$ instead of $A \cup \{i\}$, and so on;

(xvii) The complement of a set A is denoted by A^c whenever the universal set is understood. Set inclusion is denoted by $\subseteq$, while $\subset$ denotes proper inclusion. Whenever convenient, cardinalities of sets A, B, C are denoted by corresponding small letters a, b, c;

(xviii) The set interval $[A, B]$ with $A \subseteq B \subseteq X$ is defined by $[A, B] = \{C \subseteq X : A \subseteq C \subseteq B\}$. We use as well the variants $[A, B[$, $]A, B]$, and $]A, B[$ as for real intervals;

(xix) A, B being subsets of X, $A \setminus B = \{x \in A : x \notin B\}$ is the set difference. The symmetric difference of sets is denoted by $A \Delta B = (A \setminus B) \cup (B \setminus A) = (A \cup B) \setminus (A \cap B)$;

(xx) $\vee, \wedge$ are lattice supremum and infimum. When applied to real numbers, the usual ordering on real numbers is meant, hence they reduce to maximum and minimum respectively when there is a finite number of arguments;

(xxi) Useful conventions: $\sum_{i\in\varnothing} x_i = 0$, $\prod_{i\in\varnothing} x_i = 1$, where the x_i's are real numbers. Considering quantities $x_1, x_2, \ldots$ defined on an interval $I \subseteq \mathbb{R}$, we set $\wedge_{i\in\varnothing} x_i = \bigvee I$, $\vee_{i\in\varnothing} x_i = \bigwedge I$, where $\bigvee I, \bigwedge I$ are respectively the supremum and infimum of I. Also, $0! = 1$.

1.2 General Technical Results

We begin by some useful combinatorial formulas.

Lemma 1.1 *Let X be any finite nonempty set.*

(i) For every set interval $[A, B]$, $A, B \subseteq X$

$$\sum_{C\in[A,B]} (-1)^{|C\setminus A|} = \sum_{C\in[A,B]} (-1)^{|B\setminus C|} = \begin{cases} 0, & \text{if } A \subset B \\ 1, & \text{if } A = B. \end{cases} \quad (1.1)$$

(ii) For every positive integer n

$$\sum_{\ell=0}^{k} (-1)^{\ell} \binom{n}{\ell} = (-1)^k \binom{n-1}{k} \quad (k < n). \quad (1.2)$$

For $k = n$, $\sum_{\ell=0}^{n} (-1)^{\ell}\binom{n}{\ell} = (1-1)^n = 0$.

(iii) For every set interval $[A, B]$ in X, any integer k such that $|A| \leqslant k < |B|$:

$$\sum_{\substack{C\in[A,B]\\ |C|\leqslant k}} (-1)^{|C\setminus A|} = (-1)^{k-|A|} \binom{|B\setminus A| - 1}{k - |A|}.$$

(iv) For all integers $n, k \geqslant 0$

$$\sum_{j=0}^{n} \binom{n}{j} (-1)^j \frac{1}{k+j+1} = \frac{n!k!}{(n+k+1)!}.$$

(v) For all integers $n \geqslant 0, k > n$

$$\sum_{j=0}^{n} \binom{n}{j} (-1)^j \frac{1}{k-j} = (-1)^n \frac{n!(k-n-1)!}{k!}.$$

(vi) For every set interval $[A, B]$, $A, B \subseteq X$

$$\sum_{C\in[A,B]} (-2)^{|C\setminus A|} = \sum_{C\in[A,B]} (-2)^{|B\setminus C|} = (-1)^{|B\setminus A|}.$$

Proof

(i) The result is clear if $A = B$. Suppose $A \neq B$ and set $|B \setminus A| =: n$.

$$\begin{aligned}\sum_{C\in[A,B]} (-1)^{|C\setminus A|} &= \sum_{C\in[\varnothing,B\setminus A]} (-1)^{|C|} \\ &= \sum_{\ell=0}^{n} \sum_{\substack{C\in[\varnothing,B\setminus A]\\ |C|=\ell}} (-1)^{\ell} \\ &= \sum_{\ell=0}^{n} \binom{n}{\ell}(-1)^{\ell} \\ &= (1-1)^n = 0.\end{aligned}$$

The second formula holds by $\binom{n}{\ell} = \binom{n}{n-\ell}$.

(ii) Suppose that k is odd. The formula is obviously true for $k = 1$. Assuming the formula is valid till some $k < n-1$, let us prove it for $k+1$. We get:

$$\begin{aligned}\sum_{\ell=0}^{k+1}(-1)^{\ell}\binom{n}{\ell} &= -\binom{n-1}{k} + \binom{n}{k+1} \\ &= \frac{(n-1)!}{k!(n-k-1)!}\Big(-1 + \frac{n}{k+1}\Big) \\ &= \binom{n-1}{k+1}.\end{aligned}$$

For k even, the proof is similar.

(iii) Apply (ii).

(iv) (Denneberg and Grabisch [82]) Consider the shift and identity operators S, E acting on any function f defined on $\mathbb{N}_0$: $Sf(k) = f(k+1)$ and $Ef(k) = f(k)$, and the special function $f(k) = \frac{1}{k+1}$. Since S and E commute, the binomial formula applies:

$$((E-S)^n f)(k) = \sum_{j=0}^{n}\binom{n}{j}(-1)^j(S^j f)(k) = \sum_{j=0}^{n}\binom{n}{j}(-1)^j\frac{1}{k+j+1}.$$

It remains to prove that

$$((E-S)^n f)(k) = \frac{n!k!}{(n+k+1)!}. \tag{1.3}$$

We do it by induction on n. The case $n = 0$ is obvious. Suppose (1.3) holds. Applying $E - S$ to (1.3) gives Eq. (1.3) for $n + 1$:

$$\begin{aligned}((E-S)^{n+1} f)(k) &= \frac{n!k!}{(n+k+1)!} - \frac{n!(k+1)!}{(n+(k+1)+1)!}\\ &= \frac{(n+1)!k!}{((n+1)+k+1)!}.\end{aligned}$$

(v) The formula can be proved similarly to (iv), by considering the negative shift operator N defined by $Nf(k) = f(k-1)$, applied at most $k-1$ times on the function $f(k) = \frac{1}{k-1}$, $k > 1$.

(vi) Proceeding as for (i), this amounts to showing that

$$\sum_{\ell=0}^{n} \binom{n}{\ell} (-2)^\ell = (-1)^n.$$

But this is immediate because $\sum_{\ell=0}^{n} \binom{n}{\ell}(-2)^\ell = (1-2)^n$.

□

Lemma 1.2 *Let $n \in \mathbb{N}$ be fixed. For any $A \subseteq [n]$, the following holds*

$$\int_{[0,1]^n} \bigwedge_{i\in A} x_i \, \mathrm{d}x = \frac{1}{|A|+1}.$$

Proof Observe first that we can assume $A = [n]$. Next, defining the simplices

$$\mathcal{B}_\sigma = \{x \in [0,1]^n : x_{\sigma(1)} \leqslant \cdots \leqslant x_{\sigma(n)}\}$$

for all permutations σ on $[n]$, we obtain

$$\begin{aligned}\int_{[0,1]^n} \bigwedge_{i\in[n]} x_i \, \mathrm{d}x &= \sum_{\sigma\in\mathfrak{S}(n)} \int_{\mathcal{B}_\sigma} x_{\sigma(1)} \, \mathrm{d}x\\ &= \sum_{\sigma\in\mathfrak{S}(n)} \int_0^1 \int_0^{x_{\sigma(n)}} \cdots \int_0^{x_{\sigma(2)}} x_{\sigma(1)} \, \mathrm{d}x_{\sigma(1)} \cdots \mathrm{d}x_{\sigma(n)}\\ &= \sum_{\sigma\in\mathfrak{S}(n)} \frac{1}{(n+1)!} = \frac{1}{n+1}.\end{aligned}$$

□

The *Bernoulli numbers*[1] are defined recursively by

$$B_n = -\frac{1}{n+1}\sum_{k=0}^{n-1}\binom{n+1}{k}B_k \qquad (n \in \mathbb{N}), \tag{1.4}$$

starting with $B_0 = 1$. The first elements of the sequence are

$$B_0 = 1, B_1 = -\frac{1}{2}, B_2 = \frac{1}{6}, B_3 = 0, B_4 = -\frac{1}{30}, B_5 = 0, B_6 = \frac{1}{42}, \ldots,$$

and $B_{2m+1} = 0$ for $m \geqslant 1$. The *Bernoulli polynomials* are defined by

$$B_n(x) = \sum_{k=0}^{n}\binom{n}{k}B_k x^{n-k} \qquad (n \in \mathbb{N}_0, x \in \mathbb{R}).$$

These polynomials satisfy the following properties (see, e.g., Abramowitz and Stegun [1]):

$$B_n(0) = B_n \qquad (n \in \mathbb{N}_0) \tag{1.5}$$

$$B_n(1) = (-1)^n B_n \qquad (n \in \mathbb{N}_0) \tag{1.6}$$

$$B_n\Big(\frac{1}{2}\Big) = \Big(\frac{1}{2^{n-1}} - 1\Big)B_n \qquad (n \in \mathbb{N}_0) \tag{1.7}$$

$$B_n(x+y) = \sum_{k=0}^{n}\binom{n}{k}B_k(x)y^{n-k} \qquad (n \in \mathbb{N}_0, x, y \in \mathbb{R}) \tag{1.8}$$

$$\int_a^x B_n(t)\,\mathrm{d}t = \frac{B_{n+1}(x) - B_{n+1}(a)}{n+1} \qquad (n \in \mathbb{N}) \tag{1.9}$$

$$B_n(x+1) - B_n(x) = nx^{n-1} \qquad (n \in \mathbb{N}_0). \tag{1.10}$$

Lemma 1.3 *(Grabisch et al. [178]) For all $S, K \subseteq [n]$ such that $S \subseteq K$, we have*

$$\sum_{T\in[S,K]}\frac{1}{k-t+1}B_{t-s}(x) = x^{k-s} \qquad (x \in [0,1]).$$

[1]These numbers were discovered by the Swiss mathematician Jakob Bernoulli (1654–1705), born in Basel. The Bernoulli family has produced many famous mathematicians, artists and scientists, whose Jakob is the oldest representant. The Bernoulli numbers appear in the formulas for the sum of powers of the first positive integers, and they were posthumously published in 1713 in his *Ars Conjectandi*. They were also discovered independently by the Japanese mathematician Seki Kōwa (1642–1708), and published, also posthumously, in 1712.

Proof We have

$$
\begin{aligned}
\sum_{T\in[S,K]} \frac{1}{k-t+1} B_{t-s}(x) &= \sum_{t=s}^{k} \binom{k-s}{t-s} \frac{1}{k-t+1} B_{t-s}(x) \\
&= \sum_{\ell=0}^{k-s} \binom{k-s}{\ell} \frac{1}{k-s-\ell+1} B_\ell(x) \\
&= \int_0^1 \sum_{\ell=0}^{k-s} \binom{k-s}{\ell} B_\ell(x) y^{k-s-\ell}\,\mathrm{d}y \\
&= \int_0^1 B_{k-s}(x+y)\,\mathrm{d}y \qquad \text{(by 1.8)} \\
&= \frac{B_{k-s+1}(x+1) - B_{k-s+1}(x)}{k-s+1} \qquad \text{(by 1.9)} \\
&= x^{k-s} \qquad \text{(by 1.10).}
\end{aligned}
$$

□

1.3 Mathematical Prerequisites

1.3.1 Binary Relations and Orders

Let A be a nonempty set. A *binary relation* on A is a subset R of $A \times A$. The fact that $(a, b) \in R$ for some $a, b \in A$ is also denoted by aRb. A binary relation R is *reflexive* if aRa for all $a \in A$. It is *complete* if for any $a, b \in A$, either aRb or bRa holds, or both. Note that completeness implies reflexivity. A binary relation R is *antisymmetric* if aRb and bRa imply $a = b$, and it is *transitive* if for all $a, b, c \in A$, aRb and bRc imply aRc.

A *preorder* is a reflexive and transitive binary relation, often denoted by $\preceq$, or $\preccurlyeq$, or $\leqslant$. A common usage is to write $a \not\preceq b$ if $a \preceq b$ is not true (and similarly $a \not\preccurlyeq b$ or $a \not\leqslant b, \ldots$). If a and b in A are such that $a \not\preceq b$ and $b \not\preceq a$, then a and b are said to be *incomparable*. A *partial order* is an antisymmetric preorder. If the partial order is in addition complete, then it is called a *linear order* or *total order*. Taking $A = \mathbb{N}$, we have for example the following binary relations:

(i) $a \preceq b$ if $a \equiv r \mod (5)$, $b \equiv s \mod (5)$ and $r \leqslant s$, is a preorder, but not a partial order;
(ii) $a \leqslant b$ where $\leqslant$ is the usual comparison between numbers, is a linear order;
(iii) aRb if a divides b is a partial order, but not a linear order.

A binary relation R is *symmetric* if for any $a, b \in A$, aRb implies bRa. It is *asymmetric* if for all $a, b \in A$, aRb implies that bRa does not hold.

Taking any preorder $\preccurlyeq$ on A, we write $a \prec b$ if $a \preccurlyeq b$ and $b \not\preccurlyeq a$. The binary relation $\prec$ is called the *asymmetric part* of $\preccurlyeq$. Similarly, we write $a \sim b$ if $a \preccurlyeq b$ and $b \preccurlyeq a$ both hold, and $\sim$ is the *symmetric part* of $\preccurlyeq$. The symmetric part is reflexive, symmetric and transitive; such a binary relation is called an *equivalence relation*. Taking any $a \in A$, the *equivalence class* of a, denoted by $[a]$, is the set $\{b \in A : b \sim a\}$. Then A can be partitioned into equivalence classes, and the set of all equivalence classes, called the *quotient set*, is denoted by $A/_{\sim}$.

1.3.2 Partially Ordered Sets and Lattices

(See, e.g., Birkhoff [30], Caspard et al. [44], Davey and Priestley [67], Grätzer [185].) A *partially ordered set*, or *poset* for short, is a pair $(P, \preceq)$ where P is a nonempty set and $\preceq$ a partial order on P. The *dual order* of $\preceq$ is denoted by $\succeq$ or $\preceq^{\partial}$, and is defined by $x \succeq y$ if and only if $y \preceq x$. The notation P^{∂} means the *dual poset*; i.e., the set P endowed with the dual order $\preceq^{\partial}$. A function f from a poset $(P, \preceq)$ to another poset $(P', \preceq')$ is said to be *isotone* if for every $x, y \in P$, $x \preceq y$ implies $f(x) \preceq' f(y)$. It is *antitone* if the reverse inequality holds. An *isomorphism* between two posets $(P, \preceq)$ and $(P', \preceq')$ is a bijection (one-to-one and onto) $f : P \to P'$ such that $x \preceq y \Leftrightarrow f(x) \preceq' f(y)$ for every $x, y \in P$. If an isomorphism exists between two posets, they are said to be *isomorphic*. When P is isomorphic to P^{∂}, P is said to be *autodual*.

Minimal Elements, Lower Bounds and the Like $x \in P$ is a *minimal element* (respectively, a *maximal element*) if there is no $y \in P$ such that $y \prec x$ (respectively, $x \prec y$). An element $x \in P$ is the (necessarily unique) *greatest element* (respectively, the *least element*) of P if for every $y \in P$, we have $y \preceq x$ (respectively, $x \preceq y$). An *upper bound* (respectively, a *lower bound*) of a subset $Q \subseteq P$ is an element $x \in P$ satisfying $y \preceq x$ (respectively, $x \preceq y$) for all $y \in Q$. The *supremum* (respectively, the *infimum*) of $Q \subseteq P$, if it exists, is the least upper bound (i.e., the least element of the set of all upper bounds) of Q (respectively, the greatest lower bound of Q). There are denoted respectively by $\bigvee Q$, $\bigwedge Q$. The supremum (a.k.a. *join*) and infimum (a.k.a. *meet*) of $x, y \in P$ are denoted by $x \vee y, x \wedge y$ respectively, and similarly for any finite number of elements. When $\preceq$ is set inclusion, and supremum and infimum happen to be union and intersection, the notation $\bigcup Q, \bigcap Q$ can be used as well.

Lattices A poset $(L, \preceq)$ such that the supremum and the infimum of any two elements in L exist is called a *lattice*. If only the supremum (respectively, the infimum) exists, it is called an *upper semilattice* (respectively, a *lower semilattice*). The lattice is *complete* if $\bigvee Q, \bigwedge Q$ exist for every $Q \subseteq L$. Any finite lattice is complete. If L is complete, then it has a greatest and a least element, respectively called the *top* and *bottom* of the lattice. We give two remarkable examples of lattices, useful in the sequel:

(i) Let X be any finite nonempty set. Then $(2^X, \subseteq)$, the set of all subsets of X endowed with the inclusion relation, is a lattice (more precisely, a *Boolean lattice*[2]). Its top and bottom elements are respectively X and $\varnothing$, and $\vee, \wedge$ are simply $\cup, \cap$.

(ii) Consider again a finite nonempty set $X = \{x_1, \dots, x_n\}$. A *partition* of X is a family of nonempty subsets $A_1, \dots, A_q$ of X, called the *blocks* of the partition, satisfying $A_i \cap A_j = \varnothing$ for all $i, j \in \{1, \dots, q\}$, $i \neq j$, and $\bigcup_{i=1}^{q} A_i = X$. The set of all partitions of X is denoted by $\Pi(X)$, with generic elements $\pi, \pi', \dots$. We endow $\Pi(X)$ with the *coarsening order relation* $\leqslant$ defined as follows. For any two partitions $\pi = \{A_1, \dots, A_m\}$, $\pi' = \{A'_1, \dots, A'_{m'}\}$, π is *coarser* than π' (written $\pi' \leqslant \pi$) if for each $j' \in \{1, \dots, m'\}$, there exists $j \in \{1, \dots, m\}$ such that $A'_{j'} \subseteq A_j$. The top element is $\{X\}$ while the bottom element is $\{\{x_1\}, \dots, \{x_n\}\}$. The supremum and infimum of $\pi = \{A_1, \dots, A_m\}$ and $\pi' = \{A'_1, \dots, A'_{m'}\}$ are given by $\pi \wedge \pi' = \{A_j \cap A'_{j'} \mid A_j \cap A'_{j'} \neq \varnothing\}$ and

$$x(\pi \vee \pi')y \Leftrightarrow$$
$$\exists x = u_0, u_1, \dots, u_t = y \text{ such that } u_i(\pi)u_{i+1} \text{ or } u_i(\pi')u_{i+1} \text{ for all } i,$$

for any $x, y \in X$, where $x(\pi)y$ means that x, y belongs to the same block in π (see, e.g., Aigner [2]).

A subset $Q \subseteq L$ of a lattice $(L, \preceq)$ is a *sublattice* of it if for all $x, y \in Q$, $x \vee y$ and $x \wedge y$ belong to Q.

Covering Relation, Chains Let $(P, \preceq)$ be a poset. For $x, y \in P$, we say that x *covers* y, and we denote it by $y \prec\cdot\, x$, if $y \prec x$ and there is no $z \in P$ satisfying $y \prec z \prec x$. Alternatively, we may denote this by $x \succ\cdot\, y$. When the partial order is set inclusion, we use $\subset\cdot$ and $\supset\cdot$.

An *antichain* K in P is a subset of P such that any two distinct elements in K are incomparable. On the other hand, a *chain* C in P is a subset of P where all elements are pairwise comparable; i.e., for any two elements $x, y \in C$, either $x \preceq y$ or $y \preceq x$. The chain is *maximal* if no other chain contains it, or equivalently, if $C = \{x_1, \dots, x_q\}$, with $x_1 \prec\cdot\, x_2 \prec\cdot\, \cdots \prec\cdot\, x_q$, and x_1 is a minimal element of P, while x_q is a maximal element. The *length* of C is $q - 1$. The *height of an element* $x \in P$, denoted by $h(x)$, is the length of a longest chain from a minimal element of P to x. The *height of a poset* $(P, \preceq)$, denoted by $h(P, \preceq)$, is the maximum of $h(x)$ taken over all elements $x \in P$. Equivalently, it is the length of a longest (maximal) chain in P.

---

[2] A lattice L is Boolean if it is distributive (see below) and *complemented*, i.e., every element $x \in L$ has a complement $x' \in L$, which is an element satisfying $x \wedge x' = \bot$ and $x \vee x' = \top$, where $\bot, \top$ are the bottom and top elements of L. In the finite case, every Boolean lattice is isomorphic to 2^X for some X.

A poset $(P, \preceq)$ satisfies the *Jordan-Dedekind chain condition* if for any $x, y \in P$ s.t. $x \preceq y$, any maximal chain from x to y has the same length (if $(P, \preceq)$ has a least element, an equivalent definition is: $x \prec\cdot\, y$ implies $h(x) + 1 = h(y)$, $\forall x, y \in P$).

The *Hasse diagram* of a poset $(P, \preceq)$ is a graphic representation of it, with nodes figuring elements of P, and links figuring the covering relation, i.e., a link relates x and y if $x \prec\cdot\, y$, with x placed below y. The Hasse diagrams of the lattices $(2^X, \subseteq)$ and $(\Pi(X), \leqslant)$ with $X = \{1, 2, 3, 4\}$ are given on Fig. 1.1. Sets are given without commas and brackets, i.e., 123 stands for $\{1, 2, 3\}$ and 12,34 for $\{\{1, 2\}, \{3, 4\}\}$, and this convention will be used throughout the book.

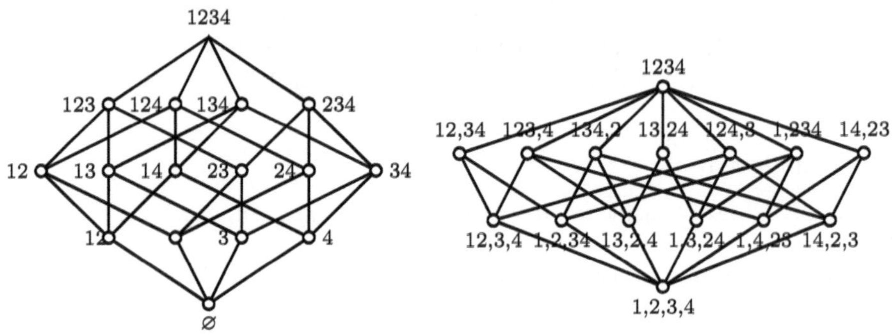

Fig. 1.1 Hasse diagram of the lattices $(2^X, \subseteq)$ (*left*) and $(\Pi(X), \leqslant)$ (*right*) with $X = \{1, 2, 3, 4\}$

Downsets and Ideals A *downset* of a poset $(P, \preceq)$ is a subset Q of P such that $x \in Q$ and $y \preceq x$ imply $y \in Q$. We denote by $\mathcal{O}(P, \preceq)$ or simply $\mathcal{O}(P)$ the set of downsets of $(P, \preceq)$. Finite unions and intersections of downsets are downsets. If a downset is nonempty and closed under $\vee$, it is called an *ideal*. For any $x \in P$, the downset $\downarrow x$ defined by $\downarrow x = \{y \in P : y \preceq x\}$ is an ideal called the *principal ideal* of x. In a finite poset, all ideals are principal ideals. For any $x, y \in P$,

$$\downarrow x \cup \downarrow y \subseteq \downarrow(x \vee y)\,, \quad \downarrow x \cap \downarrow y = \downarrow(x \wedge y)\,.$$

Also, $x \preceq y$ implies $\downarrow x \subseteq \downarrow y$.

Dual definitions of *upsets, filters, principal filters* denoted by $\uparrow x$, and results are obtained when $\preceq$ is replaced by $\succeq$, the dual order.

For any $Q \subseteq P$,

$$\downarrow Q = \{y \in P : y \preceq x \text{ for some } x \in Q\} = \bigcup_{x \in Q} \downarrow x$$

is the downset generated by Q (dually for $\uparrow Q$).

Join- and Meet-Irreducible Elements In a lattice $(L, \preceq)$, an element x is said to be *join-irreducible* (respectively, *meet-irreducible*) if it cannot be written as the supremum (respectively, the infimum) of other elements, and it is not the bottom (respectively, the top) element. When L is finite, this is equivalent to say that a join-irreducible (respectively, meet-irreducible) element covers (respectively, is covered by) only one element. The set of join-irreducible elements is denoted by $\mathcal{J}(L)$, while $\mathcal{M}(L)$ denotes the set of meet-irreducible elements. An *atom* is a join-irreducible element covering the bottom element. Dually, a *coatom* is a meet-irreducible element covered by the top element.

We introduce the mapping $\eta : L \to \mathcal{O}(\mathcal{J}(L))$, defined by $\eta(x) = \{j \in \mathcal{J}(L) : j \preceq x\}$. This mapping is one-to-one, satisfies $x = \bigvee \eta(x)$ for all $x \in X$, and

$$\eta(x \vee y) \supseteq \eta(x) \cup \eta(y), \quad \eta(x \wedge y) = \eta(x) \cap \eta(y) \qquad (x, y \in L). \tag{1.11}$$

Similarly, the mapping $\lambda : L \to \mathcal{O}(\mathcal{M}(L))$ defined by $\lambda(x) = \{m \in \mathcal{M}(L) : m \succeq x\}$ possessing dual properties can be introduced as well.

Modular and Distributive Lattices A lattice $(L, \preceq)$ is *lower semimodular* if for all $x, y \in L$, $x \prec\!\!\cdot\; x \vee y$ and $y \prec\!\!\cdot\; x \vee y$ imply $x \wedge y \prec\!\!\cdot\; x$ and $x \wedge y \prec\!\!\cdot\; y$. It is *upper semimodular* if $x \wedge y \prec\!\!\cdot\; x$ and $x \wedge y \prec\!\!\cdot\; y$ imply $x \prec\!\!\cdot\; x \vee y$ and $y \prec\!\!\cdot\; x \vee y$, and it is *modular* if it is both lower and upper semimodular. A lattice is modular if and only if it does not contain the lattice N_5 as a sublattice (Fig. 1.2). A lattice of finite height is upper semimodular if and only if it satisfies the Jordan-Dedekind chain condition and its height function is submodular; i.e., $h(x \vee y) + h(x \wedge y) \leqslant h(x) + h(y)$ for every $x, y \in L$.

A lattice $(L, \preceq)$ is *distributive* if $\vee, \wedge$ obey distributivity; i.e., $x \vee (y \wedge z) = (x \vee y) \wedge (x \vee z)$ for every $x, y, z \in L$ (equivalently, $\vee, \wedge$ can be inverted). If a lattice is distributive, then it is modular. A modular lattice is distributive if and only if it does not contain M_3 as a sublattice (see Fig. 1.2). As a consequence, a lattice is distributive if and only if it neither contains N_5 nor M_3.

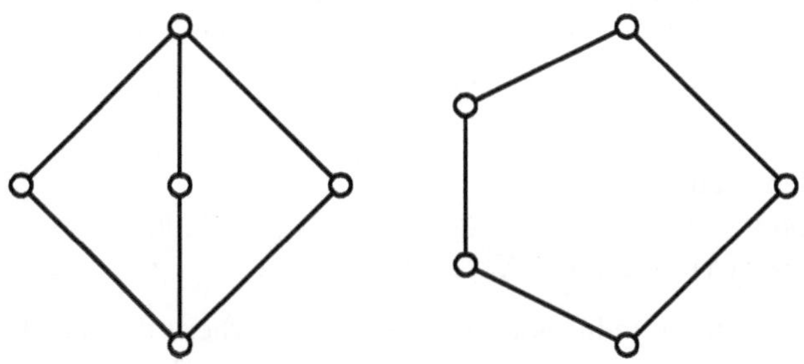

Fig. 1.2 The lattices M_3 (*left*) and N_5 (*right*)

In a distributive lattice $(L, \preceq)$, $|\mathcal{J}(L)| = |\mathcal{M}(L)| = h(L)$. In addition, when L is finite, the set of join-irreducible elements alone permits to reconstruct the whole lattice, and for this reason can be considered to be a kind of basis. Formally, this is expressed in the following theorem.

Theorem 1.4 (Representation of a distributive lattice by its join-irreducible elements) *(Birkhoff [29]) Let L be a finite distributive lattice. Then the mapping* $\eta : L \to \mathcal{O}(\mathcal{J}(L)); x \mapsto \{i \in \mathcal{J}(L) \,:\, i \preceq x\}$ *is an isomorphism from* $(L, \preceq)$ *to* $(\mathcal{O}(\mathcal{J}(L)), \subseteq)$.

As a consequence, equality holds in (1.11). Another way of expressing this theorem is to say that the set of finite distributive lattices is in bijection with the set of finite partially ordered sets. A similar result holds with meet-irreducible elements. Another important property is that in a distributive lattice, there is a unique (in the sense of inclusion) minimal representation of an element $x \in L$ by join-irreducible elements. Due to the importance of this theorem in the sequel, we illustrate it by an example (Fig. 1.3).

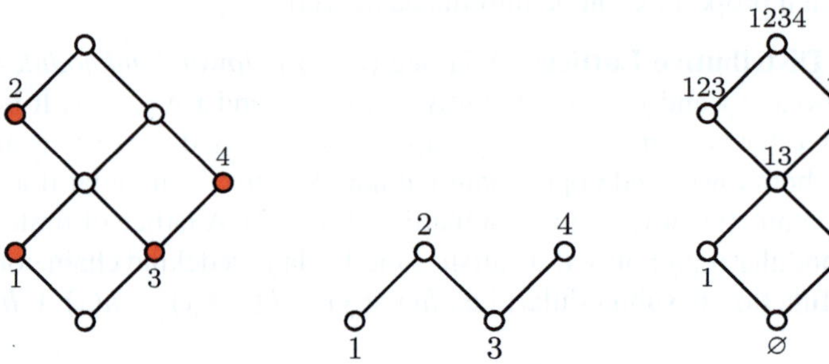

Fig. 1.3 *Left*: a distributive lattice $(L, \preceq)$. Join-irreducible elements are in *red* and labelled 1, 2, 3, 4. *Middle*: the poset $(\mathcal{J}(L), \preceq)$ of join-irreducible elements. *Right*: the distributive lattice $(\mathcal{O}(\mathcal{J}(L)), \subseteq)$ generated by the poset

Closure Systems and Operators Let X be a nonempty set. A collection $\mathcal{C} \subseteq 2^X$ is a *closure system* if $X \in \mathcal{C}$ and it is closed under intersection. Elements of $\mathcal{C}$ are called *closed sets*. A mapping $\tau : (2^X, \subseteq) \to (2^X, \subseteq)$ is a *closure operator* if it satisfies:

(i) $A \subseteq \tau(A)$ for all $A \subseteq X$;
(ii) τ is isotone;
(iii) $\tau(\tau(A)) = \tau(A)$ for all $A \subseteq X$.

Any closure operator τ induces a closure system, where the set of closed sets is the set of fixed points of τ: $\tau(A) = A$. Conversely, any closure system $\mathcal{C}$ induces a closure operator defined by $\tau(A) = \bigcap\{B \,:\, A \subseteq B, B \in \mathcal{C}\}$. A closure system is a complete lattice, with $A \wedge B = A \cap B$ and $A \vee B = \tau(A \cup B)$.

1.3.3 Cones and Convex Sets

(See, e.g., Aliprantis and Border [7], Fujishige [149, Chap. 1].) A *convex set* Z is a subset of a vector space X closed under convex combinations; i.e., for all $z, t \in Z$ and $\alpha \in [0, 1]$, $\alpha z + (1 - \alpha)t \in Z$. The *convex hull* of a set of points $x^1, \ldots, x^n \in X$ is defined by

$$\mathbf{conv}(x^1, \ldots, x^n) = \Big\{\alpha_1 x^1 + \cdots + \alpha_n x^n \ :\ \alpha_1, \ldots, \alpha_n \in [0, 1],\quad \sum_{i=1}^{n} \alpha_i = 1\Big\}.$$

A *cone* is a subset C of a vector space X such that $x \in C$ implies that $\alpha x \in C$ for all $\alpha \geqslant 0$. Note that a cone is a convex set that always contains $\mathbf{0}$. A cone is *pointed* if it does not contain a line, i.e., a set of the form $\{x + \alpha v : \alpha \in \mathbb{R}\}$. Equivalently, a cone is pointed if $C \cap (-C) = \{\mathbf{0}\}$.

The *conic hull* of a set of points $x^1, \ldots, x^n \in X$ is defined by

$$\mathbf{cone}(x^1, \ldots, x^n) = \{\alpha_1 x^1 + \cdots + \alpha_n x^n \ :\ \alpha_1, \ldots, \alpha_n \in \mathbb{R}_+\}.$$

A *ray* is a "half-line;" i.e., a set of the form $\{x + \alpha v : \alpha \in \mathbb{R}_+\}$, where $v \neq \mathbf{0}$ is the supporting vector of the ray. A cone can be seen as a set of rays passing through $\mathbf{0}$. A ray in a cone C is *extremal* if it cannot be expressed as a conic combination of other rays of C. Then any cone is the conic combination of its extremal rays.

The following result is well-known.

Theorem 1.5 (Theorem of the separating hyperplane) *Let $Z \subseteq \mathbb{R}^n$ be a convex closed set, and $x \in \mathbb{R}^n \setminus Z$. Then there exists $y \in \mathbb{R}^n$ such that*

$$\langle y, z\rangle > \langle y, x\rangle \qquad (z \in Z).$$

1.3.4 Linear Inequalities and Polyhedra

(See, e.g., Faigle et al. [136, Chap. 2, Sect. 4 and Chap. 3], Fujishige [149, Chap. 1], Schrijver [290, Chap. 8], Ziegler [360].) We consider a set of linear inequalities and equalities in n variables with real constants

$$\sum_{j=1}^{n} a_{ij} x_j \leqslant b_i \quad (i \in I) \tag{1.12}$$

$$\sum_{j=1}^{n} a_{ij} x_j = b_i \quad (i \in E), \tag{1.13}$$

I, E being two disjoint index sets. This system defines an intersection of half-spaces and hyperplanes, called a (closed convex) *polyhedron* because it is a convex set.

An *extreme point* or *vertex* of a polyhedron P is a point in P that cannot be expressed as a convex combination of other points in P. Equivalently, x is an extreme point of P if it is a solution of the system (1.12), (1.13), and it is the unique solution of a subsystem of n equations

$$\sum_{j=1}^{n} a_{ij}x_j = b_i \quad (i \in V)$$

with $V \subset I \cup E$ and $|V| = n$. A polyhedron is *pointed* if it contains an extreme point.

The *recession cone* $C(P)$ of a polyhedron P defined by (1.12) and (1.13) is a cone defined by

$$\sum_{j=1}^{n} a_{ij}x_j \leqslant 0 \quad (i \in I) \tag{1.14}$$

$$\sum_{j=1}^{n} a_{ij}x_j = 0 \quad (i \in E). \tag{1.15}$$

The recession cone is either a pointed cone (possibly reduced to $\{\mathbf{0}\}$) or it contains a line. The following basic properties are fundamental:

(i) P has rays (but no line) if and only if $C(P)$ is a pointed cone different from $\{\mathbf{0}\}$. Any non-zero solution of (1.14) and (1.15) is a ray;
(ii) P is pointed if and only if $C(P)$ does not contain a line, or equivalently, if the system (1.15) and

$$\sum_{j=1}^{n} a_{ij}x_j = 0 \quad (i \in I)$$

has $\mathbf{0}$ as unique solution.
(iii) P is a *polytope* (i.e., a bounded polyhedron) if and only if $C(P) = \{\mathbf{0}\}$.

A ray r in $C(P)$ is extremal if it satisfies (1.14) and (1.15) and the subsystem of tight inequalities and equalities satisfied by r has rank $n - 1$ (more on rays in Sect. 1.3.6).

The fundamental theorem of polyhedra asserts that any pointed polyhedron P defined by a system (1.12) and (1.13) is the Minkowski sum of its recession cone (generated by its extremal rays; this is the conic part of P) and the convex hull of its extreme points (the convex part of P):

$$P = \mathbf{cone}(r^1, \ldots, r^k) + \mathbf{conv}(\mathbf{ext}(P))$$

where $r_1, \ldots, r_k$ are the extremal rays of $C(P)$, and **ext**() is the set of extreme points of some convex set.

If P is not pointed, then it reduces to its recession cone up to a translation.

Finally, suppose that in the system (1.12) and (1.13) defining a polyhedron P, the equalities in (1.13) are independent (i.e., P is $(n - |E|)$-dimensional). A *p-dimensional face* $(0 \leq p \leq n - |E|)$ of P is a set of points in P satisfying in addition $q = n - |E| - p$ independent equalities in (1.12). In particular, P itself is a face of P $(q = 0)$, a *facet* is a $(n - |E| - 1)$-dimensional face $(q = 1)$, and a vertex is a 0-dimensional face $(q = n - |E|)$. Clearly, no vertex can exist (i.e., P is not pointed) if $|I| < n - |E|$.

Another fundamental result in linear inequalities is Farkas' Lemma, which characterizes the nonemptiness of a polyhedron.

Theorem 1.6 (Farkas' lemma I) *Consider a system of inequalities $Ax \leqslant b$, with A a $m \times n$ matrix with real coefficients and $b \in \mathbb{R}^m$. Then the system has a solution (i.e., the corresponding polyhedron is nonempty) if and only if for every nonnegative vector $y \in \mathbb{R}^m_+$ such that $y^\top A = \mathbf{0}^\top$, we have $y^\top b \geqslant 0$.*

If there are some equalities in the system, then the corresponding coordinates of y are real-valued.

There exist several equivalent avatars of Farkas' lemma. The following one is very useful because it gives a sufficient and necessary condition to find redundant inequalities in a system. Formally, we say that an inequality $a_0^\top x \leqslant b_0$ is *implied by* or *redundant in* the system $Ax \leqslant b$, where $A \in \mathbb{R}^{m \times n}$ and $b \in \mathbb{R}^m$, if for every $x \in \mathbb{R}^n$ satisfying $Ax \leqslant b$, the inequality $a_0^\top x \leqslant b_0$ holds.

Theorem 1.7 (Farkas' lemma II) *Assume that the system $Ax \leqslant b$, with $A \in \mathbb{R}^{m \times n}$ and $b \in \mathbb{R}^m$, has a solution. Then the inequality $a_0^\top x \leqslant b_0$ is implied by $Ax \leqslant b$ if and only if there exists a vector $y \in \mathbb{R}^m_+$ such that*

$$y^\top A = a_0^\top \text{ and } y^\top b \leqslant b_0.$$

Again, if there are some equalities in the system, then the corresponding coordinates of y are real-valued.

1.3.5 Linear Programming

(See, e.g., Chvátal [54], Matoušek and Gärtner [238], Schrijver [290].) A *linear program* (P) is an optimization problem whose objective function and constraints are linear, for example:

$$\begin{aligned} \max \quad & z = c^\top x \\ \text{s.t.} \quad & Ax \leqslant b \\ & x \geqslant \mathbf{0}, \end{aligned} \tag{1.16}$$

with $c, x \in \mathbb{R}^n$, $b \in \mathbb{R}^m$, and A is an $m \times n$ dimensional real matrix. Variants without the constraint $x \geqslant \mathbf{0}$, or with equality constraints $Ax = b$ exist, and all possible mixed cases (that is, some variables are nonnegative, and some inequalities are equalities). Note that in any case, one can express any linear program in the form (1.16), by expressing a real variable as the difference of two nonnegative variables, and an equality $a^\top x = b$ (with $a \in \mathbb{R}^n$) as two inequalities $a^\top x \leqslant b$ and $-a^\top x \leqslant -b$. We say that $x \in \mathbb{R}^n$ is a *feasible solution* if x satisfies all inequalities in (1.16). Supposing (P) has a feasible solution, we say that (P) is *unbounded* if z can be unbounded over the feasible domain $Ax \leqslant b, x \geqslant \mathbf{0}$.

The *dual program* (D) of the linear program (1.16) (called the *primal program*) is the linear program

$$\begin{aligned} \min \quad & w = b^\top y \\ \text{s.t.} \quad & A^\top y \geqslant c \\ & y \geqslant \mathbf{0}, \end{aligned} \tag{1.17}$$

with $y \in \mathbb{R}^m$. The dual programs of the above-mentioned variants can be found easily: for any equality constraint, the corresponding variable y_i is a real variable, while for a real variable x_j, the corresponding constraint in the dual program is an equality. The duality theorem asserts that w is always greater or equal to z, and at the optimum, objective functions of the dual and primal programs are equal.

Theorem 1.8 (Duality theorem of linear programming) *Consider the linear program (1.16) and its dual program (1.17), with objective functions $z = c^\top x$ and $w = b^\top y$ respectively. Then the following statements hold:*

(i) ***(Duality theorem, weak form)*** *For every feasible $x \in \mathbb{R}^n, y \in \mathbb{R}^m$, $b^\top y \geqslant c^\top x$;*
(ii) ***(Duality theorem, strong form)*** *The primal program has an optimal solution x^* if and only if the dual program has an optimal solution y^*, and $b^\top y^* = c^\top x^*$ holds.*

Note that if one of the program is unbounded, the other one has no solution.

A convenient way to check the optimality of a solution is to use complementary slackness (a particular case of the Karush-Kuhn-Tucker conditions; see Sect. 1.3.8).

Theorem 1.9 (Complementary slackness) *Consider a linear program (1.16) and a solution x^* to this program. Then x^* is optimal if and only if there exists y^* solution of the dual program (1.17) such that $y_i^* = 0$ for all i such that $a_i x^* < b_i$, where a_i is the ith row of matrix A, and*

$$\alpha_j^\top y^* = c_j, \quad \forall j \text{ s.t. } x_j^* > 0,$$

where α_j is the jth column of A.

1.3.6 Cone Duality

The results in this section gives some more insight on cones (especially the recession cone of polyhedra), and provides a geometric interpretation to linear programming duality (Faigle et al. [136, Sect. 3.2]). Consider a matrix $A \in \mathbb{R}^{m\times n}$ and the cone generated by the conic combination of the m rows of A, that is:

$$\mathbf{cone}(A) = \{A^\top y : y \in \mathbb{R}^m_+\}.$$

Observe that for every $x \in \mathbb{R}^n$

$$Ax \leqslant \mathbf{0} \Leftrightarrow [w^\top x \leqslant 0, \quad \forall w \in \mathbf{cone}(A)]. \tag{1.18}$$

The *dual* of a cone $C \subseteq \mathbb{R}^n$ is the cone

$$C^* = \{x \in \mathbb{R}^n : w^\top x \leqslant 0, \forall w \in C\}.$$

Taking $C = \mathbf{cone}(A)$, (1.18) yields

$$\mathbf{cone}(A)^* = \{x \in \mathbb{R}^n : Ax \leqslant \mathbf{0}\}.$$

(see Fig. 1.4 for an illustration).

Consider now the linear program $\max w^\top x$ s.t. $Ax \leqslant \mathbf{0}$ and its dual program $\min \mathbf{0}^\top y$ s.t. $A^\top y = w, y \geqslant \mathbf{0}$. The primal program has always $\mathbf{0}$ as feasible solution, but it may be unbounded. Observe that the feasibility of the dual program expresses that $w \in \mathbf{cone}(A)$. Hence, we have shown:

Lemma 1.10 *Let $A \in \mathbb{R}^{m\times n}$ and $w \in \mathbb{R}^n$. The following statements are equivalent:*

- *$w^\top x$ is bounded over the dual cone $\mathbf{cone}(A)^* = \{x : Ax \leqslant \mathbf{0}\}$;*
- *$w \in \mathbf{cone}(A)$;*
- *$w^\top x \leq 0$ for all $x \in \mathbf{cone}(A)^*$.*

The two first equivalences come from linear programming duality. The third equivalence is obtained from the first one, by observing that $x = \mathbf{0}$ is the optimal solution if and only if the primal is bounded, in which case $w^\top x = 0$. It can be also obtained from the second one, by applying (1.18) to w (see Fig. 1.4 for a geometrical interpretation).

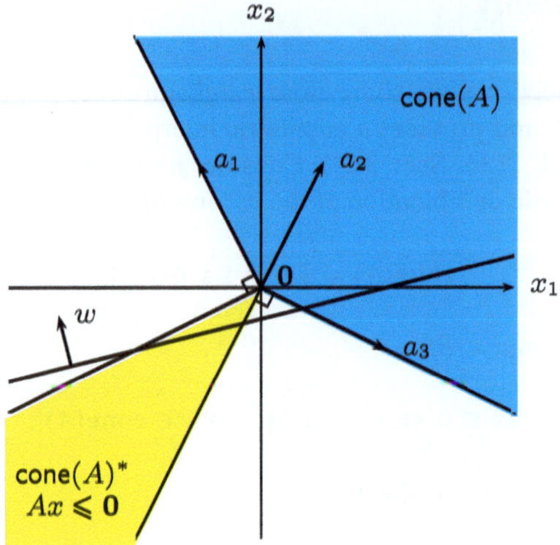

Fig. 1.4 Cone duality and linear programming with $n = 2, m = 3$. Rows of the matrix A are denoted by a_1, a_2, a_3. The vector w is such that $\max w^\top x$ is bounded

1.3.7 Support Functions of Convex Sets

(See, e.g., Aliprantis and Border [7, Sect. 7.10], Schneider [288, Sect. 1.7.1] for a more general exposition.)

Definition 1.11 A function $f : \mathbb{R}^n \to \mathbb{R}$ is said to be:

(i) *positively homogeneous* if $f(\alpha x) = \alpha f(x)$, for every $x \in \mathbb{R}^n$ and $\alpha \geqslant 0$;
(ii) *superadditive* if $f(x + y) \geqslant f(x) + f(y)$ for all $x, y \in \mathbb{R}^n$ (*subadditive* if the reverse inequality holds);
(iii) *convex* if

$$f(\lambda x + (1 - \lambda)y) \leqslant \lambda f(x) + (1 - \lambda)f(y)$$

for every $x, y \in \mathbb{R}^n$ and $\lambda \in [0, 1]$ (*concave* if the reverse inequality holds).

Observe that under positive homogeneity, superadditivity (resp., subadditivity) is equivalent to concavity (resp., convexity). Indeed, if f is superadditive,

$$f(\lambda x + (1 - \lambda)y) \geqslant f(\lambda x) + f((1 - \lambda)y) = \lambda f(x) + (1 - \lambda)f(y),$$

while f concave yields

$$f(x) + f(y) = \frac{1}{2}f(2x) + \frac{1}{2}f(2y) \leqslant f\Big(\frac{1}{2}2x + \frac{1}{2}2y\Big) = f(x + y).$$

Let $f : \mathbb{R}^n \to \mathbb{R}$ be a concave function. The set

$$\partial f(x) = \{v \in \mathbb{R}^n : f(y) \leqslant f(x) + \langle v, y - x\rangle \quad \forall y \in \mathbb{R}^n\} \tag{1.19}$$

is called the *superdifferential* of f at x. The *subdifferential* of f at x is defined with the reverse inequality. Note that if f is positively homogeneous, $\partial f(\mathbf{0}) = \{v \in \mathbb{R}^n : f(y) \leqslant \langle v, y\rangle \quad \forall y \in \mathbb{R}^n\}$.

Let $C \subseteq \mathbb{R}^n$ be nonempty. The *support function*[3] of C is a function $h_C : \mathbb{R}^n \to [-\infty, +\infty[$ defined by

$$h_C(x) = \inf_{x' \in C} \langle x, x'\rangle \qquad (x \in \mathbb{R}^n).$$

The following result shows the duality between the support function and the superdifferential.

Theorem 1.12 *Let C be a nonempty closed convex subset of $\mathbb{R}^n$. Then its support function h_C is positively homogeneous and concave, and $C = \partial h_C(\mathbf{0})$.*

Conversely, if $f : \mathbb{R}^n \to [-\infty, +\infty[$ is positively homogeneous and concave, then its superdifferential $\partial f(\mathbf{0})$ is a nonempty closed convex subset of $\mathbb{R}^n$, and $f = h_{\partial f(\mathbf{0})}$.

1.3.8 Convex Optimization and Quadratic Programming

(See, e.g., Faigle et al. [136].) We consider the following optimization problem

$$\begin{aligned} &\min_{x \in \mathbb{R}^n} f(x) \\ &\text{s.t. } \; a_i^\top x = b_i \;\; (i \in I) \\ &\qquad a_j^\top x \leqslant b_j \;\; (j \in J) \end{aligned} \tag{1.20}$$

where f is a continuously differentiable convex function on $\mathbb{R}^n$, and $a_i, a_j \in \mathbb{R}^n$, $i \in I, j \in J$. Then a feasible point $\overline{x} \in \mathbb{R}^n$ is an optimal solution of (1.20) if and only if it satisfies the *Karush-Kuhn-Tucker (KKT) conditions*:

$$\nabla^\top f(\overline{x}) + \sum_{i \in I} \lambda_i a_i + \sum_{j \in J(\overline{x})} \mu_j a_j = \mathbf{0} \tag{1.21}$$

for some $\lambda_i \in \mathbb{R}$ and $\mu_j \geqslant 0$, where $\nabla f(x) = (\frac{\partial f}{\partial x_1}(x), \dots, \frac{\partial f}{\partial x_n}(x))$ is the gradient of f, and $J(\overline{x}) = \{j \in J : a_j^\top \overline{x} = b_j\}$ corresponds to tight constraints.

[3]Usually, the support function is defined with sup instead of inf. In this case, in Theorem 1.12, the subdifferential must be used.

We apply this result to quadratic programming. A *quadratic programming problem* is a problem of the type

$$\begin{aligned} \min_{x \in \mathbb{R}^n} & f(x) = \tfrac{1}{2} x^\top Q x - c^\top x \\ \text{s.t.} \quad & Ax \leqslant b \end{aligned} \tag{1.22}$$

with $Q \in \mathbb{R}^{n \times n}$ is a symmetric matrix, $c \in \mathbb{R}^n$, $b \in \mathbb{R}^m$, and $A \in \mathbb{R}^{m \times n}$. We say that Q is *positive semidefinite* if $x^\top Q x \geqslant 0$ for all $x \in \mathbb{R}^n$, and *positive definite* if in addition $x^\top Q x = 0$ if and only if $x = \mathbf{0}$ (and in this case Q is invertible). If Q is positive semidefinite (respectively, positive definite), then its eigenvalues are nonnegative (respectively, positive) and $f(x)$ is a convex (resp., strictly convex) function.

The problem is easy to solve in case of equality constraints $Ax = b$. Then an optimal solution is a solution of the system

$$\begin{aligned} \nabla f(x) + \lambda^\top A &= \mathbf{0}^\top \\ Ax &= b \end{aligned} \qquad \text{or} \qquad \begin{bmatrix} Q & A^\top \\ A & \mathbf{0} \end{bmatrix} \begin{pmatrix} x \\ \lambda \end{pmatrix} = \begin{pmatrix} c \\ b \end{pmatrix}. \tag{1.23}$$

If Q is positive definite and $Ax = b$ has a solution, then there is a unique solution to the KKT system, hence a unique optimal solution x^*, given by

$$\begin{pmatrix} x^* \\ \lambda \end{pmatrix} = \begin{bmatrix} Q & A^\top \\ A & \mathbf{0} \end{bmatrix}^{-1} \begin{pmatrix} c \\ b \end{pmatrix}. \tag{1.24}$$

If Q is only positive semidefinite, the problem maybe unbounded, a situation that is characterized by the existence of $d \in \mathbb{R}^n$ satisfying

$$Qd = \mathbf{0}, \quad Ad = \mathbf{0}, \quad \text{and } c^\top d = 1.$$

1.3.9 Totally Unimodular Matrices and Polyhedron Integrality

(See, e.g., Schrijver [290, Chap. 19], [291, Chap. 5, Sects. 15, 16], [289, Chap. 8].) A $m \times n$ matrix is *totally unimodular* if each square submatrix of A has determinant equal to -1, 0 or $+1$. In particular, this implies that each entry is either -1, 0, or 1.

The main interest of totally unimodular matrices is that they characterize polyhedron integrality. A polyhedron is *integer* if all its extreme points have integer coordinates.

Theorem 1.13 *Let A be a totally unimodular $m \times n$ matrix. Then for all integer vectors $b \in \mathbb{R}^m$, $c \in \mathbb{R}^n$ the polyhedra*

$$P = \{x : x \geqslant \mathbf{0}, \quad Ax \leqslant b\}, \quad D = \{y : y \geqslant \mathbf{0}, \quad A^\top y \geqslant c\}$$

are integer.

The reciprocal (Hoffman-Kruskal theorem) says that if P is integer for every integer vector b, then A is totally unimodular.

Total unimodularity is however not easy to check. An easy particular case arises from directed graphs. Let $G = (V, A)$ be a *directed graph*; i.e., a set of vertices V and a set of arcs A, where an arc is an ordered pair $a = (x, y)$ that links two vertices x, y. Its *vertex-arc incidence matrix* M is defined as follows:

$$M_{x,a} = \begin{cases} +1, & \text{if a leaves } x, \\ -1, & \text{if a enters } x, \\ 0, & \text{otherwise.} \end{cases}$$

Thus, each column of M has exactly one $+1$ and one -1, the rest on the entries being 0. Conversely, any such matrix defines a directed graph, and the next theorem shows that these matrices are totally unimodular.

Theorem 1.14 *The vertex-arc incidence matrix of any directed graph is totally unimodular.*

1.3.10 Riesz Spaces

(See, e.g., Aliprantis and Border [7, Chap. 8].) Let X be a vector space, and consider a pointed cone C in X. Any pointed cone induces a partial order $\geqslant$ on X as follows: $x \geqslant y$ if $x - y \in C$. This order is said to be *compatible* with X if:

(i) $x \geqslant y$ implies $x + z \geqslant y + z$ for every $z \in X$; and
(ii) $x \geqslant y$ implies $\alpha x \geqslant \alpha y$ for all $\alpha \geqslant 0$.

An *ordered vector space* X is a real vector space with an order relation $\geqslant$ that is compatible in the above sense. The set $\{x \in X : x \geqslant \mathbf{0}\}$ is a pointed cone, called the *positive cone* of X, and denoted by X^+ or X_+. Any vector in X^+ is called *positive*.

A *Riesz space* is an ordered vector space that is also a lattice. If x is an element of a Riesz space, its *positive part* x^+, *negative part* x^-, and its *absolute value* $|x|$ are defined by

$$x^+ = x \vee 0, \quad x^- = (-x) \vee 0, \text{ and } |x| = x \vee (-x),$$

where $\vee$ is the supremum of the lattice. A norm $\|\cdot\|$ on a Riesz space X is an *L-norm* if $\|x + y\| = \|x\| + \|y\|$ for all $x, y \geqslant \mathbf{0} \in X$. A Riesz space equipped with a L-norm is called an *AL-space* (abstract Lebesgue space).

1.3.11 Laplace and Fourier Transforms

Let f be a real-valued function whose support is $\mathbb{R}_+$. Its *Laplace transform*[4] (see, e.g., Abramowitz and Stegun [1], Gradshteyn and Ryzhik [183]) $F = \mathcal{L}(f)$ is a function of the complex variable s defined by

$$F(s) = \int_0^{\infty} e^{-st} f(t)\,\mathrm{d}t \qquad (s \in \mathbb{C}, \Re s > s_0), \tag{1.25}$$

where $\Re s$ denotes the real part of s, and $s_0 \in \mathbb{R}$ is defined as the lower bound of the real numbers β such that

$$\lim_{\substack{a \to 0 \\ b \to +\infty}} \int_a^b e^{-\beta t} f(t)\,\mathrm{d}t$$

is convergent. The Laplace transform is a linear invertible operator, which is extremely useful for solving partial differential equations, due to the following property:

$$\mathcal{L}(f') = s\mathcal{L}(f) - f(0) \tag{1.26}$$

where f' is the derivative of f. Another remarkable property is that the convolution product of two functions f, g, defined by

$$(f * g)(t) = \int_0^t f(\tau) g(t-\tau)\,\mathrm{d}\tau, \tag{1.27}$$

is turned into an ordinary product:

$$\mathcal{L}(f * g) = \mathcal{L}(f)\mathcal{L}(g). \tag{1.28}$$

We give some examples of Laplace transform of usual functions in Table 1.1. Recall that functions are supposed to be zero-valued for negative real numbers. δ is the Dirac function, that is, $\delta(0) = 1$ and $\delta(t) = 0$ for every $t > 0$.

The *Fourier transform* (see p.99 for a bibliographical notice) is a particular case of the *bilateral Laplace transform* (i.e., where the integral in (1.25) is taken on $[-\infty, +\infty]$), where s has the form $s = i\xi$ ($\xi \in \mathbb{R}$) or $i2\pi\nu$ (ν is interpreted as the frequency).

---

[4] Pierre-Simon de Laplace (Beaumont-en-Auge, 1749 – Paris, 1827), French mathematician, astronom and physicist. He gave the mathematical foundations of astronomy, contributed to probability theory, and was the advocate of the determinism. What is called today the Laplace transform was in fact discovered by Leonhard Euler.

$f(t)$ $(t \geqslant 0)$		$F(s) = \mathcal{L}(f)$	s_0
1		$\frac{1}{s}$	0
t^n	$(n \in \mathbb{N}_0)$	$\frac{n!}{s^{n+1}}$	0
$t^n e^{-at}$	$(n \in \mathbb{N}_0)$	$\frac{n!}{(s+a)^{n+1}}$	$-a$
$\delta(t-\tau)$	$(\tau \geqslant 0)$	$e^{-\tau s}$	$-\infty$

Table 1.1 Laplace transform of usual functions

Chapter 2
Set Functions, Capacities and Games

This chapter opens the fundamental part of the book, focusing on mathematical properties. It presents the main classes of set functions that will be considered throughout the book, namely games (set functions vanishing on the empty set) and capacities (monotone games). In the main part of this chapter (Sects. 2.1–2.18), we consider set functions defined on a nonempty finite set X, where the domain of set functions is the power set 2^X. In Sect. 2.19 we briefly address the case of arbitrary spaces and domains.

The length of the chapter and the variety of the topics addressed made the task of organizing the chapter a bit difficult, and we hope that the readers will not be lost into this forest of definitions and theorems. In order to let the readers find more easily their way, we give a brief description of its content. After a short introduction of several application domains and interpretations of games and capacities (Sect. 2.4), we introduce the main properties (Sect. 2.7) and the main families of capacities (Sect. 2.8). The fundamental concept of Möbius transform is introduced in Sect. 2.10, as well as other important transforms, namely, the interaction transform, the Banzahf interaction transform and the co-Möbius transform in Sect. 2.11. A table giving all conversion formulas between these transforms is given in Appendix A. Section 2.12 gives a formal analysis of the class of linear invertible transforms, and can be ignored at first reading. The important concept of k-additive game is presented in Sect. 2.13. The geometrical and algebraic structure (polytope, cone, vector space, etc.) of the most important classes of games and capacities is studied in Sect. 2.15. In particular two bases of the vector space of games are exhibited. The long Sect. 2.16 takes another point of view for set functions and consider them as pseudo-Boolean functions, which permits to have a polynomial representation of set functions. There, the Walsh basis (which is orthonormal) is introduced, as well as the Fourier transform, which is fundamental in computer sciences. The polynomial form of a set function is adequate to address the problem of approximation (of a fixed degree), and of extension of a set function. Section 2.17 returns to the topic of bases and transforms, and show that the two notions are

M. Grabisch, *Set Functions, Games and Capacities in Decision Making*,
Theory and Decision Library C 46, DOI 10.1007/978-3-319-30690-2_2

intimately related. A table summarizing all known bases and transforms is given in Appendix A. Inclusion-exclusion coverings (Sect. 2.18) is related to the problem of decomposing a game into a sum of simpler games. Lastly, Sect. 2.19 considers games on infinite and finite universal sets X, whose domain is a subcollection of 2^X (called a set system).

2.1 Set Functions and Games

A *set function on* X is a mapping $\xi : 2^X \to \mathbb{R}$, assigning a real number to any subset of X. A set function can be

(i) *Additive* if $\xi(A \cup B) = \xi(A) + \xi(B)$ for every disjoint $A, B \in 2^X$;
(ii) *Monotone* if $\xi(A) \leqslant \xi(B)$ whenever $A \subseteq B$;
(iii) *Grounded* if $\xi(\varnothing) = 0$;
(iv) *Normalized* if $\xi(X) = 1$.

Note that an additive set function is uniquely determined by its value on elements of X, because $\xi(A) = \sum_{x \in A} \xi(\{x\})$.

Definition 2.1 A *game* $v : 2^X \to \mathbb{R}$ is a grounded set function.

As far as possible, throughout the book we distinguish by their notation the type of set functions (ξ for general set functions, v for games and μ for capacities, see Definition 2.5 below).

We denote the set of games on X by $\mathcal{G}(X)$. The set of set functions on X is simply $\mathbb{R}^{(2^X)}$.

A game v is *zero-normalized* if $v(\{x\}) = 0$ for every $x \in X$. We can already notice the following properties:

(i) If $\xi \geqslant 0$ (nonnegative) and additive, then ξ is monotone;
(ii) If ξ is additive, then $\xi(\varnothing) = \xi(\varnothing) + \xi(\varnothing)$, which entails $\xi(\varnothing) = 0$;
(iii) To any game v one can associate a zero-normalized game $v_0 = v - \beta$, with β an additive game defined by $\beta(\{x\}) = v(\{x\})$ for every $x \in X$.

To any set function ξ we associate its *conjugate* (a.k.a. *dual*) $\overline{\xi}$, which is a set function defined by

$$\overline{\xi}(A) = \xi(X) - \xi(A^c) \qquad (A \in 2^X). \tag{2.1}$$

Note that $\overline{\xi}(\varnothing) = \xi(X) - \xi(X) = 0$. The following properties are easy to show (try!).

Theorem 2.2 *Let* ξ *be a set function on* X.

(i) If $\xi(\varnothing) = 0$, *then* $\overline{\xi}(X) = \xi(X)$ *and* $\overline{\overline{\xi}} = \xi$;
(ii) If ξ *is monotone, then so is* $\overline{\xi}$.
(iii) If ξ *is additive, then* $\overline{\xi} = \xi$ *(*ξ *is self-conjugate).*

Remark 2.3 The term "game" may appear strange, although it is commonly used in decision theory and capacity theory. It comes from cooperative game theory (see, e.g., Owen [263], Peleg and Sudhölter [267], Peters [268]). A game v (in its full name, a *transferable utility game in characteristic function form*) represents the gain that can be achieved by cooperation of the players (more on this in Sect. 2.4). ◇

2.2 Measures

A *measure* is a nonnegative and additive set function. A normalized measure is called a *probability measure*. A *signed measure* is an additive set function, that is, it may take negative values. Measures are usually denoted by m, and $\mathcal{M}(X)$ denotes the set of measures on X.

Example 2.4 Let us give some easy examples of measures, apart from probability measures.

(i) The *counting measure* m_c just counts the elements in sets: $m_c(A) = |A|$ for all $A \in 2^X$.
(ii) Measure of length, volume, mass, etc., can be considered to be measures because they are additive and nonnegative. In $\mathbb{R}^n$, the *Lebesgue measure*[1] of a Cartesian product of real intervals $[a_1, b_1] \times \cdots \times [a_n, b_n]$ is its volume $(b_1 - a_1)(b_2 - a_2)\cdots(b_n - a_n)$.
(iii) Let $x_0 \in X$. The *Dirac measure centered at* x_0 is defined by

$$\delta_{x_0}(A) = \begin{cases} 1, & \text{if } x_0 \in A \\ 0, & \text{otherwise.} \end{cases}$$

◇

2.3 Capacities

Definition 2.5 A *capacity* $\mu : 2^X \to \mathbb{R}$ is a grounded monotone set function; i.e., $\mu(\varnothing) = 0$ and $\mu(A) \leqslant \mu(B)$ whenever $A \subseteq B$.

Note that the constant function 0 is a capacity. Also, a capacity is a monotone game, and takes only nonnegative values. A capacity is *normalized* if in addition $\mu(X) = 1$. Note that an additive normalized capacity is a probability measure. The set of

[1] Henri-Léon Lebesgue (Beauvais, 1875 – Paris, 1941), French mathematician, famous for his major contributions to measure and integration theory.

capacities on X is denoted by $\mathcal{MG}(X)$, while $\mathcal{MG}_0(X)$ denotes the set of normalized capacities.

The term "capacity" was coined by Choquet [53]. The same notion was proposed independently by Sugeno [319] under the name *fuzzy measure*. Other terms are also employed, like *nonadditive measure* (Denneberg [80]), or *monotonic measures*.

2.4 Interpretation and Usage

Set functions, and more particularly games and capacities, are used in many different fields, with different interpretations and aims. We review the most prominent ones in this section.

2.4.1 In Decision and Game Theory

The fields of decision theory and game theory seem to be the privileged area for the application of games and capacities. There are two main interpretations and usage of them here. The first one is to consider *capacities/games as a means to represent the importance/power/worth of a group*. Let X represent a set of persons, usually called players, agents, voters, experts, decision makers, etc., depending on the situation that is described and the area of decision making that is concerned: game theory (players, agents), social choice (voters), multi-person decision making (experts, decision makers). Consider a group $A \subseteq X$ of individuals, usually called a *coalition*—especially in game theory and social choice—because it is supposed that the individuals in A cooperate in some sense to achieve some common goal. One can quantify to which extent the group A is able to achieve or has achieved its goal, a quantity which is denoted by $\mu(A)$. This interpretation of capacities and games is the core of the theory of cooperative games, and it is also the basic ingredient of Chap. 6.

We now give some examples to illustrate various situations.

Example 2.6 Let X be a set of firms. Certain firms may form a coalition in order to control the market for a given product. Then $\mu(A)$ may be taken as the annual benefit of the coalition A. ◇

Example 2.7 Let X be a set of voters in charge of electing a candidate for some important position (president, director, etc.) or voting a bill by a yes/no decision. Before the election, groups of voters may agree to vote for the same candidate (or for yes or no). In many cases (presidential elections, parliament, etc.), these coalitions correspond to the political parties or to alliances among them. If the result of the election is in accordance with the wish of coalition A, the coalition is said to be winning, and we set $\mu(A) = 1$, otherwise it is loosing and $\mu(A) = 0$. ◇

Example 2.8 Let X be a set of workers in a factory, producing some goods. The aim is to produce these goods as much as possible in a given time (say, in 1 day). Then $\mu(A)$ is the number of goods produced by the group A in a given time period. Since the production needs in general the collaboration of several workers with different skills, it is likely that $\mu(A) = 0$ if A is a singleton or a too small group. ◇

The above interpretation extends to more abstract objects and other interpretations for μ, like monetary value, selling price, etc., as shown by the next example.

Example 2.9 Let X be a set of goods sold in a shop. Then $\mu(A)$ is the selling price of the set of goods A. In most cases, the selling price is a measure, that is, the price of a set is the sum of the prices of its elements. However, there are situations where the price is not additive. For example, it is common practice to reduce prices when many identical objects are bought together (e.g., 3 for the price of 2). In these cases, the price is a subadditive function (see Definition 2.18). On the other hand, it could be superadditive if the considered objects are rare or precious. For example, the complete collection of the original edition of the Encyclopedia of Diderot and D'Alembert is much more expensive that the sum of the prices of all single volumes. ◇

The readers can check that in all these examples, the function μ satisfies $\mu(\varnothing) = 0$ and monotonicity. However, the latter property may be violated if the collaboration is not beneficial. In Example 2.6, it may be the case that if the coalition A contains a firm i that is close to bankruptcy, we have $\mu(A \setminus i) > \mu(A)$. The same phenomenon may occur in Example 2.8, if the group of workers contains a worker who disturbs the others. Hence, for the sake of generality, one may abandon monotonicity and use games (Definition 2.1) instead of capacities.

The second interpretation concerns the representation of uncertainty. In an abstract way, X is the set of possible outcomes of some experiment, and it is supposed that X is exhaustive (i.e., any experiment produces an outcome that belongs to X), and that each experiment produces a single outcome. Any subset A of X is called an *event*, and $\mu(A)$ quantifies the uncertainty that the event A contains the outcome of an experiment, with the following convention: $\mu(A) = 0$ indicates total uncertainty, and $\mu(A) = 1$ indicates that there is no uncertainty. We use here the word "uncertainty" as a neutral word, which does not preclude any specific interpretation or precise meaning of the type of "non-certainty" of some event, the language having a very rich palette of words to designate the fact that some event may not occur. This framework is exactly the same as the one used for probability theory, and therefore μ can be taken as a probability measure. The use of capacities however allows more flexibility in the representation of uncertainty, and the use of some specific subfamilies of capacities permits to distinguish among the different nuances of the language. This is studied in detail in Chaps. 5 and 7.

Another way of considering the second interpretation, which does not refer to experiments, is to say that X is the set of possible answers to a question. Here also, it is assumed that only one answer is correct and that X contains that correct answer. Then $\mu(A)$ quantifies the uncertainty that A contains the correct answer.

Whatever the type of uncertainty that is represented, it is a reasonable assumption to say that if an event A is included in B, it is more likely to find the outcome of the experiment in B than in A. Therefore, the inequality $\mu(A) \leqslant \mu(B)$ makes sense and justifies the use of capacities in this context. We give some examples.

Example 2.10 David throws a dice, and wonders which number will show. Here $X = \{1, 2, 3, 4, 5, 6\}$, and $\mu(\{1, 3, 5\})$ quantifies the uncertainty of obtaining an odd number. ◇

Example 2.11 A murder was committed. After some investigation, it is found that the guilty is either Alice, Bob or Charles. Then $X = \{\text{Alice, Bob, Charles}\}$, and $\mu(\{\text{Bob, Charles}\})$ quantifies the degree to which it is "certain" (the precise meaning of this word being conditional on the type of capacity used) that the guilty is Bob or Charles. ◇

Example 2.12 Glenn is an amateur of antique Chinese porcelain. He enters a shop and sees a magnificent vase, wondering how old (and how expensive) this vase could be. Then X is the set of numbers from, say -3000 to 2012; i.e., the possible date expressed in years A.C. when the vase was created. For example, $\mu([1368, 1644])$ indicates to what degree it is certain that it is a vase of the Ming period. ◇

Example 2.13 Leonard is planning to go to the countryside tomorrow for a picnic. He wonders if the weather will be favorable or not. Here X is the set of possible states of the weather, like "sunny," "rainy," "cloudy," and so on. For example, $\mu(\{\text{sunny, cloudy}\})$ indicates to what degree of certainty it will not rain, and therefore if the picnic is conceivable or not. ◇

Except in Example 2.10, the experiment referred to is not salient in the above examples, although it is always underlying. The experiment consists in repeating the same situation and to see what happens. For example, Glenn gould enter another shop and see another vase, and so on. However, on those examples, it is more natural to interpret X as the set of possible answers to a given question. Who has committed the murder? What is the age of this vase? What will be the weather tomorrow? In all cases, there is a unique (and unknown) true answer, and this answer lies in X.

2.4.2 In Operations Research

Operations Research is another vast field where set functions are applied. They are often viewed as "pseudo-Boolean functions;" i.e., real-valued functions on $\{0, 1\}^n$ (Sect. 2.16). They have numerous applications (see the monograph of Crama and Hammer [63, Ch. 13]), e.g., in graph theory, computer sciences, data mining, production management, etc.

We mention also the following more specific fields, where games and capacities are hidden under different names.

Combinatorial Optimization

Set functions appear in combinatorial optimization. More specifically, submodular capacities are considered (Definition 2.18), under the name of the *rank function of a polymatroid* (Edmonds [122]), a polymatroid being simply a pair (X, μ), where μ is a submodular capacity, that is, satisfying the inequality

$$\mu(A \cup B) + \mu(A \cap B) \leqslant \mu(A) + \mu(B).$$

This name comes from the generalization of the following notions. A *matroid* on X is a family $\mathcal{M} \subseteq 2^X$ containing the empty set, and satisfying (1) $A \in \mathcal{M}$ and $B \subseteq A$ imply $B \in \mathcal{M}$, and (2) the property:

$$A, B \in \mathcal{M}, |A| < |B| \Rightarrow \exists x \in B \setminus A, A \cup \{x\} \in \mathcal{M}.$$

Matroids are an abstraction of the notion of sets of independent vectors in a vector space, as it can be checked from the properties. Maximal sets in $\mathcal{M}$ are called *bases* of the matroid, and correspond to the usual notion of basis of a vector space. Now, the *rank function of a matroid* is a function $\rho : 2^X \to \mathbb{N}$ defined by

$$\rho(A) = \max\{|B| : B \subseteq A, B \in \mathcal{M}\} \qquad (A \in 2^X).$$

Then it is easy to see that ρ is a submodular capacity on X. There are many useful examples of matroids, apart from those induced by matrices: the family of sets of edges without cycle in a graph, those arising from bipartite graphs and matching problems, etc. We refer the readers to the monograph of Fujishige [149] for full detail.

Reliability

Set functions model the structure of a multicomponent system and permit to study its lifetime. Consider a n-component system, whose state (0 or 1, depending whether the system is functioning or not) is ruled by a Boolean function $\phi : \{0,1\}^n \to \{0,1\}$ whose variables are the states of the n components. It is often assumed that the system is *semicoherent*, which means that ϕ is nondecreasing in each place, and satisfies $\phi(0,\dots,0) = 0$ and $\phi(1,\dots,1) = 1$. Then ϕ is a 0-1-valued capacity, a.k.a. simple game (Sect. 2.8.1).

Denoting by $X_1,\dots,X_n$ the random variables giving the lifetime of each component, the *Barlow-Proschan index* of component k is the probability that the failure of the system is provoked by the failure of component k: $I_k^{\text{BP}} = \Pr(T = X_k)$, where T is the lifetime of the system. As remarked by Marichal and Mathonet [233], when $X_1,\dots,X_n$ are continuous and i.i.d. random variables, the Barlow-Proschan index is nothing but the Shapley value of ϕ (Remark 2.43).

Another quantity of interest is the *signature* $s \in [0,1]^N$ of the system, where $s_k = \Pr(T = X_{(k)})$, the probability that the kth component failure provokes the failure of the system ($X_{(k)}$ denotes the kth smallest lifetime). When the lifetimes are continuous and i.i.d., the signature depends solely on ϕ:

$$s_k = \sum_{\substack{A\subseteq[n]\\|A|=n-k+1}} \frac{1}{\binom{n}{|A|}}\phi(A) - \sum_{\substack{A\subseteq[n]\\|A|=n-k}} \frac{1}{\binom{n}{|A|}}\phi(A) \qquad (k \in [n]).$$

Moreover, what is known in reliability theory as the *reliability function* of the system corresponds to the Owen extension of ϕ (Sect. 2.16.4).

2.5 Derivative of a Set Function

By analogy with real-valued functions, the derivative of a set function is its variation when an element is added or removed from a set.

Definition 2.14 Let ξ be a set function on X, and consider $A \subseteq X$, $i \in X$. The *derivative of* ξ *at* A *w.r.t.* i is defined by

$$\Delta_i\xi(A) = \xi(A \cup \{i\}) - \xi(A \setminus \{i\}).$$

Note that $\Delta_i\xi(A \cup \{i\}) = \Delta_i\xi(A \setminus \{i\})$ for all $A \subseteq X$ (the presence of i in A is irrelevant for $\Delta_i\xi(A)$), and that $\Delta_i\xi : 2^X \to \mathbb{R}$ is itself a set function $\forall x \in X$.

As it will become evident, the derivative of a set function possesses many properties close to derivatives of usual real-valued functions. It is very useful in cooperative game theory, where it is used to define the *marginal vectors* [see (3.8)]. An interpretation directly in terms of the usual derivative of a function will be given in Sect. 2.16, Definition 2.67.

As a first illustration, let us remark that a set function is monotone if and only if its derivative w.r.t. any element is nonnegative everywhere: $\Delta_i\xi(A) \geqslant 0$ for every $i \in X, A \subseteq X$.

Higher order derivatives can be defined as well. Consider distinct $i,j \in X$ and $A \subseteq X$, and observe that because $\Delta_i\xi$, $\Delta_j\xi$ are itself set functions, one can take their derivatives and obtains:

$$\Delta_i(\Delta_j\xi(A)) = \xi(A\cup\{i,j\}) - \xi(A\cup\{i\}\setminus\{j\}) - \xi(A\cup\{j\}\setminus\{i\}) + \xi(A\setminus\{i,j\}) = \Delta_j(\Delta_i\xi(A)).$$

The *(2nd order) derivative of* ξ *at* A *w.r.t. to* i,j is defined by

$$\Delta_{ij}\xi(A) = \Delta_i(\Delta_j\xi(A)) = \Delta_j(\Delta_i\xi(A)).$$

Continuing the process, one can take derivatives with respect to any number of elements of X, in short, we can consider the derivative with respect to a set $K \subseteq X$.

Definition 2.15 Consider subsets $A, K \subseteq X$ and a set function ξ on X. The *derivative* of ξ at A w.r.t. K is defined inductively by

$$\Delta_K \xi(A) = \Delta_{K \setminus \{i\}}(\Delta_{\{i\}} \xi(A)),$$

with the convention $\Delta_{\varnothing} \xi = \xi$, and $\Delta_{\{i\}} \xi = \Delta_i \xi$.

Obviously $\Delta_{\{i,j\}} \xi = \Delta_{ij} \xi$. Note that $\Delta_K \xi$ is a set function. When $K \cap A = \varnothing$, it is easy to obtain that $\Delta_K \xi(A) = \sum_{L \subseteq K} (-1)^{|K \setminus L|} \xi(A \cup L)$. More generally, if K and A intersect, we find

$$\Delta_K \xi(A) = \sum_{L \subseteq K} (-1)^{|K \setminus L|} \xi((A \setminus K) \cup L). \tag{2.2}$$

The next theorem expresses the derivative of the conjugate of a game.

Theorem 2.16 *For any game v on X, any set $K \subseteq X$*

$$\Delta_K \overline{v}(A) = \Delta_K v(A^c) \qquad (A \subseteq X).$$

Proof We show it by induction on $|K|$. For $K = i \in X$, we find $\Delta_i \overline{v}(A) = -v((A \cup \{i\})^c) + v((A \setminus i)^c) = -v(A^c \setminus \{i\}) + v(A^c \cup \{i\}) = \Delta_i \overline{v}(A^c)$. Assuming the formula holds for K and taking $i \in K^c$, we find:

$$\Delta_{K \cup i} \overline{v}(A) = \Delta_i(\Delta_K \overline{v}(A)) = \Delta_i(\Delta_K v(A^c)) = \Delta_{K \cup i} v(A^c).$$

□

2.6 Monotone Cover of a Game

Let v be a game on X. The *monotone cover* of v is the smallest monotone game (capacity) μ such that $\mu \geqslant v$. We denote it by $\mathbf{mc}(v)$. It is given by

$$\mathbf{mc}(v)(A) = \max_{B \subseteq A} v(B) \qquad (A \subseteq X). \tag{2.3}$$

It follows that **mc**, seen as a mapping on the lattice $(\mathcal{G}(X), \leqslant)$, where $\leqslant$ is the usual ordering of functions, is a closure operator (see Sect. 1.3.2). The set of closed games (that is, the fixed points of **mc**) is exactly the set of capacities, which forms therefore also a lattice.

The following result is easy to obtain (we leave the proof to the readers).

Theorem 2.17 *Let v be a $\{0,1\}$-valued game (i.e., whose range is $\{0,1\}$). Then*

$$\mathbf{mc}(v)(A) = 1 \text{ if and only if } A \in \uparrow \mathcal{B}_0$$

where $\mathcal{B}_0$ is the set of minimal subsets of $\mathcal{B} = \{B : v(B) = 1\}$.

We recall that $\uparrow \mathcal{B}_0$ is the upset generated by $\mathcal{B}_0$ (Sect. 1.3.2).

2.7 Properties

We give the main properties of capacities and games.

Definition 2.18 Let v be a game on X. We say that v is

(i) *superadditive* if for any $A, B \in 2^X$, $A \cap B = \varnothing$,

$$v(A \cup B) \geqslant v(A) + v(B).$$

The game is said to be *subadditive* if the reverse inequality holds;

(ii) *supermodular* if for any $A, B \in 2^X$,

$$v(A \cup B) + v(A \cap B) \geqslant v(A) + v(B).$$

The game is said to be *submodular* if the reverse inequality holds. A game that is both supermodular and submodular is said to be *modular*. Supermodular games are often improperly called *convex* games, while submodular games are called *concave* (see Remark 2.24);

(iii) *k-monotone* (for a fixed integer $k \geqslant 2$) if for any family of k sets $A_1, \ldots, A_k \in 2^X$,

$$v\Big(\bigcup_{i=1}^{k} A_i\Big) \geqslant \sum_{\substack{I \subseteq \{1,\ldots,k\} \\ I \neq \varnothing}} (-1)^{|I|+1} v\Big(\bigcap_{i \in I} A_i\Big).$$

v is *totally monotone* (or *∞-monotone*) if it is k-monotone for any $k \geqslant 2$;

(iv) *k-alternating* (for a fixed integer $k \geqslant 2$) if for any family of k sets $A_1, \ldots, A_k \in 2^X$,

$$v\Big(\bigcap_{i=1}^{k} A_i\Big) \leqslant \sum_{\substack{I \subseteq \{1,\ldots,k\} \\ I \neq \varnothing}} (-1)^{|I|+1} v\Big(\bigcup_{i \in I} A_i\Big).$$

v is *totally alternating* (or *∞-alternating*) if it is k-alternating for any $k \geqslant 2$;

(v) *maxitive* if for any $A, B \in 2^X$,

$$v(A \cup B) = v(A) \vee v(B);$$

(vi) *minitive* if for any $A, B \in 2^X$,

$$v(A \cap B) = v(A) \wedge v(B).$$

Remark 2.19

(i) All of the above properties can be applied to set functions as well.
(ii) Super(sub)modularity implies super(sub)additivity but not the converse. Also, it is easy to check that additivity and modularity are equivalent properties. This is no longer true when games are defined on subcollections of 2^X (see Sect. 2.19.2).
(iii) 2-monotonicity corresponds to supermodularity, while the 2-alternating property corresponds to submodularity.
(iv) If $2 \leqslant k' \leqslant k$, then k-monotonicity implies k'-monotonicity.
(v) k-monotonicity and the k-alternating properties generalize the following equality, valid for any measure:

$$m\Big(\bigcup_{i=1}^{k} A_i\Big) = \sum_{\substack{I \subseteq \{1,\dots,k\} \\ I \neq \varnothing}} (-1)^{|I|+1} m\Big(\bigcap_{i \in I} A_i\Big).$$

This equality comes directly from the well-known identity

$$\Big|\bigcup_{i=1}^{k} A_i\Big| = \sum_{\substack{I \subseteq \{1,\dots,k\} \\ I \neq \varnothing}} (-1)^{|I|+1} \Big|\bigcap_{i \in I} A_i\Big|. \tag{2.4}$$

The above equality for measures is related to the well-known *sieve formula* or *principle of inclusion-exclusion* (see, e.g., Berge [20, Chap. 3, Sect. 3] and Aigner [2, Chap. IV, Sect. 2.B]): Let $A_1, \dots, A_k \subseteq X$. The measure of the set of elements of X that belong to exactly p of the sets A_i is

$$T_k^p = \sum_{\ell=p}^{k} (-1)^{\ell-p} \binom{\ell}{p} \sum_{K \subseteq \{1,\dots,k\}, |K|=\ell} \underline{m}(K),$$

with $\underline{m}(K) = m\Big(\bigcap_{i \in K} A_i\Big)$ if $K \neq \varnothing$, and $\underline{m}(\varnothing) = m(X)$.

(vi) k-monotonicity and the k-alternating properties were introduced by Choquet [53, Chap. 3], although the original definitions slightly differ. Choquet introduced *difference functions* of a set function ξ as follows. For any $k \in \mathbb{N}_0$ and $A, B_1, \ldots, B_k \in 2^X$, define recursively

$$\nabla_k \xi(A, B_1, \ldots, B_k) = \nabla_{k-1} \xi(A, B_1, \ldots, B_{k-1}) - \nabla_{k-1} \xi(A \cap B_k, B_1, \ldots, B_{k-1}),$$

and $\nabla_0 \xi(A) = \xi(A)$, $A \in 2^X$. This yields

$$\nabla_k \xi(A, B_1, \ldots, B_k) = \sum_{I \subseteq [k]} (-1)^{|I|} \xi(A \cap \bigcap_{i \in I} B_i). \tag{2.5}$$

Then ξ is said to be *k-monotone in the sense of Choquet* if $\nabla_k \xi(A, B_1, \ldots, B_k) \geqslant 0$ for every family $A, B_1, \ldots, B_k \in 2^X$. Observe that 1-monotonicity is nothing but monotonicity (take $A \supseteq B$). In order to get the definition of k-alternating set function, replace in the above $\cap$ by $\cup$ and $\geqslant$ by $\leqslant$.

The exact relation between the two definitions is simply the following, for a fixed $k \geqslant 2$: ξ is k-monotone in the sense of Choquet if and only if ξ is k-monotone and monotone. Some authors, as Chateauneuf and Jaffray [50], and Barthélemy [18], use the name *weak k-monotonicity* instead of our terminology (and similarly for the k-alternating property), but it seems that our terminology follows the current usage. See also Sect. 7.2.3 for a more general presentation and other properties.

◇

The next theorem gathers some elementary properties.

Theorem 2.20 *Let v be a game on X. The following holds.*

(i) v superadditive $\Rightarrow \overline{v} \geqslant v$;
(ii) Let $k \geqslant 2$. v is k-monotone (respectively, k-alternating) if and only if $\overline{v}$ is k-alternating (respectively, k-monotone). In particular, v is supermodular (respectively, submodular) if and only if $\overline{v}$ is submodular (respectively, supermodular);
(iii) $v \geqslant 0$ and supermodular implies that v is monotone;
(iv) v maxitive or minitive $\Rightarrow$ v monotone;
(v) v is maxitive $\Leftrightarrow$ $\overline{v}$ is minitive.

Proof

(i) Suppose v is superadditive. Then for any $A \in 2^X$, $\overline{v}(A) = v(X) - v(A^c) \geqslant v(A) + v(A^c) - v(A^c)$, hence the result.

(ii) For any $A_1, \ldots, A_k \in 2^X$,

$$v\Big(\bigcup_{i=1}^k A_i\Big) \geqslant \sum_{\substack{I\subseteq\{1,\ldots,k\}\\ I\neq\varnothing}} (-1)^{|I|+1} v\Big(\bigcap_{i\in I} A_i\Big) \Leftrightarrow$$

$$\overline{v}(X) - \overline{v}\Big(\Big(\bigcup_{i=1}^k A_i\Big)^c\Big) \geqslant \sum_{\substack{I\subseteq\{1,\ldots,k\}\\ I\neq\varnothing}} (-1)^{|I|+1} \Big(\overline{v}(X) - \overline{v}\Big(\Big(\bigcap_{i\in I} A_i\Big)^c\Big)\Big) \Leftrightarrow$$

$$\overline{v}\Big(\bigcap_{i=1}^k A_i^c\Big) \leqslant \sum_{\substack{I\subseteq\{1,\ldots,k\}\\ I\neq\varnothing}} (-1)^{|I|+1} \overline{v}\Big(\bigcup_{i\in I} A_i^c\Big).$$

The result for k-alternating games holds by $\overline{\overline{v}} = v$.

(iii) Take $A \subset B \subseteq X$, and apply supermodularity to $A, B \setminus A$. We find $v(B) \geqslant v(A) + v(B \setminus A) \geqslant v(A)$ by nonnegativity of v.

(iv) Suppose $A \subseteq B$. Then, supposing v maxitive, $v(B) = v(A \cup B) = v(A) \vee v(B)$, which implies $v(A) \leqslant v(B)$ (similar if v is minitive).

(v) Suppose v is maxitive. Then

$$\begin{aligned}\overline{v}(A \cap B) &= v(X) - v(A^c \cup B^c) = v(X) - (v(A^c) \vee v(B^c)) \\ &= (v(X) - v(A^c) \wedge (v(X) - v(B^c) = \overline{v}(A) \wedge \overline{v}(B).\end{aligned}$$

The reverse implication can be obtained similarly.

□

An important question is how to test k-monotonicity and total monotonicity. A direct application of the definitions seems to be computationally intractable. The following theorem ensures a test of minimal size.

Theorem 2.21 *Let v be a game on X. The following holds.*

(i) v is k-monotone for some $k \geqslant 2$ if and only if for every disjoint $S, K \subseteq X$ with $2 \leqslant |K| \leqslant k$,

$$\Delta_K v(S) = \sum_{T\subseteq K} (-1)^{|K\setminus T|} v(S \cup T) \geqslant 0;$$

(ii) v is totally monotone if and only if v is $(2^n - 2)$-monotone, with $|X| = n$.

Proof

(i) ($\Rightarrow$) Consider the family $\{S \cup (K \setminus \{x\})\}_{x\in K}$, with $|K| = k'$, $2 \leqslant k' \leqslant k$. Since k-monotonicity implies k'-monotonicity, applying k'-monotonicity to this family yields $\Delta_K v(S) \geqslant 0$.

($\Leftarrow$) Let $A_1, \dots, A_k$ be a family of distinct sets in 2^X, and put $[k] = \{1, \dots, k\}$. First, assume that some set in the family, say A_1, is included into another one. Define $K = \{i \in \{2, \dots, k\} : A_1 \subset A_i\} \neq \varnothing$. We show that in this case the test of k-monotonicity reduces to the test of $(k-1)$-monotonicity on the family $A_2, \dots, A_k$. Indeed, clearly $\bigcup_{i=1}^k A_i = \bigcup_{i=2}^k A_i$. Moreover,

$$\sum_{\substack{I \subseteq [k] \\ I \neq \varnothing}} (-1)^{|I|+1} v\Big(\bigcap_{i \in I} A_i\Big) = \sum_{\substack{I \subseteq [k] \\ I \not\ni 1 \\ I \neq \varnothing}} (-1)^{|I|+1} v\Big(\bigcap_{i \in I} A_i\Big) + \sum_{\substack{I \subseteq [k] \\ I \ni 1}} (-1)^{|I|+1} v\Big(\bigcap_{i \in I} A_i\Big).$$

We show that the 2nd term in the right-hand side is 0, which proves the desired result.

$$\begin{aligned}
\sum_{\substack{I \subseteq [k] \\ I \ni 1}} (-1)^{|I|+1} v\Big(\bigcap_{i \in I} A_i\Big) &= \sum_{\substack{I \subseteq [k] \\ I \ni 1}} (-1)^{|I|+1} v\Big(A_1 \cap \bigcap_{i \in I \setminus K} A_i\Big) \\
&= \sum_{J \subseteq [k] \setminus (K \cup 1)} v\Big(A_1 \cap \bigcap_{i \in J} A_i\Big) \underbrace{\sum_{L \subseteq K} (-1)^{|J|+|L|+2}}_{=0 \text{ by } (1.1)} \\
&= 0.
\end{aligned}$$

Therefore, we can assume that no set is included into another one. Observe then that the family can be rewritten in the form $S \cup K_1, \dots, S \cup K_k$, with $S = \bigcap_{i=1}^k A_i$, with all K_i being distinct and nonempty and forming an antichain. The test of k-monotonicity reads

$$v(S \cup \bigcup_{i \in [k]} K_i) - \sum_{\substack{I \subseteq [k] \\ I \neq \varnothing}} (-1)^{|I|+1} v\Big(S \cup \bigcap_{i \in I} K_i\Big) \geqslant 0, \tag{2.6}$$

which we denote for convenience by $D(v; S \cup K_1, \dots, S \cup K_k) \geqslant 0$. We prove k-monotonicity by induction on $\Big|\bigcup_{i=1}^k A_i \setminus \bigcap_{i=1}^k A_i\Big| = \Big|\bigcup_{i=1}^k K_i\Big|$. Since the K_i's are distinct, nonempty and form an antichain, we have $|\cup_{i=1}^k K_i| \geqslant k$. Let us prove (2.6) when $|\cup_{i=1}^k K_i| = k$. We can put w.l.o.g. $K_i = \{i\}$ for $i = 1, \dots, k$, hence (2.6) reduces to (with a slight abuse of notation)

$$v(S \cup [k]) - \sum_{i \in [k]} v(S \cup \{i\}) - \underbrace{\sum_{\substack{I \subseteq [k] \\ |I| \geqslant 2}} (-1)^{|I|+1} v(S)}_{(-k+1)v(S)} \geqslant 0. \tag{2.7}$$

Now, $\Delta_K v(S) \geqslant 0$ for every $K \subseteq [k]$, $|K| \geqslant 2$, yields

$$\sum_{\substack{K\subseteq[k]\\|K|\geqslant 2}} \sum_{T\subseteq K} (-1)^{|K\setminus T|} v(S\cup T) \geqslant 0.$$

Rearranging and using (1.1) in Lemma 1.1, we find

$$\begin{aligned}
&\sum_{\substack{K\subseteq[k]\\|K|\geqslant 2}} \sum_{T\subseteq K} (-1)^{|K\setminus T|} v(S\cup T)\\
&= \sum_{K\subseteq[k]} \sum_{T\subseteq K} (-1)^{|K\setminus T|} v(S\cup T) - k(-1)v(S) - \sum_{i\in[k]} v(S\cup i) - v(S)\\
&= \sum_{T\subseteq[k]} v(S\cup T) \sum_{K\in[T,[k]]} (-1)^{|K\setminus T|} + (k-1)v(S) - \sum_{i\in[k]} v(S\cup i)\\
&= v(S\cup[k]) + (k-1)v(S) - \sum_{i\in[k]} v(S\cup i) \geqslant 0.
\end{aligned}$$

Hence (2.7) holds.

Assume (2.6) holds, and let us prove it still holds with the family $S\cup K_1\cup\{x\}, S\cup K_2, \ldots, S\cup K_k$, with $x \notin K_1\cup K_2\cup\cdots\cup K_k$. The k-monotonicity test becomes

$$\begin{aligned}
&D(v; S\cup K_1\cup\{x\}, S\cup K_2, \ldots, S\cup K_k)\\
&= v(S\cup \bigcup_{i\in[k]} K_i \cup\{x\}) - v(S\cup K_1\cup\{x\}) - \sum_{i\in[k]\setminus\{1\}} v(S\cup K_i)\\
&\quad - \sum_{\substack{I\subseteq[k]\\|I|\geqslant 2}} (-1)^{|I|+1} v\Big(S\cup \bigcap_{i\in I} K_i\Big)\\
&= D(v; S\cup K_1, \ldots, S\cup K_k) + v(S\cup \bigcup_{i\in[k]} K_i\cup\{x\}) - v(S\cup K_1\cup\{x\})\\
&\quad - v(S\cup \bigcup_{i\in[k]} K_i) + v(S\cup K_1)\\
&= D(v; S\cup K_1, \ldots, S\cup K_k) + D(v; S'\cup\{x\}, S'\cup K')
\end{aligned}$$

with $S' = S\cup K_1$ and $K' = K_2\cup K_3\cup\cdots\cup K_k$. By induction hypothesis, both terms in the right-hand side are nonnegative, hence $D(v; S\cup K_1\cup\{x\}, S\cup K_2, \ldots, S\cup K_k) \geqslant 0$.

(ii) It is enough to prove that $(2^n - 2)$-monotonicity implies $(2^n - 1)$- and 2^n-monotonicity.

Let us prove $(2^n - 1)$-monotonicity. Consider a family $A_1, \dots, A_{2^n-1}$ of subsets of X. If some subsets are equal, then $(2^n - 2)$-monotonicity applies and the required inequality is satisfied. If all subsets are different, then necessarily either $\varnothing$ or X occurs in the family. Suppose $A_1 = \varnothing$. Then we have

$$v\Big(\bigcup_{i=1}^{2^n-1} A_i\Big) = v\Big(\bigcup_{i=2}^{2^n-1} A_i\Big) \geqslant \sum_{\substack{I\subseteq\{2,\dots,2^n-1\}\\ I\neq\varnothing}} (-1)^{|I|+1} v\Big(\bigcap_{i\in I} A_i\Big)$$
$$= \sum_{\substack{I\subseteq\{1,\dots,2^n-1\}\\ I\neq\varnothing}} (-1)^{|I|+1} v\Big(\bigcap_{i\in I} A_i\Big),$$

applying $(2^n - 2)$-monotonicity and the fact that $\bigcap_{i\in I} A_i = \varnothing$ if $I \ni 1$. Now, suppose $A_1 = X$. Set $K = \{1, \dots, 2^n - 1\}$, and $\mathcal{K}' = \{I \subseteq K, I \ni 1\}$, $\mathcal{K}'' = \{I \subseteq K, I \not\ni 1, I \neq \varnothing\}$. Observe that the mapping $I \mapsto I\setminus\{1\}$ realizes a bijection between $\mathcal{K}' \setminus \{1\}$ and $\mathcal{K}''$. Moreover, for any $I \in \mathcal{K}'$,

$$(-1)^{|I|+1} v\Big(\bigcap_{i\in I} A_i\Big) = (-1)^{|I\setminus\{1\}|+2} v\Big(\bigcap_{i\in I\setminus\{1\}} A_i\Big).$$

Hence

$$\sum_{\substack{I\subseteq K\\ I\neq\varnothing}} (-1)^{|I|+1} v\Big(\bigcap_{i\in I} A_i\Big) = \sum_{I\in\mathcal{K}'} (-1)^{|I|+1} v\Big(\bigcap_{i\in I} A_i\Big) + \sum_{I\in\mathcal{K}''} (-1)^{|I|+1} v\Big(\bigcap_{i\in I} A_i\Big)$$
$$= v(X) = v\Big(\bigcup_{i=1}^{2^n-1} A_i\Big).$$

We prove now 2^n-monotonicity. We consider a family of 2^n subsets of X. If two or more subsets are equal, the desired equality holds by $(2^n - 1)$-monotonicity. If all subsets differ, then X belongs to the family, and the preceding argument can be applied.

□

Assertion (i) will be expressed more simply through the Möbius transform (Theorem 2.33).

Remark 2.22

(i) The proof of (i) is rather technical and we are not aware of a simpler proof, although the result is simple to show for 2-monotonicity or for total monotonicity (for the latter result, see, e.g., Crama et al. [64]). Another version of (i) expressed via the Möbius transform [Theorem 2.33(iii)] was proved by Chateauneuf and Jaffray [49]. A stronger result (in the sense that an equivalent condition is given for the derivatives w.r.t. K to be positive for every K of a given size) was shown by Foldes and Hammer [144].
(ii) Result (ii) was established by Barthélemy [18].

◇

Applying (i) for $k = 2$ gives a very simple test for supermodularity (condition (i) in the next corollary).

Corollary 2.23 *Let v be a game on X. The following holds.*

(i) v is supermodular if and only if for any $A \subseteq X$, $i, j \notin A$, $i \neq j$, we have

$$\Delta_{ij}v(A) = v(A \cup \{i,j\}) - v(A \cup \{i\}) - v(A \cup \{j\}) + v(A) \geqslant 0; \tag{2.8}$$

(ii) v is supermodular if and only if for every $A \subseteq B \subseteq X$, for every $i \in X \setminus B$, we have

$$\Delta_i v(A) \leqslant \Delta_i v(B). \tag{2.9}$$

Proof (ii) Clearly, the condition is necessary. Suppose it holds with $B = A \cup \{j\}$. Then (ii) reduces to (i), which proves supermodularity. □

Remark 2.24 Condition (ii) is often used in cooperative game theory, where it is often taken as the definition of a *convex* game. This name is justified by analogy with convex real functions. Indeed, condition (ii) says that the higher the argument of the function, the higher the derivative. For set functions defined on 2^X with X finite, the corollary shows that the notions of convexity and supermodularity coincide. But this is not true in general, for instance, if $X = \mathbb{N}$ (see Example 2.109 and Theorem 2.110), or if the domain is a subfamily of 2^N (see Theorem 2.114 for a generalization, and Example 2.115 for a counterexample). Hence, we should make a distinction between these two concepts, although it is common, especially in decision theory, to use the term "convex" for supermodular functions.[2] In combinatorial optimization, and generally in discrete mathematics, only the term "supermodular" is used. ◇

---

[2]It seems that this term was introduced by Shapley [301].

2.8 Main Families of Capacities

We present the best known families of capacities. Most of them play an important rôle in this monograph.

2.8.1 0-1-Capacities

A *0-1-capacity* is a capacity valued on $\{0, 1\}$. Apart from the null capacity 0, all 0-1-capacities are normalized. In game theory, 0-1-capacities are called *simple games*. They correspond exactly to the situation depicted in Example 2.7 (voting games).

There are two remarkable normalized 0-1-capacities, the smallest one $\mu_{\min}$ and the greatest one $\mu_{\max}$, given by

$$\mu_{\min}(A) = 0 \text{ for all } A \subset X, \quad \mu_{\max}(A) = 1 \text{ for all } \varnothing \neq A \subseteq X.$$

A 0-1-capacity μ is uniquely determined by the antichain of its minimal winning coalitions. A set A is a *winning coalition* for μ if $\mu(A) = 1$. It is *minimal winning* if in addition $\mu(B) = 0$ for all $B \subset A$. Note that two minimal winning coalitions are incomparable for inclusion, hence the collection of minimal winning coalition is an antichain in 2^X. For example, the antichains giving rise to $\mu_{\min}$ and $\mu_{\max}$ are respectively $\{X\}$ and $\{\{i\} : i \in X\}$.

The number of antichains in 2^X with $|X| = n$, or equivalently, the number of monotonic Boolean functions of n variables is the Dedekind[3] number $M(n)$ [75]. There is no known closed-form formula for an exact computation of these numbers. Up to now, only the nine first numbers are known. By the fact that a 0-1-capacity must satisfy $\mu(\varnothing) = 0$, one antichain is eliminated, namely, the antichain $\{\varnothing\}$, giving rise to the constant set function 1. Thus, the number of 0-1-capacities on X is $M(|X|) - 1$ (Table 2.1).

2.8.2 Unanimity Games

Let $A \subseteq X, A \neq \varnothing$. The *unanimity game centered on A* is the game u_A defined by

$$u_A(B) = \begin{cases} 1, & \text{if } B \supseteq A \\ 0, & \text{otherwise.} \end{cases}$$

[3]Richard Dedekind (Braunschweig, 1831 – Braunschweig, 1916) is a German mathematician. He brought important contributions in abstract algebra, algebraic number theory and the foundations of the real numbers.

n	M(n)
0	2
1	3
2	6
3	20
4	168
5	7581
6	7828354
7	2414682040998
8	56130437228687557907788

Table 2.1 The Dedekind numbers $M(n)$ for $0 \leqslant n \leqslant 8$

Note that unanimity games are 0-1-valued capacities. Dirac measures (see Example 2.4) are unanimity games (exactly the additive unanimity games) (why?).

We will see in Sect. 2.15 that unanimity games play a central rôle. They are also called *simple support functions* by Shafer [296, p. 75].

2.8.3 Possibility and Necessity Measures

A *possibility measure* on X is a normalized capacity Π on X that is maxitive; i.e., satisfying $\Pi(A \cup B) = \max(\Pi(A), \Pi(B))$ for all $A, B \in 2^X$. A *necessity measure* is a normalized capacity Nec that is minitive; i.e., satisfying $\text{Nec}(A \cap B) = \min(\text{Nec}(A), \text{Nec}(B))$. The conjugate of a possibility measure (respectively, a necessity measure) is a necessity measure (respectively, a possibility measure). Note that Π is entirely determined by its value on singletons: $\Pi(A) = \bigvee_{x \in A} \Pi(\{x\})$ for any A in 2^X, like a probability measure. The function $\pi : X \to [0, 1]$ defined by $\pi(x) = \Pi(\{x\})$ is called a *possibility distribution*. Since $\Pi(X) = 1$, we have $\max_{x \in X} \pi(x) = 1$.

Note that unanimity games are necessity measures (why?).

The concept of possibility measure was proposed initially by Zadeh [357]. Later Dubois and Prade enriched the concept by adding the dual notion of necessity measure, and developed a whole theory of representation of uncertainty [106]. Possibility measures are, however, not restricted to this usage, and arise naturally in many contexts, even in more physical ones where it is generally thought that classical additive measures are suitable. Consider, e.g., the transportation of tree trunks by a truck. Additive measures are suitable to measure the total weight of the tree trunks, and one can check if the total weight does not exceed the limit. Now, consider the length of each trunk. Clearly, the length of the bunch of trunks is determined by the length of the longest trunk, which should be within some limits. In the latter case, a maxitive measure is suitable.

Possibility and necessity measures are studied in Chap. 7.

2.8.4 Belief and Plausibility Measures

A *belief measure* is a totally monotone normalized capacity, usually denoted by Bel.

This is the usual definition, although it is enough to define a belief measure as a nonnegative normalized and totally monotone game, owing to Theorem 2.20(iii). We denote the set of belief measures on X by $\mathcal{B}(X)$.

A *plausibility measure* is a totally alternating normalized capacity, usually denoted by Pl. By Theorem 2.20(ii), the conjugate of a belief measure is a plausibility measure, and vice versa. Possibility and necessity measures are particular cases of belief and plausibility measures, as will be shown in Sect. 2.10.2. Belief measures were introduced by Shafer [296], rephrasing and developing ideas of Dempster [77], and forming what he called *evidence theory* (see also Chap. 7).

2.8.5 Decomposable Measures

They generalize the idea that measures can be defined by distributions, like probability and possibility measures, by means of some operator playing the rôle of the addition. This operator is in general a triangular conorm, although other definitions are possible.

A *triangular conorm* or *t-conorm* for short is a function $\mathbf{S} : [0,1]^2 \to [0,1]$ satisfying for all $x, y, z \in [0,1]$,

(i) associativity: $\mathbf{S}(x, \mathbf{S}(y,z)) = \mathbf{S}(\mathbf{S}(x,y), z)$;
(ii) symmetry: $\mathbf{S}(x,y) = \mathbf{S}(y,x)$;
(iii) nondecreasingness: $\mathbf{S}(x', y) \leqslant \mathbf{S}(x,y)$ for all $x' \leqslant x$;
(iv) neutral element: $\mathbf{S}(0,x) = x$.

By associativity, a t-conorm is unambiguously defined for any number of arguments. If necessary, $\mathbf{S}$ for n arguments is denoted by $\mathbf{S}^{(n)}$. Common examples of t-conorms are the maximum, the *probabilistic sum*,

$$\mathbf{S_P}(x,y) = 1 - (1-x)(1-y) = x + y - xy$$

and the *Łukasiewicz t-conorm*, a.k.a. *bounded sum*

$$\mathbf{S_L}(x,y) = \min(x+y, 1).$$

Note that by (iii) and (iv) the smallest t-conorm is the maximum, and that 1 is the annihilator element; i.e., $\mathbf{S}(1,x) = 1$ (why?).

A continuous t-conorm is *Archimedean*[4] if it satisfies $\mathbf{S}(x, x) > x$ for every $x \in]0, 1[$. Obviously, the maximum is not Archimedean but the probabilistic and bounded sums are.

A fundamental fact is that Archimedean continuous t-conorms can be represented as distorted additions. Specifically, $\mathbf{S}$ is continuous and Archimedean if and only if there exists a continuous and increasing mapping $s : [0, 1] \to [0, \infty]$, with $s(0) = 0$ (called the *additive generator*), such that

$$\mathbf{S}(x, y) = s^{-1}\big(\min(s(1), s(x) + s(y))\big) \qquad (x, y \in [0, 1]),$$

and s is unique up to a multiplicative positive constant. The additive generator of the probabilistic sum is $s_{\mathbf{P}}(u) = -\log(1 - u)$, while for the bounded sum it is simply $s_{\mathbf{L}}(u) = u$, for $u \in [0, 1]$.

An important distinction arises whether $s(1)$ is finite or not. If $s(1) = \infty$, then the t-conorm is said to be *strict*, otherwise it is said to be *nilpotent*. Thus, the probabilistic sum is strict, and the bounded sum is nilpotent. It turns out that any strict t-conorm $\mathbf{S}$ is isomorphic to the probabilistic sum, in the sense that there exists a bijection $\varphi : [0, 1] \to [0, 1]$ such that $\varphi^{-1}(\mathbf{S}(\varphi(x), \varphi(y)) = \mathbf{S_P}(x, y)$. Similarly, any nilpotent t-conorm is isomorphic to the bounded sum. Note that strict t-conorms are strictly increasing, while nilpotent t-conorms are not.

Remark 2.25 T-conorms were introduced by Schweizer and Sklar [292] for probabilistic metric spaces, and are dual to *triangular norms* (t-norms), which are binary operators on $[0, 1]$ satisfying associativity, symmetry, nondecreasingness and having neutral element 1. Both are widely used in many fields, for instance, artificial intelligence, computer sciences, many-valued logic, etc. For a thorough study and a proof of the above statements we refer the readers to the monographs of Klement et al. [210], Schweizer and Sklar [293] and to Grabisch et al. [177, Chap. 3]. ◇

Definition 2.26 Let $\mathbf{S}$ be a t-conorm. A normalized capacity μ is said to be *decomposable for* $\mathbf{S}$ or $\mathbf{S}$*-decomposable* if it satisfies

$$\mu(A \cup B) = \mathbf{S}(\mu(A), \mu(B))$$

for every disjoint $A, B \subseteq X$.

---

[4]The Archimedean property can be defined without continuity. It says that for every $x, y \in]0, 1[$, there exists $n \in \mathbb{N}$ such that $\mathbf{S}^{(n)}(x, \dots, x) > y$, or, equivalently, $\lim_{n\to\infty} \mathbf{S}^{(n)}(x, \dots, x) = 1$ for every $x \in]0, 1[$. Under continuity, it reduces to the above simple definition.

This property is a generalization of the classical Archimedean property of linearly ordered groups. Given an ordered group G and $x, y \in G$, x is said to be *infinitesimal* w.r.t. y if there is no integer n such that $\underbrace{x + \cdots + x}_{n} \geqslant y$. Then G is said to be *Archimedean* if there is no pair $x, y \in G$ such that x is infinitesimal w.r.t. y.

Clearly, a probability measure is a $\mathbf{S_L}$-decomposable measure, while a possibility measure is max-decomposable. Note that any function $\alpha : X \to [0, 1]$ defines a $\mathbf{S}$-decomposable measure μ by putting $\mu(\{x\}) = \alpha(x)$, provided $\mathbf{S}^{(n)}(\alpha(x) : x \in X) = 1$, with $n = |X|$. Then, letting $\mu(\varnothing) = 0$, nondecreasingness of the t-conorm ensures that μ is a normalized capacity. The function α is called the *distribution* of μ.

When $\mathbf{S}$ is a continuous and Archimedean t-conorm with additive generator s, decomposable measures may be distortions of (additive) measures. Specifically, for every $A \subseteq N$, we have

$$\mu(A) = s^{-1}\Big(\min(s(1), \sum_{x \in A} s \circ \mu(\{x\}))\Big). \tag{2.10}$$

Note that the condition $\mu(X) = 1$ is equivalent to $\sum_{x \in X} s \circ \mu(\{x\}) \geqslant s(1)$. Equation (2.10) shows that $s \circ \mu$ can be seen as a measure truncated by $s(1)$. This leads to the following classification of decomposable capacities:

(i) S type: $\mathbf{S}$ is strict. Then $s \circ \mu$ is additive and $s \circ \mu(X) = s(1) = \infty$; i.e., $s \circ \mu$ is an *infinite measure*.
(ii) NSA type: $\mathbf{S}$ is nilpotent and $\sum_{x \in X} s \circ \mu(\{x\}) = s(1)$. Then $s \circ \mu$ is additive and $s \circ \mu(X) = s(1) < \infty$; i.e., $s \circ \mu$ is a *finite measure* on $[0, s(1)]$.
(iii) NSP type: $\mathbf{S}$ is nilpotent and $\sum_{x \in X} s \circ \mu(\{x\}) > s(1)$. Then $s \circ \mu$ is not additive, and therefore not a measure.

$\mathbf{S_P}$-decomposable measures are of the S type.

Decomposable measures have been first introduced by Dubois and Prade [101], and independently by Weber [345]. The classification S, NSA, NSP is due to Weber.

2.8.6 λ-Measures

Let $\lambda > -1$. A normalized capacity μ is a λ-*measure* if it satisfies

$$\mu(A \cup B) = \mu(A) + \mu(B) + \lambda\mu(A)\mu(B) \tag{2.11}$$

for every disjoint $A, B \subseteq X$. Any λ-measure is a decomposable measure where the underlying t-conorm is a member of the family of *Sugeno-Weber t-conorms*, defined by:

$$\mathbf{S}^{\mathrm{SW}}_{\lambda}(x, y) = \min(1, x + y + \lambda xy)$$

for $\lambda \geqslant -1$, and $\mathbf{S}^{\mathrm{SW}}_{\infty}(x, y) = \max(x, y)$ if x or y is 0, and 1 otherwise (called the *drastic t-conorm* because it is the greatest one). Note that $\mathbf{S}^{\mathrm{SW}}_{-1} = \mathbf{S_P}$ and $\mathbf{S}^{\mathrm{SW}}_{0} = \mathbf{S_L}$.

It is a family of Archimedean t-conorms, whose additive generator is

$$s_\lambda^{\mathrm{SW}}(u) = \begin{cases} -\log(1-u), & \text{if } \lambda = -1 \\ u, & \text{if } \lambda = 0 \\ \frac{\log(1+\lambda u)}{\log(1+\lambda)}, & \text{if } \lambda \in]-1,\infty[\setminus \{0\}, \end{cases}$$

and $s_\infty^{\mathrm{SW}}(u) = 1 + u$ for $u \in]0,1]$, and $s_\infty^{\mathrm{SW}}(0) = 0$. Therefore, except for $\lambda = -1$, all $\mathbf{S}_\lambda^{\mathrm{SW}}$ are nilpotent. This explains why $\lambda = -1$ is excluded for λ-measures: since the underlying t-conorm is strict, it would lead to a decomposable measure of the S type, hence $s_{-1}^{\mathrm{SW}} \circ \mu$ would be an infinite measure.

Then, a λ-measure μ is a distorted probability (i.e., $s_\lambda^{\mathrm{SW}} \circ \mu$ is a probability) if and only if

$$\sum_{x \in X} \frac{\log(1 + \lambda\mu(\{x\})}{\log(1+\lambda)} = 1,$$

which leads to the normalization condition

$$1 + \lambda = \prod_{x \in X}(1 + \lambda\mu(\{x\}). \tag{2.12}$$

This equation has a unique solution (see, e.g., Wang and Klir [343, Theorem 4.7]), which is positive if $\sum_{x \in X} \mu(\{x\}) < 1$, 0 in case of equality, and in the interval $]-1,0[$ when $\sum_{x \in X} \mu(\{x\}) > 1$.

Formula (2.11) for $\lambda \neq 0$ can be extended to an arbitrary number of disjoint sets, using the additive generator:

$$\mu\Big(\bigcup_{i \in I} A_i\Big) = s^{-1}\Big(\min(1, \sum_{i \in I} s \circ \mu(A_i))\Big),$$

with

$$s^{-1}(u) = \frac{1}{\lambda}((1+\lambda)^u - 1), \tag{2.13}$$

acting as a distortion function. This yields, assuming the normalization condition (2.12) is fulfilled and $\lambda \neq 0$,

$$\mu\Big(\bigcup_{i \in I} A_i\Big) = \frac{1}{\lambda}\Big(\prod_{i \in I}(1 + \lambda\mu(A_i)) - 1\Big).$$

From the above formula, one obtains the expression of μ in terms of its distribution, for $\lambda \neq 0$:

$$\mu(A) = \frac{1}{\lambda}\Big(\prod_{x\in A}(1 + \lambda\mu(\{x\})) - 1\Big), \tag{2.14}$$

for any $A \subseteq X$. Be careful that this expression is valid only if (2.12) holds.

Theorem 2.27 *The conjugate of a λ-measure is a λ'-measure with $\lambda' = -\frac{\lambda}{\lambda+1}$.*

Proof (Wang and Klir [343, Corollary 4.5]) μ being a λ-measure, we have immediately by (2.11)

$$\mu(A^c) = \frac{1 - \mu(A)}{1 + \lambda\mu(A)}.$$

It follows that

$$\begin{aligned}
&\overline{\mu}(A) + \overline{\mu}(B) - \frac{\lambda}{\lambda+1}\overline{\mu}(A)\overline{\mu}(B) \\
&= 1 - \mu(A^c) + 1 - \mu(B^c) - \frac{\lambda}{\lambda+1}(1 - \mu(A^c))(1 - \mu(B^c)) \\
&= \frac{(\lambda+1)\mu(A)}{1 + \lambda\mu(A)} + \frac{(\lambda+1)\mu(B)}{1 + \lambda\mu(B)} - \lambda\frac{(\lambda+1)\mu(A)\mu(B)}{(1 + \lambda\mu(A))(1 + \lambda\mu(B))} \\
&= \frac{(\lambda+1)(\mu(A) + \mu(B) + \lambda\mu(A)\mu(B))}{(1 + \lambda\mu(A))(1 + \lambda\mu(B))} \\
&= \frac{(\lambda+1)\mu(A\cup B)}{1 + \lambda\mu(A\cup B)} = 1 - \mu((A\cup B)^c) = \overline{\mu}(A\cup B).
\end{aligned}$$

□

Clearly, a λ-measure is superadditive if $\lambda \geqslant 0$, and subadditive otherwise. A stronger result holds.

Theorem 2.28 *A λ-measure is a belief measure if and only if $\lambda \geqslant 0$, and is a plausibility measure otherwise ($\lambda \in]-1, 0]$).*

(For a proof, see Corollary 2.38.)

Remark 2.29 λ-Measures were introduced by Sugeno [319, 320]. Due to their simplicity, they are often used in applications. ◇

2.9 Summary

In Sects. 2.7 and 2.8, we have seen that many notions are related to each other. The diagram on Fig. 2.1 provides a clear view of most notions encountered so far. It is valid in the finite case and gives a classification of various families of normalized capacities.

2.10 The Möbius Transform

Definition 2.30 Let ξ be a set function on X. The *Möbius transform* or *Möbius inverse* of ξ is a set function m^ξ on X defined by

$$m^\xi(A) = \sum_{B\subseteq A}(-1)^{|A\setminus B|}\xi(B) \tag{2.15}$$

for every $A \subseteq X$.

Note that $m^\xi(\varnothing) = \xi(\varnothing)$. Given m^ξ, it is possible to recover ξ by the formula

$$\xi(A) = \sum_{B\subseteq A} m^\xi(B) \qquad (A \subseteq X). \tag{2.16}$$

Indeed, we have, using (1.1):

$$\sum_{B\subseteq A} m^\xi(B) = \sum_{B\subseteq A}\sum_{C\subseteq B}(-1)^{|B\setminus C|}\xi(C) = \sum_{C\subseteq A}\xi(C)\sum_{B\in[C,A]}(-1)^{|B\setminus C|} = \xi(A).$$

The following is a useful technical lemma.

Lemma 2.31 *For any set function ξ on X, any disjoint sets $K, A \subseteq X$, we have*

$$\Delta_K\xi(A) = \sum_{L\in[K,A\cup K]} m^\xi(L).$$

In particular, $m^\xi(A) = \Delta_A\xi(\varnothing)$, which is (2.15).

Proof We show it by induction on $|A|$. As remarked, the lemma holds with $|A| = 0$. Suppose it holds till $|A| = a$ for some $a > 0$ and let us prove it still holds for $A \cup \{i\}$, with $i \in X \setminus A$. We have

$$\begin{aligned}\sum_{L\in[K,A\cup K\cup\{i\}]} m^\xi(L) &= \sum_{L\in[K,A\cup K]} m^\xi(L) + \sum_{L\in[K\cup\{i\},A\cup K\cup\{i\}]} m^\xi(L)\\ &= \sum_{L\subseteq K}(-1)^{|K\setminus L|}\xi(A\cup L) + \sum_{L\subseteq K\cup\{i\}}(-1)^{|K\cup\{i\}\setminus L|}\xi(A\cup L)\end{aligned}$$

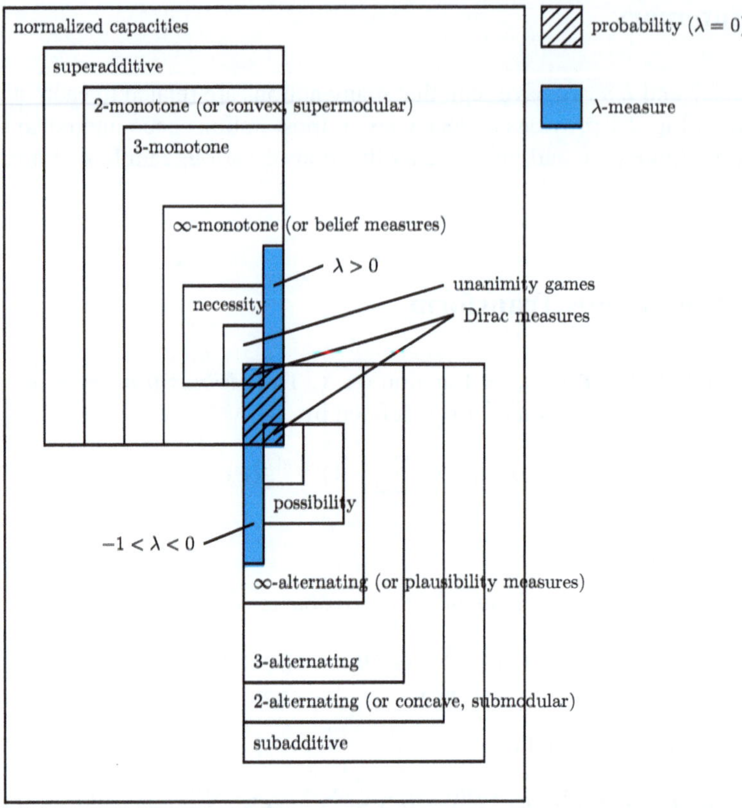

Fig. 2.1 Various families of normalized capacities on a finite set. All families are rectangular, except λ-measures. All capacities in the *left upper corner* are conjugate of those placed symmetrically in the *right lower corner*, except super/subadditive capacities. Probability measures are self-conjugate, therefore the two squares of Dirac measures coincide in reality

$$= \sum_{L \subseteq K} (-1)^{|K \setminus L|} \xi(A \cup L) - \sum_{\substack{L \subseteq K \cup \{i\} \\ L \not\ni i}} (-1)^{|K \setminus L|} \xi(A \cup L)$$

$$+ \sum_{\substack{L \subseteq K \cup \{i\} \\ L \ni i}} (-1)^{|K \cup \{i\} \setminus L|} \xi(A \cup L)$$

$$= \sum_{L' \subseteq K} (-1)^{|K \setminus L'|} \xi(A \cup \{i\} \cup L') = \Delta_K \xi(A \cup \{i\}).$$

□

Remark 2.32

(i) The Möbius transform is a widely used concept in combinatorics, capacity theory, nonadditive models of decision, and cooperative game theory. In the latter domain, the Möbius transform m^v of a game v is known under the name of *Harsanyi dividends* [193]. It has been rediscovered many times, and this notion arises naturally in many places of this monograph, because it occupies a central position.

(ii) The origin of the Möbius transform goes back to the German mathematician Möbius.[5] Later, Rota gave a general theory of Möbius functions [277] on partially ordered sets. We give below a flavor of it. A thorough treatment can be found in Aigner [2, Chap. 4], see also Berge [20, Sect. 3.2].

Let f, g be real-valued functions on some partially ordered set $(P, \leqslant)$, being locally finite (i.e., all intervals $[x, y]$ are finite) and having a least element 0. Consider the system of equations

$$f(x) = \sum_{y \leqslant x} g(y) \qquad (x \in P). \tag{2.17}$$

Knowing f, the problem is to solve (2.17), that is, to express g in terms of f. The function g is called the *Möbius inverse* of f. Rota proved that there always exists a unique solution to (2.17), given by

$$g(x) = \sum_{y \leqslant x} \mu(y, x) f(y) \tag{2.18}$$

where μ, called the *Möbius function*, is defined inductively by

$$\mu(x, y) = \begin{cases} 1, & \text{if } x = y \\ -\sum_{x \leqslant t < y} \mu(x, t), & \text{if } x < y \\ 0, & \text{otherwise.} \end{cases} \tag{2.19}$$

Note that μ depends solely on the structure of $(P, \leqslant)$.

The original definition of Möbius was given for the set of integers ordered by the integer division; i.e., $a \leqslant b$ if b is a multiple of a. Now, letting $(P, \leqslant) = (2^X, \subseteq)$, (2.17) reduces to (2.16). Let us find (2.15) using recursion formula (2.19). It suffices to show that the Möbius function is $\mu(A, B) = (-1)^{|B \setminus A|}$, for any $A \subseteq B \subseteq X$ and 0 otherwise. The latter is clear from (2.19) and we show the case $A \subseteq B \subseteq X$ by induction on the size of $B \setminus A$. For $|B \setminus A| = 1$, we

[5] August Ferdinand Möbius (Schulpforta, 1790 – Leipzig, 1868) is a German mathematician and astronomer. He discovered the famous Möbius strip, a non-orientable two-dimensional surface with only one side when embedded in three-dimensional Euclidean space.

have:

$$\mu(A, B) = -\mu(A, A) = -1.$$

Suppose the formula holds till $|B \setminus A| = k$ for some k, and consider that $|B \setminus A| = k + 1$. We have

$$\mu(A, B) = -\sum_{C \in [A,B[} \mu(A, C) = -\sum_{C \in [A,B[} (-1)^{|C \setminus A|} = -(0 - (-1)^{|B \setminus A|})$$
$$= (-1)^{|B \setminus A|},$$

using (1.1). In Sect. 2.12, we will find that this result is a particular case of a more general relation (Theorem 2.44).

(iii) The fact that the Möbius transform of a set function is closely linked to its derivatives is not surprising if one sees (2.17) as the discrete version of the integral equation

$$f(x) = \int_0^x g(y)\,\mathrm{d}y,$$

whose solution is $g(x) = f'(x)$, assuming $f(0) = 0$.

(iv) A last remark on terminology. The standard name for m^v is the *Möbius inverse* of v. The term "Möbius transform," although convenient and popular in decision making and game theory as it indicates that the original information (capacity, game) has been transformed, is however unfortunate because it conveys in the field of combinatorics exactly the inverse meaning; i.e., our Möbius transform is in fact the inverse Möbius transform (see, e.g., Björklund et al. [33]).

◇

2.10.1 Properties

Theorem 2.33 *Let v be a game on X. Then*

(i) v is additive if and only if $m^v(A) = 0$ for all $A \subseteq X$, $|A| > 1$. Moreover, we have $m^v(\{i\}) = v(\{i\})$ for all $i \in X$;

(ii) v is monotone if and only if

$$\sum_{i \in L \subseteq K} m^v(L) \geqslant 0 \qquad (K \subseteq X, \quad i \in K);$$

(iii) Let $k \geqslant 2$ be fixed. v is k-monotone if and only if

$$\sum_{L \in [A,B]} m^v(L) \geqslant 0 \qquad (A, B \subseteq X, \quad A \subseteq B, \quad 2 \leqslant |A| \leqslant k);$$

(iv) If v is k-monotone for some $k \geqslant 2$, then $m^v(A) \geqslant 0$ for all $A \subseteq X$ such that $2 \leqslant |A| \leqslant k$;
(v) v is a nonnegative totally monotone game if and only if $m^v \geqslant 0$;
(vi) A set function ξ is constant over 2^X if and only if $m^\xi(A) = 0$ for all $A \neq \varnothing$. In this case, $m^\xi(\varnothing) = \xi(A)$, $A \in 2^X$.

Proof

(i) It is easy to see that $m^v(\{i\}) = v(\{i\})$ for all $i \in X$, and $m^v = 0$ otherwise is a solution of (2.16). By uniqueness of the solution (Remark 2.32(ii)), this is the Möbius transform.
(ii) It suffices to check whether $v(A \cup i) \geqslant v(A)$ for all $i \notin A$ and all $A \subset X$. We have:

$$0 \leqslant v(A \cup i) - v(A) = \sum_{B \subseteq A \cup i} m^v(B) - \sum_{B \subseteq A} m^v(B) = \sum_{i \in B \subseteq A} m^v(B).$$

(iii) We know by Theorem 2.21 (i) that v is k-monotone for some k if and only if for every disjoint $A, K \subseteq X$ with $|K| \leqslant k$,

$$\Delta_K v(A) \geqslant 0.$$

By Lemma 2.31, we find immediately the desired result.
(iv) Take $A = B$ in (iii).
(v) Since a nonnegative and supermodular game is also monotone by Theorem 2.20(iii), from (iv) and (ii), we deduce that $m^v(A) \geqslant 0$ for every $A \subseteq X$, $|A| \geqslant 1$. Besides, $m^v(\varnothing) = 0$.
(vi) Suppose ξ is constant. Then by Lemma 1.1(i) the result is immediate. The converse is also immediate, using (2.16).

□

(iii) is due to Chateauneuf and Jaffray [49], however our proof is different.

Later, we will give upper and lower bounds of the Möbius transform of a normalized capacity (Theorem 2.63).

The following is useful.

Lemma 2.34 *Let v be a game, and consider its conjugate $\overline{v}$. The Möbius transform of $\overline{v}$ is given by*

$$m^{\overline{v}}(A) = (-1)^{|A|+1} \sum_{B \supseteq A} m^v(B).$$

Proof We use the fact that

$$\check{m}^{\overline{v}}(A) = (-1)^{|A|+1} m^{v}(A) \qquad (A \subseteq X)$$

where $\check{m}^{v}$ is the co-Möbius transform (see Sects. 2.11 and 2.12.4, (2.39) and Theorem 2.48), given by

$$m^{v}(A) = \sum_{B \supseteq A} m^{v}(B).$$

Since $\overline{\overline{v}} = v$, we obtain the desired result. □

2.10.2 Möbius Transform of Remarkable Games and Capacities

Unanimity Games

We begin by remarking that the Möbius transform of unanimity games is particularly simple. For any $A \subseteq X, A \neq \varnothing$,

$$m^{u_A}(B) = \begin{cases} 1, & \text{if } B = A \\ 0, & \text{otherwise.} \end{cases} \tag{2.20}$$

Indeed, using (2.20) in (2.16), we immediately obtain u_A (see also Theorem 2.56).

0-1-Capacities

We now provide the Möbius transform of 0-1-capacities.

Theorem 2.35 *Let μ be a 0-1-capacity, generated by the antichain $\mathcal{A} = \{A_1, \ldots, A_k\}$. Then*

$$m^{\mu}(B) = \begin{cases} (-1)^{|I|+1}, & \text{if } B = \bigcup_{i \in I} A_i \text{ for some } I \subseteq \{1, \ldots, k\}, I \neq \varnothing \\ 0, & \text{otherwise.} \end{cases}$$

Proof We prove that

$$\mu = \sum_{\substack{I \subseteq [k] \\ I \neq \varnothing}} (-1)^{|I|+1} u_{\bigcup_{i \in I} A_i},$$

from which the desired result follows. This amounts to showing that

$$\sum_{\substack{I\subseteq[k]\\ I\neq\varnothing}} (-1)^{|I|+1} u_{\bigcup_{i\in I} A_i}(S) = 1 \tag{2.21}$$

for all S such that $S \supseteq A_i$ for some $A_i \in \mathcal{A}$, and 0 otherwise. For a fixed S, define the collection $\mathcal{S} = \{A_i \in \mathcal{A} : A_i \subseteq S\}$, and let $L = \{i \in [k] : A_i \subseteq S\}$. If $\mathcal{S} = \varnothing$, then $S \not\supseteq \bigcup_{i\in I} A_i$, for all $I \subseteq [k]$, $I \neq \varnothing$. Therefore, the left term in (2.21) is equal to 0, as desired. Suppose now that $\mathcal{S} \neq \varnothing$. Then

$$\sum_{\substack{I\subseteq[k]\\ I\neq\varnothing}} (-1)^{|I|+1} u_{\bigcup_{i\in I} A_i}(S) = \sum_{\substack{I\subseteq L\\ I\neq\varnothing}} (-1)^{|I|+1} = 1,$$

using (1.1). □

Belief and Plausibility Measures

We know that belief measures are characterized by a nonnegative Möbius transform [Theorem 2.33(v)].

Let Bel be a belief measure with Möbius transform m. We know that its conjugate $\overline{\mathrm{Bel}}$ is a plausibility measure, which we denote by Pl. We first express Pl in terms of m. We have for every $A \subseteq X$,

$$\begin{aligned}
\mathrm{Pl}(A) &= \mathrm{Bel}(X) - \mathrm{Bel}(A^c)\\
&= \sum_{B\subseteq X} m(B) - \sum_{B\subseteq A^c} m(B)\\
&= \sum_{B\subseteq X} m(B) - \sum_{B\cap A=\varnothing} m(B)\\
&= \sum_{B\cap A\neq\varnothing} m(B).
\end{aligned} \tag{2.22}$$

Now, we can obtain the Möbius transform of Pl in terms of m by application of Lemma 2.34:

$$m^{\mathrm{Pl}}(A) = (-1)^{|A|+1}\check{m}(A) = (-1)^{|A|+1}\sum_{B\supseteq A} m(B), \tag{2.23}$$

where $\check{m}$ is the co-Möbius transform of Bel (see Sects. 2.11 and 2.12.4).

Possibility and Necessity Measures

Theorem 2.36 *Let* Π *be a possibility measure on* X*, and* Nec *its conjugate (necessity measure). Then the Möbius transform of* Nec *is nonnegative and lives on a chain* $A_1 \subset A_2 \subset \cdots \subset A_q$.

Proof We first prove by contradiction that the Möbius transform of Nec lives on a chain. Suppose there exist $A, B \in 2^X$ with $m^{\text{Nec}}(A), m^{\text{Nec}}(B) \neq 0$ such that $A \setminus B$ and $B \setminus A$ are nonempty, and choose such A, B that are both smallest w.r.t. inclusion. We prove that necessarily $\text{Nec}(A \cap B) \neq \text{Nec}(A) \wedge \text{Nec}(B)$.

Suppose $A \cap B = \varnothing$. Then there is no $A' \subset A$ such that $m^{\text{Nec}}(A') \neq 0$, and similarly for B, so that

$$\text{Nec}(A) \wedge \text{Nec}(B) = m^{\text{Nec}}(A) \wedge m^{\text{Nec}}(B) \neq 0 = \text{Nec}(\varnothing) = \text{Nec}(A \cap B).$$

Suppose then that $A \cap B \neq \varnothing$. We have

$$\text{Nec}(A) = m^{\text{Nec}}(A) + \sum_{C \subseteq A \cap B} m^{\text{Nec}}(C)$$

$$\text{Nec}(B) = m^{\text{Nec}}(B) + \sum_{C \subseteq A \cap B} m^{\text{Nec}}(C)$$

$$\text{Nec}(A \cap B) = \sum_{C \subseteq A \cap B} m^{\text{Nec}}(C).$$

Because $m^{\text{Nec}}(A), m^{\text{Nec}}(B)$ are both nonzero, we have $\text{Nec}(A \cap B) \neq \text{Nec}(A) \wedge \text{Nec}(B)$.

Second, we prove nonnegativity. Suppose $m^{\text{Nec}}(A) < 0$ for some A in the chain, and call A', A'' the "neighbors" of A in the chain; i.e., $A' \subset A \subset A''$. Then $\text{Nec}(A) < \text{Nec}(A')$, which contradicts the monotonicity of Nec. □

As a consequence, by Theorem 2.33(v), necessity measures are monotone and totally monotone normalized capacities (i.e., belief measures), and possibility measures are special cases of plausibility measures.

The precise determination of the Möbius transform of Π and Nec can be readily done from previous Theorem 2.36. Let π be the possibility distribution generating Π, with $\mathbf{ran}\,\pi \setminus \{0\} = \{\pi^1, \pi^2, \ldots, \pi^q\}$ its range after exclusion of the 0 value, supposing $1 = \pi^1 > \pi^2 > \cdots > \pi^q > 0$. We claim that the sets in the support of m^{Nec} are given by

$$A_i = \{x \in X : \pi(x) \geqslant \pi^i\} \qquad (i = 1, \ldots, q) \tag{2.24}$$

and

$$m^{\text{Nec}}(A_i) = \pi^i - \pi^{i+1} \qquad (i = 1, \ldots, q), \tag{2.25}$$

with the convention $\pi^{q+1} = 0$. By (2.22), it suffices to check that the plausibility measure Pl generated by m^{Nec} coincides with Π. We have

$$\text{Pl}(A) = \sum_{\substack{i\in\{1,\dots,q\}\\ A\cap A_i\neq\varnothing}} m^{\text{Nec}}(A_i) = \sum_{i\geqslant k_0} m^{\text{Nec}}(A_i) = \pi^{k_0},$$

with k_o such that $\pi^{k_0} = \max_{x\in A}\pi(x)$. On the other hand, we have $\Pi(A) = \pi^{k_0}$.

The Möbius transform of Π can be obtained from Lemma 2.34:

$$m^{\Pi}(A) = (-1)^{|A|+1}\sum_{B\supseteq A} m^{\text{Nec}}(B) = (-1)^{|A|+1}\sum_{i\geqslant \ell_0} m^{\text{Nec}}(A_i) = (-1)^{|A|+1}\pi^{\ell_0}$$

with ℓ_0 such that $\pi^{\ell_0} = \min_{x\in A}\pi(x)$. Hence,

$$m^{\Pi}(A) = (-1)^{|A|+1}\min_{x\in A}\pi(x). \tag{2.26}$$

λ-Measures

Theorem 2.37 *The Möbius transform of a λ-measure μ is given by*

$$m^{\mu}(A) = \lambda^{|A|-1}\prod_{i\in A}\mu(\{i\}) \qquad (\varnothing\neq A\subseteq X)$$

for any $\lambda > -1$, $\lambda\neq 0$.

Proof We have, using (2.14)

$$\begin{aligned}\mu(A) &= \frac{1}{\lambda}\Big(\prod_{x\in A}(1+\lambda\mu(\{x\}))-1\Big)\\ &= \frac{1}{\lambda}\sum_{\substack{B\subseteq A\\ B\neq\varnothing}}\Big(\lambda^{|B|}\prod_{x\in B}\mu(\{x\})\Big)\\ &= \sum_{\substack{B\subseteq A\\ B\neq\varnothing}}\Big(\lambda^{|B|-1}\prod_{x\in B}\mu(\{x\})\Big) = \sum_{B\subseteq A} m^{\mu}(B).\end{aligned}$$

□

Corollary 2.38 *A λ-measure is a belief measure if $\lambda\geqslant 0$, and a plausibility measure otherwise.*

Proof If $\lambda = 0$, μ is a probability measure, therefore a belief measure. Clearly, m^{μ} is nonnegative when $\lambda > 0$, and because μ is monotone, it follows that it is a belief measure by Theorem 2.33(v). Now, if $-1 < \lambda < 0$, we put $\lambda' = -\frac{\lambda}{\lambda+1} > 0$. By Theorem 2.27, we deduce that μ is a plausibility measure. □

(See Wang and Klir [343, Theorem 4.21] for these two results.)

2.11 Other Transforms

We may take a more general viewpoint by considering the Möbius transform as a *transformation* acting on the space of set functions on X, denoted by $\mathbb{R}^{(2^X)}$, that is, a mapping $T : \mathbb{R}^{(2^X)} \to \mathbb{R}^{(2^X)}$, assigning to any set function ξ its *transform* $T(\xi)$ by T.

A transformation T is *linear* if $T(\alpha\xi_1+\xi_2) = \alpha T(\xi_1)+T(\xi_2)$ for any set functions ξ_1, ξ_2 and $\alpha \in \mathbb{R}$. It is *invertible* if T^{-1} exists.

Note: Although this is not mathematically correct, we will in the whole book not distinguish any more between "transform" and "transformation" and speak only of "transform". The fact is that this distinction causes more intricacies than it clarifies the things, and also it seems to be a widespread usage.

We observe that the Möbius transform is linear and invertible. It is the usage to call the inverse Möbius transform the *Zeta transform*.

We introduce three other natural transforms that are useful. We set in all this section $|X| = n$.

Definition 2.39 (Grabisch et al. [178]) The *co-Möbius transform* $\check{m}$ is a linear invertible transform, defined for any set function ξ by

$$\check{m}^{\xi}(A) = \sum_{B \supseteq X \setminus A} (-1)^{n-|B|} \xi(B) = \sum_{B \subseteq A} (-1)^{|B|} \xi(X \setminus B) \tag{2.27}$$

for all $A \subseteq X$.

The inverse transform is given by

$$\xi(A) = \sum_{B \subseteq X \setminus A} (-1)^{|B|} \check{m}^{\xi}(B). \tag{2.28}$$

Remark 2.40 It is known in Dempster-Shafer theory under the name of "commonality function" (Shafer [296]), and in possibility theory under the name of "guaranteed possibility measure;" see, e.g., Dubois and Prade [109], and Chap. 7. ◇

Definition 2.41 (Grabisch [163]) The *interaction transform* I is a linear invertible transform, defined for any set function ξ by

$$I^{\xi}(A) = \sum_{B \subseteq X \setminus A} \frac{(n-b-a)!b!}{(n-a+1)!} \Delta_A \xi(B) = \sum_{K \subseteq X} \frac{|X \setminus (A \cup K)|!|K \setminus A|!}{(n-a+1)!} (-1)^{|A \setminus K|} \xi(K) \tag{2.29}$$

for all $A \subseteq X$, where a, b, k are cardinalities of subsets A, B, K, respectively.

The second equality is obtained from the first one as follows:

$$I^{\xi}(A) = \sum_{B\subseteq X\setminus A} \frac{(n-b-a)!b!}{(n-a+1)!} \sum_{L\subseteq A} (-1)^{|A\setminus L|}\xi(B\cup L)$$
$$= \sum_{K\subseteq X} \frac{|X\setminus (A\cup K)|!|K\setminus A|!}{(n-a+1)!}(-1)^{|A\setminus K|}\xi(K),$$

letting $K = B \cup L$.

The inverse transform will be given in Sect. 2.12 [Eq. (2.43)].

Definition 2.42 (Roubens [278]) The *Banzhaf interaction transform* is a linear invertible transform, defined for any set function ξ by

$$I_{\mathrm{B}}^{\xi}(A) = \left(\frac{1}{2}\right)^{n-a} \sum_{B\subseteq X\setminus A} \Delta_A\xi(B) = \left(\frac{1}{2}\right)^{n-a} \sum_{K\subseteq X} (-1)^{|A\setminus K|}\xi(K) \tag{2.30}$$

for all $A \subseteq X$, where a is the cardinality of A. The second equality is obtained as for the interaction transform.

The inverse transform is given by

$$\xi(A) = \sum_{K\subseteq X} \left(\frac{1}{2}\right)^{k} (-1)^{|K\setminus A|} I_{\mathrm{B}}^{\xi}(K) \tag{2.31}$$

for all $A \subseteq X$.

Remark 2.43 The two interaction transforms presented above have their origin in cooperative game theory and multicriteria decision making, as an extension of the *Shapley value* [298] and *Banzhaf value* [17] (or *Banzhaf power index*) (see Sect. 3.5 for a formal definition of a value). In the context of cooperative games and voting games, the Shapley and Banzhaf values, which we denote by ϕ^{Sh} and ϕ^{B} respectively, assign to any game v a vector $\phi^{\mathrm{Sh}}(v)$ (respectively, $\phi^{\mathrm{B}}(v)$) in $\mathbb{R}^X$, whose coordinate i is the "average" marginal contribution of player $i \in X$. The marginal contribution of i in coalition S is simply $\Delta_i v(S)$, and the way the average is computed over all coalitions differs for the Shapley and Banzhaf values. More precisely,

$$\phi_i^{\mathrm{Sh}}(v) = \sum_{S\subseteq X\setminus i} \frac{s!(n-s-1)!}{n!}\big(v(S\cup i) - v(S)\big) = I^{v}(\{i\}) \tag{2.32}$$

$$\phi_i^{\mathrm{B}}(v) = \frac{1}{2^{n-1}} \sum_{S\subseteq X\setminus i} \big(v(S\cup i) - v(S)\big) = I_{\mathrm{B}}^{v}(\{i\}), \tag{2.33}$$

for any $i \in X$. The Banzhaf value is often used as a power index in voting games, while the Shapley value has many applications in economics and computer sciences (see, e.g., Moretti and Patrone [247]) because, due to its property $\sum_{i=1}^{n} \phi_i^{\mathrm{Sh}}(v) = v(N)$ (see Theorem 2.45(ii) hereafter), it can be used as a rule of sharing of the total worth $v(N)$. We will return to the Shapley value and interaction transform in Chap. 6.

Finally, we mention that the interaction and Banzhaf interaction have been axiomatized in the framework of cooperative games by Grabisch and Roubens [180] (see also Fujimoto et al. [146]). ◇

All conversion formulas between these transforms are established in the subsequent sections, and are summarized in Appendix A, Tables A.2 and A.3. Still some other transforms are introduced in Sect. 2.16.

2.12 Linear Invertible Transforms

We investigate the case of linear invertible transforms, called hereafter operators. This will permit us to derive in a simple way all conversion formulas between the different representations. This section is mainly based on Denneberg and Grabisch [82]. Throughout it we assume $|X| = n$.

2.12.1 Definitions and Examples

An *operator* is a two-place set function $\Phi : 2^X \times 2^X \longrightarrow \mathbb{R}$. The multiplication $\star$ between operators and set functions is defined as follows, for every $A, B \subseteq X$

$$(\Phi \star \Psi)(A, B) = \sum_{C \subseteq X} \Phi(A, C)\Psi(C, B),$$

$$(\Phi \star \xi)(A) = \sum_{C \subseteq X} \Phi(A, C)\xi(C),$$

$$(\xi \star \Psi)(B) = \sum_{C \subseteq X} \xi(C)\Psi(C, B).$$

Defining a linear order on 2^X that is an extension of the partial order induced by $\subseteq$ (i.e., $A \subseteq B$ implies that A is ranked before B), we can identify 2^X with $\{1, 2, \ldots, 2^n\}$, and $\star$ becomes simply the ordinary multiplication of square matrices or matrices and vectors.

The Kronecker's delta

$$\Delta(A,B) = \begin{cases} 1, & \text{if } A = B \\ 0, & \text{otherwise} \end{cases}$$

is the unique neutral element from the left and from the right. If Φ is invertible, the inverse of Φ is denoted by Φ^{-1}, satisfying $\Phi \star \Phi^{-1} = \Delta,\ \Phi^{-1} \star \Phi = \Delta$.

Noting that the family

$$\mathcal{G} = \{\Phi : 2^X \times 2^X \to \mathbb{R} \,:\, \Phi(A,A) = 1\ \forall A \subseteq X,\ \Phi(A,B) = 0 \text{ if } A \not\subseteq B\}$$

of functions of two variables corresponds to triangular matrices with 1 on the diagonal (and therefore invertible), it follows that $(\mathcal{G}, \star)$ forms a group, and the inverse $\Phi^{-1} \in \mathcal{G}$ of $\Phi \in \mathcal{G}$ can be computed recursively using

$$\Phi^{-1}(A,B) = \begin{cases} 1, & \text{if } A = B \\ -\sum\limits_{A \subseteq C \subset B} \Phi^{-1}(A,C)\Phi(C,B), & \text{if } A \subset B. \end{cases} \tag{2.34}$$

(Berge [20, Chap. 3, Sect. 2]).

We introduce a first fundamental operator, the *Zeta operator* $Z(A,B)$, defined by

$$Z(A,B) = \begin{cases} 1, & \text{if} \quad A \subseteq B \\ 0, & \text{otherwise,} \end{cases}$$

and its inverse, the *Möbius operator*. Indeed, if we compare it with (2.16), we find

$$\xi = m^\xi \star Z, \tag{2.35}$$

and therefore $m^\xi = \xi \star Z^{-1}$ with

$$Z^{-1}(A,B) = \begin{cases} (-1)^{|B \setminus A|}, & \text{if } A \subseteq B \\ 0, & \text{otherwise.} \end{cases}$$

With the help of the Zeta operator, we introduce the co-Möbius transform

$$\check{m}^\xi = Z \star m^\xi = Z \star \xi \star Z^{-1}.$$

We will prove later that this is indeed the co-Möbius transform introduced in Definition 2.39.

The next fundamental operator we introduce is the *inverse Bernoulli operator* Γ (the reason for such a name will become clear later):

$$\Gamma(A,B) = \begin{cases} \frac{1}{|B\setminus A|+1}, & \text{if } A \subseteq B \\ 0, & \text{otherwise.} \end{cases}$$

2.12.2 Generator Functions, Cardinality Functions

We turn now to a special class of operators in $\mathcal{G}$, satisfying

$$\Phi(A,B) = \Phi(\varnothing, B \setminus A) \quad \text{for } A \subseteq B, \tag{2.36}$$

i.e., they can be represented by an ordinary set function $\varphi(A) = \Phi(\varnothing, A)$, denoted with the corresponding small letter. We call it the *generator function*. In fact, the set of such operators forms an Abelian group, as well as the corresponding set of generator functions:

$$\mathbf{g} = \{\varphi : 2^X \to \mathbb{R} : \varphi(\varnothing) = 1\}$$

with operation $\star$ defined by

$$\varphi \star \psi(A) = \sum_{C \subseteq A} \varphi(C)\psi(A \setminus C), \quad A \subseteq X.$$

The neutral element δ of $\mathbf{g}$ is

$$\delta(A) = \begin{cases} 1, & \text{if } A = \varnothing \\ 0, & \text{otherwise} \end{cases},$$

and the inverse of φ is denoted by $\varphi^{\star -1}$. Since Z and Γ have property (2.36), we can introduce the corresponding *Zeta generator function* and *Bernoulli generator functions*:

$$\zeta(A) = 1 \text{ for all } A \in 2^X,$$

$$\gamma(A) = \frac{1}{|A|+1}, \quad A \in 2^X.$$

If moreover φ is a function only of the cardinality of sets, then we call it a *cardinality function*, and the corresponding Φ a *cardinality operator*. Note that Z and Γ have also this property. It is convenient to associate to any cardinality function φ its *cardinal representation* $\widehat{\varphi} : \mathbb{N}_0 \to \mathbb{R}$ defined by $\widehat{\varphi}(m) = \varphi(A)$ for some A such that $|A| = m$, with $0 \leqslant m \leqslant n$.

2.12.3 Inverse of Cardinality Operators

The inverse of a cardinality function is computed as follows.

Theorem 2.44 (Inverse of a cardinality function) *Let* $\varphi \in \mathbf{g}$ *be a cardinality function. Then* $\varphi^{\star -1}$ *is a cardinality function with cardinal representation* $\widehat{\varphi^{\star -1}}$ *given by*

$$\widehat{\varphi^{\star -1}}(m) = \begin{cases} 1, & \text{if } m = 0 \\ -\sum_{k=0}^{m-1} \binom{m}{k} \widehat{\varphi}(m-k) \widehat{\varphi^{\star -1}}(k), & \text{if } m \in \mathbb{N}. \end{cases} \tag{2.37}$$

Proof We apply the recursion formula (2.34) with Φ defined by $\Phi(A, B) = \varphi(B \setminus A)$ for $A \subseteq B$, which yields

$$\varphi^{\star -1}(A) = \begin{cases} 1, & \text{if } A = \varnothing \\ -\sum_{C \subset A} \varphi^{\star -1}(C) \varphi(A \setminus C), & \text{otherwise.} \end{cases}$$

Using $\widehat{\varphi}$ we get

$$\varphi^{\star -1}(A) = -\sum_{k=0}^{m-1} \sum_{\substack{C \subset A \\ |C| = k}} \varphi^{\star -1}(C) \widehat{\varphi}(m-k) \qquad (|A| = m > 0).$$

We show by induction that $\varphi^{\star -1}(A)$ is a cardinality function. The assertion obviously holds for $|A| = 0$. Supposing that $\varphi^{\star -1}(C) = g(|C|)$ for $|C| < m$, we get for $|A| = m$,

$$\varphi^{\star -1}(A) = -\sum_{k=0}^{m-1} \binom{m}{k} g(k) \widehat{\varphi}(m-k),$$

which clearly depends only on m. Hence we can put $\varphi^{\star -1}(A) = g(m) = \widehat{\varphi^{\star -1}}(m)$, and the proof is complete. □

One can apply this formula to the Zeta generator function, and the readers can easily check that the Möbius generator function $\zeta^{\star -1}$ is

$$\zeta^{\star -1}(A) = (-1)^{|A|} \qquad (A \in 2^X).$$

Applying it to the Bernoulli function, we find for its cardinal representation $\widehat{\gamma}$ that $\widehat{\gamma}(0) = 1$, and for any $m \in \mathbb{N}$,

$$\widehat{\gamma}(m) = -\sum_{k=0}^{m-1} \binom{m}{k} \frac{1}{m-k+1} \widehat{\gamma}(k) = -\frac{1}{m+1} \sum_{k=0}^{m-1} \binom{m+1}{k} \widehat{\gamma}(k). \qquad (2.38)$$

We recognize in $\widehat{\gamma}(m)$, $m \in \mathbb{N}_0$, the recursive expression of the Bernoulli numbers B_m [see (1.4)]. The sequence of Bernoulli numbers starts with $B_0 = 1, B_1 = -\frac{1}{2}, B_2 = \frac{1}{6}, B_3 = 0, B_4 = -\frac{1}{30}, \ldots$, and $B_{2m+1} = 0$ for $m \geqslant 1$.

2.12.4 The Co-Möbius Operator

We recall that $\check{m}^\xi = Z \star m^\xi = Z \star \xi \star Z^{-1}$. This yields

$$\check{m}^\xi(A) = \sum_{B \supseteq A} m^\xi(B). \qquad (2.39)$$

We introduce the *co-Möbius operator* C by $\check{m}^\xi = \xi \star C$. Let us express $C(A, B)$. We have

$$\begin{aligned}
\check{m}^\xi(A) &= \sum_{B \subseteq X} Z(A, B) \sum_{D \subseteq X} \xi(D) Z^{-1}(D, B) \\
&= \sum_{B \supseteq A} \sum_{D \subseteq B} \xi(D)(-1)^{|B \setminus D|} \\
&= \sum_{D \subseteq X} \xi(D) \sum_{B \supseteq A \cup D} (-1)^{|B \setminus D|} \\
&= \sum_{D \subseteq X} \xi(D)(-1)^{|A| - |A \cap D|} \sum_{B \supseteq A \cup D} (-1)^{|B \setminus (A \cup D)|} \\
&= \sum_{D \supseteq X \setminus A} (-1)^{|X \setminus D|} \xi(D) = \sum_{D \subseteq A} (-1)^{|D|} \xi(X \setminus D),
\end{aligned}$$

where we have used (1.1). We recognize (2.27). Hence the co-Möbius operator reads

$$C(A, B) = \begin{cases} (-1)^{|X \setminus A|}, & \text{if } X \setminus A \subseteq B \\ 0, & \text{otherwise.} \end{cases}$$

Clearly, it does not belong to $\mathcal{G}$. The inverse formula is $\xi = \breve{m}^\xi \star C^{-1}$, and can be found similarly:

$$\xi(A) = \sum_{B \subseteq X \setminus A} (-1)^{|B|} \breve{m}^\xi(B),$$

which gives

$$C^{-1}(A, B) = \begin{cases} (-1)^{|B|}, & \text{if } B \subseteq X \setminus A \\ 0, & \text{otherwise.} \end{cases}$$

We remark that, in a way similar to the Möbius transform (compare with Lemma 2.31), we have the relation

$$\breve{m}^\xi(A) = \Delta_A \xi(X) \qquad (A \subseteq X), \tag{2.40}$$

as it can be easily checked by comparing (2.27) with (2.2).

2.12.5 The Interaction Operator

The interaction transform of a set function ξ is defined by

$$I^\xi = \Gamma \star m^\xi = \Gamma \star (\xi \star Z^{-1})$$

(we prove hereafter that this is indeed the transform introduced in Definition 2.41). This gives immediately

$$I^\xi(A) = \sum_{K \supseteq A} \frac{1}{k - a + 1} m^\xi(K) \qquad (A \subseteq X). \tag{2.41}$$

Let us express the Möbius transform of a set function ξ in terms of its interaction transform. We have

$$\begin{aligned} m^\xi(A) = \Gamma^{-1} \star I^\xi(A) &= \sum_{C \supseteq A} \Gamma^{-1}(A, C) I^\xi(C) \\ &= \sum_{C \supseteq A} \gamma^{\star -1}(C \setminus A) I^\xi(C) \\ &= \sum_{m=0}^{|A^c|} B_m \sum_{\substack{C \supseteq A \\ |C \setminus A| = m}} I^\xi(C), \end{aligned}$$

or more simply, letting $|C| = c$

$$m^{\xi}(A) = \sum_{C \supseteq A} B_{c-a} I^{\xi}(C). \tag{2.42}$$

We can now compute ξ in terms of its interaction transform.

$$\begin{aligned}
\xi(A) &= \sum_{C \subseteq A} m^{\xi}(C) \\
&= \sum_{C \subseteq A} \sum_{D \supseteq C} B_{|D \setminus C|} I^{\xi}(D) \\
&= \sum_{D \subseteq X} \Big(\sum_{C \subseteq A \cap D} B_{|D \setminus C|} \Big) I^{\xi}(D) \\
&= \sum_{D \subseteq X} \left(\sum_{j=0}^{|A \cap D|} \binom{|A \cap D|}{j} B_{|D|-j} \right) I^{\xi}(D).
\end{aligned}$$

Hence we find

$$\xi(A) = \sum_{D \subseteq X} \beta^{|D|}_{|A \cap D|} I^{\xi}(D) \qquad (A \subseteq X), \tag{2.43}$$

with the coefficients β^l_k defined by

$$\beta^l_k = \sum_{j=0}^{k} \binom{k}{j} B_{l-j} \qquad (k \leqslant l). \tag{2.44}$$

The first values of β^l_k are given in Table 2.2. The β^l_k numbers have remarkable properties, in particular they satisfy the property of Pascal's triangle; i.e.,

$$\beta^{l+1}_{k+1} = \beta^l_k + \beta^{l+1}_k \qquad (0 \leqslant k \leqslant l).$$

They show also the following symmetry

$$\beta^l_k = (-1)^l \beta^l_{l-k} \qquad (0 \leqslant k \leqslant l).$$

The numbers β^l_0 are the Bernoulli numbers, so that by the above symmetry and $B_{2m+1} = 0$ for $m \geqslant 1$ we obtain that the diagonal elements $\beta^l_l = (-1)^l \beta^l_0 =$

$k \backslash l$	0	1	2	3	4
0	1	$-\frac{1}{2}$	$\frac{1}{6}$	0	$-\frac{1}{30}$
1		$\frac{1}{2}$	$-\frac{1}{3}$	$\frac{1}{6}$	$-\frac{1}{30}$
2			$\frac{1}{6}$	$-\frac{1}{6}$	$\frac{2}{15}$
3				0	$-\frac{1}{30}$
4					$-\frac{1}{30}$

Table 2.2 The coefficients β^l_k

$(-1)^l B_l$ are the Bernoulli numbers except for $l = 1$, where $\beta^1_1 = -B_1$. Furthermore, the columns sum to zero:

$$\sum_{k=0}^{l} \beta^l_k = 0 \qquad (l > 0).$$

(see Denneberg and Grabisch [82], Grabisch [162] for further details).

We express properties of a set function in terms of its interaction transform.

Theorem 2.45 *Let ξ be a set function on X. Then*

(i) $\xi(\varnothing) = \sum_{K \subseteq X} B_k I^\xi(K)$;
(ii) $\xi(X) = \sum_{i \in X} I^\xi(\{i\}) + \xi(\varnothing)$;
(iii) *ξ is monotone if and only if* $\sum_{K \subseteq X\setminus\{i\}} \beta^{|K|}_{|K\cap L|} I^\xi(K\cup\{i\}) \geqslant 0,\ \forall i \in X,\ \forall L \subseteq X\setminus\{i\}$;
(iv) *ξ is constant if and only if $I^\xi(A) = 0$ for all $A \neq \varnothing$. In this case, $I^\xi(\varnothing) = \xi(A)$, $A \in 2^X$;*
(v) *If ξ is k-monotone for some $2 \leqslant k \leqslant n$, then $I^\xi(A) \geqslant 0$ for all $A \subseteq X$ such that $2 \leqslant |A| \leqslant k$.*

Proof (i) is clear from (2.43). For (ii), use (2.43) again, (i) and properties of β^l_l.

(iii) By (2.16) we find

$$\xi(A \cup \{i\}) = \sum_{B \subseteq A} m^\xi(B) + \sum_{B \subseteq A} m^\xi(B \cup \{i\}) \qquad (i \notin A).$$

Then ξ is monotone if and only if the second sum is nonnegative for all $i \in X$ and all $A \subseteq X \setminus \{i\}$. Transforming this sum like we did for proving (2.43) yields the desired result.

(iv) Obvious from Theorem 2.33(vi) and (2.41).

(v) Clear from Definition 2.41 and Theorem 2.21(i). □

The converse of Theorem 2.45(v) is false, as shown by the next example.

Example 2.46 Take $n = 3$ and consider the game v defined by $v(S) = 1$ if $|S| > 1$, and 0 otherwise. Then $I^v(S) = 0$ for any S s.t. $|S| = 2$, however, v is not 2-monotone because $v(123) + v(1) \not\geq v(12) + v(13)$. ◇

We return to the interaction transform and express it as an operator. From the relation $I^\xi = \Gamma \star (\xi \star Z^{-1})$ we define the *interaction operator* $\mathcal{I}$ by

$$I^\xi = \xi \star \mathcal{I}.$$

Clearly, $\mathcal{I}$ is a linear operator on the set of set functions $\mathbb{R}^{(2^X)}$, but it does not belong to the group $\mathcal{G}$. Indeed, we know already by (2.43) that the inverse operator is $\mathcal{I}^{-1}(B,A) = \beta^{|B|}_{|B\cap A|}$, showing that $\mathcal{I}^{-1}$ and consequently $\mathcal{I}$, do not belong to $\mathcal{G}$. It remains to express $\mathcal{I}$. From $I^\xi = \Gamma \star (\xi \star Z^{-1})$ we get

$$
\begin{aligned}
I^\xi(A) = & \sum_{B\supseteq A}\sum_{C\subseteq B} \frac{1}{|B\setminus A|+1}(-1)^{|B\setminus C|}\xi(C) \\
& \text{substituting } D = B\setminus A, C_A = C\cap A, C_D = C\cap D \\
= & \sum_{D\subseteq A^c} \frac{1}{|D|+1} \sum_{C_D\subseteq D}(-1)^{|D\setminus C_D|}\sum_{C_A\subseteq A}(-1)^{|A\setminus C_A|}\xi(C_A\cup C_D) \\
& \text{substituting } C = C_D, E = D\setminus C, B = C_A \\
= & \sum_{C\subseteq A^c}\sum_{E\subseteq A^c\setminus C}(-1)^{|E|}\frac{1}{|C|+|E|+1}\sum_{B\subseteq A}(-1)^{|A\setminus B|}\xi(B\cup C) \\
& \text{letting } m = |A^c| \\
= & \sum_{k=0}^{m}\sum_{\substack{C\subseteq A^c\\|C|=k}}\sum_{j=0}^{m-k}\sum_{\substack{E\subseteq A^c\setminus C\\|E|=j}}\frac{(-1)^j}{k+j+1}\sum_{B\subseteq A}(-1)^{|A\setminus B|}\xi(B\cup C) \\
= & \sum_{k=0}^{m}\sum_{\substack{C\subseteq A^c\\|C|=k}}\sum_{j=0}^{m-k}\binom{m-k}{j}\frac{(-1)^j}{k+j+1}\sum_{B\subseteq A}(-1)^{|A\setminus B|}\xi(B\cup C).
\end{aligned}
$$

Applying Lemma 1.1(iv) with $n = m-k$, we get finally

$$I^\xi(A) = \sum_{C\subseteq X\setminus A}\frac{(m-k)!k!}{(m+1)!}\sum_{B\subseteq A}(-1)^{|A\setminus B|}\xi(B\cup C)$$

with $m = |X \setminus A|$, $k = |C|$, which is Eq. (2.29). The operator $\mathcal{I}$ is therefore, using the second equation in (2.29),

$$\mathcal{I}(D,A) = (-1)^{|A\setminus D|}\frac{|X \setminus (A \cup D)|!|D \setminus A|!}{m-n+1}.$$

Note the symmetry

$$\mathcal{I}(D^c,A) = (-1)^{|A|}\mathcal{I}(D,A). \tag{2.45}$$

2.12.6 The Banzhaf Interaction Operator

The Banzhaf interaction transform introduced in Sect. 2.11 can be treated in the same way as the interaction operator, although computations are much simpler. First, we introduce the operator Θ, playing the same rôle as Γ in Sect. 2.12.1:

$$\Theta(A,B) = \begin{cases} \left(\frac{1}{2}\right)^{|B\setminus A|}, & \text{if } A \subseteq B\\ 0, & \text{otherwise.}\end{cases}$$

Since $\Theta(A,B) = \Theta(\varnothing, B \setminus A)$ and its value depends only on the cardinality of A and B, it is a cardinality operator with generator function $\theta(A) = \left(\frac{1}{2}\right)^{|A|}$, hence its inverse can be found via Theorem 2.44. Denoting by $\widehat{\theta}, \widehat{\theta^{\star-1}}$ the cardinal representations of θ and its inverse $\theta^{\star-1}$, we find

$$\widehat{\theta^{\star-1}}(m) = \begin{cases} 1, & \text{if } m = 0\\ -\sum_{k=0}^{m-1}\binom{m}{k}\left(\frac{1}{2}\right)^{m-k}\widehat{\theta^{\star-1}}(k), & \text{if } m \in \mathbb{N}.\end{cases}$$

Hence, unless $m = 0$, we have $\sum_{k=0}^{m}\binom{m}{k}\left(\frac{1}{2}\right)^{m-k}\widehat{\theta^{\star-1}}(k) = 0$. Comparing with (1.1), the solution immediately appears to be $\widehat{\theta^{\star-1}}(k) = \left(-\frac{1}{2}\right)^k$. It follows that

$$\Theta^{-1}(A,B) = \begin{cases} \left(-\frac{1}{2}\right)^{|B\setminus A|}, & \text{if } A \subseteq B\\ 0, & \text{otherwise.}\end{cases} \tag{2.46}$$

We define the Banzhaf interaction transform of a set function ξ by

$$I_{\mathrm{B}}^{\xi} = \Theta \star m^{\xi} = \Theta \star (\xi \star Z^{-1})$$

which yields

$$I^\xi_{\mathrm{B}}(A) = \sum_{K\supseteq A} \Big(\frac{1}{2}\Big)^{k-a} m^\xi(K) \qquad (A\subseteq X). \tag{2.47}$$

The inverse relation $m^\xi = \Theta^{-1} \star I^\xi$ reads, using (2.46),

$$m^\xi(A) = \sum_{K\supseteq A} \Big(-\frac{1}{2}\Big)^{k-a} I^\xi_{\mathrm{B}}(K) \qquad (A\subseteq X). \tag{2.48}$$

It remains to express the Banzhaf interaction transform in terms of the set function and vice versa. We have

$$\begin{aligned}
\xi(A) &= \sum_{C\subseteq A} m^\xi(C) = \sum_{C\subseteq A}\sum_{D\supseteq C}\Big(-\frac{1}{2}\Big)^{d-c} I^\xi_{\mathrm{B}}(D)\\
&= \sum_{D\subseteq X} I^\xi_{\mathrm{B}}(D) \sum_{C\subseteq A\cap D}\Big(-\frac{1}{2}\Big)^{d-c}\\
&= \sum_{D\subseteq X}(-1)^d\Big(\frac{1}{2}\Big)^d I^\xi_{\mathrm{B}}(D) \underbrace{\sum_{C\subseteq A\cap D}(-2)^c}_{(-1)^{|A\cap D|} \text{ by Lemma 1.1(vi)}}\\
&= \sum_{D\subseteq X}(-1)^{|D\setminus A|}\Big(\frac{1}{2}\Big)^d I^\xi_{\mathrm{B}}(D).
\end{aligned} \tag{2.49}$$

From $I^\xi_{\mathrm{B}} = \Theta \star (\xi \star Z^{-1})$ we find

$$\begin{aligned}
I^\xi_{\mathrm{B}}(A) &= \sum_{B\supseteq A}\Big(\frac{1}{2}\Big)^{b-a}\sum_{C\subseteq B}(-1)^{|B\setminus C|}\xi(C)\\
&= \sum_{C\subseteq X}\xi(C)\sum_{B\supseteq A\cup C}\Big(\frac{1}{2}\Big)^{b-a}(-1)^{b-c}\\
&= \sum_{C\subseteq X}(-1)^{n-c}\Big(\frac{1}{2}\Big)^{n-a}\xi(C)\sum_{B\supseteq A\cup C}(-1)^{b-n}\Big(\frac{1}{2}\Big)^{b-n}\\
&= \sum_{C\subseteq X}(-1)^{n-c}\Big(\frac{1}{2}\Big)^{n-a}\xi(C) \underbrace{\sum_{B\supseteq A\cup C}(-2)^{n-b}}_{(-1)^{n-|A\cup C|} \text{ by Lemma 1.1(vi)}}\\
&= \Big(\frac{1}{2}\Big)^{n-a}\sum_{C\subseteq X}(-1)^{|A\setminus C|}\xi(C).
\end{aligned}$$

Thus, the *Banzhaf interaction operator* $\mathcal{I}_{\mathrm{B}}$, defined by

$$I_{\mathrm{B}}^{\xi} = \xi \star \mathcal{I}_{\mathrm{B}}$$

is given by

$$\mathcal{I}_{\mathrm{B}}(C, A) = \left(\frac{1}{2}\right)^{n-a} (-1)^{|A \setminus C|},$$

and we note as for the interaction operator the symmetry

$$\mathcal{I}_{\mathrm{B}}(C^c, A) = (-1)^{|A|} \mathcal{I}_{\mathrm{B}}(C, A) \qquad (A, C \subseteq X). \tag{2.50}$$

We express properties of the set function in terms of its Banzhaf interaction transform.

Theorem 2.47 *Let ξ be a set function on X. Then*

(i) $\xi(\varnothing) = \sum_{B \subseteq X} \left(-\frac{1}{2}\right)^b I_{\mathrm{B}}^{\xi}(B)$;

(ii) $\xi(X) = \sum_{B \subseteq X} \left(\frac{1}{2}\right)^b I_{\mathrm{B}}^{\xi}(B)$;

(iii) *ξ is monotone if and only if* $\sum_{K \subseteq X, K \ni i} (-1)^{|K \setminus A|} \left(\frac{1}{2}\right)^k I_{\mathrm{B}}^{\xi}(K) \geqslant 0$, $\forall i \in X$, $\forall A \subseteq X \setminus i$;

(iv) *ξ is constant if and only if $I_{\mathrm{B}}^{\xi}(A) = 0$ for all $A \neq \varnothing$. In this case, $I_{\mathrm{B}}^{\xi}(\varnothing) = \xi(A)$, $A \in 2^X$;*

(v) *If ξ is k-monotone for some $2 \leqslant k \leqslant n$, then $I_{\mathrm{B}}^{\xi}(A) \geqslant 0$ for all $A \subseteq X$ such that $2 \leqslant |A| \leqslant k$.*

Proof (i) and (ii) are obvious.

(iii) Monotonicity of ξ is equivalent to

$$\xi(A \cup i) - \xi(A) = \sum_{B \subseteq A} m^{\xi}(B \cup i) \geqslant 0 \qquad (i \in X, A \subseteq X \setminus i).$$

Proceeding as for (2.49) with the above inequalities yields the desired result.

(iv) Obvious by Theorem 2.33(vi) and (2.47).

(v) Clear from Definition 2.42 and Theorem 2.21(i). □

2.12.7 Transforms of Conjugate Set Functions

The Möbius transform of a set function is closely related to the co-Möbius transform of its conjugate.

Theorem 2.48 *For any set function ξ on X, we have*

$$\check{m}^{\bar{\xi}}(A) = (-1)^{|A|+1} m^{\xi}(A) \tag{2.51}$$

for all $A \subseteq X$, $A \neq \varnothing$.

Proof From (2.27), we have

$$\begin{aligned}
\check{m}^{\bar{\xi}}(A) &= \sum_{B \subseteq A} (-1)^{|B|} \bar{\xi}(B^c) \\
&= \sum_{B \subseteq A} (-1)^{|B|} (\xi(X) - \xi(B)) \\
&= -\sum_{B \subseteq A} (-1)^{|B|} \xi(B) = (-1)^{|A|+1} m^{\xi}(A).
\end{aligned}$$

□

Let us now express the interaction and Banzhaf interaction transforms for the conjugate set function.

Theorem 2.49 *For any set function ξ,*

$$I^{\bar{\xi}}(A) = \begin{cases} \xi(X) - I^{\xi}(\varnothing), & \text{if } A = \varnothing \\ (-1)^{|A|+1} I^{\xi}(A), & \text{otherwise.} \end{cases}$$

$$I_{\mathrm{B}}^{\bar{\xi}}(A) = \begin{cases} \xi(X) - I_{\mathrm{B}}^{\xi}(\varnothing), & \text{if } A = \varnothing \\ (-1)^{|A|+1} I_{\mathrm{B}}^{\xi}(A), & \text{otherwise.} \end{cases}$$

Proof From Theorem 2.45(iv) and the linearity of I, we obtain

$$I^{\bar{\xi}} = \xi(X)\delta_{\varnothing} - \xi^c \star \mathcal{I},$$

with the shorthand $\xi^c(\cdot) = \xi(\cdot^c)$, and $\delta_{\varnothing}$ is the set function defined by $\delta_{\varnothing}(\varnothing) = 1$ and $\delta_{\varnothing}(A) = 0$ if $A \neq \varnothing$. Now,

$$\begin{aligned}
\xi^c \star \mathcal{I}(A) &= \sum_{C \subseteq X} \xi(C^c)\mathcal{I}(C, A) \\
&= \sum_{D \subseteq X} \xi(D)\mathcal{I}(D^c, A) \\
&= (-1)^{|A|} \sum_{D \subseteq X} \xi(D)\mathcal{I}(D, A) \text{ using (2.45)} \\
&= (-1)^{|A|} \xi \star \mathcal{I}(A),
\end{aligned}$$

which proves the result. Proceed exactly in the same way for the Banzhaf interaction transform. □

2.13 k-Additive Games

Definition 2.50 A game[6] v on X is said to be *k-additive* for some integer $k \in \{1,\ldots,|X|\}$ if $m^v(A) = 0$ for all $A \subseteq X$, $|A| > k$, and there exists some $A \subseteq X$ with $|A| = k$ such that $m^v(A) \neq 0$.

A game v is *at most k-additive* for some $1 \leqslant k \leqslant |X|$ if it is k'-additive for some $k' \in \{1,\ldots,k\}$ (equivalently, if m^v vanishes for subsets of more than k elements). The notion of k-additivity can also be applied to subclasses of games as well, like capacities, belief measures, etc. The set of k-additive games on X (respectively, capacities, etc.) is denoted by $\mathcal{G}^k(X)$ (respectively, $\mathcal{MG}^k(X)$, etc.), while we denote by $\mathcal{G}^{\leqslant k}(X), \mathcal{MG}^{\leqslant k}(X)$ the set of at most k-additive games and capacities. Clearly,

$$\mathcal{G}(X) = \mathcal{G}^1(X) \cup \mathcal{G}^2(X) \cup \cdots \cup \mathcal{G}^{|X|}(X) = \mathcal{G}^1(X) \cup \mathcal{G}^{\leqslant 2}(X) \cup \cdots \cup \mathcal{G}^{\leqslant |X|}(X)$$

and similarly for capacities and other subclasses. The first expression is a disjoint union, while for the second note that $\mathcal{G}^1(X) = \mathcal{G}^{\leqslant 1}(X) \subset \mathcal{G}^{\leqslant 2}(X) \subset \cdots \subset \mathcal{G}^{\leqslant |X|}(X) = \mathcal{G}(X)$.

Remark 2.51

(i) By Theorem 2.33(i) we see that $\mathcal{G}^1(X)$ is the set of additive games. Hence, k-additivity can indeed be seen as a generalization of additivity.
(ii) By (2.41), one can see that m^v can be replaced by I^v [or by I^v_{B}, using (2.47)] without any change in Definition 2.50.
(iii) A k-additive game needs $\binom{|X|}{1} + \binom{|X|}{2} + \cdots + \binom{|X|}{k}$ coefficients to be defined, while a game needs in general $2^{|X|} - 1$ coefficients.
(iv) k-additive games were introduced under this name by the author [163]. We will see in Sect. 2.16 that they correspond to games with a polynomial representation of degree k. The same idea was also proposed by Vassil'ev [331].

◇

We mention some elementary properties.

Lemma 2.52 *The following holds:*

(i) If v is a k-additive game for some $1 \leqslant k \leqslant n$, then

$$m^v(A) = \check{m}^v(A) = I^v(A) = I^v_{\mathrm{B}}(A) \qquad (A \subseteq X, |A| = k);$$

[6] The definition applies to any set function as well.

(ii) If v is a 2-additive game, then

$$v(A) = \sum_{\{i,j\}\subseteq A} v(\{i,j\}) - (|A|-2)\sum_{i\in A} v(\{i\}) \qquad (A \subseteq X, |A| \geqslant 2),$$

and moreover, $I^v(\{i,j\}) = v(\{i,j\}) - v(\{i\}) - v(\{j\})$ for any distinct $i,j \in X$.

The proof is immediate and left to the readers.

2.14 p-Symmetric Games

As k-additivity generalizes the concept of additivity, p-symmetry generalizes the concept of symmetric games. A game v on X is *symmetric* if $v(A) = v(B)$ whenever $|A| = |B|$. In other words, v depends only on the cardinality of sets.

Consider a game v on X. A nonempty subset $A \subseteq X$ is a *subset of indifference for v* if for all $B_1, B_2 \subseteq A$ such that $|B_1| = |B_2|$, we have $v(C \cup B_1) = v(C \cup B_2)$, for all $C \subseteq X \setminus A$.

Observe that any nonempty subset of a subset of indifference is also a subset of indifference, and that any singleton is a subset of indifference. These two properties imply that for any game v, it is possible to partition X into subsets of indifference of v, $A_1, \dots, A_\ell$; i.e., they satisfy $A_i \cap A_j = \varnothing$ for $i \neq j$, and $\bigcup_{i=1}^{\ell} A_i = X$. We recall that a partition $\pi = \{A_1, \dots, A_\ell\}$ of X is *coarser* than another partition $\pi' = \{A'_1, \dots, A'_{\ell'}\}$ of X if every block A_i of π is a union of blocks of π'. One is interested in finding a coarsest partition into subsets of indifference; i.e., with the biggest possible blocks. Such a partition is unique (see below) and is called the *basis* of v.

Definition 2.53 A game v on X is *p-symmetric* for some integer $p \in \{1, \dots, |X|\}$ if its basis contains p subsets.

As for k-additivity, any game is p-symmetric for some p, and has a basis. Therefore, p-symmetry induces a partition of the set of games. The notion of *at most p-symmetry* can be introduced as well.

Example 2.54 A symmetric game has basis $\{X\}$; i.e., it is a 1-symmetric game. Any unanimity game u_A, $\varnothing \neq A \subseteq X$, is a 2-symmetric game with basis $\{A, X \setminus A\}$. ◇

The basis of a game is closely related to notion of symmetry among elements of X. This notion is particularly meaningful in game theory, where elements of X are players. We say that $i,j \in X$ are *symmetric* for the game v (denoted by $i \sim_v j$) if for all $A \subseteq X \setminus \{i,j\}$, $v(A \cup \{i\}) = v(A \cup \{j\})$.

Theorem 2.55 *The basis of a game v is the collection of equivalence classes of the equivalence relation $\sim_v$.*

Proof We first prove that $\sim_v$ is an equivalence relation on X. Reflexivity and symmetry are obvious. To see transitivity, suppose that $i \sim_v j$ and $j \sim_v k$, and let us prove that $i \sim_v k$. We have immediately that $v(A \cup i) = v(A \cup j) = v(A \cup k)$ for all $A \subseteq X \setminus \{i,j,k\}$. It remains to prove the equality $v(A \cup i) = v(A \cup k)$ for all $A \subseteq X \setminus \{i,k\}$ such that $A \ni j$, or equivalently, $v(A \cup \{i,j\}) = v(A \cup \{j,k\})$ for all $A \subseteq X \setminus \{i,j,k\}$. We have for any such A, $v(A \cup \{i,j\}) = v(A \cup \{i,k\})$ by $j \sim_v k$, and $v(A \cup \{i,k\}) = v(A \cup \{j,k\})$ by $i \sim_v j$, and the desired equality is proved.

Now, any equivalence class of $\sim_v$ is a subset of indifference. Indeed, consider $[i]$ the equivalence class of $i \in X$, supposing $|[i]| \geqslant 2$ (otherwise we are done), and take two distinct $B, B' \subseteq [i]$, with $B = \{i_1, \ldots, i_b\}$, $B' = \{j_1, \ldots, j_b\}$. Since $i_1 \sim_v j_1, \ldots, i_b \sim_v j_b$, we find successively $v(A \cup i_1) = v(A \cup j_1)$, $v(A \cup \{i_1, i_2\}) = v(A \cup \{j_1, j_2\})$, and eventually $v(A \cup B) = v(A \cup B')$, for all $A \subseteq X \setminus [i]$.

Finally, there cannot be larger subsets of indifference than the equivalence classes, because if $v(A \cup i) = v(A \cup j)$ for some $j \notin [i]$ and all $A \subseteq X \setminus ([i] \cup j)$, this would imply $j \sim_v i$, a contradiction. □

Consider a p-symmetric game v, with basis $\{A_1, \ldots, A_p\}$, and a subset $B \subseteq X$. Clearly, the value $v(B)$ depends uniquely on the numbers $b_1, \ldots, b_p$, with $b_i = |A_i \cap B|$. Because $0 \leqslant b_i \leqslant |A_i|$, it follows that v needs $\prod_{i=1}^{p}(|A_i| + 1)$ coefficients to be defined.

p-symmetric games have been introduced by Miranda and Grabisch [245, 246].

2.15 Structure of Various Sets of Games

We study the structure of the main families of games.

2.15.1 *The Vector Space of Games*

Considering the addition of functions and scalar multiplication, it is plain that $\mathcal{G}(X)$ is a vector space. Two popular bases for $\mathcal{G}(X)$ are the set of Dirac games and the set of unanimity games. Take any $A \subseteq X$, $A \neq \varnothing$. The *Dirac game* δ_A *centered at* A (also called *identity game*, see, e.g., Bilbao [21]) is the 0-1-game defined by

$$\delta_A(B) = \begin{cases} 1, & \text{if } A = B \\ 0, & \text{otherwise.} \end{cases}$$

Theorem 2.56 (Bases of $\mathcal{G}(X)$) *The set of Dirac games $\{\delta_A\}_{A \in 2^X \setminus \{\varnothing\}}$ and the set of unanimity games $\{u_A\}_{A \in 2^X \setminus \{\varnothing\}}$ are bases of $\mathcal{G}(X)$ of dimension $2^{|X|} - 1$.*

Proof Since for any $v \in \mathcal{G}(X)$ we can write $v = \sum_{A \subseteq X, A \neq \varnothing} v(A)\delta_A$, it is clear that $\{\delta_A\}_{A \in 2^X \setminus \{\varnothing\}}$ spans $\mathcal{G}(X)$. Let us prove it is an independent set, that is, $\sum_{A \in 2^X \setminus \{\varnothing\}} \lambda_A \delta_A = 0$ if and only if $\lambda_A = 0$ for all $A \subseteq X$, $A \neq \varnothing$. Suppose on the contrary that this not true, and consider some nonnull coefficient λ_{A_0}. Then $\sum_{A \in 2^X \setminus \{\varnothing\}} \lambda_A \delta_A(A_0) = \lambda_{A_0}$, a contradiction.

Let us prove that the set of unanimity games is a basis. Because there are $2^{|X|} - 1$ unanimity games, it suffices to prove that it forms an independent set. Assume on the contrary that this is not true; i.e., $\sum_{A \in 2^X \setminus \{\varnothing\}} \lambda_A u_A = 0$ with some $\lambda_A \neq 0$, and consider A_0, a minimal set in the collection $\{A \subseteq X : \lambda_A \neq 0\}$. Then $\sum_{A \in 2^X \setminus \{\varnothing\}} \lambda_A u_A(A_0) = \lambda_{A_0}$, a contradiction. □

In the basis of Dirac games, the coordinates of a game v are simply $\{v(A)\}_{A \in 2^X \setminus \{\varnothing\}}$.

Let us consider the basis of unanimity games. We have for any game $v \in \mathcal{G}(X)$

$$v(B) = \sum_{A \in 2^X \setminus \{\varnothing\}} \lambda_A u_A(B) = \sum_{A \subseteq B, A \neq \varnothing} \lambda_A \qquad (B \subseteq X).$$

Comparing with (2.16), it follows that the coefficients of a game v in the basis of unanimity games are nothing other than its Möbius transform: $\lambda_A = m^v(A)$ for all $A \subseteq X, A \neq \varnothing$ [see also (2.20)].

It is easy to see that the set of conjugate of unanimity games is a basis as well. We have for any $A \neq \varnothing$,

$$\overline{u_A}(B) = 1 - u_A(B^c) = \begin{cases} 1, & \text{if } B \cap A \neq \varnothing \\ 0, & \text{otherwise.} \end{cases}$$

Let us determine the coefficients of a game in this basis. We observe that for all $A \subseteq X$,

$$\overline{v}(A) = v(X) - v(A^c) = \sum_{B \subseteq X} m^v(B) - \sum_{B \subseteq A^c} m^v(B) = \sum_{B \cap A \neq \varnothing} m^v(B).$$

On the other hand, the decomposition in the basis of the conjugate of unanimity games reads

$$v(A) = \sum_{B \subseteq X, B \neq \varnothing} \lambda_B \overline{u_B}(A) = \sum_{B \cap A \neq \varnothing} \lambda_A \qquad (A \subseteq X),$$

from which we deduce that $\lambda_B = m^{\overline{v}}(B)$, $B \in 2^X \setminus \{\varnothing\}$. Using Theorem 2.48 and (2.39), we find

$$\lambda_B = (-1)^{|B|+1} \sum_{A \supseteq B} m^v(A). \tag{2.52}$$

Remark 2.57

(i) The fact that the set of unanimity games is a basis of $\mathcal{G}(X)$ was first shown by Shapley [298, Lemma 3].
(ii) The above two bases can serve also as bases for the $2^{|X|}$-dim vector space of set functions, by adding respectively $\delta_\varnothing$ and $u_\varnothing$. We introduce other bases of set functions in Sects. 2.16.1 and 2.17, see in particular the latter for the relation between transforms and bases.

◇

We end this section by presenting two natural orders for listing the coordinates of a game. Indeed, if one wants to use vectors and matrices for representing games and their transforms, it is necessary to fix an ordering of all subsets of X. The most popular one is the lexicographic ordering subordinated to cardinality; i.e., considering smaller subsets first, and in case of equal cardinality, the lexicographic ordering. This gives for $X = \{1, 2, 3, 4\}$, omitting braces and commas:

$$\varnothing, 1, 2, 3, 4, 12, 13, 14, 23, 24, 34, 123, 124, 134, 234, 1234.$$

Although simple, this ordering does not lead to nicely structured matrices. Also, if a new element is added to X, the whole list has to be rebuilt. The *binary order* does not have these drawbacks, because it has a recursive structure. Each subset is coded by an integer, whose binary code is precisely the characteristic vector of the subset. For example, again with $X = \{1, 2, 3, 4\}$, the subset $\{1, 3\}$ has characteristic vector $(1, 0, 1, 0)$, written as 0101 (rightmost position corresponds to element 1 of X), which is the binary code of 5. Then, subsets are ordered according to the ordering of integers. We obtain:

$$\varnothing, 1, 2, 12, 3, 13, 23, 123, 4, 14, 24, 124, 34, 134, 234, 1234.$$

The fundamental property of this list is that in order to get the list for $\{1, \ldots, n, n+1\}$, we add to the list obtained for $\{1, \ldots, n\}$ a duplicate of it, where to each subset the element $n + 1$ is added. In [178], it is shown how this order permits to have a simple matrix representation of most of the transforms given in Sect. 2.12. These matrices are called "fractal" because of their replicative structure induced by the binary order.

2.15.2 The Cone of Capacities

Since the null set function 0 is a capacity, the set of capacities $\mathcal{MG}(X)$ is a pointed cone. Indeed, it is plain that it is closed under multiplication by a nonnegative scalar. Moreover, $-\mathcal{MG}(X)$ is the set of antitone games (i.e., $v(A) \geqslant v(B)$ whenever $A \subseteq B$), therefore only 0 can be both in $\mathcal{MG}(X)$ and $-\mathcal{MG}(X)$.

Also, note that unanimity games are extremal rays of $\mathcal{MG}(X)$, that is, they cannot be written as a conic combination of other capacities. To see this, consider the unanimity game u_A for some $A \neq \varnothing$, and write

$$u_A = \sum_{i \in I} \alpha_i \mu_i$$

with $\alpha_i > 0$, and $\mu_i \in \mathcal{MG}(X)$, for all $i \in I$. Then nonnegativity of the α_i's and μ_i's entails $\mu_i(B) = 0$ for all $B \not\supseteq A$, $\forall i \in I$. By monotonicity of the capacities μ_i and nonnegativity of the α_i's, we have for any $B \supseteq A$

$$1 = u_A(B) = \sum_{i \in I} \alpha_i \mu_i(B) \geqslant \sum_{i \in I} \alpha_i \mu_i(A) = u_A(A) = 1,$$

forcing equality throughout, and by nonnegativity again, this yields $\mu_i(B) = \mu_i(A)$ for all $B \supseteq A$. It follows that μ_i and u_A are colinear for all $i \in I$. However, unanimity games are not the only extremal rays, because one cannot express any capacity as a conic combination of unanimity games: linear combinations are necessary.

2.15.3 The Cone of Supermodular Games

The set of supermodular games (as well as the set of submodular games) is a cone. However, it is not pointed as it contains the set of additive games, which form a linear space.

We consider instead the cone of zero-normalized supermodular games, which is pointed because the only zero-normalized additive game is the null game 0. We know from Sect. 2.1 that any game v can be zero-normalized by considering $v_0 = v - \beta$, where β is an additive game defined by $\beta(\{i\}) = v(\{i\})$ for all $i \in X$. Moreover, v is supermodular if and only if v_0 is, hence $v \mapsto v_0$ defines a surjective mapping from the cone of supermodular games to the cone of supermodular zero-normalized games. Let us denote the latter by $\mathcal{G}_\diamond(X)$.

The problem of finding the extremal rays of $\mathcal{G}_\diamond(X)$ is difficult. In his 1971 paper [301], Shapley gives the 37 extremal rays of $\mathcal{G}_\diamond(X)$ with $|X| = 4$, computed by S. A. Cook, and says that "for larger n little is known about the set of all extremals". Later, Rosenmüller and Weidner [276] find all extremal rays of the cone of nonnegative supermodular games (it differs from $\mathcal{G}_\diamond(X)$ only by the addition of the extremal rays $u_{\{i\}}$, $i \in X$) by representing each such game as a maximum over shifted additive games:

$$v = \max(m^1 - \alpha_1, \ldots, m^t - \alpha_t)$$

where $m^1, \ldots, m^t$ are additive games and $\alpha_1, \ldots, \alpha_t \in \mathbb{R}_+$. Up to some additional conditions, the above representation can be put into a unique canonical form. For a given v and some $H \subseteq \{1, \dots, t\}$, $|H| \geqslant 2$, where t pertains to the canonical form, consider the collection $\mathcal{S}_H$ of sets such that $v(S) = m^i(S) - \alpha_i$ for every $i \in H$. Then v is an extremal ray if and only if $m^1, \dots, m^t$ are the unique solution of the systems

$$m^i(S) - \alpha^i = v(S) \qquad (i \in H, S \in \mathcal{S}_H)$$

for all $H \subseteq \{1, \dots, t\}$, $|H| \geqslant 2$. We refer the readers to [276] for the details.

Later, Studený and Kroupa [317] find out another criterion for finding the extremal rays of $\mathcal{G}_\diamond(X)$, based on the fact that supermodular games are exact and their core coincides with the Weber set (see Chap. 3 and especially Sect. 3.4 for the necessary concepts in what follows); i.e., any supermodular v can be written as a minimum over the marginal vectors of v (which are the extreme points of its core)

$$v = \min_{\sigma \in \mathfrak{S}(X)} x^{\sigma,v},$$

where $\mathfrak{S}(X)$ is the set of permutations on X (observe the duality with the approach of Rosenmüller and Weidner). Then v is an extremal ray of $\mathcal{G}_\diamond$ if and only if every solution $\{y^\sigma\}_{\sigma \in \mathfrak{S}(X)}$ of the system

$$\begin{aligned} y_i^\sigma &= 0 && (i \in X \text{ s.t. } x_i^{\sigma,v} = 0, \sigma \in \mathfrak{S}(X)) \\ \sum_{i \in S} y_i^\sigma &= \sum_{i \in S} y_i^\tau && (S \subseteq N, \sigma, \tau \in \mathfrak{S}(X) \text{ s.t. } v(S) = \sum_{i \in S} x_i^\sigma = \sum_{i \in S} x_i^\tau) \end{aligned}$$

is equal to $\{x^{\sigma,v}\}_{\sigma \in \mathfrak{S}(X)}$ up to a real multiplicative constant.

2.15.4 The Cone of Totally Monotone Nonnegative Games

We know from Theorem 2.20(iii) that totally monotone nonnegative games are exactly totally monotone capacities. Therefore, they constitute a subset of the cone of capacities, which we denote by $\mathcal{G}_+(X)$ for reasons that will become clear in the next paragraph. By Theorem 2.33(v), we know that totally monotone capacities have a nonnegative Möbius transform. Therefore, they form a cone too, which is pointed.

As above, unanimity games are extremal rays of $\mathcal{G}_+(X)$, but in this case, there are no extremal rays other than unanimity games, because any totally monotone capacity can be expressed as a conic combination of unanimity games (due to nonnegativity of the Möbius transform). In summary:

Theorem 2.58 *$\mathcal{G}_+(X)$ is a pointed cone, whose extremal rays are the unanimity games u_A, $A \in 2^X \setminus \{\varnothing\}$.*

2.15.5 The Riesz Space of Games

(See Marinacci and Montrucchio [235], Gilboa and Schmeidler [156], and Sect. 1.3.10 for definitions) We can endow $\mathcal{G}(X)$ with a richer structure thanks to the cone $\mathcal{G}_+(X)$. We introduce the partial order $\succeq$ on $\mathcal{G}(X)$ as follows: $v \succeq v'$ if $v - v' \in \mathcal{G}_+(X)$. Note that $\{v \in \mathcal{G}(X) : v \succeq 0\} = \mathcal{G}_+(X)$.

Endowed with $\succeq$, $\mathcal{G}(X)$ becomes an ordered vector space, whose positive cone is $\mathcal{G}_+(X)$, hence the notation. Moreover, under the lattice operations $\vee, \wedge$ induced by $\succeq$, it is a Riesz space.

Theorem 2.59 *The ordered vector space* $(\mathcal{G}(X), \succeq)$ *is a Riesz space, with lattice operations given by*

$$v_1 \vee v_2 = \sum_{A \in 2^X \setminus \{\varnothing\}} (m^{v_1}(A) \vee m^{v_2}(A)) u_A$$

$$v_1 \wedge v_2 = \sum_{A \in 2^X \setminus \{\varnothing\}} (m^{v_1}(A) \wedge m^{v_2}(A)) u_A.$$

Proof We only prove the result for $\vee$, a similar argument can be used for $\wedge$. We set

$$\widehat{v} = \sum_{A \in 2^X \setminus \{\varnothing\}} (m^{v_1}(A) \vee m^{v_2}(A)) u_A$$

and want to prove that $\widehat{v} = v_1 \vee v_2$. We have for $i = 1, 2$,

$$\widehat{v} - v_i = \sum_{A \in 2^X \setminus \{\varnothing\}} \big((m^{v_1}(A) \vee m^{v_2}(A)) - m^{v_i}(A)\big) u_A,$$

which yields $\widehat{v} - v_i \in \mathcal{G}_+(X)$ because $(m^{v_1}(A) \vee m^{v_2}(A)) - m^{v_i}(A) \geqslant 0$ for all $\varnothing \neq A \subseteq X$. It follows that $\widehat{v} \succeq v_i$, $i = 1, 2$. It remains to show that $\widehat{v}$ is the least upper bound of v_1, v_2, that is, for every $\widehat{v}'$ such that $\widehat{v}' \succeq v_i$, $i = 1, 2$, we have $\widehat{v}' \succeq \widehat{v}$.

As $\widehat{v}' - v_i \in \mathcal{G}_+(X)$, we have $m^{\widehat{v}' - v_i}(A) = m^{\widehat{v}'}(A) - m^{v_i}(A) \geqslant 0$ for each A, and $i = 1, 2$. This yields $m^{\widehat{v}'} \geqslant m^{v_1} \vee m^{v_2}$ pointwise, which implies that

$$\widehat{v}' - \widehat{v} = \sum_{A \in 2^X \setminus \{\varnothing\}} \big(m^{\widehat{v}'}(A) - (m^{v_1}(A) \vee m^{v_2}(A))\big) u_A$$

belongs to $\mathcal{G}_+(X)$. We conclude that $\widehat{v}' \succeq \widehat{v}$, as desired. □

The *positive part* and *negative part* of a game v are given by

$$v^+ = \sum_{A \in 2^X \setminus \{\varnothing\}} (m^v(A) \vee 0) u_A \tag{2.53}$$

$$v^- = \sum_{A \in 2^X \setminus \{\varnothing\}} (-m^v(A) \vee 0) u_A. \tag{2.54}$$

The *absolute value* of a game v is defined by $|v| = v^+ + v^-$, and is given by

$$|v| = \sum_{A \in 2^X \setminus \{\varnothing\}} |m^v(A)| u_A.$$

The associated norm $\|\cdot\|_c$, called the *composition norm* (Gilboa and Schmeidler [157]), is defined by $\|v\|_c = |v|(X) = v^+(X) + v^-(X) = \sum_{A \in 2^X \setminus \{\varnothing\}} |m^v(A)|$. Because $\|v_1 + v_2\|_c = \|v_1\|_c + \|v_2\|_c$ holds for any $v_1, v_2 \in \mathcal{G}_+(X)$, it follows that $\|\cdot\|_c$ is a L-norm, and $\mathcal{G}(X)$ an AL-space.

We end this paragraph by showing a decomposition theorem for games.

Theorem 2.60 *For any game $v \in \mathcal{G}(X)$, the games v^+, v^- are the unique totally monotone capacities such that*

$$v = v^+ - v^-$$

and

$$\|v\|_c = \|v^+\|_c + \|v^-\|_c.$$

Proof We use a well-known result of decomposition for vectors in $\mathbb{R}^m$: given $z \in \mathbb{R}^m$, the positive and negative parts of z, z^+ and z^-, are the unique vectors in $\mathbb{R}^m$ such that $z = z^+ - z^-$ and $\|z\|_1 = \|z^+\|_1 + \|z^-\|_1$, where $\|\cdot\|_1$ is the L_1-norm, that is, $\|z\|_1 = \sum_{i=1}^m |z_i|$.

Clearly, by Theorem 2.56, to any game v corresponds a vector m in $\mathbb{R}^{2^n-1}$ containing the Möbius transform of v, and $\|v\|_c = \|m\|_1$. The correspondence being a bijection, the desired result follows from the decomposition of real vectors. □

2.15.6 The Polytope of Normalized Capacities

Taking for simplicity the basis of Dirac games, where coordinates of a game v are simply $\{v(A)\}_{A \in 2^X \setminus \{\varnothing\}}$, the set of normalized capacities reads

$$\mathcal{MG}_0(X) = \Big\{\mu \in \mathbb{R}^{2^{|X|}-1} : \begin{array}{ll} \mu(A) \geqslant \mu(B), & \varnothing \neq B \subseteq A \subseteq X \\ \mu(A) \geqslant 0, & A \subseteq X \\ \mu(X) = 1 & \end{array}\Big\}.$$

Removing redundant inequalities, we find

$$\mathcal{MG}_0(X) = \Big\{\mu \in \mathbb{R}^{2^{|X|}-1} : \begin{array}{lll} \mu(A \cup \{i\}) & \geqslant \mu(A), \varnothing \neq A \subset X, & i \notin A \\ \mu(\{i\}) & \geqslant 0, & i \in X \\ \mu(X) & = 1 & \end{array}\Big\}. \tag{2.55}$$

It follows that $\mathcal{MG}_0(X)$ is a polyhedron. Because for any set $A \in 2^X \setminus \{\varnothing\}$, we have $0 \leqslant \mu(A) \leqslant 1$, the polyhedron is bounded, hence it is a polytope.

Theorem 2.61 (The extreme points of $\mathcal{MG}_0(X)$) *(Stanley [316], Radojevic [272]) The set $\mathcal{MG}_0(X)$ of normalized capacities on X is a $(2^{|X|}-2)$-dimensional polytope, whose extreme points are all 0-1-capacities, except the null capacity 0. Moreover, each inequality and equality in (2.55) defines a facet.*

Proof We prove that every 0-1-capacity (different from 0) is an extreme point of $\mathcal{MG}_0(v)$; i.e., it cannot be expressed as a convex combination of other capacities. Take $\mu \neq 0$ a 0-1-capacity, and suppose that $\mu = \alpha\nu + (1-\alpha)\nu'$ for some $\alpha \in]0, 1[$, $\nu, \nu' \in \mathcal{MG}_0(X)$. Consider $A \subseteq X$ such that $\mu(A) = 1$. We have

$$1 = \mu(A) = \alpha\nu(A) + (1-\alpha)\nu'(A).$$

Since $\alpha \neq 0, 1$ and ν, ν' are normalized, it follows that necessarily $\nu(A) = \nu'(A) = 1$. Similarly, if A is such that $\mu(A) = 0$, the equality $0 = \alpha\nu(A) + (1-\alpha)\nu'(A)$ entails by nonnegativity of ν, ν' that $\nu(A) = \nu'(A) = 0$. This yields $\mu = \nu = \nu'$.

Conversely, suppose that μ is an extreme point, but not a 0-1-capacity. Put

$$\epsilon = \min(1 - \max_{A:\mu(A)<1} \mu(A), \min_{A:\mu(A)>0} \mu(A)),$$

and define $\mu'(A) = \mu(A) + \epsilon$ for all A such that $\mu(A) \neq 0, 1$, and $\mu''(A) = \mu(A) - \epsilon$ for all A such that $\mu(A) \neq 0, 1$, and $\mu'(A) = \mu(A) = \mu''(A)$ otherwise. Then μ', μ'' are normalized capacities and $\mu = \frac{\mu'+\mu''}{2}$, a contradiction. □

Recalling that 0-1-capacities are in bijection with antichains in $(2^X, \subseteq)$ (Sect. 2.8), we see that the number of extreme points of $\mathcal{MG}_0(X)$ is extremely large: it is $M(|X|) - 2$ (do not forget that the null capacity 0, corresponding to the empty antichain, has been removed), where $M(n)$ is the nth Dedekind number.

Let us prove a stronger result than Theorem 2.61, from which the latter is obtained as a simple corollary. We let $|X| = n$ for simplicity and write the inequalities in (2.55) under a matrix form $M\mu \geqslant b$, where μ is a $(2^n - 1)$-dim vector with $\mu_A = \mu(A)$ for all $\varnothing \neq A \subseteq X$, M is a matrix with a number of rows equal to

$$n + n(n-1) + \binom{n}{2}(n-2) + \cdots + \binom{n}{k}(n-k) + \cdots n \cdot 1 =$$

$$n\left(1 + \binom{n-1}{1} + \cdots + \binom{n-1}{k} + \cdots + 1\right) = n2^{n-1}$$

and $2^n - 1$ columns, with entry

$$M_{A,i;B} = \begin{cases} -1, & \text{if } B = A \\ 1, & \text{if } B = A \cup \{i\} \\ 0, & \text{otherwise} \end{cases}$$

for $A \subset X$, $i \in X \setminus A$, and $B \subseteq X$, $B \neq \varnothing$, and b is a $n2^{n-1}$ dimensional zero vector. Let us give an example with $n = 3$ and the following order on subsets: $\varnothing, 1, 2, 3, 12, 13, 23, 123$. The matrix M is given by

$$M = \begin{array}{r} \\ \varnothing,1 \\ \varnothing,2 \\ \varnothing,3 \\ 1,2 \\ 1,3 \\ 2,1 \\ 2,3 \\ 3,1 \\ 3,2 \\ 12,3 \\ 13,2 \\ 23,1 \end{array} \begin{array}{ccccccc} 1 & 2 & 3 & 12 & 13 & 23 & 123 \\ \hline 1 & & & & & & \\ & 1 & & & & & \\ & & 1 & & & & \\ -1 & & & 1 & & & \\ -1 & & & & 1 & & \\ & -1 & & 1 & & & \\ & -1 & & & & 1 & \\ & & -1 & & 1 & & \\ & & -1 & & & 1 & \\ & & & -1 & & & 1 \\ & & & & -1 & & 1 \\ & & & & & -1 & 1 \end{array}.$$

Theorem 2.62 *The matrix M is totally unimodular.*

Proof We follow the argument in Miranda et al. [243, Theorem 2]. We prove that $M^\top$ is totally unimodular, which is equivalent to the desired result. We remark that $M^\top = (I, B)$, where the submatrix I is a submatrix of the $(2^n - 1)$-dimensional identity matrix Id_{2^n-1}, and each column of B contains exactly one entry $+1$ and one entry -1. It follows from Theorem 1.14 that B is totally unimodular. Now, it is easy to see that (Id_{2^n-1}, B) is also totally unimodular, and so is $M^\top$ because it is a submatrix of (Id_{2^n-1}, B). □

The vector b being integer, it follows from Theorem 1.13 that the set of capacities such that $\mu(X)$ is integer is an integer polytope. In particular, if $\mu(X) = 1$, we recover Theorem 2.61.

Normalized capacities take their value in $[0, 1]$. One may think that the Möbius transform of normalized capacities take values in the symmetrized interval $[-1, 1]$. The following result shows that this is far from being true: the Möbius transform grows exponentially fast with $|X|$.

Theorem 2.63 (Exact bounds of the Möbius transform) *For any normalized capacity μ, its Möbius transform satisfies for any $A \subseteq X$, $|A| > 1$:*

$$-\binom{|A|-1}{l'_{|A|}} \leqslant m^{\mu}(A) \leqslant \binom{|A|-1}{l_{|A|}},$$

with

$$l_{|A|} = 2\left\lfloor \frac{|A|}{4} \right\rfloor, \quad l'_{|A|} = 2\left\lfloor \frac{|A|-1}{4} \right\rfloor + 1 \tag{2.56}$$

and for $|A| = 1 < n$:

$$0 \leqslant m^{\mu}(A) \leqslant 1,$$

and $m^{\mu}(A) = 1$ if $|A| = n = 1$. These upper and lower bounds are attained by the normalized capacities μ^*_A, μ_{A*}, respectively:

$$\mu^*_A(B) = \begin{cases} 1, & \text{if } |A| - l_{|A|} \leqslant |B \cap A| \leqslant |A| \\ 0, & \text{otherwise} \end{cases}$$

$$\mu_{A*}(B) = \begin{cases} 1, & \text{if } |A| - l'_{|A|} \leqslant |B \cap A| \leqslant |A| \\ 0, & \text{otherwise} \end{cases}$$

for any $B \subseteq X$.

We give in Table 2.3 the first values of the bounds.

Proof Let us prove the result for the upper bound when $A = X$. For any permutation $\sigma \in \mathfrak{S}(X)$ and any capacity $\mu \in \mathcal{MG}_0(X)$, we define the capacity $\sigma(\mu) \in \mathcal{MG}_0(X)$ by $\sigma(\mu)(B) = \mu(\sigma^{-1}(B))$ for any $B \subseteq X$.

We observe that $m^{\mu}(X)$ is invariant under permutation. Indeed,

$$m^{\sigma(\mu)}(X) = \sum_{B \subseteq X} (-1)^{n-|B|} \mu(\sigma^{-1}(B))$$

$$= \sum_{B' \subseteq X} (-1)^{n-|B'|} \mu(B') \qquad \text{(letting } B' = \sigma^{-1}(B))$$

$$= m^{\mu}(X).$$

$\vert A\vert$	1	2	3	4	5	6	7
u.b. of $m^{\mu}(A)$	1	1	1	3	6	10	15
l.b. of $m^{\mu}(A)$	1(0)	−1	−2	−3	−4	−10	−20

$\vert A\vert$	8	9	10	11	12
u.b. of $m^{\mu}(A)$	35	70	126	210	462
l.b. of $m^{\mu}(A)$	−35	−56	−126	−252	−462

Table 2.3 Lower and upper bounds for the Möbius transform of a normalized capacity

For every set function μ on X, define its symmetric part $\mu^s = \frac{1}{n!}\sum_{\sigma\in\mathfrak{S}(X)}\sigma(\mu)$, which is a symmetric function. By convexity of $\mathcal{MG}_0(X)$, if $\mu \in \mathcal{MG}_0(X)$, then so is μ^s, and by linearity of the Möbius inverse, we have

$$m^{\mu^s}(X) = \frac{1}{n!}\sum_{\sigma\in\mathfrak{S}(X)} m^{\sigma(\mu)}(X) = \frac{1}{n!}\sum_{\sigma\in\mathfrak{S}(X)} m^{\mu}(X) = m^{\mu}(X).$$

It is therefore sufficient to maximize $m^{\mu}(X)$ on the set of symmetric normalized capacities. But this set is also a convex polytope, whose extreme points are the following $\{0, 1\}$-valued capacities μ_k defined by

$$\mu_k(B) = 1 \text{ iff } |B| \geqslant n-k \qquad (k = 0, \ldots, n-1).$$

Indeed, if μ is symmetric, it can be written as a convex combination of these capacities:

$$\mu = \mu(\{1\})\mu_{n-1} + \sum_{k=2}^{n}\big(\mu(\{1,\ldots,k\}) - \mu(\{1,\ldots,k-1\})\big)\mu_{n-k}.$$

It follows that the maximum of $m^{\mu}(X)$ is attained on one of these capacities, say μ_k. We compute

$$\begin{aligned} m^{\mu_k}(X) &= \sum_{B\subseteq X}(-1)^{|X\setminus B|}\mu_k(B) = \sum_{i=n-k}^{n}(-1)^{n-i}\binom{n}{i} \\ &= \sum_{i'=0}^{k}(-1)^{i'}\binom{n}{n-i'} = (-1)^k\binom{n-1}{k}, \end{aligned} \tag{2.57}$$

where the third equality is obtained by letting $i' = n - i$ and the last one follows from (1.2). Therefore k must be even. If $n-1$ is even, the maximum of $\binom{n-1}{k}$ for k even is attained for $k = \frac{n-1}{2}$ if this is an even number, otherwise $k = \frac{n-3}{2}$. If $n-1$ is odd, the maximum of $\binom{n-1}{k}$ is reached for $k = \lceil\frac{n-1}{2}\rceil$ and $k-1 = \lfloor\frac{n-1}{2}\rfloor$, among which the even one must be chosen. As it can be checked (see Table 2.4 below), this amounts to taking

$$k = 2\left\lfloor\frac{n}{4}\right\rfloor,$$

that is, $k = l_n$ as defined in (2.56), and we have defined the capacity

$$\mu^*(B) = 1 \text{ if } n - l_n \leqslant |B| \leqslant n,$$

which is μ^*_X as defined in the theorem.

For establishing the upper bound of $m^{\mu}(A)$ for any $A \subset X$, remark that the value of $m^{\mu}(A)$ depends only on the subsets of A. It follows that applying the above result to the sublattice 2^A, the set function ξ^*_A defined on 2^X by

$$\xi^*_A(B) = 1 \text{ if } B \subseteq A \text{ and } |A| - l_{|A|} \leqslant |B| \leqslant |A|, \text{ and 0 otherwise},$$

yields an optimal value for $m^{\mu}(A)$. It remains to turn this set function into a capacity on X, without destroying optimality. This can be done since ξ^*_A is monotone on 2^A, so that taking the monotonic cover of ξ^*_A yields an optimal capacity, given by

$$\mathbf{mc}(\xi^*_A)(B) = \max_{C\subseteq B}\xi^*_A(C) = 1 \text{ if } |A| - l_{|A|} \leqslant |B\cap A| \leqslant |A|, \text{ and 0 otherwise},$$

which is exactly μ^*_A as desired. Note however that this is not the only optimal solution in general, since values of the capacity on the sublattice $2^{X\setminus A}$ are irrelevant.

One can proceed in a similar way for the lower bound. In this case however, as it can be checked, the capacity must be equal to 1 on the $l'_n + 1$ first lines of the lattice 2^X, with $l'_n = 2\left\lfloor\frac{n-1}{4}\right\rfloor + 1$ (see Table 2.4). □

n/k	0	1	2	3	4	5	6	7	8	9	10	11
$n = 1$	1											
$n = 2$	1	−1										
$n = 3$	1	−2	1									
$n = 4$	1	−3	3	−1								
$n = 5$	1	−4	6	−4	1							
$n = 6$	1	−5	10	−10	5	−1						
$n = 7$	1	−6	15	−20	15	−6	1					
$n = 8$	1	−7	21	−35	35	−21	7	−1				
$n = 9$	1	−8	28	−56	70	−56	28	−8	1			
$n = 10$	1	−9	36	−84	126	−126	84	−36	9	−1		
$n = 11$	1	−10	45	−120	210	−252	210	−120	45	−10	1	
$n = 12$	1	−11	55	−105	330	−462	462	−330	165	−55	11	−1

Table 2.4 Computation of the upper (red) and lower (blue) bounds

The value of the capacity μ is 1 for the $k + 1$ first lines of the lattice 2^X. Each entry (n, k) equals $m^\mu(X)$, as given by (2.57)

This result appears in Grabisch and Miranda [179] (corrected version of [244]). It can be shown that the exact bounds of I^μ, $I^\mu_{\rm B}$ are the same, attained by the same capacities [179]. Using the well-known Stirling's approximation $\binom{2n}{n} \simeq \frac{4^n}{\sqrt{\pi n}}$ for $n \to \infty$, we obtain the asymptotic behavior of the bounds:

$$-\frac{4^{\frac{n}{2}}}{\sqrt{\frac{\pi n}{2}}} \leqslant m^\mu(X) \leqslant \frac{4^{\frac{n}{2}}}{\sqrt{\frac{\pi n}{2}}}. \tag{2.58}$$

Remark 2.64

(i) The polytope of normalized capacities is a particular case of *order polytopes* studied among others by Stanley [316]. Considering a poset $(P, \leqslant)$ of n elements, the order polytope of P is defined by

$$\mathcal{O}(P) = \{f : P \to \mathbb{R} \text{ s.t. } 0 \leqslant f(x) \leqslant 1, \quad f(x) \leqslant f(y) \text{ if } x \leqslant y\}.$$

$\mathcal{MG}_0(X)$ is recovered with $(P, \leqslant) = (2^X, \subseteq)$. Theorem 2.61 still holds *mutatis mutandis* in this general framework. In particular, the extreme points of $\mathcal{O}(P)$ are bijectively associated to the antichains of P.

Stanley also introduces a related polytope, called the *chain polytope* $\mathcal{C}(P)$ of P. It is defined as

$$\mathcal{C}(P) = \{g : P \to \mathbb{R} \text{ s.t. } g(x) \geqslant 0, \quad g(y_1) + g(y_k) \leqslant 1$$
$$\text{for all maximal chains } y_1, \ldots, y_k \text{ in } P\}.$$

It is shown that the extreme points of $\mathcal{C}(P)$ are the characteristic functions of antichains of P, showing that $\mathcal{C}(P)$ and $\mathcal{O}(P)$ have the same number of extreme points. However, in general faces are different. The following function ϕ is a continuous piecewise linear bijection between $\mathcal{O}(P)$ and $\mathcal{C}(P)$:

$$\phi : \mathcal{O}(P) \to \mathcal{C}(P); \quad \phi \circ f(x) = \min\{f(x) - f(y), \quad x \text{ covers } y\}.$$

Applied to our case, the mapping ϕ is related to the derivative of a capacity. For a capacity μ, $\phi \circ \mu(A) = \min_{i \in A} \Delta_i \mu(A \setminus \{i\})$, for $A \in 2^X$.

(ii) A detailed study of the polytope of normalized capacities was done by Combarro and Miranda [56]. They studied the adjacency of extreme points in $\mathcal{MG}_0(X)$, where two extreme points are adjacent if they belong to the same edge (face of dimension 1). In particular, they showed that an extreme point chosen at random is adjacent to u_X with probability tending to 1 when $|X|$ tends to infinity. The same result holds for the conjugate game $\overline{u_X}$. Moreover, the diameter of $\mathcal{MG}_0(X)$, i.e., the maximal number of adjacency relations necessary to relate any two extreme points, is 3 when $|X| \geqslant 3$.

◇

2.15.7 *The Polytope of Belief Measures*

Let us express the set $\mathcal{B}(X)$ using the basis of unanimity games:

$$\mathcal{B}(X) = \left\{ m \in \mathbb{R}^{2^{|X|}-1} : \begin{array}{ll} m(A) & \geqslant 0, \varnothing \neq A \subseteq X \\ \sum_{A \subseteq X, A \neq \varnothing} m(A) & = 1 \end{array} \right\}. \tag{2.59}$$

This is a $(2^{|X|} - 2)$-dimensional polytope, because it is clearly bounded. Each (in)equality defines a facet. This polytope is very simple, because it is merely the intersection of the hyperplane $\sum_{A \subseteq X, A \neq \varnothing} m(A) = 1$ with the positive orthant (or, with the cone of totally monotone nonnegative games). Its extreme points are therefore all $2^{|X|} - 1$ points of the form $(0, \ldots, 0, 1, 0, \ldots, 0)$. These correspond exactly to the unanimity games u_A, $A \subseteq X, A \neq \varnothing$.

The result can be found also by simply remarking that the (unique) decomposition of a belief measure on the basis of unanimity games reduces to a convex sum.

2.15.8 The Polytope of At Most k-Additive Normalized Capacities

The study of $\mathcal{MG}^{\leqslant k}(X)$ reveals to be more difficult. We begin by studying the case $k = 1$, that is, the set of probability measures. Proceeding as for belief measures, the set

$$\mathcal{MG}^1(X) = \left\{ m \in \mathbb{R}^{|X|} \,:\, \begin{array}{ll} m_i & \geqslant 0,\, i \in X \\ \sum_{i \in X} m_i & = 1 \end{array} \right\}$$

is a $(|X|-1)$-dimensional polytope, whose extreme points are all vectors of the form $(0,\ldots,0,1,0,\ldots,0)$. These correspond to the unanimity games $u_{\{i\}}$, $i \in X$, or put differently, the set of Dirac measures on X.

Let us study $\mathcal{MG}^{\leqslant 2}(X)$, and write this set using the basis of unanimity games. Discarding all null coordinates, we obtain

$$\mathcal{MG}^{\leqslant 2}(X) = \left\{ m \in \mathbb{R}^{\kappa(2)} \,:\, \begin{array}{ll} m_i & \geqslant 0,\, i \in X \\ m_i + \sum_{j \in K} m_{ij} & \geqslant 0,\, i \in X, \varnothing \neq K \subseteq X \setminus i \\ \sum_{i \in X} m_i + \sum_{\{i,j\} \subseteq X} m_{ij} & = 1 \end{array} \right\} \tag{2.60}$$

with $\kappa(2) = \binom{|X|}{1} + \binom{|X|}{2}$, and where we have written for simplicity m_i, m_{ij} instead of $m(\{i\}), m(\{i,j\})$. The two first sets of inequalities come from monotonicity [Theorem 2.33(ii)].

Theorem 2.65 (The extreme points of $\mathcal{MG}^{\leqslant 2}(X)$) *(Miranda et al. [243, Proposition 11]) $\mathcal{MG}^{\leqslant 2}(X)$ is a $(\kappa(2) - 1)$-dimensional polytope, whose extreme points are all at most 2-additive normalized 0-1-capacities. These 0-1-capacities are of three different types:*

(i) The unanimity games $u_{\{i\}}$, $i \in X$ (these are the extreme points of $\mathcal{MG}^1(X)$);
(ii) The unanimity games $u_{\{i,j\}}$, $\{i,j\} \subseteq X$;
(iii) The conjugate of the unanimity games $u_{\{i,j\}}$, $\{i,j\} \subseteq X$, given by

$$\overline{u_{\{i,j,\}}} = u_i + u_j - u_{\{i,j\}}.$$

Moreover, the convex decomposition of any $\mu \in \mathcal{MG}^{\leqslant 2}(X)$ is

$$\mu = \sum_{\{i,j\} \subseteq X \,:\, m^\mu_{ij} < 0} (-m^\mu_{ij}) \overline{u_{\{i,j\}}} + \sum_{\{i,j\} \subseteq X \,:\, m^\mu_{ij} > 0} m^\mu_{ij} u_{\{i,j\}} + \sum_{i \in X} \Big(m^\mu_i + \sum_{j \in X \setminus i \,:\, m^\mu_{ij} < 0} m^\mu_{ij} \Big) u_{\{i\}}, \tag{2.61}$$

using the above shorthand.

Proof By Theorem 2.61, we know that any normalized 0-1-capacity is an extreme point of $\mathcal{MG}_0(X)$. It is therefore plain that any 2-additive normalized 0-1-capacity is an extreme point of $\mathcal{MG}^{\leqslant 2}(X)$. Then we have to check that the above candidates are indeed 2-additive, normalized and 0-1-valued, and that there is no other extreme point.

The first point is easy to check and is left to the readers. As for the second point, we show that any $\mu \in \mathcal{MG}^{\leqslant 2}(X)$ can be written as (2.61), and that this decomposition is indeed a convex combination. By 2-additivity, we have:

$$\begin{aligned}
\mu &= \sum_{i\in X} m_i^\mu u_i + \sum_{\{i,j\}\subseteq X} m_{ij}^\mu u_{\{i,j\}} \\
&= \sum_{i\in X} m_i^\mu u_i + \sum_{i,j\,:\,m_{ij}^\mu>0} m_{ij}^\mu u_{\{i,j\}} + \sum_{i,j\,:\,m_{ij}^\mu<0} m_{ij}^\mu u_{\{i,j\}} \\
&= \sum_{i\in X} m_i^\mu u_i + \sum_{i,j\,:\,m_{ij}^\mu>0} m_{ij}^\mu u_{\{i,j\}} + \sum_{i,j\,:\,m_{ij}^\mu<0} m_{ij}^\mu (u_i + u_j - \overline{u_{\{i,j\}}}) \\
&= \sum_{i,j\,:\,m_{ij}^\mu>0} m_{ij}^\mu u_{\{i,j\}} + \sum_{i,j\,:\,m_{ij}^\mu<0} (-m_{ij}^\mu)\overline{u_{\{i,j\}}} + \sum_{i\in X}\Big(m_i^\mu + \sum_{j\,:\,m_{ij}^\mu<0} m_{ij}^\mu\Big) u_{\{i\}},
\end{aligned}$$

the desired expression. We check now that this is convex combination. The coefficients of terms in $u_{\{i\}}$ are nonnegative by monotonicity of μ [second set of inequalities in (2.60)], so that all coefficients are nonnegative. It remains to prove that they sum up to 1. This sum is:

$$\begin{aligned}
-\sum_{m_{ij}^\mu<0} m_{ij}^\mu + \sum_{m_{ij}^\mu>0} m_{ij}^\mu + \sum_{i\in X} m_i^\mu + \sum_{i\in X}\sum_{j\neq i\,:\,m_{ij}^\mu<0} m_{ij}^\mu &= \\
\sum_{i\in X} m_i^\mu + \sum_{m_{ij}^\mu>0} m_{ij}^\mu + \sum_{m_{ij}^\mu<0} m_{ij}^\mu = \mu(X) = 1.
\end{aligned}$$

This completes the proof. □

The above result permits to know the number of extreme points of $\mathcal{MG}^{\leqslant 2}(X)$. Summing the cardinality of all three types of extreme points, we find $|X| + \binom{|X|}{2} + \binom{|X|}{2} = |X|^2$.

Remark 2.66

(i) Up to now, no other remarkable result is known for $\mathcal{MG}^{\leqslant k}(X)$ when $k > 2$. Miranda et al. [243] have found an example of extreme point in $\mathcal{MG}^{\leqslant 3}(X)$ with $|X| = 4$, which is not a 0-1-capacity. It seems that apart from at most k-

additive 0-1-capacities, many other extreme points exist, and that the structure of k-additive capacities is very complex. The readers can consult Combarro and Miranda [57] for further results and an algorithm to compute extreme points.

(ii) The polytope of p-symmetric normalized capacities was studied by Miranda et al. [243]. Its structure is close to the one of normalized capacities because it is an order polytope [Remark 2.64(i)].

◇

2.16 Polynomial Representations

We set $|X| = n$ for convenience in the whole section, and denote the set $\{1, \ldots, n\}$ by $[n]$.

Set functions can be seen as functions on the vertices of the hypercube $[0, 1]^n$, by means of the characteristic function of sets $1\colon A \mapsto 1_A$, which is an isomorphism between 2^X and $\{0, 1\}^n$. A *pseudo-Boolean function* is any function $f : \{0, 1\}^n \to \mathbb{R}$; $x \mapsto f(x)$. To any set function ξ corresponds a unique pseudo-Boolean function f_ξ defined by

$$f_\xi(x) = \xi(\{i \in X : x_i = 1\}) \qquad (x \in \{0, 1\}^n).$$

Conversely, to any pseudo-Boolean function f corresponds a unique set function ξ_f defined by

$$\xi_f(A) = f(1_A) \qquad (A \subseteq X).$$

In short, $\xi_f = f \circ 1$, and $f_\xi = \xi \circ 1^{-1}$. The mapping 1 can be seen as a coding function. Other coding functions are possible, see Remark 2.68(iii). We denote by $\mathcal{PB}(n)$ the set of pseudo-Boolean functions on $\{0, 1\}^n$. It is a vector space.

Definition 2.67 The *derivative* with respect to coordinate i of a pseudo-Boolean function is defined as for ordinary functions by

$$\Delta_i f(x) = f(x_1, \ldots, x_{i-1}, 1, x_{i+1}, \ldots, x_n) - f(x_1, \ldots, x_{i-1}, 0, x_{i+1}, \ldots, x_n).$$

Derivatives w.r.t. several coordinates are defined recursively as follows: for any $A \subseteq [n], A \neq \varnothing$,

$$\Delta_A f(x) = \Delta_{\{i\}}(\Delta_{A\setminus\{i\}} f(x)) \text{ for some } i \in A,$$

with $\Delta_{\{i\}}f = \Delta_i f$, and $\Delta_\varnothing f = f$. Recalling the derivative of set functions (Definitions 2.14 and 2.15), it is easy to see that they correspond:

$$\Delta_A \xi_f(B) = \Delta_A f(1_B) \qquad (A, B \subseteq [n]).$$

Remark 2.68

(i) Pseudo-Boolean functions were introduced by Hammer and Rudeanu [191], and have many applications in operations research, optimization (Boros and Hammer [36]), circuit design, coding theory, and many other topics in theoretical computer sciences (Crama and Hammer [63, Chap. 13], de Wolf [73], O'Donnell [258, 259]).

(ii) Any definition or property established for set functions can be transposed to pseudo-Boolean functions and vice versa. In particular, grounded pseudo-Boolean functions (that is, vanishing at $\mathbf{0}$) correspond to games, while capacities are grounded nondecreasing pseudo-Boolean functions. The interest of pseudo-Boolean functions is that they lead to polynomial representations and their extensions on the hypercube $[0, 1]^n$, which will be seen to be integrals, and also to approximation problems.

(iii) The coding of a subset A of $[n]$ by 0 and 1 can be changed to -1 and $+1$ if this happens to be more convenient. The latter is often used in theoretical computer sciences (O'Donnell [258, 259]); more on this in Sect. 2.16.2.

◇

2.16.1 Bases of $\mathcal{PB}(n)$

Let f be a pseudo-Boolean function. It is easy to check that it can be written as

$$f(x) = \sum_{A\subseteq[n]} f(1_A) \prod_{i\in A} x_i \prod_{i\in A^c} (1 - x_i), \tag{2.62}$$

for every $x \in \{0, 1\}^n$, and with the convention $\prod_{i\in\varnothing} x_i = 1$. Indeed, if $x = 1_B$ for some $B \subseteq [n]$, the term $\prod_{i\in A} x_i \prod_{i\in A^c}(1-x_i)$ is nonnull if and only if $A = B$. Hence, the pseudo-Boolean function $\prod_{i\in A} x_i \prod_{i\in A^c}(1 - x_i)$ corresponds to the Dirac game δ_A (with the exception that $\delta_\varnothing$ is not a game!). We know from Theorem 2.56 that the set of Dirac games forms a basis of the set of games $\mathcal{G}(X)$. Hence, the polynomials $\prod_{i\in A} x_i \prod_{i\in A^c}(1 - x_i)$ form a 2^n-dimensional basis of the vector space $\mathcal{PB}(n)$.[7] We call (2.62) the *standard representation* of f.

From Theorem 2.56 again, we know that the unanimity games form a basis of $\mathcal{G}(X)$. It is easily seen that unanimity games correspond to the monomials $\prod_{i\in A} x_i$.

[7] Our proof of Theorem 2.56 extends easily if $\delta_\varnothing$ is considered.

Hence, another polynomial representation of pseudo-Boolean functions is given by

$$f(x) = \sum_{T\subseteq[n]} a_T \prod_{i\in T} x_i, \tag{2.63}$$

for every $x \in \{0,1\}^n$, where the coefficients a_T (this is the traditional notation) form the Möbius transform of ξ_f, the set function associated to f. We call this representation the *Möbius representation* of f.

Let us endow $\mathcal{PB}(n)$ with the inner product

$$\langle f, g\rangle = \frac{1}{2^n}\sum_{x\in\{0,1\}^n} f(x)g(x) \qquad (f,g \in \mathcal{PB}(n)).$$

(note that this defines an inner product for the vector space of games: $\langle v, w\rangle = \frac{1}{2^n}\sum_{A\subseteq X} v(A)w(A)$). Accordingly, we define the associated (L_2) norm by $\|f\| = \sqrt{\langle f,f\rangle}$. The question is to find an orthonormal basis of $\mathcal{PB}(n)$. It is easily seen that the first basis (corresponding to the Dirac games) is orthogonal but not orthonormal, while the second is even not orthogonal. An orthonormal basis exists, denoted by $\{w_T\}_{T\subseteq[n]}$, with

$$w_T(x) = \prod_{i\in T}(2x_i - 1) \qquad (T \subseteq [n]).$$

We call any w_T a *Walsh function*, with some abuse (Remark 2.70). The set function corresponding to w_T is denoted by the same symbol w_T, and is expressed as

$$w_T(S) = (-1)^{|T\setminus S|} \qquad (S \subseteq [n]). \tag{2.64}$$

We call such set functions *Walsh set functions*. These are *not* games because $w_T(\varnothing) \neq 0$.

Theorem 2.69 *The family $\{w_T\}_{T\subseteq[n]}$ is an orthonormal basis of $\mathcal{PB}(n)$.*

Proof We start by proving that the Walsh functions are orthonormal; i.e., $\langle w_S, w_T\rangle = 1$ if $S = T$, and 0 otherwise. Setting $z_i = 2x_i - 1$ for $i = 1,\dots,n$, we have $w_S(z) = \prod_{i\in S} z_i$, denoted for simplicity by z^S, for any $z \in \{-1,1\}^{[n]}$. Then we find

$$\langle w_S, w_T\rangle = \frac{1}{2^n}\sum_{z\in\{-1,1\}^{[n]}} z^S z^T = \frac{1}{2^n}\sum_{z\in\{-1,1\}^{[n]}} z^{S\Delta T}.$$

If $S = T$, we get $\langle w_S, w_T\rangle = 1$, owing to the convention $\prod_{i\in\varnothing} z_i = 1$. If $S \neq T$, we have:

$$\sum_{z\in\{-1,1\}^{[n]}} z^{S\Delta T} = 2^{|[n]\setminus(S\Delta T)|}\sum_{z\in\{-1,1\}^{S\Delta T}} z^{S\Delta T} = 2^{|[n]\setminus(S\Delta T)|}(1-1)^{|S\Delta T|} = 0.$$

Now, because the Walsh functions are pairwise orthogonal, they form an independent subset of $\mathcal{PB}(n)$. Since there are $2^n = \dim\mathcal{PB}(n)$ Walsh functions, this independent set must be a basis.

□

Let us express any pseudo-Boolean function f in this basis. Setting $z_i = 2x_i - 1 \in \{-1, +1\}$, the Walsh functions reduce to monomials $\prod_i z_i$. From $x_i = \frac{1}{2}z_i + \frac{1}{2}$, we get

$$\begin{aligned} f(z) &= \sum_{T\subseteq[n]} a_T \prod_{i\in T} \frac{1}{2}(z_i + 1) \\ &= \sum_{T\subseteq[n]} \frac{a_T}{2^{|T|}} \Big(\sum_{S\subseteq T} \prod_{i\in S} z_i\Big) \\ &= \sum_{T\subseteq[n]} \Big(\sum_{S\supseteq T} \frac{a_S}{2^{|S|}}\Big) \prod_{i\in T} z_i. \end{aligned}$$

In summary, a pseudo-Boolean function reads as follows in the basis of Walsh functions:

$$f(x) = \sum_{T\subseteq[n]} \Big(\sum_{S\supseteq T} \frac{a_S}{2^{|S|}}\Big) w_T(x) = \sum_{T\subseteq[n]} \frac{1}{2^{|T|}} I^f_{\mathrm{B}}(T) w_T(x), \tag{2.65}$$

using (2.47). Note that up to a factor, the coordinates in the basis of the Walsh functions are merely the Banzhaf interaction coefficients. We call this expression the *Walsh representation* of f.

Remark 2.70

(i) The basis w_T is closely related to the *Walsh functions* [342], well-known in signal processing (Hurst et al. [199]). The original Walsh functions are defined as follows, for any $k \in \mathbb{N}_0$ and any $x \in [0, 1]$:

$$W_k(x) = (-1)^{\sum_{j=0}^{\infty} k_j x_{j+1}}, \tag{2.66}$$

with $k = k_0 + k_1 2 + k_2 2^2 + \cdots k_m 2^m$, $k_i \in \{0, 1\}$ for all i, and $x = x_1 2^{-1} + x_2 2^{-2} + x_3 2^{-3} + \cdots$, $x_i \in \{0, 1\}$ for all i, the binary representations of k and x. They form an orthonormal basis of the set of square integrable functions on $[0, 1]$. The connection with our Walsh functions is that the latter have a discretized domain $0, \frac{1}{2^n}, \frac{2}{2^n}, \frac{3}{2^n}, \ldots, \frac{2^n-1}{2^n}$ of 2^n points, corresponding to the 2^n subsets of $[n]$. More precisely, $w_S(x)$ corresponds to $W_k(x')$ such that S and k have same binary coding (see Sect. 2.15.1), and $x'_1 = 1 - x_1, \ldots, x'_n = 1 - x_n$, $x'_j = 0$ for $j > n$. For illustration, we take $n = 3$, and represent graphically the Walsh functions on Fig. 2.2. The Walsh functions are also closely related to the Fourier transform, see Sect. 2.16.2.

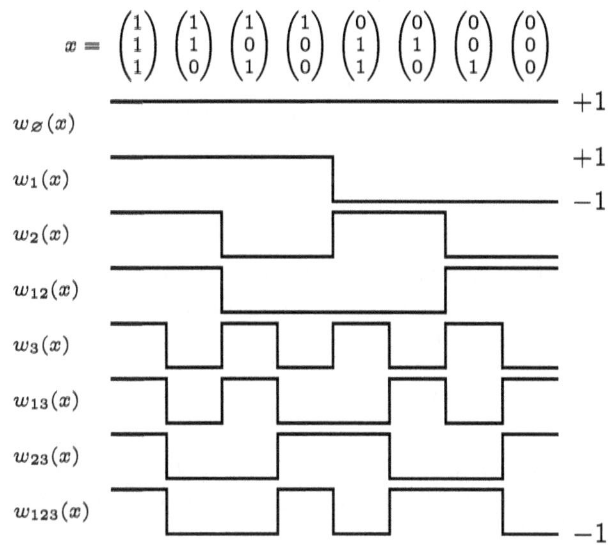

Fig. 2.2 The Walsh functions $w_T(x)$ for $n = 3$

(ii) Some authors have proposed to use a more general inner product (Ding et al. [90, 91] and Marichal and Mathonet [232]), starting from probabilities $p_1, \ldots, p_n$, where p_i indicates the probability that $x \in \{0,1\}^n$ has coordinate $x_i = 1$. Considering that coordinates are statistically independent, this induces a probability distribution over $\{0,1\}^n$ given by

$$p(x) = \prod_{i\in[n]} p_i^{x_i}(1-p_i)^{1-x_i}.$$

Assuming that $p_i > 0$ for all $i \in [n]$, the inner product is defined by

$$\langle f, g\rangle = \sum_{x\in\{0,1\}^n} p(x)f(x)g(x). \tag{2.67}$$

Note that our previous inner product is recovered with $p_i = \frac{1}{2}$ for all $i \in [n]$ (uniform distribution on the hypercube). It is shown in Ding et al. [91] that there still exists an orthonormal basis of pseudo-Boolean functions, given by:

$$w_T^p(x) = \prod_{i\in T} \frac{x_i - p_i}{\sqrt{p_i(1-p_i)}}.$$

◇

2.16.2 The Fourier Transform

Let us introduce another orthonormal basis of pseudo-Boolean functions, through the parity functions. For any subset $S \subseteq [n]$, the *parity function* associated to S is the function

$$\chi_S(x) = (-1)^{1_S^\top x} = (-1)^{\sum_{i \in S} x_i} \qquad (x \in \{0,1\}^n), \tag{2.68}$$

where $1_S^\top x$ is the inner product between the two vectors $1_S, x$. The parity function outputs 1 if the number of variables in S having value 1 is even, and -1 if it is odd. In terms of set functions, the parity function reads

$$\chi_S(T) = (-1)^{|S \cap T|} \qquad (T \in 2^{[n]}).$$

An important remark is that the parity functions are, up to a recoding, identical to the Walsh functions: consider the coding function ε defined by $\varepsilon(1) = 0$ and $\varepsilon(-1) = 1$. Recall that the Walsh functions are monomials:

$$w_S(z) = \prod_{i \in S} z_i \qquad (z \in \{-1,+1\}^n).$$

Then it holds that $w_S(z) = \chi_S \circ \varepsilon(z)$. Indeed,

$$w_S(z) = \prod_{i \in S} z_i = (-1)^{\sum_{i \in S} \varepsilon(z_i)} = \chi_S(\varepsilon(z)).$$

[see also Remark 2.70(i)] It follows that the parity functions inherit the properties of the Walsh functions (and vice versa), in particular, $\{\chi_S\}_{S \subseteq [n]}$ is an orthonormal basis of $\mathcal{PB}(n)$: $\langle \chi_S, \chi_T \rangle = 1$ if and only if $S = T$, and 0 otherwise. As other interesting property, we have, as the readers can check,

$$\chi_S(x \oplus y) = \chi_S(x)\chi_S(y) \qquad (x, y \in \{0,1\}^n), \tag{2.69}$$

where $\oplus$ denotes the coordinatewise binary addition:

$$1 \oplus 1 = 0 = 0 \oplus 0, \quad 1 \oplus 0 = 0 \oplus 1 = 1.$$

Following the tradition, let us denote by $\widehat{f}(S)$, $S \subseteq [n]$, the coordinates of a pseudo-Boolean function f in the basis of parity functions:

$$f = \sum_{S \subseteq [n]} \widehat{f}(S) \chi_S. \tag{2.70}$$

The basis being orthonormal, it follows from (2.70) that $\widehat{f}(S)$ is simply given by

$$\widehat{f}(S) = \langle f, \chi_S \rangle = \frac{1}{2^n} \sum_{x \in \{0,1\}^n} (-1)^{1_S^\top x} f(x) \qquad (S \subseteq [n]), \tag{2.71}$$

or, in terms of set functions,

$$\widehat{\xi}(S) = \frac{1}{2^n} \sum_{T \subseteq [n]} (-1)^{|S \cap T|} \xi(T) \qquad (S \subseteq [n]). \tag{2.72}$$

The set of coordinates $\{\widehat{f}(S)\}_{S \subseteq [n]}$ is the *Fourier transform*[8] or *Fourier spectrum* of f. We may also call the basis of parity functions the *Fourier basis*.

Note that Formula (2.70) gives the inverse Fourier transform; i.e., how to recover f from $\widehat{f}$.

We gather in the next theorem the most basic properties of the Fourier transform. We first introduce some notation and definitions. Considering $x \in \{0,1\}^n$ as a random variable with uniform distribution, the expected value and the variance of a pseudo-Boolean function f are

$$\mathbb{E}[f] = \frac{1}{2^n} \sum_{x \in \{0,1\}^n} f(x); \quad \mathrm{Var}[f] = \mathbb{E}\big[(f - \mathbb{E}[f])^2\big] = \mathbb{E}[f^2] - \mathbb{E}^2[f].$$

Next, given two pseudo-Boolean functions f, g, their *convolution product* $f * g$ is a pseudo-Boolean function defined by

$$(f * g)(x) = \frac{1}{2^n} \sum_{y \in \{0,1\}^n} f(x \oplus y) g(y) \qquad (x \in \{0,1\}^n). \tag{2.73}$$

Theorem 2.71 *Let $f, g \in \mathcal{PB}(n)$. The following holds.*

(i) $f(\mathbf{0}) = \sum_{S \subseteq [n]} \widehat{f}(S)$*;*
(ii) $\widehat{f}(\varnothing) = \mathbb{E}[f]$*;*
(iii) (Parseval's identity) $\|f\|^2 = \sum_{S \subseteq [n]} \widehat{f}^2(S)$*;*
(iv) $\sum_{S \in 2^{[n]} \setminus \{\varnothing\}} \widehat{f}^2(S) = \mathrm{Var}[f]$*;*
(v) f is constant if and only if $\widehat{f}(S) = 0$ for all $S \neq \varnothing$;
(vi) $\widehat{(f * g)}(S) = \widehat{f}(S)\widehat{g}(S)$ *for all $S \in 2^{[n]}$.*

---

[8]The Fourier transform indeed induces a transformation or transform in the sense of Sect. 2.11, which to any set function ξ maps another set function $\widehat{\xi}$. It is linear and invertible (Sect. 2.17.1).

Proof (i) and (ii) are obvious.

(iii) By orthonormality of the Fourier basis,

$$\|f\|^2 = \langle f, f\rangle = \sum_{S\subseteq[n]} \widehat{f}(S) \sum_{T\subseteq[n]} \widehat{f}(T)\langle \chi_S, \chi_T\rangle = \sum_{S\subseteq[n]} \widehat{f}^2(S).$$

(iv) Immediate by definition of the variance, (ii) and (iii).

(v) Supposing f constant and using the set function notation, it suffices to show that $\sum_{T\subseteq[n]}(-1)^{|S\cap T|} = 0$ for every nonempty S. We have, for $S \neq \varnothing$

$$\sum_{T\subseteq[n]} (-1)^{|S\cap T|} = \sum_{K\subseteq[n]\setminus S} \sum_{L\subseteq S} (-1)^{|L|} = 0$$

by applying Lemma 1.1(i). The reverse implication comes from (iv).

(vi) We have by (2.71)

$$\begin{aligned}
\widehat{f * g}(S) &= \frac{1}{2^n} \sum_x (f * g)(x)\chi_S(x) \\
&= \frac{1}{2^{2n}} \sum_x \sum_y f(x\oplus y)g(y)\chi_S(x) \\
&= \frac{1}{2^n} \sum_y g(y)\chi_S(y)\Big(\frac{1}{2^n}\sum_x f(x\oplus y)\chi_S(x\oplus y)\Big) \\
&= \widehat{f}(S)\widehat{g}(S),
\end{aligned}$$

where we have used (2.69) in the third equality. □

Finally, we establish the relation between the Fourier, Möbius and Banzhaf transforms [176]. Taking any set function ξ, we have

$$\begin{aligned}
\widehat{\xi}(S) = \frac{1}{2^n} \sum_{T\subseteq[n]} (-1)^{|S\cap T|} \xi(T) &= \frac{1}{2^n} \sum_{T\subseteq[n]} (-1)^{|S\cap T|} \sum_{K\subseteq T} m^{\xi}(K) \\
&= \frac{1}{2^n} \sum_{K\subseteq[n]} m^{\xi}(K) \sum_{T\supseteq K} (-1)^{|S\cap T|}.
\end{aligned}$$

Now,

$$\sum_{T\supseteq K}(-1)^{|S\cap T|} = (-1)^{|K\cap S|}2^{n-|K\cup S|} + (-1)^{|K\cap S|+1}2^{n-|K\cup S|}\binom{|S\setminus K|}{1} + $$
$$(-1)^{|K\cap S|+2}2^{n-|K\cup S|}\binom{|S\setminus K|}{2} + \cdots + (-1)^{|S|}2^{n-|K\cup S|}$$
$$= (-1)^{|K\cap S|}2^{n-|K\cup S|}\underbrace{\left(1 - \binom{|S\setminus K|}{1} + \binom{|S\setminus K|}{2} + \cdots + (-1)^{|S\setminus K|}\right)}_{=0 \text{ except if } |S\setminus K|=0}.$$

It follows that

$$\widehat{\xi}(S) = \frac{1}{2^n}\sum_{K\supseteq S} m^{\xi}(K)(-1)^{|K\cap S|}2^{n-|K\cup S|}$$
$$= (-1)^{|S|}\sum_{K\supseteq S}\frac{1}{2^k}m^{\xi}(K). \tag{2.74}$$

Now, using (2.47), we obtain

$$\widehat{\xi}(S) = \left(\frac{-1}{2}\right)^s I_{\mathrm{B}}^{\xi}(S). \tag{2.75}$$

Finally, from (2.75) and (2.48), we obtain

$$m^{\xi}(S) = (-2)^s\sum_{T\supseteq S}\widehat{\xi}(T). \tag{2.76}$$

Remark 2.72

(i) The name "Fourier transform" comes from the initial work of Fourier[9] on the representation of periodic functions by trigonometric series (Fourier series), later generalized to any integrable function (Sect. 1.3.11). The Fourier transform of a function, viewed as a time function or *signal*, gives its frequency representation, and is a fundamental tool in signal processing. Exactly the same results as those given in Theorem 2.71 hold for the original Fourier transform, for instance, a constant signal has a null spectrum for all frequencies, except

[9]Joseph Fourier (Auxerre, 1768 – Paris 1830) is a French mathematician and physicist. He created the Royal University of Grenoble and he is especially known for his work on heat propagation, where his famous trigonometric series were used.

for frequency 0, the well-known Parseval[10] identity says that the energy of a signal in time and frequency domains is the same, and most importantly, convolution of signals in time domain is transformed into their product in frequency domain. It is also a particular case of the bilateral Laplace transform.

Note however that all these properties, except the last one on convolution, are immediate consequences of the orthonormality of the basis, so that the name "Fourier transform" for pseudo-Boolean functions is more an analogy than an exact mathematical correspondence. This is common usage in theoretical computer sciences, however to the opinion of the author, this transform should be rather called the *Walsh transform*, because the definition of the Walsh function is an infinite version of the parity function used here, as a comparison of (2.66) and (2.68) reveals.

(ii) As we have noted in Theorem 2.33(vi), the Möbius transform also satisfies (v). The same property is valid for the interaction transform [Theorem 2.45(iv)] and the Banzhaf interaction transform [Theorem 2.47(iv)].

(iii) The original definition of a convolution of two functions f, g (often viewed as signals; i.e., time functions) is [see (1.27)]

$$(f * g)(x) = \int f(x-y)g(y)dy.$$

The counterpart for pseudo-Boolean functions proposed above is indeed very close, because binary addition $\oplus$ and binary subtraction $\ominus$ coincide: $1 \ominus 1 = 0 = 0 \ominus 0$, and $1 \ominus 0 = 1 = 0 \ominus 1$.

(iv) We have seen that the Walsh functions on $\{-1, +1\}^n$ are identical to the Fourier basis, up to a recoding. Another variant exists, where the normalization of the inner product is $1/2^{n/2}$. Then with this normalization, the Fourier transform (a.k.a. *Hadamard transform*) becomes its own inverse, and the Parseval identity would read $\|f\| = \|\widehat{f}\|$.

(v) The Fourier transform of pseudo-Boolean functions is a central tool in theoretical computer sciences, with many applications (the readers should consult the surveys by de Wolf and O'Donnell [73, 258, 259] and the references therein), for instance, list decoding, learning, random parities, influence of variables, threshold phenomena, and also social choice theory.

◇

[10]Marc-Antoine Parseval des Chênes (1755 – 1836) is a French mathematician born in Rosières-aux-Salines. It seems that in all his life he published only five articles. This shows, in our times where bibliometrics exerts its tyranny, that it is not necessary to publish a lot to remain famous for centuries!

2.16.3 Approximations of a Fixed Degree

When dealing with polynomials, a natural question is the approximation of a given polynomial by a polynomial of a fixed degree. Considering a pseudo-Boolean function f, we would like to find a best approximation by a pseudo-Boolean function f_k^* of degree at most k, where the degree of a pseudo-Boolean function is the highest degree of its monomials, and the degree of a monomial $\prod_{i\in T} x_i$ is simply $|T|$. Let us denote by $\mathcal{PB}^{\leqslant k}(n)$ the vector space of pseudo-Boolean functions of degree at most k. Considering (2.63) and recalling Definition 2.50, we see that the above approximation problem amounts to approximating a set function by an at most k-additive set function.

The best approximation is defined with respect to the (normalized) Euclidean distance, which is the norm associated to the inner product introduced in Sect. 2.16.1; i.e.:

$$d(f,g) = \sqrt{\frac{1}{2^n}\sum_{x\in\{0,1\}^n} (f(x)-g(x))^2} = \|f-g\|$$

with $\|f\| = \sqrt{\langle f,f\rangle}$. Hence the solution of the approximation problem is simply the orthogonal projection of f onto $\mathcal{PB}^{\leqslant k}(n)$, which can be very easily expressed if one uses an orthonormal basis w.r.t the above inner product: it is just the set of coordinates of f of degree at most k in this basis. From Sect. 2.16.1 we know that the basis of Walsh functions is suitable for this. We obtain:

Theorem 2.73 *Let $f \in \mathcal{PB}(n)$. The best approximation of f in $\mathcal{PB}^{\leqslant k}(n)$ is given by*

$$f_k^*(x) = \sum_{\substack{T\subseteq[n]\\ |T|\leqslant k}} a_T^{(k)} \prod_{i\in T} x_i,$$

with

$$a_T^{(k)} = a_T + (-1)^{k-|T|} \sum_{\substack{S\supseteq T\\ |S|>k}} \binom{|S|-|T|-1}{k-|T|} \left(\frac{1}{2}\right)^{|S\setminus T|} a_S \qquad (T\subseteq N, |T|\leqslant k).$$

Proof By (2.65), the best approximation of degree at most k reads:

$$
\begin{aligned}
f_k^*(x) &= \sum_{\substack{T\subseteq[n]\\|T|\leqslant k}} \underbrace{\Big(\sum_{S\supseteq T}\frac{a_S}{2^{|S|}}\Big)}_{\beta_T}\prod_{i\in T}(2x_i-1)\\
&= \sum_{\substack{T\subseteq[n]\\|T|\leqslant k}} \beta_T\Big(\sum_{S\subseteq T}(-1)^{|T\setminus S|}2^{|S|}\prod_{i\in S}x_i\Big)\\
&= \sum_{\substack{S\subseteq N\\|S|\leqslant k}} \underbrace{\Big(\sum_{\substack{T\supseteq S\\|T|\leqslant k}}(-1)^{|T\setminus S|}2^{|S|}\beta_T\Big)}_{\gamma_S}\prod_{i\in S}x_i.
\end{aligned}
$$

We have

$$
\begin{aligned}
\gamma_S &= \sum_{\substack{T\supseteq S\\|T|\leqslant k}}(-1)^{|T\setminus S|}2^{|S|}\Big(\sum_{L\supseteq T}\frac{a_L}{2^{|L|}}\Big)\\
&= \sum_{L\supseteq S}\frac{a_L}{2^{|L\setminus S|}}\Big(\sum_{\substack{T\in[S,L]\\|T|\leqslant k}}(-1)^{|T\setminus S|}\Big).
\end{aligned}
$$

From Lemma 1.1(i) and (iii), we get:

$$
\sum_{\substack{T\in[S,L]\\|T|\leqslant k}}(-1)^{|T\setminus S|} = \begin{cases} 1, & \text{if } S=L\\ 0, & \text{if } S\subset L \text{ and } |L|\leqslant k\\ (-1)^{k-|S|}\binom{|L\setminus S|-1}{k-|S|}, & \text{otherwise}\end{cases}
$$

which yields the desired result. □

Remark 2.74

(i) This result was first established by Hammer and Holzman for $k = 1$ and $k = 2$ [190], then later for any k by Grabisch et al. [178].
(ii) A generalization of this result exists for the weighted distance built from the weighted inner product w.r.t. probabilities $p_1, \dots, p_n$ [see Eq. (2.67)]. The best approximation in this case has the following coefficients:

$$
a_T^{(k)} = a_T + (-1)^{k-|T|}\sum_{\substack{S\supseteq T\\|S|>k}}\binom{|S|-|T|-1}{k-|T|}\Big(\prod_{i\in S\setminus T}p_i\Big)a_S \qquad (T\subseteq N, |T|\leqslant k).
$$

(see Ding et al. [91], and Marichal and Mathonet [232]).

◇

An important fact is that the best approximation f_k^* has the same Banzhaf interaction transform as the original function (up to truncation, of course).

Theorem 2.75 *Let $f \in \mathcal{PB}(n)$. The best approximation of f in $\mathcal{PB}^{\leqslant k}(n)$ satisfies*

$$I_{\mathrm{B}}^{f_k^*}(S) = I_{\mathrm{B}}^{f}(S) \qquad (S \subseteq [n], |S| \leqslant k).$$

This was remarked by Marichal and Mathonet [232], who generalized this result to the weighted distance.

Proof First remark that by (2.65), we have

$$I_{\mathrm{B}}^{f}(S) = 2^s \langle f, w_S \rangle.$$

Then we have for all $S \subseteq [n]$ such that $|S| \leqslant k$, $I_{\mathrm{B}}^{f_k^*}(S) = I_{\mathrm{B}}^{f}(S)$ if and only if $\langle f - f_k^*, w_S \rangle = 0$, but this characterizes the projection of f onto the basis of Walsh functions.[11] □

Another type of approximation is called *faithful approximation* by Hammer and Holzman [190]. It consists in finding a function $g \in \mathcal{PB}^{\leqslant k}(n)$ that minimizes the

[11]Direct proof using (2.47): For every $S \subseteq [n]$, $|S| \leqslant k$, we have, denoting the cardinality of sets $L, T, \ldots$ by corresponding small letters $l, t, \ldots$:

$$\begin{aligned} I_{\mathrm{B}}^{f_k^*}(S) &= \sum_{\substack{L \supseteq S \\ l \leqslant k}} \left(\frac{1}{2}\right)^{l-s} \left(a_L + (-1)^{k-l} \sum_{\substack{T \supseteq L \\ t > k}} \binom{t-l-1}{k-l} \left(\frac{1}{2}\right)^{t-l} a_T \right) \\ &= \sum_{\substack{L \supseteq S \\ l \leqslant k}} \left(\frac{1}{2}\right)^{l-s} a_L + \sum_{\substack{L \supseteq S \\ l \leqslant k}} \left(\frac{1}{2}\right)^{l-s} (-1)^{k-l} \sum_{\substack{T \supseteq L \\ t > k}} \binom{t-l-1}{k-l} \left(\frac{1}{2}\right)^{t-l} a_T \\ &= \sum_{\substack{L \supseteq S \\ l \leqslant k}} \left(\frac{1}{2}\right)^{l-s} a_L + \sum_{\substack{T \supseteq S \\ t > k}} \left(\frac{1}{2}\right)^{t-s} a_T \sum_{\substack{L \in [S,T] \\ l \leqslant k}} (-1)^{k-l} \binom{t-l-1}{k-l}. \end{aligned}$$

It suffices to prove that the summation on $L \in [S, T]$ equals 1. We have

$$\begin{aligned} \sum_{\substack{L \in [S,T] \\ l \leqslant k}} (-1)^{k-l} \binom{t-l-1}{k-l} &= \sum_{l=s}^{k} (-1)^{k-l} \binom{t-s}{l-s} \binom{t-l-1}{k-l} \\ &= \sum_{l=s}^{k} (-1)^{k-l} \binom{t-s}{k-s+1} \binom{k-s}{l-s} \frac{k-s+1}{t-l} \\ &= (-1)^k (k-s+1) \binom{t-s}{k-s+1} \underbrace{\sum_{l=s}^{k} (-1)^l \binom{k-s}{l-s} \frac{1}{t-l}}_{\sigma}. \end{aligned}$$

Euclidean distance to f, under the constraints $f(\mathbf{0}) = g(\mathbf{0})$ and $f(\mathbf{1}) = g(\mathbf{1})$, where $\mathbf{0}, \mathbf{1}$ are the n-dimensional vectors $(0, \dots, 0)$ and $(1, \dots, 1)$. We mention here the result found by Ding et al., without proof [91]. There is a unique such function g whose coordinates $\{\beta_S\}_{S\subseteq[n]}$ in the basis of Walsh functions are

$$\beta_S = \begin{cases} b_S + \dfrac{\sum_{\substack{T\subseteq[n]\\ |T|>k, |T| \text{ even}}} b_T}{D_0(n,k)}, & \text{if } |S| \text{ is even and } |S| \leqslant k \\ b_S + \dfrac{\sum_{\substack{T\subseteq[n]\\ |T|>k, |T| \text{ odd}}} b_T}{D_1(n,k)}, & \text{if } |S| \text{ is odd and } |S| \leqslant k \end{cases} \quad (S \subseteq [n])$$

where $\{b_S\}_{S\subseteq[n]}$, are the coordinates of f in the basis of Walsh functions, and

$$D_0(n,k) = \sum_{\substack{0\leqslant j\leqslant k\\ j \text{ even}}} \binom{n}{j}, \quad D_1(n,k) = \sum_{\substack{1\leqslant j\leqslant k\\ j \text{ odd}}} \binom{n}{j}.$$

Letting $k = 1$, it is possible (after some algebra...) to recover the result from Hammer and Holzman [190]: the best linear faithful approximation of f is the function $g^*(x) = f(\mathbf{0}) + \sum_{i\in[n]} v_i x_i$, with

$$v_i = \sum_{i\in T} \frac{a_T}{2^{|T|-1}} + \frac{1}{n}\Big(f(\mathbf{1}) - f(\mathbf{0}) - \sum_{T\subseteq[n]} \frac{|T|a_T}{2^{|T|-1}}\Big).$$

Approximation of Degree 1: Approximation of a Game by an Additive Game

The simplest possible approximation is to approximate a pseudo-Boolean function (or a set function) on $[n]$ by a polynomial of degree 1. Let us consider a slightly more restricted problem: approximating a game v by an additive game ψ, defined by a n-dimensional vector ψ as $\psi(S) = \sum_{i\in S} \psi_i$ (with some abuse of notation). This is a kind of faithful approximation in the above sense because $v(\varnothing) = \psi(\varnothing) = 0$, however in general nothing is imposed on $\psi(X)$. This problem deserves some interest, because it can be seen as an equivalent to the linear regression problem, or, in cooperative game theory, to find a value (like the Shapley or Banzhaf value)

The summation σ reads, letting $l' = l - s$,

$$\sigma = (-1)^s \sum_{l'=0}^{k-s} (-1)^{l'} \binom{k-s}{l'} \frac{1}{t-l'-s} = (-1)^k \frac{(k-s)!(t-k-1)!}{(t-s)!}$$

by applying Lemma 1.1(v). Therefore, the summation on L equals 1, and the proof is complete.

so that the resulting additive game is as close as possible to the original game. Similarly, in decision making, it amounts to finding a probability measure as close as possible to the original capacity.

We know already from Theorem 2.75 (with $k = 1$) that the best approximation is nothing other than the Banzhaf value of v; i.e., $\phi^{\mathrm{B}}(v) = I_{\mathrm{B}}^{v}(\{i\})$ (originally shown by Hammer and Holzman [190]). Another remarkable result is due to Charnes et al. [45], who give the solution of the faithful approximation when the distance is weighted with positive symmetric coefficients. In particular, this permits to find the Shapley value as the best faithful approximation for a certain set of weights. We give hereafter a more general result, where weights need not be positive nor symmetric (Faigle and Grabisch [129]).

We consider the following optimization problem (expressed under the game form):

$$\min_{\phi\in\mathbb{R}^X} \sum_{S\subseteq X} \alpha_S(v(S) - \phi(S))^2 \text{ subject to } \phi(X) = v(X), \tag{2.77}$$

with the convention $\phi(S) = \sum_{i\in S}\phi_i$, and $\alpha_S \in \mathbb{R}$ for all S. The objective function can be rewritten as

$$\sum_{S\subseteq X}\alpha_S\Big(v^2(S) + 2\sum_{\{i,j\}\subseteq S}\phi_i\phi_j - 2v(S)\sum_{i\in S}\phi_i + \sum_{i\in S}\phi_i^2\Big).$$

Hence (2.77) amounts to minimizing the following quantity subject to $\phi(X) = v(X)$:

$$\sum_{i\in X}\phi_i^2\sum_{S\ni i}\alpha_S + 2\sum_{\{i,j\}\subseteq X}\phi_i\phi_j\sum_{S\supseteq\{i,j\}}\alpha_S - 2\sum_{i\in X}\phi_i\sum_{S\ni i}v(S)\alpha_S.$$

Then the optimization problem can be put into the standard quadratic form

$$\min_{\phi\in\mathbb{R}^X} \phi^\top Q\phi - 2c^\top\phi \text{ subject to } \sum_{i\in X}\phi_i = v(X), \tag{2.78}$$

with Q a symmetric $n\times n$ matrix with component $q_{ij} = \sum_{S\supseteq\{i,j\}}\alpha_S$ for $i\neq j$, $q_{ii} = \sum_{S\ni i}\alpha_S$, and the n-dimensional vector c with component $c_i = \sum_{S\ni i}v(S)\alpha_S$. The constraint being an equality constraint and linear in ϕ, and supposing Q to be positive semidefinite, we are exactly in the case described in Sect. 1.3.8: any optimal solution satisfies the KKT conditions (1.23) in the $n+1$ variables $\phi_1,\ldots,\phi_n,z$:

$$\begin{bmatrix} Q & \mathbf{1}\\ \mathbf{1}^\top & 0\end{bmatrix}\begin{pmatrix}\phi\\ z\end{pmatrix} = \begin{pmatrix}c\\ v(X)\end{pmatrix}, \text{ or } \begin{cases} Q\phi - c = -z\mathbf{1}\\ \mathbf{1}^\top\phi \quad = v(X),\end{cases} \tag{2.79}$$

with $\mathbf{1}^\top = (1, \ldots, 1) \in \mathbb{R}^n$. Clearly, $\mathbf{1}^\top \phi = v(X)$ has a solution, therefore if Q is positive definite, there is a unique optimal solution given by

$$\begin{pmatrix} \phi^* \\ z^* \end{pmatrix} = \begin{bmatrix} Q & \mathbf{1} \\ \mathbf{1}^\top & 0 \end{bmatrix}^{-1} \begin{pmatrix} c \\ v(X) \end{pmatrix}.$$

The explicit expression of the optimal solution can be obtained if the matrix Q is sufficiently simple, for example, if all diagonal elements are identical (let us denote them by q), as well as all off-diagonal elements (let us denote them by p). Let us call *regular* such a matrix. It can be checked that if the coefficients α_S are symmetric, i.e., $\alpha_S = \alpha_T =: \alpha_s$ whenever $|S| = |T| = s$, then Q is regular (this is however not the only case). Indeed, one finds

$$q_{ij} = \begin{cases} \sum_{s=2}^{n} \binom{n-2}{s-2} \alpha_s, & \text{if } i \neq j \\ \sum_{s=1}^{n} \binom{n-1}{s-1} \alpha_s, & \text{otherwise.} \end{cases}$$

Lemma 2.76 *Let Q be regular, with $q = q_{ii}$ and $p = q_{ij}$ for $i \neq j$. Then Q is positive definite if and only if $q > p \geqslant 0$.*

Proof For any $\phi \in \mathbb{R}^X$, we have, after some algebra

$$\phi^\top Q \phi = (q - p) \sum_{i=1}^{n} \phi_i^2 + p\phi^2(X).$$

Then $\phi^\top Q \phi = 0$ only for $\phi = \mathbf{0}$ is equivalent to $q - p > 0$ and $p \geqslant 0$. □

For regular positive definite matrices, it is then easy to get the unique optimal solution.

Theorem 2.77 *If Q is regular with $q = q_{ii}$ and $p = q_{ij}$ for $i \neq j$ and satisfies the condition $q > p \geqslant 0$, then the unique optimal solution ϕ^* of (2.78) is:*

$$\phi_i^* = \frac{1}{q-p}\Big(c_i + \frac{(q-p)v(X) - C}{n}\Big),$$

with $C = \mathbf{1}^\top c$.

Proof Let us solve (2.79) under the conditions on Q. Note that

$$-z = q\phi_i + p\sum_{j \neq i} \phi_j - c_i = (q-p)\phi_i + pv(X) - c_i \qquad (i = 1, \ldots, n).$$

Summing over the i's, we obtain, letting $C = \mathbf{1}^\top c$,

$$-nz = (q-p)v(X) + npv(X) - C$$

and hence

$$-z^* = (q-p)\frac{v(X)}{n} + pv(X) - \frac{C}{n}.$$

Knowing z^*, we immediately obtain the components of ϕ^*:

$$\phi_i^* = \frac{-z^* - pv(X) + c_i}{q-p} = \frac{1}{q-p}\Big(c_i + \frac{(q-p)v(X) - C}{n}\Big).$$

□

Now, supposing that the coefficients α_S are symmetric, we find:

$$\begin{aligned} q-p &= \sum_{s=1}^{n}\binom{n-1}{s-1}\alpha_s - \sum_{s=2}^{n}\binom{n-2}{s-2}\alpha_s \\ &= \alpha_1 + \sum_{s=2}^{n}\alpha_s\left(\binom{n-1}{s-1} - \binom{n-2}{s-2}\right) = \sum_{s=1}^{n-1}\alpha_s\binom{n-2}{s-1}. \end{aligned} \tag{2.80}$$

Substituting into ϕ_i^* above yields exactly the solution given by Charnes et al. (also shown in Ruiz et al. [280]). We note however that the result given in Theorem 2.77 is slightly more general, because the coefficients need not be positive. We find now how to recover the Shapley value.

Theorem 2.78 *(Charnes et al. [45]) The Shapley value is the unique optimal solution of (2.78) when the coefficients α_S are given by*

$$\alpha_S = \frac{1}{\binom{n-2}{|S|-1}} \qquad (\varnothing \neq S \subseteq X).$$

Proof Since the Shapley value and the optimal solution of (2.78) are linear over games, it suffices to match them on any basis, for example, the identity games δ_S, $S \in 2^X \setminus \{\varnothing\}$. Taking any δ_S and supposing the coefficients α_S to be symmetric, we have

$$c_i = \begin{cases} \alpha_s, & \text{if } i \in S \\ 0, & \text{otherwise} \end{cases}$$

so that $C = s\alpha_s$, and we obtain, for $S \neq X$:

$$\phi_i^* = \begin{cases} \frac{\alpha_s(n-s)}{n(q-p)}, & \text{if } i \in S \\ -\frac{s\alpha_s}{n(q-p)}, & \text{otherwise,} \end{cases}$$

and if $S = X$, $\phi_i^* = 1/n$ for all $i \in X$. On the other hand, we find for the Shapley value:

$$\phi_i^{\mathrm{Sh}}(\delta_S) = \begin{cases} \frac{1}{n\binom{n-1}{s-1}}, & \text{if } i \in S \\ -\frac{1}{n\binom{n-1}{s}}, & \text{otherwise.} \end{cases}$$

Equating ϕ_i^* and $\phi_i^{\mathrm{Sh}}(\delta_S)$ for every $i \in X$ yields

$$\frac{\alpha_s}{q-p} = \frac{1}{s\binom{n-1}{s}} \qquad (1 \leqslant s \leqslant n-1). \tag{2.81}$$

Letting $\alpha_s = 1/\binom{n-2}{s-1}$ immediately yields $q-p = n-1$ by (2.80). Injecting $q-p = n-1$ in (2.81) shows that the proposed solution works. □

As a second application of Theorem 2.77, let us find the best faithful approximation with $\alpha_S = 1$ for every $S \subseteq X$. Using (2.80) we find that $q - p = 2^{n-2}$, hence Theorem 2.77 can be applied. We have $c_i = \sum_{S \ni i} v(S)$ and therefore $C = \sum_{S \subseteq X} |S| v(S)$. This yields

$$\begin{aligned} \phi_i^* &= \frac{1}{n2^{n-2}}\Big(n \sum_{S \ni i} v(S) + 2^{n-2} v(X) - \sum_{S \subseteq X} |S| v(S)\Big) \\ &= \frac{v(X)}{n} + \frac{1}{n2^{n-2}}\Big(\sum_{S \ni i} (n - |S|) v(S) - \sum_{S \subseteq X \setminus i} |S| v(S)\Big). \end{aligned}$$

Letting $B_i(v) = \frac{1}{2^{n-2}} \sum_{S \subseteq X \setminus i} v(S \cup i)$, an equivalent form is:

$$\phi_i^* = \frac{v(X)}{n} + B_i(v) - \frac{1}{n} \sum_{j \in X} B_j(v). \tag{2.82}$$

Remark 2.79 Recall that the best (unfaithful) approximation with $\alpha_S = 1\ \forall S$ leads to the Banzhaf value. The best faithful approximation was proposed by Ruiz et al. [279] under the name of *least square prenucleolus*, in the form (2.82). In [280], Ruiz et al. consider the family of values obtained by a faithful least square approximation and axiomatize it. ◇

2.16.4 Extensions of Pseudo-Boolean Functions

Considering any pseudo-Boolean function $f(x)$, $x \in \{0, 1\}^n$, one may let x vary over $[0, 1]^n$ or $\mathbb{R}^n$, in any of the expressions (2.62), (2.63) or (2.65). These polynomials are *extensions* of the original pseudo-Boolean function f, and more generally, any polynomial $\overline{f}$ on $\mathbb{R}^n$ coinciding with f on $\{0, 1\}^n$ is an extension of f.

The Owen Extension

Consider the standard representation (2.62) of a pseudo-Boolean function f. Letting x vary over $\mathbb{R}^n$, we obtain the *Owen extension* of f, which we denote by f^{Ow}:

$$f^{\mathrm{Ow}}(x) = \sum_{A\subseteq[n]} f(1_A)\prod_{i\in A} x_i \prod_{i\in A^c}(1-x_i) \qquad (x\in\mathbb{R}^n). \tag{2.83}$$

Rearranging terms, it can be equivalently written into the Möbius representation:

$$f^{\mathrm{Ow}}(x) = \sum_{T\subseteq[n]} a_T \prod_{i\in T} x_i \qquad (x\in\mathbb{R}^n).$$

We denote the set of all such polynomials by $\mathcal{P}(n)$. It is a vector space.

Remark 2.80

(i) Since pseudo-Boolean functions bijectively correspond to set functions, Eq. (2.83) gives the Owen extension of a set function. If the set function is a game, note that its Owen extension is a multilinear polynomial, that is, linear in each variable. Precisely, the multilinear extension of games was introduced by Owen [261] as a way to extend a game on the hypercube $[0, 1]^n$.

(ii) Usually, extensions are defined on $[0, 1]^n$ instead of $\mathbb{R}^n$. However, as remarked by Marinacci and Montrucchio [235], there is no need to restrict to $[0, 1]^n$, and moreover some interesting properties can be found when working over $\mathbb{R}^n$. They also show that f^{Ow} is the least-degree Bernstein polynomial associated to f.

◇

Theorem 2.81 *$\mathcal{P}(n)$ is a 2^n-dimensional vector space. The set of monomials $\prod_{i\in T} x_i$, $T \subseteq [n]$, is a basis of $\mathcal{P}(n)$, as well as the family of polynomials $\prod_{i\in T} x_i \prod_{i\in T^c}(1-x_i)$, and the family of Walsh functions $\prod_{i\in T}(2x_i-1)$, the latter basis being orthonormal.*

Proof Clearly, each $P \in \mathcal{P}(n)$ can be written in the form $\sum_{T\subseteq[n]} a_T \prod_{i\in T} x_i$, therefore it remains to prove uniqueness. We proceed by induction on the size of the subsets T. For $T = \varnothing$, $P(\mathbf{0}) = a_\varnothing$, therefore $a_\varnothing$ is uniquely determined. Assume next that all a_T are determined, for all T such that $|T| \leqslant k$, and consider T with $|T| = k + 1$. Since $P(1_T) = \sum_{S\subseteq T} a_S$, we have

$$a_T = P(1_T) - \sum_{S\subset T} a_S.$$

Then by induction hypothesis, a_T is uniquely determined. Because there are 2^n monomials $\prod_{i\in T} x_i$, the dimension of the basis is 2^n.

To prove that the polynomials $\prod_{i\in T} x_i \prod_{i\in T^c}(1 - x_i)$ form a basis, it suffices to show that they are linearly independent. Suppose

$$P(x) = \sum_{T\subseteq[n]} \alpha_T \prod_{i\in T} x_i \prod_{i\in T^c}(1 - x_i) = 0 \qquad (x \in \mathbb{R}^n).$$

Since $P(1_T) = \alpha_T$, we get $\alpha_T = 0$ for each T, so that the polynomials are linearly independent.

For the Walsh functions, the statement is proved exactly like in Theorem 2.69.

□

An important consequence of the theorem is that, as remarked by Owen [261], f^{Ow} is the unique multilinear extension of f when f is grounded.

The conversion formulas between the bases are the same as for $\mathcal{PB}(n)$. Denoting by a_T, α_T, β_T the coefficients in the basis of monomials, of the polynomials $\prod_{i\in T} x_i \prod_{i\in T^c}(1-x_i)$, and of the Walsh functions respectively, we have that $\{a_T\}_{T\subseteq[n]}$ is the Möbius transform of $\{\alpha_T\}_{T\subseteq[n]}$, and the coefficients β_T are given by (2.65).

Example 2.82 Define nonnegative weights $p_1, \ldots, p_n$, not all null, and consider the monotone game v defined by $v(A) = \left(\sum_{i\in A} p_i\right)^2$, $A \subseteq X$. Writing $\left(\sum_{i\in A} p_i\right)^2 = \left(\sum_{i\in[n]} p_i x_i\right)^2$ with $x = 1_A$, and remarking that $x_i^2 = x_i$ if $x_i = 0$ or 1, we directly obtain the expression in terms of monomials

$$f_v^{\mathrm{Ow}}(x) = \sum_{i\in[n]} p_i^2 x_i + 2 \sum_{\{i,j\}\subseteq[n]} p_i p_j x_i x_j.$$

This shows that this is a 2-additive game, whose Möbius transform is equal to p_i^2 on the singleton $\{i\}$, and $2p_i p_j$ on the pair $\{i,j\}$. Therefore, it is a totally monotone game. Its expression in the basis of Walsh functions is

$$f_v^{\mathrm{Ow}}(x) = \sum_{i\in[n]} \Big(\frac{p_i^2}{2} + \sum_{j\neq i} \frac{p_i p_j}{2}\Big) w_{\{i\}}(x) + \sum_{\{i,j\}\subseteq[n]} \frac{p_i p_j}{2} w_{\{i,j\}}(x).$$

Note that if $p_1, \ldots, p_n$ defines a probability distribution, then v is the square of a probability measure, and by the above result it is a belief measure. Now, if all weights are equal to 1, we find $v(A) = |A|^2$. ◇

The next lemma gives the partial derivatives of the Owen extension, and is useful for showing subsequent theorems. Following our convention 1.1 in Sect. 1.1, we use throughout this section the shorthand:

$$\frac{\partial^s f}{\partial x_{|S}}(x) = \frac{\partial^s f}{\partial x_{i_1} \cdots \partial x_{i_s}}(x)$$

for any function f of the variable $x = (x_1, \ldots, x_n)$ and any $S \subseteq [n]$, with $S = \{i_1, \ldots, i_s\}$.

Lemma 2.83 *Let $K = \{i_1, \ldots, i_k\} \subseteq [n]$. The partial derivative of the Owen extension of a set function ξ w.r.t. $x_{i_1}, \ldots, x_{i_k}$ is given by*

$$\frac{\partial^k f_\xi^{\mathrm{Ow}}}{\partial x_{|K}}(x) = \sum_{A \subseteq [n] \setminus K} \Delta_K \xi(A) \prod_{i \in A} x_i \prod_{i \in [n] \setminus (A \cup K)} (1 - x_i)$$

$$= \sum_{A \supseteq K} m^\xi(A) \prod_{i \in A \setminus K} x_i \qquad (x \in \mathbb{R}^n), \tag{2.84}$$

with m^ξ the Möbius transform of ξ. In particular, for any $A \subseteq [n]$,

$$\frac{\partial^k f_\xi^{\mathrm{Ow}}}{\partial x_{|K}}(1_A) = \Delta_K f_\xi(1_A) = \Delta_K \xi(A). \tag{2.85}$$

Proof By definitions of the partial derivative and of $\Delta_K f$, $\Delta_K \xi$, it suffices to show the result for $K = \{k\}$. We have:

$$\frac{\partial f_\xi^{\mathrm{Ow}}}{\partial x_k}(x) = \sum_{A \subseteq [n] \setminus k} \xi(A \cup k) \prod_{i \in A} x_i \prod_{i \notin A, i \neq k} (1 - x_i)$$

$$- \sum_{A \subseteq [n] \setminus k} \xi(A) \prod_{i \in A} x_i \prod_{i \notin A, i \neq k} (1 - x_i)$$

$$= \sum_{A \subseteq [n] \setminus k} (\xi(A \cup k) - \xi(A)) \prod_{i \in A} x_i \prod_{i \notin A, i \neq k} (1 - x_i),$$

the desired result. The second formula can be proved similarly. Now, for any $A \subseteq [n]$, $A \not\ni k$, we have clearly $\frac{\partial f_\xi^{\mathrm{Ow}}}{\partial x_k}(1_A) = \xi(A \cup k) - \xi(A) = \Delta_k \xi(A)$. If $A \ni k$, we have $\frac{\partial f_\xi^{\mathrm{Ow}}}{\partial x_k}(1_A) = \frac{\partial f_\xi^{\mathrm{Ow}}}{\partial x_k}(1_{A \setminus k}) = \xi(A) - \xi(A \setminus k)$. In any case, the partial derivative of f_ξ^{Ow} is equal to $\Delta_k \xi(A)$. □

Theorem 2.84 *Let v be a game on X, and consider its Owen extension. The following statements hold.*

(i) $v \geqslant 0$ if and only if $f_v^{\mathrm{Ow}}(x) \geqslant 0$ for all $x \in [0, 1]^n$;
(ii) v is monotone if and only if for all $k \in [n]$,

$$\frac{\partial f_v^{\mathrm{Ow}}}{\partial x_k}(x) \geqslant 0 \qquad (x \in [0, 1]^n);$$

(iii) *v is monotone and k-monotone for some $2 \leqslant k \leqslant n$ if and only if for each set $L \subseteq [n]$ such that $1 \leqslant |L| \leqslant k$,*

$$\frac{\partial^{|L|} f_v^{\text{Ow}}}{\partial x_{|L}}(x) \geqslant 0 \qquad (x \in [0,1]^n);$$

(iv) *v is nonnegative and totally monotone if and only if $f_v^{\text{Ow}}(x) \geqslant 0$ for all $x \in \mathbb{R}^n_+$.*

Proof

(i) If $v \geqslant 0$, its Owen extension (2.83) has nonnegative coefficients. As $\prod_{i\in A} x_i \prod_{i\in A^c}(1-x_i) \geqslant 0$ on $[0,1]^n$, we deduce that $f_v^{\text{Ow}}(x) \geqslant 0$ on $[0,1]^n$. The converse is obvious.

(ii) v is monotone if and only if $v(A \cup k) - v(A) \geqslant 0$ for every $A \subseteq [n]$. Due to Lemma 2.83, this implies that $\frac{\partial f_v^{\text{Ow}}}{\partial x_k}(x) \geqslant 0$ for all $x \in [0,1]^n$. Conversely, monotonicity of v is proved by applying nonnegativity of the partial derivative at $x = 1_A$, $A \subseteq [n] \setminus \{k\}$.

(iii) By Theorem 2.21(i), k-monotonicity is equivalent to $\Delta_L v(A) \geqslant 0$ for all disjoint A, L and $2 \leqslant |L| \leqslant k$. From Lemma 2.83 and (ii), this implies that $\frac{\partial^{|L|} f_v^{\text{Ow}}}{\partial x_{|L}}(x) \geqslant 0$ for all $x \in [0,1]^n$. Conversely, the nonnegativity of these partial derivatives on the vertices of $[0,1]^n$ implies k-monotonicity and monotonicity by Theorem 2.21(i) and (ii).

(iv) (Marinacci and Montrucchio [235]) Writing $f_v^{\text{Ow}}(x) = \sum_{T\subseteq[n]} a_T \prod_{i\in T} x_i$ and invoking Theorem 2.33(v) shows that $f_v^{\text{Ow}}(x) \geqslant 0$ on $\mathbb{R}^n_+$. Conversely, suppose $f_v^{\text{Ow}} \geqslant 0$ and that there exists $T \subseteq [n]$ such that $a_T < 0$. Consider the vector $\alpha 1_T$ with $\alpha > 0$. Then

$$f_v^{\text{Ow}}(\alpha 1_T) = a_T \alpha^{|T|} + \text{ terms of lower degree.}$$

Therefore, for α large enough, we get $f_v^{\text{Ow}}(\alpha 1_T) < 0$, a contradiction.

□

Remark 2.85 The above results (ii) and (iii) can be stated in a more precise way. Monotonicity of v implies the nonnegativity of the partial derivatives on $[0,1]^n$. But to infer monotonicity, we need only nonnegativity on the vertices of $[0,1]^n$. Another way to infer monotonicity is to impose nonnegativity of the partial derivatives on $]0,1[^n$. Then by continuity of the Owen extension, nonnegativity holds also on the vertices. The same remarks apply to k-monotonicity as well. ◇

The next result shows the connection with various transforms.

Theorem 2.86 *For any set function ξ and any $S \subseteq [n]$, we have*

$$m^{\xi}(S) = \frac{\partial^s f_{\xi}^{\mathrm{Ow}}}{\partial x_{|S}}(0, 0, \dots, 0) \tag{2.86}$$

$$\check{m}^{\xi}(S) = \frac{\partial^s f_{\xi}^{\mathrm{Ow}}}{\partial x_{|S}}(1, 1, \dots, 1) \tag{2.87}$$

$$I_{\mathrm{B}}^{\xi}(S) = \frac{\partial^s f_{\xi}^{\mathrm{Ow}}}{\partial x_{|S}}\Big(\frac{1}{2}, \frac{1}{2}, \dots, \frac{1}{2}\Big) \tag{2.88}$$

$$I_{\mathrm{B}}^{\xi}(S) = \int_{[0,1]^n} \frac{\partial^s f_{\xi}^{\mathrm{Ow}}}{\partial x_{|S}}(x)\, \mathrm{d}x \tag{2.89}$$

$$I^{\xi}(S) = \int_0^1 \frac{\partial^s f_{\xi}^{\mathrm{Ow}}}{\partial x_{|S}}(x, x, \dots, x)\, \mathrm{d}x. \tag{2.90}$$

Proof Since by (2.85) $\frac{\partial^s f_{\xi}^{\mathrm{Ow}}}{\partial x_{|S}}(0, 0, \dots, 0) = \Delta_S \xi(\varnothing)$ and $\frac{\partial^s f_{\xi}^{\mathrm{Ow}}}{\partial x_{|S}}(1, 1, \dots, 1) = \Delta_S \xi(X)$, (2.86) and (2.87) are nothing but (2.15) and (2.40).

Equation (2.88) comes immediately from (2.47) and (2.84).

For the two last formulas, using Lemma 2.83 and (2.41), we find:

$$\int_0^1 \frac{\partial^s f_{\xi}^{\mathrm{Ow}}}{\partial x_{|S}}(x, x, \dots, x)\, \mathrm{d}x = \sum_{T \supseteq S} m^{\xi}(T) \int_0^1 x^{t-s} = \sum_{T \supseteq S} m^{\xi}(T) \frac{1}{t-s+1} = I^{\xi}(S),$$

with $s = |S|, t = |T|$. Using (2.47), we find now

$$\int_{[0,1]^n} \frac{\partial^s f_{\xi}^{\mathrm{Ow}}}{\partial x_{|S}}(x)\, \mathrm{d}x = \sum_{T \supseteq S} m^{\xi}(T) \int_{[0,1]^n} \prod_{i \in T \setminus S} x_i \, \mathrm{d}x$$
$$= \sum_{T \supseteq S} m^{\xi}(T) \Big(\frac{1}{2}\Big)^{t-s} = I_{\mathrm{B}}^{\xi}(S).$$

□

Remark 2.87 These results show that the interaction transform is the integral of the partial derivative on the diagonal of the hypercube, while the Banzhaf interaction is the integral over the whole hypercube. They were proved by Grabisch et al. [178]. Owen in his seminal paper [261] proved (2.90) for $|S| = 1$; i.e., for the Shapley value. ◇.

Based on the foregoing theorem, the Taylor expansion of the Owen extension is useful to derive conversion formulas between some transforms [178]. Recall that the Taylor expansion of a real-valued function f of several variables $x_1, \ldots, x_d$ at point $a = (a_1, \ldots, a_d)$ is given by:

$$T(x) = f(a) + \sum_{j=1}^{d}(x_j - a_j)\frac{\partial f}{\partial x_j}(a) + \frac{1}{2!}\sum_{j=1}^{d}\sum_{k=1}^{d}(x_j - a_j)(x_k - a_k)\frac{\partial^2 f}{\partial x_j \partial x_k}(a)$$
$$+ \frac{1}{3!}\sum_{j=1}^{d}\sum_{k=1}^{d}\sum_{\ell=1}^{d}(x_j - a_j)(x_k - a_k)(x_\ell - a_\ell)\frac{\partial^3 f}{\partial x_j \partial x_k \partial x_\ell}(a) + \cdots$$

Because the Owen extension is a multilinear function of n variables, its Taylor expansion at any point $y \in [0,1]^n$ is exact and reduces to:

$$f^{\mathrm{Ow}}(x) = \sum_{T \subseteq [n]} \prod_{i \in T}(x_i - y_i)\frac{\partial^t f^{\mathrm{Ow}}}{\partial x_{|T}}(y) \qquad (x \in [0,1]^n). \tag{2.91}$$

Taking $x = 1_S$ and $y = (\alpha, \ldots, \alpha)$, we obtain, for f_ξ^{Ow}:

$$\xi(S) = \sum_{T \subseteq [n]} \prod_{i \in T}(1_S(i) - \alpha)\frac{\partial^t f_\xi^{\mathrm{Ow}}}{\partial x_{|T}}(\alpha, \ldots, \alpha) \qquad (\alpha \in [0,1], S \subseteq [n]). \tag{2.92}$$

By Theorem 2.86, letting $\alpha = 0, 1$ and $\frac{1}{2}$ in (2.92) gives the conversion formulas respectively from $m^\xi, \check{m}^\xi, I_{\mathrm{B}}^\xi$ to ξ, which we know already from Sect. 2.12. Now, by successive derivations of (2.91), we obtain

$$\frac{\partial^s f^{\mathrm{Ow}}}{\partial x_{|S}}(x) = \sum_{T \supseteq S} \prod_{i \in T \setminus S}(x_i - y_i)\frac{\partial^t f^{\mathrm{Ow}}}{\partial x_{|T}}(y) \qquad (x, y \in [0,1]^n, S \subseteq [n]).$$

In particular, if x, y are constant vectors we obtain

$$\frac{\partial^s f_\xi^{\mathrm{Ow}}}{\partial x_{|S}}(\alpha, \ldots, \alpha) = \sum_{T \supseteq S}(\alpha - \beta)^{t-s}\frac{\partial^t f_\xi^{\mathrm{Ow}}}{\partial x_{|T}}(\beta, \ldots, \beta) \qquad (\alpha, \beta \in [0,1], S \subseteq [n]). \tag{2.93}$$

Again by Theorem 2.86, letting α and β be 0, 1 or $\frac{1}{2}$ in (2.93) gives the conversion formulas between m, $\check{m}$ and I_{B}, in particular:

$$\check{m}^{\xi}(S) = \sum_{T\supseteq S}\left(\frac{1}{2}\right)^{t-s} I_{\mathrm{B}}^{\xi}(T)$$

$$I_{\mathrm{B}}^{\xi}(S) = \sum_{T\supseteq S}\left(-\frac{1}{2}\right)^{t-s} \check{m}^{\xi}(T).$$

Now, combining (2.90) with (2.93), we obtain

$$I^{\xi}(S) = \sum_{T\supseteq S}\left(\int_0^1 (\alpha-\beta)^{t-s}d\alpha\right)\frac{\partial^t f_{\xi}^{\mathrm{Ow}}}{\partial x_{|T}}(\beta,\dots,\beta) \tag{2.94}$$

$$= \sum_{T\supseteq S}\frac{(1-\beta)^{t-s+1}-(-\beta)^{t-s+1}}{t-s+1}\frac{\partial^t f_{\xi}^{\mathrm{Ow}}}{\partial x_{|T}}(\beta,\dots,\beta) \tag{2.95}$$

for any $\beta \in [0,1]$ and $S \subseteq [n]$. Replacing β by 0, 1, and $\frac{1}{2}$, we obtain the conversion formulas from $m, \check{m}, I_{\mathrm{B}}$ to I, in particular:

$$I^{\xi}(S) = \sum_{T\supseteq S}\frac{(1/2)^{t-s+1}-(-1/2)^{t-s+1}}{t-s+1}I_{\mathrm{B}}^{\xi}(T) = \sum_{T\supseteq S}\frac{1+(-1)^{t-s}}{2^{t-s+1}(t-s+1)}I_{\mathrm{B}}^{\xi}(T).$$

The following result is helpful to find the converse formula.

Lemma 2.88 *(Grabisch et al. [178]) For any set function ξ,*

$$\frac{\partial^s f_{\xi}^{\mathrm{Ow}}}{\partial x_{|S}}(\alpha,\dots,\alpha) = \sum_{T\supseteq S} B_{t-s}(\alpha)I^{\xi}(T) \qquad (\alpha\in[0,1], S\subseteq[n]).$$

Proof We have

$$\begin{aligned}
\sum_{T\supseteq S} B_{t-s}(\alpha)I^{\xi}(T) &= \sum_{T\supseteq S} B_{t-s}(\alpha)\sum_{K\supseteq T}\frac{1}{k-t+1}m^{\xi}(K) \qquad \text{(by (2.41))}\\
&= \sum_{K\supseteq S} m^{\xi}(K)\sum_{T\in[S,K]} B_{t-s}(\alpha)\frac{1}{k-t+1}\\
&= \sum_{K\supseteq S} m^{\xi}(K)\alpha^{k-s} \qquad \text{(by Lemma 1.3)}\\
&= \frac{\partial^s f_{\xi}^{\mathrm{Ow}}}{\partial x_{|S}}(\alpha,\dots,\alpha) \qquad \text{(by Lemma 2.83).}
\end{aligned}$$

□

Letting $\alpha = \frac{1}{2}$ in the foregoing lemma yields, using (1.7):

$$I_{\mathrm{B}}^{\xi}(S) = \sum_{T\supseteq S} \Big(\frac{1}{2^{t-s-1}} - 1\Big) B_{t-s} I^{\xi}(T).$$

The Lovász Extension

Consider now the Möbius representation of a pseudo-Boolean function f:

$$f(x) = \sum_{T\subseteq[n]} a_T \prod_{i\in T} x_i \qquad (x \in \{0,1\}^n).$$

Letting x vary over $\mathbb{R}^n$ gives the Owen extension, as we have seen. However, nothing prevents us to replace the product operator by any other operator coinciding with the product on $\{0,1\}^n$. A simple example is the minimum operator. Doing so, we obtain the *Lovász extension* f^{Lo} of f:

$$f^{\mathrm{Lo}}(x) = \sum_{T\subseteq[n]} a_T \bigwedge_{i\in T} x_i \qquad (x \in \mathbb{R}^n_+). \tag{2.96}$$

Remark 2.89 In his 1983's paper [226], Lovász considered the problem of maximizing a linear function $c \cdot x$ over the core (see Chap. 3) of a supermodular set function f.[12] Lovász remarked that the optimal solution $\hat{f}(c)$ was an extension of f on $\mathbb{R}^n_+$. The formula he gaves was not the above one, but an equivalent form. Later, Singer [307] proved that the Lovász extension is the unique affine function that interpolates f at the $n+1$ vertices of each simplex $[0,1]^n_\sigma = \{x \in [0,1]^n : x_{\sigma(1)} \leqslant x_{\sigma(2)} \leqslant \cdots \leqslant x_{\sigma(n)}\}$, for all permutations σ on $[n]$. These simplices are called the *canonical simplices of the unit hypercube*. Later, Marichal [227, 230] remarked that the Lovász extension was the Choquet integral, which is introduced in Chap. 4. ◇

As for the Owen extension, we write f_v^{Lo} to mean that it is the Lovász extension of the game v.

Since the Lovász extension is a sum of minima, it is not differentiable, and unlike the Owen extension, one cannot compute its partial derivatives. Instead, we extend the discrete derivative $\Delta_S f_v^{\mathrm{Lo}}$ to $\mathbb{R}^n_+$ in the following way:

$$\Delta_S f_v^{\mathrm{Lo}}(x) = \sum_{T\supseteq S} a_T \bigwedge_{i\in T\setminus S} x_i \qquad (x \in \mathbb{R}^n_+) \tag{2.97}$$

[12] Originally in the paper, the dual version of the core of a submodular set function was considered, which amounts to the same.

(see Lemma 2.83, and replace $\prod_{i\in T\setminus S}$ by $\bigwedge_{i\in T\setminus S}$ as they coincide on $\{0,1\}^n$). Thanks to this, we obtain the following result, showing that the interaction transform is the integral of the derivative of the Lovász extension.

Theorem 2.90 *(Grabisch et al. [178]) For any game v and any $S \subseteq [n]$, we have*

$$I^v(S) = \int_{[0,1]^n} \Delta_S f_v^{\mathrm{Lo}}(x)\,\mathrm{d}x.$$

Proof Using Lemma 1.2, (2.41) and (2.97), we find

$$\int_{[0,1]^n} \Delta_S f_v^{\mathrm{Lo}}(x)\,\mathrm{d}x = \sum_{T\supseteq S} a_T \int_{[0,1]^n} \bigwedge_{i\in T\setminus S} x_i\,\mathrm{d}x = \sum_{T\supseteq S} m^v(T)\frac{1}{t-s+1} = I^v(S).$$

□

2.17 Transforms, Bases and the Inverse Problem

2.17.1 Transforms and Bases

We consider the vector space $\mathbb{R}^{2^X}$ of set functions (of course, similar considerations can be done for the set of games: see Remark 2.92). So far, we have introduced several bases (unanimity games, Walsh functions, etc.), and studied in depth invertible linear transforms (see Sect. 2.12), separately. However, it is an obvious fact from linear algebra that these two topics are the two faces of the same medal: any basis induces a linear invertible transform, and vice versa. As a curious matter of fact, it seems than no one has taken advantage of this duality. The next result gives the explicit correspondence (we recall that δ_S denotes the Dirac game or identity game).

Lemma 2.91 (Duality between bases and transforms) *(Faigle and Grabisch [131]) For every basis $\{b_S\}_{S\in 2^X}$ of $\mathbb{R}^{2^X}$, there exists a unique linear invertible transform Ψ such that for any $\xi \in \mathbb{R}^{2^X}$,*

$$\xi = \sum_{S\in 2^X} \Psi^\xi(S) b_S, \tag{2.98}$$

whose inverse Ψ^{-1} is given by $\xi \mapsto (\Psi^{-1})^\xi = \sum_{T\in 2^X} \xi(T) b_T$.

Conversely, to any transform Ψ corresponds a unique basis $\{b_S\}_{S\in 2^X}$ such that (2.98) holds, given by $b_S = (\Psi^{-1})^{\delta_S}$.

Proof The above formulas simply express the change of basis. Considering the above functions and transforms as (column) vectors and matrices, the representation of ξ in the basis $\{b_S\}_{S\in 2^X}$, that is, $\xi = \sum_S w_S b_S$, can be rewritten as $\xi = Bw$, or $w = B^{-1}\xi$, and the mapping $\xi \mapsto w$ can be seen as a linear invertible transform, with matrix representation B^{-1}. □

Let us apply this result to all bases and transforms introduced so far, together with a new one.

(i) *The Möbius transform:* the associated basis is given by $b_S = (m^{-1})^{\delta_S}$, for any $S \in 2^X$. From (2.16) we get

$$(m^{-1})^{\xi}(T) = \sum_{B\subseteq T} \xi(B),$$

yielding $b_S(T) = \sum_{B\subseteq T} \delta_S(B)$, which is equal to 1 if and only if $T \supseteq S$, and 0 otherwise. We recover the fact that the associated basis is the basis of unanimity games, with the additional set function $u_\varnothing(S) = 1$ for all $S \in 2^X$ (Sect. 2.15.1).

(ii) *The co-Möbius transform:* proceeding similarly, we have from (2.28)

$$(\check{m}^{-1})^{\xi}(S) = \sum_{T\subseteq X\setminus S} (-1)^{|T|}\xi(T),$$

yielding the associated basis $\{\check{u}_T\}_{T\in 2^X}$:

$$\check{u}_T(S) = \sum_{B\subseteq X\setminus S} (-1)^{|B|}\delta_T(B) = \begin{cases} (-1)^{|T|} & \text{if } S\cap T = \varnothing \\ 0 & \text{otherwise.}\end{cases} \tag{2.99}$$

(iii) *The basis of conjugate unanimity games:* we recall that the conjugate unanimity games are defined by

$$\overline{u_T}(S) = 1 - u_T(S^c) = \begin{cases} 1, & \text{if } S\cap T \neq \varnothing \\ 0, & \text{otherwise.}\end{cases} \qquad (S\subseteq X)$$

(Sect. 2.15.1). We have already established in Sect. 2.15.1 the coordinates of a game in this basis [see (2.52)]. The induced transform $\overline{U}$ is then

$$\overline{U}^{\xi}(S) = m^{\overline{\xi}}(S) = (-1)^{|S|+1}\sum_{T\supseteq S} m^{\xi}(T) = (-1)^{|S|+1}\check{m}^{\xi}(S),$$

where we have used (2.39) for the last equality. Finally, we get from (2.27)

$$\overline{U}^{\xi}(S) = (-1)^{|S|+1}\sum_{T\supseteq X\setminus S} (-1)^{n-|T|}\xi(T).$$

The inverse transform is directly obtained from Lemma 2.91:

$$(\overline{U}^{-1})^{\xi}(S) = \sum_{T \cap S \neq \varnothing} \xi(T).$$

(iv) *The (Shapley) interaction transform:* From (2.43) the inverse transform is given by

$$(I^{-1})^{\xi}(S) = \sum_{K \subseteq X} \beta^{|K|}_{|S \cap K|} \xi(K),$$

with coefficients β^l_k given in (2.44), yielding the corresponding basis

$$b^I_T(S) = \beta^{|T|}_{|T \cap S|} \qquad (S, T \in 2^X). \tag{2.100}$$

(v) *The Banzhaf interaction transform:* From (2.31), we find the associated basis:

$$b^{I_B}_T(S) = \sum_{K \subseteq X} \left(\frac{1}{2}\right)^k (-1)^{|K \setminus S|} \delta_T(K) = \left(\frac{1}{2}\right)^{|T|} (-1)^{|T \setminus S|} \qquad (S, T \in 2^X). \tag{2.101}$$

(vi) *The Fourier transform:* The transform $\xi \mapsto F^{\xi}$ (denoted by $\widehat{\xi}$ in Sect. 2.16.2) is defined by

$$F^{\xi}(S) = \frac{1}{2^n} \sum_{K \subseteq X} (-1)^{|S \cap K|} \xi(K) \tag{2.102}$$

with inverse given by (2.70)

$$(F^{-1})^{\xi}(S) = \sum_{K \subseteq X} (-1)^{|S \cap K|} \xi(K).$$

The corresponding basis is therefore

$$b^F_T(S) = \sum_{K \subseteq X} (-1)^{|S \cap K|} \delta_T(K) = (-1)^{|S \cap T|} \qquad (S, T \in 2^X). \tag{2.103}$$

We recover as expected the Fourier basis, denoted by χ_T in Sect. 2.16.2.

(vii) *The Walsh basis:* We recall that this basis is defined by $w_T(S) = (-1)^{|T \setminus S|}$ [Eq. (2.64)]. Let us find the corresponding transform $\xi \mapsto W^{\xi}$. By

Lemma 2.91, the inverse transform is immediate:

$$(W^{-1})^{\xi}(S) = \sum_{T\subseteq X} \xi(T)(-1)^{|T\setminus S|}.$$

The direct transform can be discerned by solving the linear system

$$\xi(S) = \sum_{T\subseteq X} W^{\xi}(T)(-1)^{|T\setminus S|} \quad (S \in 2^X),$$

or by simply noticing that $w_T(S) = 2^{|T|}b_T^{I_B}(S)$, which from

$$\xi(S) = \sum_{T\subseteq X} I_B^{\xi}(T)b_T^{I_B}(S) = \sum_{T\subseteq X} W^{\xi}(T)w_T(S)$$

yields the components of W^{ξ} as

$$W^{\xi}(T) = \left(\frac{1}{2}\right)^{|T|} I_B^{\xi}(T) \qquad (T \in 2^X).$$

We recover Formula (2.65). Note that the Fourier and Walsh bases are related as follows:

$$b_T^F(S) = b_S^F(T) = (-1)^{|S\cap T|} = (-1)^{|S\setminus(X\setminus T)|} = w_S(X \setminus T).$$

Also, from (2.75), we find

$$F^{\xi}(S) = (-1)^s W^{\xi}(S) \qquad (S \in 2^X). \tag{2.104}$$

(viii) *The Yokote basis*[13] *[354, 355]:* it is a basis of the set of games, which is defined by

$$\upsilon_T(S) = \begin{cases} 1, & \text{if } |S \cap T| = 1 \\ 0, & \text{otherwise} \end{cases} \qquad (S \in 2^X \setminus \varnothing). \tag{2.105}$$

Any game v reads in this basis

$$v = \sum_{T\in 2^X\setminus\varnothing} Y^v(T)\upsilon_T \tag{2.106}$$

where the coordinates $Y^v(S)$ define the *Yokote transform* Y. We give now Y^v in terms of m^v and v, as well as the inverse relations. We start with m^v.

---

[13]Reference [354] contains a more general set of bases including this one.

From (2.106) we have

$$v(S) = \sum_{T:|T\cap S|=1} Y^v(T),$$

hence by (2.15),

$$\begin{aligned}
m^v(S) &= \sum_{T\subseteq S}(-1)^{|S\setminus T|}\sum_{K:|K\cap T|=1} Y^v(K)\\
&= \sum_{K\cap S\neq\varnothing} Y^v(K)\sum_{\substack{T\subseteq S\\ |T\cap K|=1}}(-1)^{|S\setminus T|}\\
&= \sum_{K\cap S\neq\varnothing} Y^v(K)\sum_{i\in K\cap S}\sum_{L\subseteq S\setminus K}(-1)^{|S\setminus(L\cup i)|}\\
&= \sum_{K\cap S\neq\varnothing} Y^v(K)\sum_{i\in K\cap S}(-1)^{|K\cap S|+1}\underbrace{\sum_{L\subseteq S\setminus K}(-1)^{|(S\setminus K)\setminus L|}}_{=0\text{ except if }S\setminus K=\varnothing}\\
&= \sum_{K\supseteq S} Y^v(K)\sum_{i\in K\cap S}(-1)^{|S|+1}
\end{aligned}$$

hence finally

$$m^v(S) = |S|(-1)^{|S|+1}\sum_{K\supseteq S} Y^v(K) \qquad (\varnothing\neq S\subseteq X). \tag{2.107}$$

Let us find the inverse relation. Putting $\mu(S) = \frac{(-1)^{|S|+1}}{|S|}m^v(S)$, and using the relation between m and $\check{m}$ in Table A.2 (μ and Y playing the rôle of $\check{m}$ and m respectively), we get

$$Y^v(S) = \sum_{K\supseteq S}(-1)^{|K\setminus S|}\mu(K) = \sum_{K\supseteq S}(-1)^{|K\setminus S|+|K|+1}\frac{1}{|K|}m^v(K),$$

which yields

$$Y^v(S) = (-1)^{|S|+1}\sum_{K\supseteq S}\frac{1}{|K|}m^v(K) \qquad (\varnothing\neq S\subseteq X). \tag{2.108}$$

We remark that $Y^v(\{i\}) = \phi_i^{\mathrm{Sh}}(v)$, the Shapley value of v [see Remark 2.43, Eq. (2.32)].

Let us find the expression of Y^v in terms of v. Using (2.108) and (2.15), we find

$$\begin{aligned}Y^v(S) &= (-1)^{|S|+1}\sum_{K\supseteq S}\frac{1}{|K|}\sum_{L\subseteq K}(-1)^{|K\setminus L|}v(L)\\ &= (-1)^{|S|+1}\sum_{L\subseteq X}v(L)\sum_{K\supseteq S\cup L}\frac{1}{|K|}(-1)^{|K\setminus L|}\\ &= (-1)^{|S|+1}\sum_{L\subseteq X}(-1)^{|S\setminus L|}v(L)\sum_{K\supseteq S\cup L}\frac{1}{|K|}(-1)^{|K\setminus(L\cup S)|}.\end{aligned}$$

Now, putting $S \cup L = T$, we have

$$\begin{aligned}\sum_{K\supseteq S\cup L}\frac{1}{|K|}(-1)^{|K\setminus(L\cup S)|} &= \sum_{K\in[T,X]}\frac{1}{|K|}(-1)^{|K\setminus T|} = \sum_{K\in[\varnothing,X\setminus T]}\frac{1}{|K|+|T|}(-1)^{|K|}\\ &= \sum_{k=0}^{n-t}\frac{1}{k+t}(-1)^k\binom{n-t}{k}\end{aligned}$$

with $t = |T|$, $k = |K|$. Using Lemma 1.1(iv) we find

$$\sum_{k=0}^{n-t}\frac{1}{k+t}(-1)^k\binom{n-t}{k} = \frac{(n-t)!(t-1)!}{n!}.$$

Substituting in the above, we get

$$Y^v(S) = (-1)^{|S|+1}\sum_{L\subseteq X}(-1)^{|S\setminus L|}v(L)\frac{(n-s-l)!(s+l-1)!}{n!}$$

which finally gives

$$Y^v(S) = \sum_{L\subseteq X}(-1)^{|S\cap L|+1}\frac{(n-s-l)!(s+l-1)!}{n!}v(L). \tag{2.109}$$

Table A.1 in Appendix A summarizes all the foregoing results.

Remark 2.92 The above results can be easily adapted to the vector space of games, even if most of the foregoing bases are not composed of games. Since for any game $v(\varnothing) = 0$, given a basis $\{b_S\}_{S\in 2^X}$ of the space of set functions, it can be turned into a basis of games $\{b'_S\}_{S\in 2^X\setminus\{\varnothing\}}$ by letting

$$b'_S(T) = \begin{cases} b_S(T), & \text{if } T \neq \varnothing \\ 0, \text{otherwise} \end{cases} \qquad (S \in 2^X \setminus \{\varnothing\}). \tag{2.110}$$

◇

2.17.2 The Inverse Problem

The duality between transforms and bases permits to easily solve the "inverse problem," well known in game theory: For a given game $v \in \mathcal{G}(X)$, find all games $v' \in \mathcal{G}(X)$ with same Shapley value, or Banzhaf value, or any other linear value ϕ (see Remark 2.43 and Eqs. (2.32) and (2.33) for their definition); i.e., such that

$$\phi(v) = \phi(v') \quad \text{or, equivalently,} \quad \phi(v - v') = \mathbf{0}.$$

Hence, the solution of the problem amounts to finding the *null space* or *kernel* of the value ϕ, viewed as a linear mapping:

$$\ker(\phi) = \{\xi \in \mathbb{R}^{2^X} \mid \phi(\xi) = \mathbf{0}\}.$$

Kleinberg and Weiss [209] exhibited a basis of the kernel for the Shapley value. Other solutions were given by Dragan [95], who solved this problem for the Shapley value [93] and later for all semivalues[14] [94] in a simpler way than Kleinberg and Weiss, and more recently by Yokote et al. [355].

Lemma 2.91 provides an easy solution for the inverse problem if a transform that extends the value in question is available. We illustrate the method with the Shapley value ϕ^{Sh}. Since the Shapley interaction transform $\xi \mapsto I^\xi$ extends ϕ^{Sh} in the sense that $I^\xi(\{i\}) = \phi_i^{\mathrm{Sh}}(\xi)$, we have

$$\xi = \sum_{S\in 2^X} I^\xi(S) b_S^I = \sum_{i\in X} \phi_i^{\mathrm{Sh}}(\xi) b_{\{i\}}^I + \sum_{\substack{S\in 2^X \\ |S|\neq 1}} I^\xi(S) b_S^I,$$

[14]A *semivalue* (Dubey et al. [97]) assigns to each game v a vector $\psi(v)$ defined by

$$\psi_i(v) = \sum_{S\subseteq X\setminus i} p_s(v(S\cup i) - v(S)) \qquad (i \in X)$$

where $\{p_s\}_{s=0,\dots,|X|-1}$ is a probability distribution on the size of the sets, satisfying $\sum_{j=0}^{|X|-1}\binom{|X|-1}{s}p_s = 1$. If the distribution depends on i, then it is called a *probabilistic value* (Weber [344]). The Shapley value is a particular case of semivalue.

which implies

$$\xi \in \ker(\phi^{\mathrm{Sh}}) \iff \xi = \sum_{\substack{S\in 2^X\\ |S|\neq 1}} I^{\xi}(S) b^I_S$$

i.e.,

$$\ker(\phi^{\mathrm{Sh}}) = \Big\{ \sum_{\substack{S\in 2^X\\ |S|\neq 1}} \lambda_S b^I_S \mid \lambda_S \in \mathbb{R} \Big\}. \tag{2.111}$$

Example 2.93 Let us give an explicit form of the kernel for $n = 3$, using Table 2.2. We obtain that any member ξ of the kernel has the form

$$\begin{aligned}
\xi(\varnothing) &= \lambda_\varnothing + \frac{1}{6}(\lambda_{12} + \lambda_{13} + \lambda_{23})\\
\xi(1) &= \lambda_\varnothing - \frac{1}{3}\lambda_{12} - \frac{1}{3}\lambda_{13} + \frac{1}{6}\lambda_{23} + \frac{1}{6}\lambda_{123}\\
\xi(2) &= \lambda_\varnothing - \frac{1}{3}\lambda_{12} + \frac{1}{6}\lambda_{13} - \frac{1}{3}\lambda_{23} + \frac{1}{6}\lambda_{123}\\
\xi(3) &= \lambda_\varnothing + \frac{1}{6}\lambda_{12} - \frac{1}{3}\lambda_{13} - \frac{1}{3}\lambda_{23} + \frac{1}{6}\lambda_{123}\\
\xi(12) &= \lambda_\varnothing + \frac{1}{6}\lambda_{12} - \frac{1}{3}\lambda_{13} - \frac{1}{3}\lambda_{23} - \frac{1}{6}\lambda_{123}\\
\xi(13) &= \lambda_\varnothing - \frac{1}{3}\lambda_{12} + \frac{1}{6}\lambda_{13} - \frac{1}{3}\lambda_{23} - \frac{1}{6}\lambda_{123}\\
\xi(23) &= \lambda_\varnothing - \frac{1}{3}\lambda_{12} - \frac{1}{3}\lambda_{13} + \frac{1}{6}\lambda_{23} - \frac{1}{6}\lambda_{123}\\
\xi(123) &= \lambda_\varnothing + \frac{1}{6}(\lambda_{12} + \lambda_{13} + \lambda_{23}),
\end{aligned}$$

where $\lambda_\varnothing, \lambda_{12}, \lambda_{13}, \lambda_{23}, \lambda_{123}$ are arbitrary real constants.

This method was already proposed by the author in [163, Sect. 7], and applied to find all k-additive capacities having the same Shapley value. A more general method for finding the kernel of the Shapley value can be found in [131].

2.18 Inclusion-Exclusion Coverings

Set functions being exponentially complex in the size of X (defined to be n in this section), it is important to find subfamilies that can be represented in a simpler way. Additive games are extreme examples needing only n real values to be defined, while k-additive games and p-symmetric games are other examples. We have seen in Sect. 2.13 that k-additivity is a natural way to generalize additivity. There exists another natural way via partitions: take any partition π of X, and say that π is an *interadditive partition* for a game v if for all sets $A \subseteq X$,

$$v(A) = \sum_{P \in \pi} v(A \cap P). \tag{2.112}$$

For an additive game, any partition is interadditive, in particular the finest one (see Sect. 1.3.2 for the definition of the poset of partitions). Conversely, if the finest partition is interadditive, then v is an additive game.

If (2.112) holds, v can be represented via the restrictions of v to each $P \in \pi$; i.e.,

$$v(A) = \sum_{P \in \pi} v_P(A \cap P) \qquad (A \subseteq X) \tag{2.113}$$

with $v_P \in \mathcal{G}(P)$, defined by $v_P(B) = v(B)$ for any $B \subseteq P$. Therefore, v needs only $\sum_{P \in \pi}(2^{|P|} - 1)$ coefficients instead of 2^n to be defined.

We may try to be more general and use coverings instead of partitions. A *covering* $\mathcal{C} = \{C_1, \dots, C_k\}$ of X is a collection of nonempty subsets of X such that $\bigcup_{i=1}^k C_i = X$. We denote by $\mathfrak{C}(X)$ the set of coverings of X. Then Eq. (2.112) does not make sense any more because overlapping between sets may occur. It can be generalized as follows. We say that $\mathcal{C} = \{C_1, \dots, C_k\}$ is an *inclusion-exclusion covering* for v if for all $A \subseteq X$,

$$v(A) = \sum_{\substack{I \subseteq [k] \\ I \neq \varnothing}} (-1)^{|I|+1} v(\bigcap_{i \in I} C_i \cap A). \tag{2.114}$$

Similarities with the notions of k-monotone/alternating properties explain the name [Remark 2.19(v)]. The set of inclusion-exclusion coverings for v on X is denoted by $\mathfrak{IEC}(v, X)$.

Inclusion-exclusion coverings are intimately related to the Möbius transform.

Lemma 2.94 *Let v be a game on X, which has no null set (see Definition 2.107 hereafter). Then $\mathcal{M}(v) = \{C \subseteq X : m^v(C) \neq 0\}$ is a covering of X, called the Möbius covering for v.*

Proof Assume that $\mathcal{M}(v)$ is not a covering of X. Then there exists $x \in X$ such that for all $A \ni x$, $m^v(A) = 0$. Consider any set $B \subseteq X \setminus \{x\}$. Then

$$v(B \cup \{x\}) = \sum_{A \subseteq B \cup \{x\}} m^v(A) = \sum_{A \subseteq B} m^v(A) = v(B),$$

proving that $\{x\}$ is a null set, a contradiction. □

Theorem 2.95 *Consider a game v on X such that $\mathcal{M}(v)$ is a covering*[15] *of X. Then the Möbius covering $\mathcal{M}(v)$ is an inclusion-exclusion covering for v.*

Proof We denote by $\mathcal{M}(v) = \{C_1, \ldots, C_k\}$ the Möbius covering.

We show (2.114) for some $A \subseteq X$, that is

$$\sum_{D \in \mathcal{M}_A(v)} m^v(D) = v(A) = \sum_{\substack{I \subseteq [k] \\ I \neq \varnothing}} (-1)^{|I|+1} v\Big(\bigcap_{i \in I} C_i \cap A\Big),$$

introducing $\mathcal{M}_A(v) = \{C \in \mathcal{M}(v) \,:\, C \subseteq A\}$. We have

$$\sum_{\substack{I \subseteq [k] \\ I \neq \varnothing}} (-1)^{|I|+1} v\Big(\bigcap_{i \in I} C_i \cap A\Big) = \sum_{\substack{I \subseteq [k] \\ I \neq \varnothing}} (-1)^{|I|+1} \sum_{\substack{D \subseteq \bigcap_{i \in I} C_i \\ D \in \mathcal{M}_A(v)}} m^v(D),$$

therefore we have to show that $m^v(D)$ for each $D \in \mathcal{M}_A(v)$ appears once and only once in the right member. Let us introduce for each $D \in \mathcal{M}_A(v)$ the collection

$$\mathcal{D}(D) = \{C \in \mathcal{M}(v) \,:\, C \supseteq D\},$$

which is nonempty because $D \in \mathcal{D}(D)$. Put $d = |\mathcal{D}(D)|$. The right member reads

$$\begin{aligned}
&\sum_{\substack{I \subseteq [k] \\ I \neq \varnothing}} (-1)^{|I|+1} \sum_{\substack{D \subseteq \bigcap_{i \in I} C_i \\ D \in \mathcal{M}_A(v)}} m^v(D) \\
&= \sum_{D \in \mathcal{M}_A(v)} m^v(D) \Big(\sum_{C \in \mathcal{D}(D)} 1 - \sum_{\substack{C, C' \in \mathcal{D}(D) \\ C \neq C'}} 1 + \sum_{\substack{C, C', C'' \in \mathcal{D}(D) \\ C \neq C' \neq C''}} 1 - \cdots \Big) \\
&= \sum_{D \in \mathcal{M}_A(v)} m^v(D) \left(\binom{d}{1} - \binom{d}{2} + \binom{d}{3} - \cdots + (-1)^{d+1} \binom{d}{d} \right) \\
&= \sum_{D \in \mathcal{M}_A(v)} m^v(D)
\end{aligned}$$

by applying (1.1). □

The Möbius covering is by no means the only inclusion-exclusion covering. However, it plays a fundamental rôle because in a sense it permits to find all of them. To this end, let us endow the set of coverings $\mathfrak{C}(X)$ with a preorder. Let $\mathcal{C}$ and $\mathcal{D}$ be two coverings of X. We write $\mathcal{C} \sqsubseteq \mathcal{D}$ if for every $C \in \mathcal{C}$, there exists $D \in \mathcal{D}$ such that $C \subseteq D$. Note that $\sqsubseteq$, although reflexive and transitive, is *not* a partial order

[15]By Lemma 2.94, a sufficient condition is that v has no null set.

because it is not antisymmetric: consider $|X| > 1$ and $\mathcal{C} = \{X, \{x\}\}$, $\mathcal{D} = \{X, \{y\}\}$ with $x \neq y$. Then $\mathcal{C} \sqsubseteq \mathcal{D}$ and $\mathcal{D} \sqsubseteq \mathcal{C}$ both hold, although they are different.

There is a simple way to make $\sqsubseteq$ a partial order: it suffices to restrict to irreducible coverings. The *reduction* of a covering $\mathcal{C}$ is the covering denoted by $\mathcal{C}^\circ$, which is the greatest antichain in $\mathcal{C}$: $\mathcal{C}^\circ$ is obtained from $\mathcal{C}$ by removing any set $C \in \mathcal{C}$ such that there exists $D \in \mathcal{C}$ that contains C. A covering $\mathcal{C}$ is *irreducible* if $\mathcal{C}^\circ = \mathcal{C}$; i.e., if it is an antichain. With some abuse, we denote by $\mathfrak{C}^\circ(X)$ the set of irreducible coverings of X. Now, $(\mathfrak{C}^\circ(X), \sqsubseteq)$ is a partially ordered set.

Consider again the Möbius covering $\mathcal{M}(v)$ of some game v, and a covering $\mathcal{C} \in \mathfrak{C}(X)$ such that $\mathcal{M}(v) \sqsubseteq \mathcal{C}$. It is easy to see by inspection of the proof of Theorem 2.95 that $\mathcal{C}$ is also an inclusion-exclusion covering for v. Indeed, exactly the same proof holds putting $\mathcal{C} = \{C_1, \dots, C_k\}$ and $\mathcal{D}(D) = \{C \in \mathcal{C} : C \supseteq D\}$, because $\mathcal{D}(D)$ is never empty for each $D \in \mathcal{M}_A(v)$ since $\mathcal{M}(v) \sqsubseteq \mathcal{C}$.

On the other hand, suppose there exist $C_i, C_j \in \mathcal{M}(v)$ such that $C_i \subset C_j$ (say $i = k$ for simplicity). Then $\mathcal{M}(v) \setminus \{C_k\}$ is still an inclusion-exclusion covering. Again, the proof of Theorem 2.95 still works, replacing $[k]$ by $[k-1]$ and letting $\mathcal{D}(D) = \{C \in \mathcal{M}(v) \setminus \{C_k\} : C \supseteq D\}$, because $\mathcal{D}(D)$ is never empty for each $D \in \mathcal{M}_A(v)$. Of course, the same reasoning applies each time two sets are comparable by inclusion in $\mathcal{M}(v)$, however, the removal of a set C_i that would not be included into another one in $\mathcal{M}(v)$ is forbidden, because $\mathcal{D}(D)$ would be empty for $D = C$. In summary, we have shown:

Lemma 2.96 *Consider a game v on X such that $\mathcal{M}(v)$ is a covering. Then:*

(i) Any covering $\mathcal{C}$ such that $\mathcal{M}(v) \sqsubseteq \mathcal{C}$ is an inclusion-exclusion covering for v;
(ii) The irreducible covering $\mathcal{M}^\circ(v)$ is an inclusion-exclusion covering for v. Moreover, it is a smallest one.

We show another simple result.

Lemma 2.97 *Consider a game v on X such that $\mathcal{M}(v)$ is a covering, and an inclusion-exclusion covering $\mathcal{C}$ for v. Then $\mathcal{M}(v) \sqsubseteq \mathcal{C}$.*

Proof Let $\mathcal{C} = \{C_1, \dots, C_k\}$. Consider $A \in \mathcal{M}(v)$ such that no $C \in \mathcal{C}$ includes A. Since $\mathcal{C}$ is an inclusion-exclusion covering, we have

$$\sum_{\substack{B \in \mathcal{M}(v) \\ B \subseteq A}} m^v(B) = v(A) = \sum_{\substack{I \subseteq [k] \\ I \neq \emptyset}} (-1)^{|I|+1} v\Big(\bigcap_{i \in I} C_i \cap A\Big) = \sum_{\substack{I \subseteq [k] \\ I \neq \emptyset}} (-1)^{|I|+1} \sum_{\substack{B \subseteq \bigcap_{i \in I} C_i \\ B \in \mathcal{M}(v) \\ B \subseteq A}} m^v(B).$$

Observe that $m^v(A)$ appears in the left member of the equation, but it cannot appear in the right member because no C_i contains A. □

Combining the above Lemma with Lemma 2.96(i), we conclude that a covering $\mathcal{C}$ is a an inclusion-exclusion covering for v if and only if $\mathcal{M}(v) \sqsubseteq \mathcal{C}$. Therefore, the irreducible covering $\mathcal{M}^\circ(v)$ is the least element of $\mathfrak{IEC}(v, X)$. In summary, we have shown:

Theorem 2.98 (The set of inclusion-exclusion coverings) *Consider a game v on X such that $\mathcal{M}(v)$ is a covering. Then*

$$\mathfrak{IEC}(v,X) = \{\mathcal{C} \in \mathfrak{C}(X) \, : \, \mathcal{M}^\circ(v) \sqsubseteq \mathcal{C}\}.$$

Consider $\mathcal{C}, \mathcal{D}$ two coverings in $\mathfrak{C}(X)$, and introduce the operations $\sqcup, \sqcap$ by

$$\begin{aligned} \mathcal{C} \sqcap \mathcal{D} &= \{C \cap D \, : \, C \in \mathcal{C}, D \in \mathcal{D}\} \\ \mathcal{C} \sqcup \mathcal{D} &= \mathcal{C} \cup \mathcal{D}. \end{aligned}$$

Clearly, $\mathcal{C} \sqcap \mathcal{D}$ and $\mathcal{C} \sqcup \mathcal{D}$ are coverings of X. Moreover, if $\mathcal{C}, \mathcal{D}$ are inclusion-exclusion coverings for v, then Theorem 2.98 shows that so are $\mathcal{C} \sqcap \mathcal{D}$ and $\mathcal{C} \sqcup \mathcal{D}$. Let us consider the poset $(\mathfrak{IEC}^\circ(v,X), \sqsubseteq)$, i.e., the set of irreducible inclusion-exclusion coverings endowed with $\sqsubseteq$, and introduce the internal operations

$$\begin{aligned} \mathcal{C} \sqcap^\circ \mathcal{D} &= \{C \cap D \, : \, C \in \mathcal{C}, D \in \mathcal{D}\}^\circ \\ \mathcal{C} \sqcup^\circ \mathcal{D} &= (\mathcal{C} \cup \mathcal{D})^\circ. \end{aligned}$$

Then it is easy to check that $(\mathfrak{IEC}^\circ(v,X), \sqsubseteq)$ is a distributive lattice, whose supremum and infimum are $\sqcup^\circ$ and $\sqcap^\circ$ respectively.

After this detailed study of the set of inclusion-exclusion coverings, let us return to our initial goal, that is, to find simpler representations, similar to (2.113). Of course, the most economic representation should be based on the smallest irreducible inclusion-exclusion covering $\mathcal{M}^\circ(v)$. We partition $\mathcal{M}(v)$ in subcollections $\mathcal{C}(B)$, $B \in \mathcal{M}^\circ(v)$, where each $\mathcal{C}(B)$ contains the subsets of B. A tie occurs for every $C \in \mathcal{M}(v)$ included in several B's in $\mathcal{M}^\circ(v)$, hence the partition is not unique. Using the basis of unanimity games (see Sect. 2.15.1), we have for any $A \subseteq X$

$$\begin{aligned} v(A) &= \sum_{B \in \mathcal{M}(v)} m^v(B) u_B(A) \\ &= \sum_{B \in \mathcal{M}^\circ(v)} \sum_{C \in \mathcal{C}(B)} m^v(C) u_C(A) \\ &= \sum_{B \in \mathcal{M}^\circ(v)} v_B(A) \qquad (2.115) \\ &= \sum_{B \in \mathcal{M}^\circ(v)} v_B(A \cap B), \qquad (2.116) \end{aligned}$$

where v_B is a game in $\mathcal{G}(B)$ for each $B \in \mathcal{M}^\circ(v)$, defined by $v_B = \sum_{C \in \mathcal{C}(B)} m^v(C) u_C$. Equation (2.116) provides the sought decomposition of v in terms of simpler games, like (2.113). Note that the decomposition may not be unique, because the partition of $\mathcal{M}(v)$ is not unique in general, and that the complexity of the representation (number of coefficients needed) is $|\mathcal{M}(v)|$, which is no surprise.

Example 2.99 Consider $X = \{1, 2, 3, 4\}$ and a game v on X with

$$\mathcal{M}(v) = \{1, 3, 4, 12, 13, 14, 24, 34, 134\}.$$

Clearly, $\mathcal{M}(v)$ is a covering of X, therefore it is an inclusion-exclusion covering. The smallest irreducible covering is $\mathcal{M}^\circ(v) = \{12, 24, 134\}$. The set $\mathfrak{IEC}(v, X)$ is very large: it contains in particular all supercollections of $\mathcal{M}^\circ(v)$ (there are 2^{12} such collections, including $\mathcal{M}^\circ(v)$), plus all collections obtained from $\mathcal{M}^\circ(v)$ (or any of its supercollections) by replacing some of the sets by a superset, e.g., $\{123, 24, 134\}$ or $\{12, 234, 1234\}$.

The decomposition of v reads

$$v(A) = v_{12}(A \cap 12) + v_{24}(A \cap 24) + v_{134}(A \cap 134),$$

with

$$\begin{aligned} v_{12} &= m^v(1)u_1 + m^v(12)u_{12} \\ v_{24} &= m^v(4)u_4 + m^v(24)u_{24} \\ v_{134} &= m^v(3)u_3 + m^v(13)u_{13} + m^v(14)u_{14} + m^v(34)u_{34} + m^v(134)u_{134}. \end{aligned}$$

If $\mathcal{M}(v)$ contains X, then $\mathcal{M}^\circ(v)$ reduces to $\{X\}$, and there is no decomposition of v any more. The set $\mathfrak{IEC}(v, X)$ is then simply the set of collections in $2^X \setminus \{\varnothing\}$ containing X. There are $2^{(2^n-2)}$ such collections. ◇

Remark 2.100 The notion of inclusion-exclusion covering was introduced and studied in depth by Fujimoto, Murofushi and Sugeno in several papers [147, 148, 323]. Our presentation, however, is different and provides alternative proofs. ◇

2.19 Games on Set Systems

So far we have assumed that set functions were defined on 2^X, with $|X| < \infty$. In many application domains, this assumption does not hold. Referring to the interpretations of capacities given in Sect. 2.4, the two main domains of application of capacities are the representation of uncertainty, and the representation of power/worth of a group (cooperative game theory and social choice theory, mainly). In the representation of uncertainty, capacities extend classical probability measures, and in this field X is most often infinite, and probability measures are defined on algebras or σ-algebras, representing the set of possible events. The whole corpus of classical measure theory deals with this framework (see, e.g., Halmos [188]). In cooperative game theory too, games with infinitely many players have been considered (see the seminal work of Aumann and Shapley [11]), and in the finite case, it is not uncommon to consider games with *restricted cooperation*, that is, defined on a proper subset of 2^X. Indeed, in many real situations, it is not reasonable to assume that any coalition or group can form, and coalitions that can actually form are called *feasible*. If X is a set of political parties, leftist and rightist parties will never form a feasible coalition. Also, if some hierarchy exists among players, feasible coalitions should correspond to sets including all subordinates, or all superiors, depending on the interpretation of what a coalition represents. A last example concerns games induced by a communication graph. A feasible coalition is then a group of players who can communicate, in other terms, it corresponds to a connected component of the graph.

In this section, we briefly address the infinite case (a complete treatment of set functions on infinite sets would take a whole monograph, including in particular classical measure theory (Halmos [188]), and nonclassical measure theory, as it can be found in Denneberg [80], König [215], Pap [264, 265], Wang and Klir [343]), and focus on the finite case. We will present several possible algebraic structures for the subcollections of 2^X.

We use the general term *set system* to denote the subcollection of 2^X where set functions are defined. Its precise definition is as follows.

Definition 2.101 A *set system* $\mathcal{F}$ on X is a subcollection of 2^X containing $\varnothing$ and such that $\bigcup_{A\in\mathcal{F}} A = X$.

$\mathcal{F}$ endowed with set inclusion is therefore a poset, and $\varnothing$ is its least element (see Sect. 1.3.2 for all definitions concerning posets and lattices). We recall that $A \subset\cdot\, B$ means that $A \subset B$ and there is no C such that $A \subset C \subset B$. Elements of $\mathcal{F}$ are *feasible sets*. Definitions of set functions, games and capacities remain unchanged, only the domain changes. In particular, a *game* v *on* $(X, \mathcal{F})$ is a mapping $v : \mathcal{F} \subseteq 2^X \to \mathbb{R}$ satisfying $v(\varnothing) = 0$. We denote by $\mathcal{G}(X, \mathcal{F})$ the set of games on $\mathcal{F}$.[16]

2.19.1 Case Where X Is Arbitrary

(see Halmos [188, Chaps. 1 and 2])

Definition 2.102

(i) A nonempty subcollection $\mathcal{F}$ of 2^X is an *algebra on* X if it is closed under finite union and complementation:

$$A, B \in \mathcal{F} \Rightarrow A \cup B \in \mathcal{F}; \qquad A \in \mathcal{F} \Rightarrow A^c \in \mathcal{F};$$

[16]This notation implies that our previous notation $\mathcal{G}(X)$ is a shorthand for $\mathcal{G}(X, 2^X)$. The omission of the set system means that we consider the Boolean lattice 2^X. We keep this convention throughout the book.

(ii) A nonempty subcollection $\mathcal{R}$ of 2^X is a *ring on X* if it is closed under finite union and set difference:

$$A, B \in \mathcal{R} \Rightarrow A \cup B \in \mathcal{R} \text{ and } A \setminus B \in \mathcal{R}.$$

Observe that:

(i) An algebra $\mathcal{F}$ is closed under finite $\cap$, and $\varnothing, X \in \mathcal{F}$. Hence an algebra is a set system;
(ii) For a ring $\mathcal{R}$, $\varnothing \in \mathcal{R}$ but X is not necessarily an element of $\mathcal{R}$;
(iii) Every algebra is a ring; Every ring containing X is an algebra.

Definition 2.103 An algebra $\mathcal{F}$ is a *σ-algebra* if it is closed under countable unions:

$$\{A_n\} \subseteq \mathcal{F} \Rightarrow \bigcup_{n=1}^{\infty} A_n \in \mathcal{F}.$$

Observe that a σ-algebra is closed under countable intersection. A similar definition exists for σ-rings.

The set of finite subsets of X with their complement is an algebra, while the set of countable subsets of X with their complement is a σ-algebra.

We introduce some additional properties of set functions.

Definition 2.104 Let $\mathcal{X}$ be a nonempty subcollection of 2^X and ξ be a set function on $(X, \mathcal{X})$.

(i) ξ is *σ-additive* if it satisfies

$$\xi\Big(\bigcup_{n=1}^{\infty} A_n\Big) = \sum_{n=1}^{\infty} \xi(A_n)$$

for any family $\{A_n\}$ of pairwise disjoint sets in $\mathcal{X}$ such that $\bigcup_{n=1}^{\infty} A_n \in \mathcal{X}$;

(ii) ξ is *continuous from below at a set $A \in \mathcal{X}$* if for every countable family $\{A_n\}$ of sets in $\mathcal{X}$ such that $A_1 \subseteq A_2 \subseteq \cdots$ and $\lim_{n\to\infty} A_n = A$, it holds

$$\lim_{n\to\infty} \xi(A_n) = \xi(A).$$

ξ is continuous from below if this holds for every $A \in \mathcal{X}$;

(iii) ξ is *continuous from above at a set $A \in \mathcal{X}$* if for every countable family $\{A_n\}$ of sets in $\mathcal{X}$ such that $A_1 \supseteq A_2 \supseteq \cdots$, $\xi(A_m) < \infty$ for some m, and $\bigcap_{n=1}^{\infty} A_n = A$, it holds

$$\lim_{n\to\infty} \xi(A_n) = \xi(A).$$

ξ is continuous from above if this holds for every $A \in \mathcal{X}$;

(iv) ξ is *continuous* if it is continuous from below and from above.

A *measure*[17] m is a nonnegative σ-additive set function on a ring, such that $m(\varnothing) = 0$. Observe that by the latter property, every measure is finitely additive. A measure m is *finite* if $m(X) < \infty$. A probability measure is a normalized measure. A *charge* is a finitely additive nonnegative set function vanishing at the empty set.

The continuity properties and σ-additivity are intimately related.

Theorem 2.105 *Let ξ be a finite, nonnegative, and finitely additive set function on a ring $\mathcal{R}$.*

(i) If ξ is either continuous from below at every $A \in \mathcal{X}$ or continuous from above at $\varnothing$, then μ is σ-additive, i.e., it is a (finite) measure;

(ii) If μ is a measure on $\mathcal{R}$, then it is continuous from below and continuous from above.

Remark 2.106

(i) In probability theory, algebras and σ-algebra are often called *fields* and σ-*fields*. σ-additivity is also called *countable additivity*, and continuity from above (respectively, below) is sometimes called *outer* (respectively, *inner*) *continuity*.

(ii) σ-additivity and σ-algebras are related to the famous *Problem of Measure* (see Aliprantis and Border [7, pp. 372–373] for a more detailed discussion): Given a set X, is there any probability measure defined on its power set so that the probability of each singleton is 0? The motivation for this question is that most often in applied sciences, to each point of the real line we assign measure zero. Returning to the Problem of Measure, if X is countable, then σ-additivity entails that no such probability measure exists, therefore sets of higher cardinality must be chosen. The *Continuum Hypothesis* asserts that the smallest uncountable cardinality is the cardinality of the interval $[0, 1]$. However, Banach and Kuratowski have shown that under this hypothesis, still no probability measure can have measure zero on singletons. It follows that in order to make probability measures satisfy this requirement, there are two choices: either σ-additivity is abandoned, or measurability of every set (that is, $\mathcal{F} = 2^X$) is abandoned. The latter choice is the most common one, and leads to σ-algebras.

◇

Null Sets

The notion of null sets is well known in classical measure theory, where it indicates a set that cannot be "seen" by a (signed) measure, in the sense that its measure, as

[17]This is the classical definition. It generalizes the definition given in Sect. 2.2 for finite sets.

well as the measure of all its subsets, is zero. For more general set functions, this notion can be extended as follows.

Definition 2.107 (Murofushi and Sugeno [252]) Let v be a game on $(X, \mathcal{F})$, where $\mathcal{F}$ is a set system. A set $N \in \mathcal{F}$ is called a *null set* w.r.t. v if

$$v(A \cup M) = v(A) \qquad (\forall M \subseteq N \text{ s.t. } A \cup M \in \mathcal{F}), (\forall A \in \mathcal{F}).$$

We give the main properties of null sets.

Theorem 2.108 *Let v be a game on $(X, \mathcal{F})$. The following holds.*

- *(i) The empty set is a null set;*
- *(ii) If N is a null set, then $v(N) = 0$;*
- *(iii) If N is a null set, then every $M \subseteq N$, $M \in \mathcal{F}$ is a null set;*
- *(iv) If $\mathcal{F}$ is closed under finite unions, the finite union of null sets is a null set;*
- *(v) If $\mathcal{F}$ is closed under countable unions and if v is continuous from below, the countable union of null sets is a null set;*
- *(vi) Assume $\mathcal{F}$ is an algebra. Then N is a null set if and only if $v(A \setminus M) = v(A)$ (equivalently, $v(A \Delta M) = v(A)$), for all $M \subseteq N$, $M \in \mathcal{F}$, and for all $A \in \mathcal{F}$;*
- *(vii) If v is monotone, N is a null set if and only if $v(A \cup N) = v(A)$ for all $A \in \mathcal{F}$;*
- *(viii) If v is additive, N is a null set if and only if $v(M) = 0$ for every $M \subseteq N$, $M \in \mathcal{F}$;*
- *(ix) If v is additive and nonnegative, N is null if and only if $v(N) = 0$.*

The proof of these statements is immediate from the definitions, and is left to the readers. Statement (viii) shows that our definition of null sets is an extension of the classical one.

Supermodular and Convex Games

The definition of supermodularity [see Definition 2.18(ii)] is left unchanged on algebras, because they are closed under finite union and intersection. When X is infinite, the equivalence between convexity and supermodularity [see Corollary 2.23(ii)] is lost in general, even if $\mathcal{F} = 2^X$ and X is countable. This is shown by the following example (Fragnelli et al. [145]).

Example 2.109 Consider $X = \mathbb{N}$ and the game v defined by:

$$v(S) = \begin{cases} 1, & \text{if } |S| = +\infty \\ 0, & \text{otherwise.} \end{cases}$$

It satisfies $v(S \cup i) - v(S) \leqslant v(T \cup i) - v(T)$ for every $S \subseteq T \subseteq \mathbb{N} \setminus \{i\}$, and therefore v is convex. However, consider S and T being respectively the set of odd and even numbers. Then $v(S) = v(T) = v(S \cup T) = 1$, and $v(S \cap T) = 0$. Therefore v is not supermodular. ◇

A remedy to this is to impose continuity from below.

Theorem 2.110 *(Fragnelli [145]) Suppose* $X = \mathbb{N}$, $\mathcal{F} = 2^{\mathbb{N}}$, *and consider a game* v *on* $(\mathbb{N}, 2^{\mathbb{N}})$ *that is continuous from below. Then* v *is convex if and only if it is supermodular.*

Proof Supermodularity implies as in the finite case convexity. Let us show the converse. Take $S, T \subseteq \mathbb{N}$, and let $S \setminus T = \{s_1, s_2, s_3, \ldots\}$, with $s_1 < s_2 < s_3 < \cdots$ (any subset of $\mathbb{N}$ has a least element). By continuity from below, we have for $r \geqslant 1$ and $s_r \in S \setminus T$:

$$v(S) - v(S \cap T) = \sum_{r \geqslant 1} \big(v((S \cap T) \cup \{s_1, \ldots, s_r\}) - v((S \cap T) \cup \{s_1, \ldots, s_{r-1}\}) \big)$$

$$v(S \cup T) - v(T) = \sum_{r \geqslant 1} \big(v(T \cup \{s_1, \ldots, s_r\}) - v(T \cup \{s_1, \ldots, s_{r-1}\}) \big).$$

Since by convexity each term in the first summand is not larger than the corresponding term in the second summand, it follows that $v(S) + v(T) \leqslant v(S \cup T) + v(S \cap T)$. □

The Variation Norm of a Game

Aumann and Shapley [11] introduced the *variation norm* of a game v as follows:

$$\|v\| = \sup \sum_{i=1}^{n} |v(A_i) - v(A_{i-1})|, \tag{2.117}$$

where the supremum is taken over all finite chains $\varnothing = A_0 \subseteq A_1 \subseteq \cdots \subseteq A_n = X$ in $\mathcal{F}$ (this should not be confused with the composition norm $\|v\|_c$; see Sect. 2.15.5). Observe that if v is monotone, then $\|v\| = v(X)$. We denote by $\mathcal{BV}(\mathcal{F})$ the set of games on $(X, \mathcal{F})$ of bounded variation.

Aumann and Shapley showed that $v \in \mathcal{BV}(\mathcal{F})$ if and only if there exist two capacities μ_1, μ_2 such that $v = \mu_1 - \mu_2$ and $\|v\| = \mu_1(X) + \mu_2(X)$. See also Murofushi et al. [255] for other properties of games of bounded variation.

2.19.2 Case Where X Is Finite

We assume here that $|X| = n$.

Set Systems Closed Under Union and Intersection

In the finite case, algebras are set systems closed under union, intersection and complementation. In many cases, in particular when studying the core (see Chap. 3), complementation is not useful and it makes sense to consider set systems closed under union and intersection. Such set systems endowed with inclusion are distributive lattices whose supremum and infimum are union and intersection respectively, and are therefore isomorphic to the set of downsets of some poset P, which can be considered to be a partition of X; i.e., the elements of P are the blocks of the partition (see Sect. 1.3.2, in particular Theorem 1.4). Hence, there is no fundamental difference between a set system generated by a partition of X or by X itself, because in the former case, it simply amounts to considering that some elements of X are glued together. For this reason, we always consider set systems closed under union and intersection as generated by a partial order $\preceq$ on X. We write, according to our notation of Sect. 1.3.2, $\mathcal{F} = \mathcal{O}(X, \preceq)$, or simply $\mathcal{F} = \mathcal{O}(X)$. Note that this class of set systems coincides with the class of distributive lattices of height n. By Birkhoff's theorem (Theorem 1.4), it is isomorphic to the class of posets on X. Figure 2.3 (reproduction of Fig. 1.3) gives an example of such a set system.

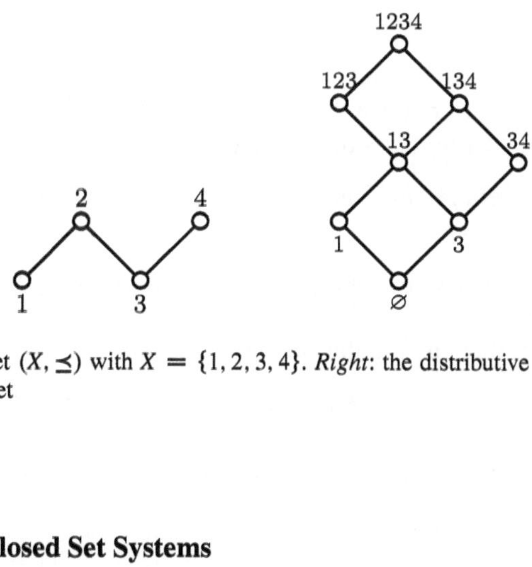

Fig. 2.3 *Left*: a poset $(X, \preceq)$ with $X = \{1, 2, 3, 4\}$. *Right*: the distributive lattice $\mathcal{F} = \mathcal{O}(X, \preceq)$ generated by the poset

Weakly Union-Closed Set Systems

One way to get a more general class of set systems is to weaken the assumption on union-closedness.

Definition 2.111 A set system $\mathcal{F}$ is *weakly union-closed* if $A, B \in \mathcal{F}$, $A \cap B \neq \varnothing$ implies $A \cup B \in \mathcal{F}$.

Note that X does not necessarily belong to $\mathcal{F}$. An important property of weakly union-closed systems is that for any $A \subseteq X$, the family $\mathcal{F}(A) = \{F \in \mathcal{F} : F \subseteq A\}$ has pairwise disjoint maximal elements.

The *basis* of $\mathcal{F}$ is the collection of sets S in $\mathcal{F}$ that cannot be written as $S = A \cup B$, with $A, B \in \mathcal{F}, A, B \neq S, A \cap B \neq \varnothing$. All singletons and pairs of $\mathcal{F}$ are in the basis. Clearly, knowing the basis permits to recover $\mathcal{F}$. Figure 2.4 illustrates these notions.

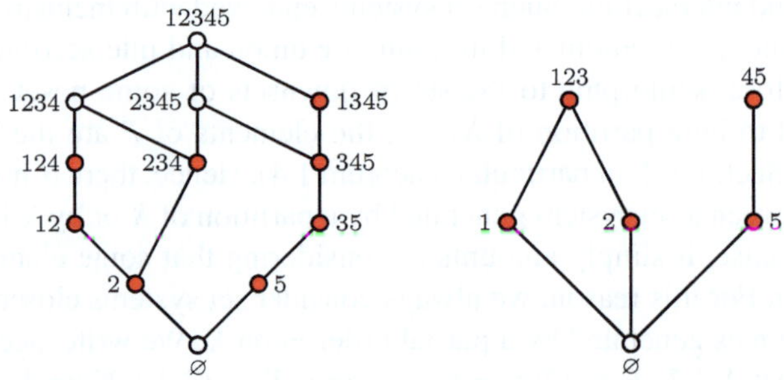

Fig. 2.4 Weakly union-closed systems on $X = \{1, 2, 3, 4, 5\}$. Elements of the basis are in *red*

Regular Set Systems

In a distributive lattice $\mathcal{O}(X)$, all maximal chains from $\varnothing$ to X have length n. This property, which does not characterize distributive lattices, can be taken as the defining property of a larger class of set systems.

Definition 2.112 A set system $\mathcal{F}$ is *regular* if it contains X and all maximal chains from $\varnothing$ to X have length n.

Equivalently, $\mathcal{F}$ is regular if and only if it contains X and for all $A, B \in \mathcal{F}$ such that $A \prec B$, we have $|B \setminus A| = 1$.

It is easy to prove that any regular set system satisfies

(i) The *one-point extension property*: if $A \in \mathcal{F}, A \neq X$, then $\exists i \in X \setminus A$ such that $A \cup \{i\} \in \mathcal{F}$;
(ii) The *accessibility property*: if $A \in \mathcal{F}, A \neq \varnothing$, then $\exists i \in A$ such that $A \setminus \{i\} \in \mathcal{F}$.

The converse is not true (Fig. 2.5).

Comparisons and Further Remarks

All the kinds of set systems presented above form distinct classes. It is apparent from the definitions that distributive lattices of the form $\mathcal{O}(X)$ are both regular set systems and weakly union-closed systems. The precise situation of these three classes is depicted in Fig. 2.6. One can find specimens of set systems in each part of this figure, even if one restricts to weakly-union closed systems containing X (see Fig. 2.4 (left)

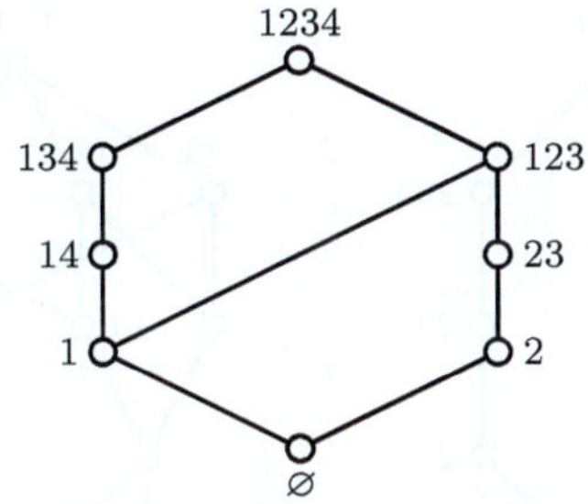

Fig. 2.5 A set system satisfying one-point extension and accessibility, but that is not regular

for an example of a weakly union-closed system containing X but being not regular, and Fig. 2.7).

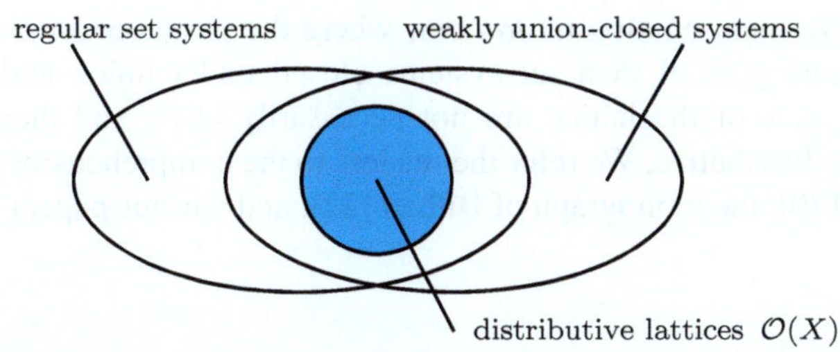

Fig. 2.6 Relations between classes of set systems

Remark 2.113

(i) The idea of games with restricted cooperation seems to go back to Myerson [256], inspired by Aumann and Drèze [10], and Owen [262]. In the latter, a fixed partition of X is given (called a *coalition structure*), which serves as a basis for defining games. No significant work seems to have emerged on this topic till the paper of Faigle [128], who coined the term "restricted cooperation."

(ii) The idea to consider a set system generated by a poset on X is due to Faigle and Kern [134]. This type of game was called *game with precedence constraint*. Regular set systems were introduced in [197, 221], while weakly union-closed sets systems were introduced by Faigle and Grabisch [130, 132], but already studied under the name *union stable systems* by Algaba [3] (summarized in Bilbao [22, Chap. 6], see also [4]).

(iii) Still many other kinds of set systems have been proposed and studied [mainly *convex geometries* (Edelman and Jamison [121]), *antimatroids* (Dilworth [89]), and *augmenting systems* (Bilbao [23])] in the context of cooperative game

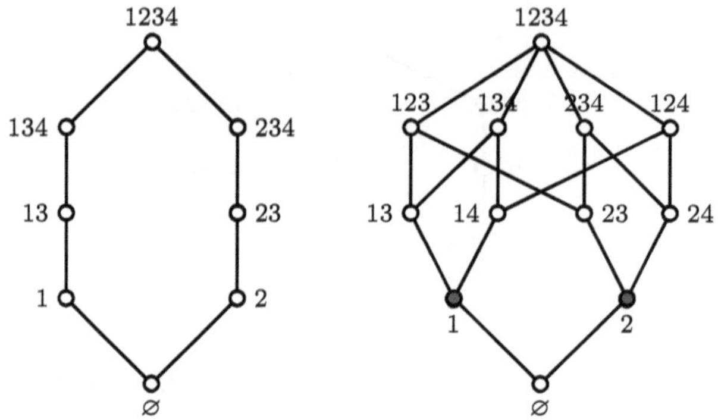

Fig. 2.7 *Left*: regular but not weakly union-closed; *Right*: regular and weakly union-closed but not a lattice, because 1 and 2 have no supremum

theory. We mention also *set lattices*, where the partial order is the inclusion order, more general than set systems closed under union and intersection, because $\vee, \wedge$ of the lattice are not necessarily $\cup, \cap$, and they may be not mutually distributive. We refer the readers to the comprehensive survey by the author [170], the monograph of Bilbao [22], and various papers [5, 6, 21, 25–27].

◇

Supermodular and Convex Games

Generally speaking, the usual notion of supermodularity makes sense if $\mathcal{F}$ is a lattice: we say that v is *supermodular* if

$$v(A \vee B) + v(A \wedge B) \geqslant v(A) + v(B) \qquad (A, B \in \mathcal{F}) \tag{2.118}$$

where $\vee, \wedge$ are the supremum and infimum of $\mathcal{F}$. For distributive lattices of the type $\mathcal{O}(X, \preceq)$, supremum and infimum are set union and intersection, so that one recovers the classical definition (same as for algebras). However, for weakly union-closed set systems and regular set systems, the definition does not make sense in general.

The following adaptation for weakly union-closed systems was proposed by Bilbao and Ordóñez [27], and Faigle et al. [132]: v on $\mathcal{F}$ is *supermodular* if

$$v(A \cup B) + \sum_{F \text{ maximal in } \mathcal{F}(A\cap B)} v(F) \geqslant v(A) + v(B), \tag{2.119}$$

for $A, B \in \mathcal{F}, A \cap B \neq \varnothing$, where $\mathcal{F}(A \cap B) = \{F \in \mathcal{F} : F \subseteq A \cap B\}$, and maximality is meant w.r.t. set inclusion.

The following result shows that for distributive lattices of the type $\mathcal{O}(X, \preceq)$, convexity and supermodularity are equivalent [generalization of Corollary 2.23(ii)].

Theorem 2.114 *Let $\mathcal{F} = \mathcal{O}(X, \preceq)$. A game v on $(X, \mathcal{F})$ is supermodular if and only if it is convex:*

$$\Delta_i v(A) \leqslant \Delta_i v(B) \qquad (i \in X, A \subseteq B \subseteq X \setminus i, A \cup i, B \in \mathcal{F}). \tag{2.120}$$

Proof (Grabisch and Sudhölter [182]) The "only if" part is easy and left to the readers. Let us show that $v(A \cup B) + v(A \cap B) \geqslant v(A) + v(B)$, assuming that $A \setminus B \neq \varnothing$ so that there exists $i_1, \ldots, i_p \in X$, where $p = |A \setminus B|$, such that $(A \cap B) \cup \{i_1, \ldots, i_m\} \in \mathcal{F}$ for all $m = 1, \ldots, p-1$ and $A \setminus B = \{i_1, \ldots, i_p\}$ (it suffices that $i_k \not\prec i_\ell$ for $k > \ell$). By (2.120),

$$\begin{aligned}
v(A) - v(A \cap B) &= \sum_{m=1}^{p} (v((A \cap B) \cup \{i_1, \ldots, i_m\}) - v((A \cap B) \cup \{i_1, \ldots, i_{m-1}\})) \\
&\leqslant \sum_{m=1}^{p} (v(B \cup \{i_1, \ldots, i_m\}) - v(B \cup \{i_1, \ldots, i_{m-1}\})) \\
&= v(A \cup B) - v(B).
\end{aligned}$$

□

Convexity and supermodularity are no longer equivalent on weakly union-closed systems and on regular set systems that are set lattices, as shown by the next examples.

Example 2.115 Consider the set system $\mathcal{F}$ on $X = \{1, 2, 3\}$ given in Fig. 2.8, which is weakly union-closed but not a distributive lattice. Observe that v on $(X, \mathcal{F})$ is

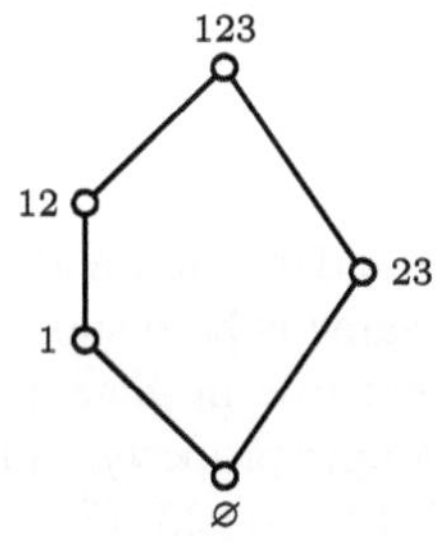

Fig. 2.8 Example of weakly union-closed not being a distributive lattice

convex if and only if $v(1) - v(\varnothing) \leqslant v(123) - v(23)$. However, supermodularity as defined in (2.119) requires that

$$v(123) + v(\varnothing) \geqslant v(12) + v(23)$$

which cannot be deduced from the former condition. ◇

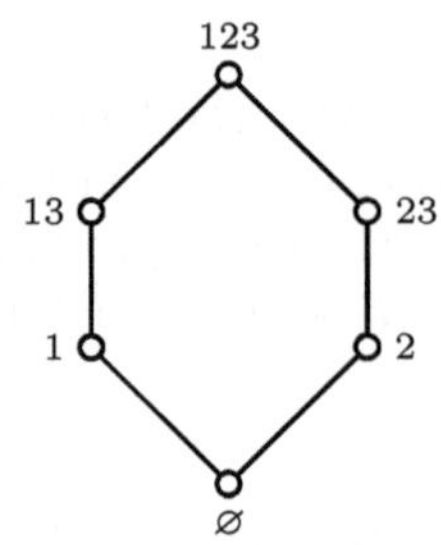

Fig. 2.9 Example of regular set lattice not being distributive

Example 2.116 Consider the set system $\mathcal{F}$ on $X = \{1, 2, 3\}$ given in Fig. 2.9, which is a regular set lattice but not distributive. v on $(X, \mathcal{F})$ is convex if and only if

$$v(1) \leqslant v(123) - v(23)$$
$$v(2) \leqslant v(123) - v(13).$$

However supermodularity requires in particular

$$v(123) \geqslant v(1) + v(2)$$

which cannot be deduced from the former conditions. ◇

Modular and Additive Games

Let us suppose that $\mathcal{F}$ is closed under union and intersection, that is, of the type $\mathcal{O}(X, \preceq)$. Recall that a modular game is both super- and submodular. Contrarily to the case where $\mathcal{F} = 2^X$, it is not true in general that additivity is equivalent to modularity, the latter being a stronger property. This is easy to see on an example. Take $n = 4$ and $\mathcal{F} = \{\varnothing, 1, 3, 13, 34, 123, 134, 1234\}$ (Fig. 2.3). Since the only disjoint pairs are the singletons $\{1\}, \{3\}$, and $\{1\}\ \{3, 4\}$, additivity is equivalent to

$$v(\{1\}) + v(\{3\}) = v(\{1, 3\})$$
$$v(\{1\}) + v(\{3, 4\}) = v(\{1, 3, 4\}).$$

However, modularity also requires for example

$$v(\{1,2,3\}) + v(\{1,3,4\}) = v(\{1,2,3,4\}) + v(\{1,3\}),$$

which cannot be deduced from the former conditions. However, if $\mathcal{F}$ is in addition closed under complementation, then the two properties are equivalent.

Theorem 2.117 *Let $\mathcal{F}$ be an algebra. Then v is modular if and only if v is additive.*

Proof We need only to prove that additivity implies modularity. Take $A, B \in \mathcal{F}$. By assumption, it follows that $A \cup B, A \cap B, X \setminus A, X \setminus B$, and therefore $A \setminus B = (X \setminus B) \cap A$, $B \setminus A$ belong to $\mathcal{F}$. Then, by additivity of v,

$$\begin{aligned} v(S \cup T) + v(S \cap T) &= v(S \setminus T) + v(T \setminus S) + 2v(S \cap T) \\ &= v(S) + v(T). \end{aligned}$$

□

k-Monotone and Totally Monotone Games

In the whole section, we suppose that $\mathcal{F}$ is a (set) lattice. Under this condition, the definition of k-monotonicity can be generalized. A game v on $(X, \mathcal{F})$ is *k-monotone* for some $k \geqslant 2$ if for any family of k sets $A_1, \ldots, A_k$ in $\mathcal{F}$,

$$v\Big(\bigvee_{i=1}^{k} A_i\Big) \leqslant \sum_{\substack{I \subseteq \{1,\ldots,k\} \\ I \neq \varnothing}} (-1)^{|I|+1} v\Big(\bigwedge_{i \in I} A_i\Big). \tag{2.121}$$

Similarly as in the case of $\mathcal{F} = 2^X$, we say that v is *totally monotone* or *∞-monotone* if it is monotone for every $k \geqslant 2$. A first fundamental result is the generalization of Theorem 2.21(ii).

Lemma 2.118 *Let v be a game on $(X, \mathcal{F})$ where $\mathcal{F}$ is a lattice. Then v is totally monotone if and only if v is $(|\mathcal{F}| - 2)$-monotone.*

This result was proved by Barthélemy [18], and the proof of Theorem 2.21(ii) can be used *mutatis mutandis*.

We turn now to k-valuations. A *k-valuation* is a function on a lattice L for which (2.121) holds with equality. Similarly, an ∞-valuation is a set function being a k-valuation for every $k \geqslant 2$. Note that a 2-valuation is a modular function, and that a probability measure is an ∞-valuation.

Interestingly, the existence of a k-valuation on a lattice L gives information on its properties. The following results are well known in lattice theory (see, e.g., Birkhoff [30, Chap. X]).

Lemma 2.119 *Let L be a lattice. The following holds.*

(i) L is modular if and only if it admits a strictly monotone 2-valuation;
(ii) L is distributive if and only if it admits a strictly monotone 3-valuation;
(iii) L is distributive if and only if it is modular and every strictly monotone 2-valuation is a 3-valuation;
(iv) L is distributive if and only if it is modular and every strictly monotone 2-valuation is an ∞-valuation.

In view of these results, a natural question is: Does the existence of a totally monotone function on a lattice L implies some property on L? The answer is negative.

Lemma 2.120 *(Barthélemy [18]) Any lattice L admits a monotone and totally monotone function vanishing at the bottom of L.*

Proof Let $\mathcal{J}(L)$ be the set of join-irreducible elements of L. We use the mapping η on L defined by $\eta(x) = \{t \in \mathcal{J}(L) \, : \, t \leqslant x\}$ (Sect. 1.3.2). We know that η is injective, and satisfies $\eta(x \wedge y) = \eta(x) \cap \eta(y)$, and $\eta(x) \cup \eta(y) \subseteq \eta(x \vee y)$. Define the function f on L by $f(x) = |\eta(x)|$. It is obviously monotone and $f(\perp) = 0$. We claim that f is also a totally monotone function on L. Since $|A| + |B| = |A \cap B| + |A \cup B|$ for any set $A, B \subseteq L$, $|\cdot|$ is a strictly monotone 2-valuation on the Boolean lattice 2^L, hence distributive. It follows from Lemma 2.119(iv) that $|\cdot|$ is an ∞-valuation,[18] hence for any $x_1, \ldots, x_k \in L$

$$\begin{aligned} f\Big(\bigvee_{i=1}^{k} x_i\Big) &\geqslant \Big|\bigcup_{i=1}^{k} \eta(x_i)\Big| \\ &= \sum_{\substack{J \subseteq \{1,\ldots,k\} \\ J \neq \varnothing}} (-1)^{|J|+1} \Big|\bigcap_{j \in J} \eta(x_j)\Big| \\ &= \sum_{\substack{J \subseteq \{1,\ldots,k\} \\ J \neq \varnothing}} (-1)^{|J|+1} f\Big(\bigwedge_{j \in J} x_j\Big), \end{aligned}$$

which proves the claim. □

This result means that belief measures (i.e., a monotone and totally monotone game) exist on any lattice. By contrast, Lemma 2.119(iv) tells us that a probability measure can live only on a distributive lattice.

The last question we address in this section is the following: In the classical case $\mathcal{F} = 2^X$, we know by Theorem 2.33(v) that there is an equivalence for a game between being monotone and totally monotone, and having a nonnegative Möbius

[18] See also (2.4).

transform. Does this equivalence still holds when $\mathcal{F}$ is any lattice? The answer is positive.

We recall that the Möbius transform is defined for functions on lattices in Remark 2.32(ii).

Theorem 2.121 *Let μ be a capacity on $(X, \mathcal{F})$, where $\mathcal{F}$ is a lattice. Then μ has a nonnegative Möbius transform if and only if μ is totally monotone.*

Proof ⇒) By assumption, we know that m^μ is nonnegative and satisfies $m^\mu(\varnothing) = 0$. For any $S \in \mathcal{F}$, we consider its principal ideal $\downarrow S = \{T \in \mathcal{F} : T \subseteq S\}$. We recall (see Sect. 1.3.2) that $S \subseteq S'$ implies $\downarrow S \subseteq \downarrow S'$, $\downarrow S \cup \downarrow S' \subseteq \downarrow(S \vee S')$, and $\downarrow S \cap \downarrow S' = \downarrow(S \wedge S')$. Consider the Boolean lattice $2^{\mathcal{F}}$ together with the function m' defined on it by $m'(A) = m^\mu(S)$ if $A = \downarrow S$ for some $S \in \mathcal{F}$, and 0 otherwise. Thus, m' considered to be the Möbius transform of some set function μ' on $2^{\mathcal{F}}$ implies that $\mu(S) = \mu'(\downarrow S)$ for every $S \in \mathcal{F}$, and μ' is totally monotone on $2^{\mathcal{F}}$. Taking a family of k sets $S_1, \ldots, S_k \in \mathcal{F}$, we have

$$\begin{aligned}\mu(S_1 \vee \cdots \vee S_k) &= \mu'(\downarrow(S_1 \vee \cdots \vee S_k))\\ &\geqslant \mu'(\downarrow S_1 \cup \cdots \cup \downarrow S_k)\\ &\geqslant \sum_{\substack{J \subseteq K\\ J \neq \varnothing}} (-1)^{|J|+1} \mu'\Big(\bigcap_{j \in J} \downarrow S_j\Big).\end{aligned}$$

However $\mu'\Big(\bigcap_{j\in J} \downarrow S_j\Big) = \mu'\Big(\Big\downarrow\Big(\bigwedge_{j\in J} S_j\Big)\Big) = \mu\Big(\bigwedge_{j\in J} S_j\Big)$, whence the result.

⇐) Taking a totally monotone capacity μ, we proceed inductively to show that $m^\mu \geqslant 0$, using the formula

$$m^\mu(S) = \mu(S) - \sum_{\substack{T \subset S\\ T \in \mathcal{F}}} m^\mu(T) \qquad (S \in \mathcal{F}). \tag{2.122}$$

We already observe that $m^\mu(\varnothing) = 0$, and from (2.122) and the nonnegativity of μ, we deduce that $m^\mu(S) \geqslant 0$ for all atoms of $\mathcal{F}$. Take $S \in \mathcal{F}$ that is not an atom, and define $\mathcal{S} = \{T \in \mathcal{F} : T \subset S\} = (\downarrow S) \setminus \{S\}$. Observe that $\mathcal{S} = \bigcup_{\substack{T \subset S\\ T \in \mathcal{F}}} \downarrow T$. Now, we introduce the set function M on $2^{\mathcal{F}}$ defined by $M(\mathcal{T}) = \sum_{T \in \mathcal{T}} m^\mu(T)$, which is an additive set function, and therefore an ∞-valuation. From the above considerations, we have

$$\begin{aligned}\sum_{T \in \mathcal{S}} m^\mu(T) &= M(\mathcal{S}) = M\Big(\bigcup_{\substack{T \subset S\\ T \in \mathcal{F}}} \downarrow T\Big)\\ &= \sum_{\substack{\mathcal{T} \subseteq \mathcal{S}\\ \mathcal{T} \neq \varnothing}} (-1)^{|\mathcal{T}|+1} \Big(\sum_{K \in \bigcap_{T \in \mathcal{T}} \downarrow T} m^\mu(K)\Big)\end{aligned}$$

$$= \sum_{\substack{\mathcal{T} \subseteq \mathcal{S} \\ \mathcal{T} \neq \varnothing}} (-1)^{|\mathcal{T}|+1} \Big(\sum_{K \in \downarrow \bigwedge_{L \in \mathcal{T}} L} m^\mu(K) \Big)$$

$$= \sum_{\substack{\mathcal{T} \subseteq \mathcal{S} \\ \mathcal{T} \neq \varnothing}} (-1)^{|\mathcal{T}|+1} \mu\Big(\bigwedge_{L \in \mathcal{T}} L \Big).$$

Moreover, since μ is totally monotone,

$$\mu(S) \geqslant \mu\Big(\bigvee_{T \in \mathcal{S}} T \Big)$$

$$\geqslant \sum_{\substack{\mathcal{T} \subseteq \mathcal{S} \\ \mathcal{T} \neq \varnothing}} (-1)^{|\mathcal{T}|+1} \mu(\Big(\bigwedge_{T \in \mathcal{T}} T \Big)$$

$$= \sum_{T \in \mathcal{S}} m^\mu(T).$$

It follows that

$$m^\mu(S) = \mu(S) - \sum_{T \in \mathcal{S}} m^\mu(T) \geqslant 0.$$

□

Remark 2.122 The “if” part was shown by Barthélemy [18], while the “only if” part was shown more recently by Zhou [359], contradicting a (wrong!) counterexample found by the author [168, Example 2]. ◇

Chapter 3
The Core and the Selectope of Games

This second fundamental chapter addresses the following problem: Given a game or a capacity, does there exist an additive game dominating it on every subset, under the constraint that both coincide on the universal set? For normalized capacities, the problem amounts to finding probability measures dominating a given capacity, while in cooperative game theory, it amounts to the problem of sharing a cake so that no coalition of players is dissatisfied. The set of additive games (or equivalently vectors, if one represents additive games by their "distribution") dominating a given game is called the *core* of that game. It is a convex polyhedron, whenever it is nonempty, and its properties have been studied in depth. In Sect. 3.2, we study the case where the game is defined on the whole power set, while Sect. 3.3 addresses the case of games on set systems. The latter case reveals to be much more complex, because the core is most often unbounded and may even have no vertices. Section 3.4 goes further in the analysis of the core of games on the power set, through the concept of exact games and large cores. Exact games are those which coincide with the lower envelope of their core and are of primary importance in decision under uncertainty. A game with large core has the property that for any vector y dominating it without constraint, it is possible to find a core element x smaller than y. The last topic addressed in this chapter is the selectope of a game (Sect. 3.5). It is the set of additive games (or vectors) obtained by sharing the Möbius transform of that game on every element. It always contains the core and equality holds if and only if the game has a nonnegative Möbius transform, except possibly on singletons.

M. Grabisch, *Set Functions, Games and Capacities in Decision Making*, Theory and Decision Library C 46, DOI 10.1007/978-3-319-30690-2_3

In the whole chapter, we consider a finite set N, with $|N| = n$.[1] The chapter makes an extensive use of Sects. 1.3.3–1.3.6 on polyhedra and linear programming.

3.1 Definition and Interpretations of the Core

Definition 3.1 Let us consider a game $v \in \mathcal{G}(N, \mathcal{F})$, where $\mathcal{F}$ is any set system on N (Definition 2.101). The *core* of v is defined by

$$\mathbf{core}(v) = \{x \in \mathbb{R}^N : x(S) \geqslant v(S), \forall S \in \mathcal{F}, \quad x(N) = v(N)\}, \tag{3.1}$$

where $x(S)$ is a shorthand for $\sum_{i \in S} x_i$. By convention, $x(\varnothing) = 0$.

The core of a game v is therefore a set of real vectors x having the property that the additive game generated by x is greater than v. Since it is defined by a set of linear inequalities plus one linear equality, it is a convex closed polyhedron of dimension at most $n - 1$, which may be empty.

The next two sections study in depth the properties of this polyhedron. Beforehand, we make some remarks on the interpretations of the core. To this end, we recall the two main interpretations of games and capacities given in Sect. 2.4.1.

In the first interpretation, N is a set of players, agents, etc., and $v(S)$ is the "worth" of coalition $S \subseteq N$. This pertains to cooperative game theory, social choice and group decision making, however the notion of core is best suited to cooperative game theory, and we therefore stick to this framework here. For a better understanding, we develop a little bit more its presentation (see Driessen [96], Owen [263], Peleg and Sudhölter [267] and Peters [268] for monographs on the topic, and Examples 2.6 and 2.8 for illustrations of this situation).

In most cases of interest, the function v represents the maximum benefit (or minimum cost, in which case inequalities in (3.1) have to be reversed) a coalition can achieve by cooperation of its members (or by using in common a resource). If all players in N cooperate, the quantity $v(N)$ represents the achieved benefit (or paid cost) in total.[2] Let us assume that the coalition N eventually forms. Then each player in N would like to be rewarded for his cooperation, for having contributed to the realization of the total benefit $v(N)$. This amounts to defining an *allocation*

[1]We apologize for the change of notation from X to N, since X is the universal set in Chaps. 2 and 4. We chose X for these chapters of general interest, as being "neutral," compared to the more specific Ω (obviously related to uncertainty), E (standing for the set of edges, which is common in combinatorial optimization), N (standard in game theory and for pseudo-Boolean functions), etc. We have chosen N in this chapter because it is more closely related to game theory. Also, throughout the chapter, vectors in $\mathbb{R}^n$ are used, more conveniently denoted by x, y, z, which could have caused some confusion with elements of X.

[2]Generally, people think benefits are positive amounts, however $v(N)$ could be negative and is considered then to be a loss. The following discussion works as well when $v(N)$ is a loss.

vector $x \in \mathbb{R}^N$ (identified as $\mathbb{R}^n$, letting $N = \{1, \ldots, n\}$), where x_i is the reward given to player i. Of course $x(N) = \sum_{i\in N} x_i$ could not exceed $v(N)$, hence the best to do is to impose $x(N) = v(N)$, in order to maximize the rewards. Now, suppose there exists a coalition $S \in \mathcal{F}$ such that $x(S) < v(S)$. Then this coalition has no interest to cooperate with the players in $N \setminus S$ to form the grand coalition N, because doing so they would get $x(S)$, strictly less than what they could achieve by themselves; i.e., $v(S)$. In other words, in such a situation, the game is "unstable" in the sense that S would leave N and form a subgame. The core is therefore the set of allocation vectors that are optimal and ensure stability of the game. For this reason, it is a central notion in cooperative game theory. It was introduced by Shapley [301], although the first definition of the core was proposed by Gillies [158] in a different form. The two forms coincide in particular for superadditive games.

In the second interpretation, N is the set of possible outcomes of some experiment (states of nature), and v has to be monotone; i.e., it is a capacity μ, which in addition is supposed to be normalized. A subset $A \in \mathcal{F}$ is an event, and $\mu(A)$ quantifies the amount of uncertainty that A realizes, in other words, that the true state of nature lies in A. If μ is not a probability measure, the usual and well-developed tools of probability theory and statistics do not apply. It is therefore tempting to try to replace μ by a probability measure, the problem being which one to choose and under which rationale. The commonly admitted interpretation is the following: if enough statistical evidence would be available on the realization of A, $\mu(A)$ would be a probability. If not, it means that our knowledge on the experiment is incomplete, and we lack evidence on the realization of A. Therefore, the amount of certainty (accumulated evidence) on the realization of A, quantified by $\mu(A)$, should be less than the (true) probability $P(A)$ of the event A. It follows that the set of probability measures compatible with the incomplete model represented by μ is, assuming that $\mathcal{F}$ is an algebra (see Definition 2.102),

$$\{P \text{ probability measure on } (N, \mathcal{F}) : P(A) \geqslant \mu(A), \forall A \in \mathcal{F}\}.$$

Because a probability measure on a finite set N is equivalently represented by a vector in $\mathbb{R}^N$ (under some mild condition on $\mathcal{F}$), and since $P(N) = \mu(N) = 1$, it follows that the set of compatible probability measures is the core of μ. This view of uncertainty is a particular case of the model of *imprecise probabilities* (Sect. 5.3.5).

Referring now to the usage of games in combinatorial optimization (see Sect. 2.4.2), we have seen that a submodular capacity is used under the name of rank function ρ of a polymatroid (N, ρ). In this context, a dual notion of the core was introduced by Edmonds [122], the *base polyhedron* associated to the polymatroid (N, ρ):

$$B(\rho) = \{x \in \mathbb{R}^N_+ : x(A) \leqslant \rho(A), \forall A \in 2^N, \quad x(N) = \rho(N)\}.$$

It is easy to see that $B(\rho) = \mathbf{core}(\overline{\rho})$, the core of the conjugate capacity [see (2.1)]. In game theory, $B(\rho)$ is sometimes called the *anticore* of ρ, and is denoted by $\mathbf{core}^*(\rho)$. Edmonds [122] has shown that, when ρ is the rank function of a matroid, the convex hull of the characteristic vectors of the bases of the matroid is precisely

$B(\rho)$. Following Edmonds, the structure of the base polyhedron was studied in depth in various papers by Fujishige and Tomizawa (see a synthesis in Fujishige [149, Sect. 3.3]).

3.2 The Core of Games on $(N, 2^N)$

The core of games whose domain is 2^N was thoroughly studied in the literature, and its structure is completely determined for supermodular games. We begin by mentioning the following relation between the cores of the game and its conjugate (we let the readers check that it is correct):

$$\mathbf{core}(v) = -\mathbf{core}(-\overline{v}), \tag{3.2}$$

and the relation between the core and the anticore:

$$\mathbf{core}^*(v) = \mathbf{core}(\overline{v}). \tag{3.3}$$

We recall that the core is always a closed convex polyhedron of dimension at most $n-1$. Moreover it is bounded, as is easy to see from the inequalities

$$x_i \geqslant v(\{i\}) \qquad (i \in N) \tag{3.4}$$

valid for any $x \in \mathbf{core}(v)$, and the equality $\sum_{i \in N} x_i = v(N)$. Therefore, the core is completely determined by its extreme points. However, the core may be empty, as shown by the following example.

Example 3.2 Let $N = \{1, 2, 3\}$ and v such that $v(\{i\}) = 1$, $i \in N$, and $v(N) < 3$. Then the inequalities $x_i \geqslant 1$ for $i \in N$, and $x_1+x_2+x_3 = v(N) < 3$ are incompatible, hence the core is empty. ◇

We begin by examining when the core is nonempty.

3.2.1 Nonemptiness of the Core

Definition 3.3 Let $\mathcal{B} \subseteq 2^N$ be a collection of nonempty sets. We say that $\mathcal{B}$ is a *balanced collection* if there exist $\lambda_A > 0$, $A \in \mathcal{B}$, such that for each $i \in N$, $\sum_{A \in \mathcal{B} : A \ni i} \lambda_A = 1$. In shorter form:

$$\sum_{A \in \mathcal{B}} \lambda_A 1_A = 1_N. \tag{3.5}$$

The quantities $\lambda_A, A \in \mathcal{B}$, form a *system of balancing weights*.

Note that any partition of N is a balanced collection with weights all equal to 1. Hence the notion of balanced collection generalizes the notion of partition.

Example 3.4 Let $N = \{1, 2, 3\}$ and consider $\mathcal{B} = \{\{1, 2\}, \{1, 3\}, \{2, 3\}\}$. This is balanced collection with weights $(\frac{1}{2}, \frac{1}{2}, \frac{1}{2})$. It can be checked that there is no other system of balancing weights. However, if we consider the collection $\mathcal{B} = \{\{1, 2\}, \{1, 3\}, \{2, 3\}, \{1, 2, 3\}\}$, it is also balanced but there exist infinitely many systems of balancing weights, of the form $(\alpha, \alpha, \alpha, 1 - 2\alpha)$, with $0 < \alpha < \frac{1}{2}$. ◇

Balanced collections have the property to be *separating*: for all $i, j \in N$, if there exists $A \in \mathcal{B}$ such that $i \in A \not\ni j$, then there exists $B \in \mathcal{B}$ such that $j \in B \not\ni i$. This can be seen from

$$1 = \sum_{A \in \mathcal{B} : i \in A} \lambda_A = \sum_{A \in \mathcal{B} : i \in A \not\ni j} \lambda_A + \sum_{A \in \mathcal{B} : i,j \in A} \lambda_A$$

$$1 = \sum_{B \in \mathcal{B} : j \in B} \lambda_B = \sum_{B \in \mathcal{B} : j \in B \not\ni i} \lambda_B + \sum_{B \in \mathcal{B} : i,j \in B} \lambda_B,$$

where $(\lambda_A)_{A \in \mathcal{B}}$ is a system of balancing weights for $\mathcal{B}$. A characterization of balanced collections is given in the next theorem.

Theorem 3.5 *A collection $\mathcal{B} \subseteq 2^N$ of nonempty sets is balanced if and only if for every vector $y \in \mathbb{R}^N$ such that $y(N) = 0$, either $y(S) = 0$ for every $S \in \mathcal{B}$ or there exist $S, T \in \mathcal{B}$ such that $y(S) > 0$ and $y(T) < 0$.*

Proof (Zumsteg [361], Derks and Peters [86][3])

$\Rightarrow$) Suppose $\mathcal{B}$ is balanced with $(\lambda_S)_{S \in \mathcal{B}}$ a system of balancing weights. Then for any $y \in \mathbb{R}^N$,

$$\sum_{S \in \mathcal{B}} \lambda_S y(S) = y(N).$$

Letting $y(N) = 0$, the positivity of λ_S, $S \in \mathcal{B}$ implies that either all $y(S)$ are zero, or there must exist $S, T \in \mathcal{B}$ such that $y(S) > 0$ and $y(T) < 0$.

$\Leftarrow$) Suppose that for all $y \in \mathbb{R}^N$ such that $y(S) \geqslant 0$, $S \in \mathcal{B}$, and $y(N) = 0$, it holds $y(S) = 0$. Putting $\mathcal{B}^* = \mathcal{B} \setminus \{N\}$, this means that the set of inequalities $\{y(S) \geqslant 0, S \in \mathcal{B}^*, y(N) = 0\}$ implies the inequalities $-y(S) \geqslant 0$ for all $S \in \mathcal{B}^*$. Hence, by Farkas' Lemma II (Theorem 1.7), for all $S \in \mathcal{B}^*$, there exist nonnegative coefficients λ^S_T, $T \in \mathcal{B}^*$ and $\lambda^S_N \in \mathbb{R}$ such that

$$\sum_{T \in \mathcal{B}^*} \lambda^S_T 1_T + \lambda^S_N 1_N = -1_S \qquad (S \in \mathcal{B}^*).$$

[3]Derks and Peters show a slightly more general result. We follow their proof.

Observe that $\lambda_N^S < 0$ for all $S \in \mathcal{B}^*$, hence letting

$$\gamma_T^S = \begin{cases} \frac{\lambda_T^S}{-\lambda_N^S}, & \text{if } T \neq S \\ \frac{\lambda_S^S + 1}{-\lambda_N^S}, & \text{otherwise} \end{cases}$$

we find

$$\sum_{T \in \mathcal{B}^*} \gamma_T^S 1_T = 1_N \qquad (S \in \mathcal{B}^*),$$

with for any $S \in \mathcal{B}^*$, $\gamma_T^S \geqslant 0$ for any $T \in \mathcal{B}^*$, $\gamma_S^S > 0$. Define

$$\lambda_T = \frac{\sum_{S \in \mathcal{B}^*} \gamma_T^S}{|\mathcal{B}^*|} \qquad (T \in \mathcal{B}^*).$$

Observe that $\lambda_T > 0$ for all $T \in \mathcal{B}^*$, and $\sum_{T \in \mathcal{B}^*} \lambda_T 1_T = 1_N$. Hence $\mathcal{B}^*$ (and therefore $\mathcal{B}$) is balanced.

□

Definition 3.6 A game v on N is *balanced* if for any balanced collection $\mathcal{B}$ it holds

$$v(N) \geqslant \sum_{A \in \mathcal{B}} \lambda_A v(A).$$

Theorem 3.7 (Bondareva-Shapley theorem, weak form) *Let* $v \in \mathcal{G}(N)$. *Then* **core**(v) *is nonempty if and only if* v *is balanced.*

Proof Consider the set of $2^n - 2$ inequalities plus one equality defining the core. By Farkas' Lemma (Theorem 1.6), the core is nonempty if and only if for every vector $y \in \mathbb{R}^{2^N \setminus \{\varnothing\}}$ such that $y_A \geqslant 0$ for $A \neq N$, and $\sum_{A \ni i} y_A = 0$ for all $i \in N$, we have $\sum_{A \in 2^N \setminus \{\varnothing\}} y_A v(A) \leqslant 0$. Take any such vector different from $\mathbf{0}$, and remark that

$$\sum_{A \ni i} y_A = \sum_{A \ni i, A \neq N} y_A + y_N = 0, \forall i \in N,$$

yielding $y_N < 0$ because not all y_A, $A \neq N$, can be 0. Hence, we may assume w.l.o.g. that $y_N = -1$; then **core**$(v) \neq \varnothing$ if and only if for every $y \in \mathbb{R}^{2^N \setminus \{\varnothing, N\}}$ with

$$y_A \geqslant 0, \forall A \subset N, A \neq \varnothing, \text{ and } \sum_{\varnothing \neq A \subset N} y_A 1_A = 1_N \tag{3.6}$$

we have

$$\sum_{\varnothing \neq A \subset N} y_A v(A) \leqslant v(N). \tag{3.7}$$

Moreover, each vector y satisfying (3.6) corresponds to the balanced collection $\mathcal{B} = \{\varnothing \neq A \subset N \, : \, y_A > 0\}$ with system of balancing weights $\lambda_A = y_A$, $A \in \mathcal{B}$, and vice versa. Therefore (3.7) means that v is balanced. □

Remark 3.8 This result was shown independently by Bondareva [34, 35] and Shapley [299]. The classical proof uses the strong form of the duality theorem in linear programming (see, e.g., Peleg and Sudhölter [267]). Our proof uses Farkas' Lemma. It shows that there is no need to consider balanced collections containing N. The sharp form of the theorem (Theorem 3.12) will discard from the analysis many more balanced collections. ◇

We say that a balanced collection is *minimal* if it does not contain a proper subcollection that is balanced.

Lemma 3.9 *A balanced collection is minimal if and only if it has a unique system of balancing weights.*

Proof ⇐) Suppose that $\mathcal{B}$ is not minimal. Then there exists $\mathcal{B}^* \subset \mathcal{B}$ that is balanced with a system of balancing weights $(\lambda^*_A)_{A \in \mathcal{B}^*}$. Then $\mathcal{B}$ has infinitely many systems of balancing weights $(\lambda^\alpha_A)_{A \in \mathcal{B}}$, defined by

$$\lambda^\alpha_A = \begin{cases} \alpha\lambda_A + (1-\alpha)\lambda^*_A, & \text{if } A \in \mathcal{B}^* \\ \alpha\lambda_A, & \text{if } A \in \mathcal{B} \setminus \mathcal{B}^* \end{cases}$$

with $0 < \alpha \leqslant 1$.

⇒) Suppose that $\mathcal{B}$ has two different systems of balancing weights $(\lambda_A)_{A \in \mathcal{B}}$ and $(\lambda'_A)_{A \in \mathcal{B}}$. Then there exists $A \in \mathcal{B}$ such that $\lambda'_A > \lambda_A$, and we put

$$\tau = \min\Big\{\frac{\lambda_A}{\lambda'_A - \lambda_A} \, : \, \lambda'_A > \lambda_A\Big\}.$$

We define the system of weights $(\tilde{\lambda}_A)_{A \in \mathcal{B}}$:

$$\tilde{\lambda}_A = (1+\tau)\lambda_A - \tau\lambda'_A \qquad (A \in \mathcal{B}).$$

Then $\mathcal{B}^* = \{A \in \mathcal{B} \, : \, \tilde{\lambda}_A > 0\}$ is a proper subcollection of $\mathcal{B}$ that is balanced with system of balancing weights $(\tilde{\lambda}_A)_{A \in \mathcal{B}^*}$. □

Let us consider the convex polytope

$$F = \Big\{\lambda \in \mathbb{R}^{2^N \setminus \{\varnothing\}} \, : \, \sum_{\varnothing \neq A \subseteq N} \lambda_A 1_A = 1_N, \quad \lambda_A \geqslant 0, \quad \forall \varnothing \neq A \subseteq N\Big\}.$$

Lemma 3.10 *Let $\lambda \in F$ and consider $\mathcal{B} = \{A \subseteq N : \lambda_A > 0\}$. Then λ is an extreme point of F if and only if $\mathcal{B}$ is a minimal balanced collection.*

Proof $\Rightarrow$) If $\mathcal{B}$ is not minimal, then there exists $\mathcal{B}^* \subset \mathcal{B}$ that is balanced, with a system of balancing weights $(\lambda^*_A)_{A\in\mathcal{B}^*}$. We set

$$\gamma_A = (1-t)\lambda_A + t\lambda^*_A$$
$$\gamma'_A = (1+t)\lambda_A - t\lambda^*_A$$

for all $A \in \mathcal{B}$, letting $\lambda^*_A = 0$ if $A \notin \mathcal{B}^*$, with $t > 0$ small enough to ensure $\gamma_A, \gamma'_A > 0$ for all $A \in \mathcal{B}$. Then $(\gamma_A)_{A\in\mathcal{B}}$, $(\gamma')_{A\in\mathcal{B}}$ are systems of balancing weights for $\mathcal{B}$ that are different, because $\gamma_A < \gamma'_A$ for all $A \in \mathcal{B} \setminus \mathcal{B}^*$. Moreover, $\lambda_A = \frac{1}{2}(\gamma_A + \gamma'_A)$ for all $A \in \mathcal{B}$, hence λ is not an extreme point.

$\Leftarrow$) Suppose that $\mathcal{B}$ is a minimal collection. If λ is not an extreme point, there exist distinct $\gamma, \gamma' \in F$ such that

$$\lambda_A = \frac{1}{2}(\gamma_A + \gamma'_A) \qquad (A \in \mathcal{B}).$$

Since γ, γ' are nonnegative, $\lambda_A = 0$ implies $\gamma_A = \gamma'_A = 0$, therefore γ, γ' define distinct systems of balancing weights for collections $\mathcal{C}, \mathcal{C}'$, subcollections of $\mathcal{B}$, which by Lemma 3.9 contradicts the minimality of $\mathcal{B}$. □

Corollary 3.11 *A minimal balanced collection contains at most n sets.*

Proof From Lemma 3.10, $\mathcal{B}$ is minimal if and only if its unique system of balancing weight corresponds to an extreme point λ of F. Therefore λ is the (unique) solution of a system of at least $2^n - 1$ equalities among the system $\{\sum_{A\ni i}\lambda_A = 1, i \in N; \lambda_A \geqslant 0, A \in 2^N \setminus \{\varnothing\}\}$. Since the number of equalities in this system is $n + 2^n - 1 - |\mathcal{B}|$, the above condition yields $|\mathcal{B}| \leqslant n$. □

Theorem 3.12 (Bondareva-Shapley theorem, sharp form) *Let $v \in \mathcal{G}(N)$. Its core is nonempty if and only if for any minimal balanced collection $\mathcal{B}$ with system of balancing weights $(\lambda_A)_{A\in\mathcal{B}}$, we have $v(N) \geqslant \sum_{A\in\mathcal{B}} \lambda_A v(A)$. Moreover, none of the inequalities is redundant, except the one for $\mathcal{B} = \{N\}$.*

Proof Every $\lambda \in F$ is a convex combination of extreme points $\lambda^1, \ldots, \lambda^k$:

$$\lambda = \alpha_1\lambda^1 + \cdots + \alpha_k\lambda^k.$$

For each λ^i, the inequality $v(N) \geqslant \sum_{A\in\mathcal{B}} \lambda^i_A v(A)$ is valid, therefore

$$\underbrace{\sum_{i=1}^k \alpha_i v(N)}_{v(N)} \geqslant \sum_{A\in\mathcal{B}} v(A) \underbrace{\sum_{i=1}^k \alpha_i \lambda^i_A}_{\lambda_A}.$$

Hence v is balanced, and by Theorem 3.7 its core is nonempty.

The converse statement is obvious.

It remains to prove that none of the inequalities in the system $\{\sum_{A\in\mathcal{B}} \lambda_A^{\mathcal{B}} v(A) \leqslant v(N), \mathcal{B}$ minimal balanced , $\mathcal{B} \neq \{N\}\}$ is redundant. From Farkas' Lemma II (Theorem 1.7), it suffices to prove that choosing any inequality $\sum_{A\in\mathcal{B}^*} \lambda_A^{\mathcal{B}^*} v(A) \leqslant v(N)$ in the system, a conic combination of the left members of the remaining ones cannot give the left member of the chosen inequality. In symbols, for all nonnegative coefficients $\gamma^{\mathcal{B}}$, with $\mathcal{B}$ minimal balanced and different from $\mathcal{B}^*$, the equalities

$$\sum_{\mathcal{B}\neq\mathcal{B}^*, \mathcal{B}\ni S} \gamma^{\mathcal{B}} \lambda_S^{\mathcal{B}} = \begin{cases} \lambda_S^{\mathcal{B}^*}, & S \in \mathcal{B}^* \\ 0, & \text{otherwise} \end{cases}$$

cannot hold simultaneously. Choose $S \in \mathcal{B}^*$. Then there exists some minimal balanced collection $\tilde{\mathcal{B}} \neq \mathcal{B}^*$ such that $\tilde{\mathcal{B}} \ni S$ and $\gamma^{\tilde{\mathcal{B}}} > 0$ (otherwise $0 < \lambda_S^{\mathcal{B}^*} = \sum_{\mathcal{B}\neq\mathcal{B}^*, \mathcal{B}\ni S} \gamma^{\mathcal{B}} \lambda_S^{\mathcal{B}}$ is not possible). Because $\tilde{\mathcal{B}} \neq \mathcal{B}^*$ and $\tilde{\mathcal{B}} \subset \mathcal{B}^*$ is impossible by minimality, there exists $T \in \tilde{\mathcal{B}}$, $T \notin \mathcal{B}^*$. Therefore

$$0 = \sum_{\mathcal{B}\neq\mathcal{B}^*, \mathcal{B}\ni T} \gamma^{\mathcal{B}} \lambda_T^{\mathcal{B}} \geqslant \gamma^{\tilde{\mathcal{B}}} \lambda_T^{\tilde{\mathcal{B}}} > 0,$$

a contradiction. □

Example 3.13 We enumerate the minimal balanced collections for $N = \{1, 2, 3, 4\}$. Every partition is obviously minimal, and there are 15 partitions of N. Apart these, the following are minimal balanced collections:

$$\mathcal{B} = \{\{1,2,3\}, \{1,2,4\}, \{1,3,4\}, \{2,3,4\}\}, \quad \text{with } \lambda = \left(\frac{1}{3}, \frac{1}{3}, \frac{1}{3}, \frac{1}{3}\right)$$

$$\mathcal{B} = \{\{1,2\}, \{1,3\}, \{1,4\}, \{2,3,4\}\}, \quad \text{with } \lambda = \left(\frac{1}{3}, \frac{1}{3}, \frac{1}{3}, \frac{2}{3}\right)$$

$$\mathcal{B} = \{\{1,2\}, \{1,3\}, \{2,3\}, \{4\}\}, \quad \text{with } \lambda = \left(\frac{1}{2}, \frac{1}{2}, \frac{1}{2}, 1\right)$$

$$\mathcal{B} = \{\{1,2\}, \{1,3,4\}, \{2,3,4\}\}, \quad \text{with } \lambda = \left(\frac{1}{2}, \frac{1}{2}, \frac{1}{2}\right)$$

$$\mathcal{B} = \{\{1,2,3\}, \{1\}, \{3,4\}, \{2,4\}\} \quad \text{with } \lambda = \left(\frac{1}{2}, \frac{1}{2}, \frac{1}{2}, \frac{1}{2}\right)$$

and those obtained by permutations. ◇

There exists an algorithm generating all minimal balanced collections (see Peleg [266]).

3.2.2 Extreme Points of the Core

Permutations on N correspond bijectively to maximal chains in the Boolean lattice $(2^N, \subseteq)$ in the following way. Letting σ be a permutation on N, we associate to it a maximal chain C^σ; i.e., a sequence of $n+1$ sets $\varnothing = A_0 \subset A_1 \subset \cdots \subset A_n = N$ defined by

$$\begin{aligned} A_1 &= \{\sigma(1)\} \\ A_2 \setminus A_1 &= \{\sigma(2)\} \\ \vdots &= \vdots \\ A_n \setminus A_{n-1} &= \{\sigma(n)\}, \end{aligned}$$

that is, $A_i = \{\sigma(1), \dots, \sigma(i)\}$. The usual convention is that i is the rank, and $\sigma(i)$ is the element of rank i. Next we associate to σ and v its *marginal vector* $x^{\sigma,v} \in \mathbb{R}^N$ defined by

$$x^{\sigma,v}_{\sigma(i)} = v(A_i) - v(A_{i-1}) \qquad (i \in N). \tag{3.8}$$

It is easy to check that this is equivalent to

$$\sum_{j=1}^{i} x^{\sigma,v}_{\sigma(j)} = x^{\sigma,v}(A_i) = v(A_i) \qquad (i \in N). \tag{3.9}$$

The next theorem asserts that the convex hull of the marginal vectors always contains the core. As we will see, this convex hull, sometimes called the *Weber set* [344], plays a central rôle in this chapter. We denote it by $\mathbf{Web}(v)$:

$$\mathbf{Web}(v) = \mathbf{conv}(x^{\sigma,v} : \sigma \in \mathfrak{S}(N)). \tag{3.10}$$

Theorem 3.14 *For any game v in $\mathcal{G}(N)$, $\mathbf{Web}(v) \supseteq \mathbf{core}(v)$.*

Proof (Derks [83]) Suppose there exists $x \in \mathbf{core}(v) \setminus \mathbf{Web}(v)$. By the separating hyperplane Theorem 1.5, there exists $y \in \mathbb{R}^n$ such that

$$\langle w, y \rangle > \langle x, y \rangle \qquad (w \in \mathbf{Web}(v)).$$

Let $\pi \in \mathfrak{S}(N)$ be a permutation such that $y_{\pi(1)} \geqslant y_{\pi(2)} \geqslant \cdots \geqslant y_{\pi(n)}$. In particular for $w = x^{\pi,v}$, we find

$$\langle x^{\pi,v}, y \rangle > \langle x, y \rangle. \tag{3.11}$$

Since $x \in \mathbf{core}(v)$, we have

$$\begin{aligned}\langle x^{\pi,v}, y\rangle &= \sum_{i=1}^{n} y_{\pi(i)}\big(v(\{\pi(1),\ldots,\pi(i)\}) - v(\{\pi(1),\ldots,\pi(i-1)\})\big)\\ &= y_{\pi(n)}v(N) - y_{\pi(1)}v(\varnothing) + \sum_{i=1}^{n-1}(y_{\pi(i)} - y_{\pi(i+1)})v(\{\pi(1),\ldots,\pi(i)\})\\ &\leqslant y_{\pi(n)}x(N) + \sum_{i=1}^{n-1}(y_{\pi(i)} - y_{\pi(i+1)})x(\{\pi(1),\ldots,\pi(i)\})\\ &= \sum_{i=1}^{n} y_{\pi(i)}x(\{\pi(1),\ldots,\pi(i)\}) - \sum_{i=2}^{n} y_{\pi(i)}x(\{\pi(1),\ldots,\pi(i-1)\})\\ &= \sum_{i=1}^{n} y_{\pi(i)}x_{\pi(i)} = \langle y, x\rangle,\end{aligned}$$

which contradicts (3.11). □

The next important theorem shows that the converse holds only for supermodular games, and makes clear the polyhedral structure of the core.

Theorem 3.15 (Structure of the core of supermodular games) *Let v be a game in $\mathcal{G}(N)$. The following propositions are equivalent.*

(i) *v is supermodular;*
(ii) *$x^{\sigma,v} \in \mathbf{core}(v)$ for all $\sigma \in \mathfrak{S}(N)$;*
(iii) *$\mathbf{core}(v) = \mathbf{Web}(v)$;*
(iv) *$\mathbf{ext}(\mathbf{core}(v)) = \{x^{\sigma,v} : \sigma \in \mathfrak{S}(N)\}$.*

Proof (i)⇔(ii): see proof of (i)⇔(ii) for Theorem 3.27, which is more general.

(iv)⇒(ii): Clear.

(ii)⇒(iii): (ii) implies that $\mathbf{conv}(\{x^{\sigma,v}\}_{\sigma\in\mathfrak{S}(N)}) \subseteq \mathbf{core}(v)$. The converse holds by Theorem 3.14.

(iii)⇒(iv): It remains to prove that every marginal vector is an extreme point of the core. Take σ a permutation and the corresponding maximal chain $\mathcal{C}^\sigma$. Observe that the n inequalities $x^{\sigma,v}(S) \geqslant v(S)$ for $S \in \mathcal{C}^\sigma$ are tight [see Eq. (3.9)]. This system of linear equalities is triangular with no zero on the diagonal, and therefore nonsingular. Hence it defines an extreme point. □

Remark 3.16

(i) Theorem 3.15 is well-known in game theory, where it was shown by Shapley [301] in 1971 (implications (i)⇒(ii) and (i)⇒(iv)) and Ichiishi [201] (implication (ii)⇒(i)). At the same time, Edmonds [122] (acknowledged in Shapley [301]) proved (i)⇒(iv) in terms of the base polyhedron (Sect. 3.1): for a monotonic submodular game, the marginal vectors are the extreme points of the base polyhedron (see also Lovász [226]). Even earlier in 1967, Dempster [77] proved that the marginal vectors are the extreme points of the core for belief measures (totally monotone and monotone games).

(ii) When games are not supermodular, the core (if nonempty) can have extreme points that are not marginal vectors. For example, taking $n = 3$ and the game defined by $v(12) = v(13) = 60$, $v(23) = 20$, $v(123) = 100$, and $v(S) = 0$ otherwise, it can be checked that $(20, 40, 40)$ is an extreme point of $\mathbf{core}(v)$ but not a marginal vector.

◇

3.2.3 Additivity Properties

The next theorem summarizes the properties relating addition of games and the (Minkovsky) sum of their cores and Weber sets.

Theorem 3.17 *The following holds.*

(i) $\mathbf{core}(v) + \mathbf{core}(v') \subseteq \mathbf{core}(v + v')$, *for all balanced games* v, v';
(ii) $\mathbf{Web}(v) + \mathbf{Web}(v') \supseteq \mathbf{Web}(v + v')$, *for all games* v, v';
(iii) *If* v, v' *are supermodular, then* $\mathbf{core}(v) + \mathbf{core}(v') = \mathbf{core}(v + v')$.

Proof

(i) Take $x \in \mathbf{core}(v)$ and $x' \in \mathbf{core}(v')$. Then for any $S \subseteq N$,

$$(x + x')(S) = x(S) + x'(S) \geqslant v(S) + v'(S) = (v + v')(S),$$

with equality if $S = N$. Therefore, $x + x' \in \mathbf{core}(v + v')$.

(ii) Take $x \in \mathbf{Web}(v + v')$. Then x is a convex sum of marginal vectors of $v + v'$:

$$x = \sum_i \alpha_i x^{\sigma_i, v+v'} = \sum_i \alpha_i x^{\sigma_i, v} + \sum_i \alpha_i x^{\sigma_i, v'} = y + y'$$

where y, y' are elements of $\mathbf{Web}(v)$ and $\mathbf{Web}(v')$ respectively, proving that $x \in \mathbf{Web}(v) + \mathbf{Web}(v')$.

(iii) If v, v' are convex, it follows from Theorem 3.15 that the core and the Weber set coincide for these games, hence the result follows by (i) and (ii).

□

The following examples show that the inclusions in the above theorem may be strict.

Example 3.18 Consider $N = \{1, 2, 3\}$ and the following games v, v': The cores of v, v' are reduced to the singletons $\{(1, 1, 0)\}$ and $\{(1, 0, 1)\}$, respectively, however $(1, 1, 0) + (1, 0, 1) \neq (3, 0, 1) \in \mathbf{core}(v + v')$.

S	1	2	3	12	13	23	123
$v(S)$	1	0	0	2	1	0	2
$v'(S)$	0	0	1	1	2	0	2
$(v + v')(S)$	1	0	1	3	3	0	4

As for the Weber set, consider the two following games: Observe that $v + v'$

S	1	2	3	12	13	23	123
$v(S)$	1	0	-1	1	0	1	1
$v'(S)$	0	0	1	0	1	-1	0
$(v + v')(S)$	1	0	0	1	1	0	1

is additive, therefore its Weber set reduces to $\{(1, 0, 0)\}$. However, $(0, 2, -1)$ and $(0, 0, 0)$ are marginal vectors of v and v', respectively, and their sum is not equal to $(1, 0, 0)$. ◇

3.3 The Core of Games on Set Systems

The study of the core for games with restricted cooperation is much more complex because most often the core is no longer a bounded polyhedron: it is often unbounded, and may even have no vertices. The readers can consult also the survey paper [170] of the author for other results.

We consider throughout this section a set system $\mathcal{F}$ (see Sect. 2.19 for all definitions of notions used hereafter) that contains N. The last assumption is necessary otherwise the condition $x(N) = v(N)$ does not make sense. For a good understanding of Sects. 3.3.2–3.3.5, we recommend to read Sect. 1.3.4 before.

3.3.1 *Nonemptiness of the Core*

The results of Sect. 3.2.1 apply without change: it suffices to replace 2^N by $\mathcal{F}$ everywhere in the results and proofs. We refer to Faigle [128] for an alternative equivalent view.

A simple result is given in the next theorem. First, we need the notion of connectedness. We say that i and j are *connected* in $\mathcal{F}$ if there is a sequence of elements $i_1, \ldots, i_k$ with $i_1 = i$ and $i_k = j$ such that for each $\ell = 1, \ldots, k-1$, either for each $S \in \mathcal{F}$, $i_\ell \in S$ implies $i_{\ell+1} \in S$, or for each $S \in \mathcal{F}$, $i_{\ell+1} \in S$ implies $i_\ell \in S$. Then $\mathcal{F}$ is *connected* if every two elements in N are connected.

When $\mathcal{F} = \mathcal{O}(N, \preceq)$, i.e., a distributive lattice of height n, the above definition simplifies as follows: i and j are connected if there exists a sequence $i_1, \ldots, i_k$ with $i_1 = i$ and $i_k = j$ such that for each $\ell = 1, \ldots, k-1$, either $i_\ell \prec i_{\ell+1}$ or $i_{\ell+1} \prec i_\ell$. In short, i and j are connected in the sense of graph theory in the Hasse diagram of $(N, \preceq)$.

Theorem 3.19 *(Grabisch and Sudhölter [181]) Let $\mathcal{F}$ be a set system containing N. The following holds.*

(i) If $\mathcal{F}$ is connected, then for any game v in $\mathcal{G}(N, \mathcal{F})$, $\mathbf{core}(v) \neq \varnothing$.
(ii) If $\mathcal{F}$ is not connected, then there exists a game v on $\mathcal{F}$ such that $\mathbf{core}(v) = \varnothing$.

Proof (i) Let $\mathcal{B} \subseteq \mathcal{F}$ be a balanced collection and $(\lambda_A)_{A\in\mathcal{B}}$ be a system of balancing weights. In view of Theorem 3.12 it suffices to show that $\mathcal{B} = \{N\}$ is the only minimal balanced collection. Let $R \in \mathcal{B}$, $R \neq \varnothing$. Then there exists $i \in R$. In order to show that $R = N$, let $j \neq i$. As $\mathcal{F}$ is connected, there exist $i_1, \ldots, i_k \in N$ such that $i_1 = i, i_k = j$, and for all $\ell = 1, \ldots, k-1$, either for all $S \in \mathcal{F}$ $i_\ell \in S$ implies $i_{\ell+1} \in S$, or for all $S \in \mathcal{F}$ $i_{\ell+1} \in S$ implies $i_\ell \in S$. We show that $i_\ell \in R$ by induction on ℓ. For $\ell = 1$ nothing has to be proved. Assume that $i_\ell \in R$ for some $\ell < k$. If $i_\ell \in S$ implies $i_{\ell+1} \in S$, then $i_{\ell+1} \in R$. If $i_{\ell+1} \in S$ implies $i_\ell \in S$, any $S \in \mathcal{B}$ with $i_{\ell+1} \in S$ also contains i_ℓ. As $\mathcal{B}$ is separating (see Sect. 3.2.1), $i_{\ell+1} \in R$.

(ii) Let $\mathcal{F}$ be non-connected and $v \in \mathcal{G}(\mathcal{F})$ that satisfies

$$v(N) < \sum\{v(A) \,:\, A \text{ is a connected component of } \mathcal{F}\},$$

where a connected component of $\mathcal{F}$ is any connected maximal subset of $\mathcal{F}$ (i.e., any superset is not connected). Clearly, $\mathbf{core}(v) = \varnothing$. □

3.3.2 Boundedness

We study in this section whether the core, whenever nonempty, is pointed (i.e., has vertices), and if yes, if it is bounded. We recall from Sect. 1.3.4 that this amounts to studying the recession cone of the core:

$$\mathbf{core}(0) = \{x \in \mathbb{R}^N \,:\, x(N) = 0, x(S) \geqslant 0, \forall S \in \mathcal{F}\}.$$

The next example shows that one cannot expect that the core is pointed or bounded in general.

Example 3.20 Consider $N = \{1, 2, 3\}$ and the set system $\mathcal{F} = \{\varnothing, 1, 123\}$. Since a vertex of the core of a game v is defined by a set of at least three equalities in $\mathbf{core}(v) = \{x \in \mathbb{R}^3 \,:\, x_1 \geqslant v(\{1\}), x(N) = v(N)\}$, there are clearly not enough sets in $\mathcal{F}$ to achieve this. Hence the core cannot be pointed.

Let us now consider $\mathcal{F}' = \{\varnothing, 1, 2, 123\}$. The recession cone **core**(0) reads

$$x_1 \geqslant 0$$
$$x_2 \geqslant 0$$
$$x_1 + x_2 + x_3 = 0.$$

The core, if nonempty, is pointed because putting equality everywhere leads to a system whose unique solution is **0**. Now, the core is not bounded because the vector $(1, 0, -1)$ belongs to **core**(0) (it is the supporting vector of a ray). ◇

From the theory of polyhedra, the general conditions for the core (when nonempty) to be pointed is that the system of linear equalities in real variables

$$x(S) = 0 \qquad (S \in \mathcal{F} \setminus \{\varnothing\}),$$

should have **0** as unique solution. Equivalently, by Gauss elimination, the condition is that for all $i \in N$, there exists a linear combination of the equations yielding $x_i = 0$:

$$\forall i \in N, \forall S \in \mathcal{F} \setminus \{\varnothing\}, \exists \alpha_S^i \in \mathbb{R}, \sum_{S \in \mathcal{F} \setminus \{\varnothing\}} \alpha_S^i 1_S = 1_i. \tag{3.12}$$

If this condition is satisfied, we say that the set system $\mathcal{F}$ is *nondegenerate*, otherwise it is said to be *degenerate*.

Several simple sufficient conditions for nondegenerateness or degenerateness are noteworthy:

(i) $\mathcal{F}$ is degenerate if $|\mathcal{F} \setminus \{\varnothing\}| < n$ (this is Example 3.20);
(ii) $\mathcal{F}$ is degenerate if there exists a *macro-element*, that is, a subset $K \subseteq N$, $|K| > 1$, such that either $K \subseteq S$ or $K \cap S = \emptyset$ for every nonempty $S \in \mathcal{F}$ [by (3.12)];
(iii) $\mathcal{F}$ is nondegenerate if $\mathcal{F}$ contains all singletons [by (3.12)];
(iv) $\mathcal{F}$ is nondegenerate if $\mathcal{F}$ contains a chain of length n (e.g., if $\mathcal{F}$ is regular, in particular, if $\mathcal{F} = \mathcal{O}(N, \preceq)$; i.e., a distributive lattice of height n). Indeed, a chain of length n is a maximal chain $\varnothing = S_0, S_1, \dots, S_n = N$, hence all $1_{\{i\}}$'s can be recovered from $1_{S_j} - 1_{S_{j-1}}$, for two consecutive sets S_j, S_{j-1} in the chain.

The following result shows that if $\mathcal{F}$ is closed under intersection, the absence of macro-element is a necessary and sufficient condition to be nondegenerate.

Theorem 3.21 (Necessary and sufficient condition to be nondegenerate) *(Faigle et al. [133]) Let $\mathcal{F}$ be closed under intersection. Then $\mathcal{F}$ is nondegenerate if and only if it contains no macro-element.*

Proof The "only if" part is obvious since the presence of a macro-element implies degeneracy.

Suppose that $\mathcal{F}$ is closed under intersection and has no macro-element. We prove by induction on $n = |N|$ that it is nondegenerate. The assertion is easily checked for $n = 1$, with the only possible set system $\{\emptyset, \{1\}\}$. Suppose the assertion holds till some value $n-1$ and let us prove it for n.

CLAIM: there exists $i \in N$ such that $\{i\} \in \mathcal{F}$.

PROOF OF THE CLAIM: Since $\mathcal{F}$ has no macro-element, necessarily every atom is a singleton. Indeed, suppose *per contra* that S is an atom, with $|S| > 1$. Because S is not a macro-element, there exists $T \in \mathcal{F}$ separating S; i.e., $j \in T \not\ni k$ for some $j, k \in S$. Since $\mathcal{F}$ is closed under intersection, it follows that $S \cap T \in \mathcal{F}$ and $\emptyset \neq S \cap T \subsetneq S$, a contradiction with the fact that S is an atom. □

Consider then $\mathcal{F}^{-i} = \{S \subseteq N \setminus i : S \text{ or } S \cup i \in \mathcal{F}\}$ on $N \setminus i$, the collection of sets obtained from $\mathcal{F}$ by removing i in every set. Note that $\emptyset \in \mathcal{F}^{-i}$. We prove that $\mathcal{F}^{-i}$ is a set system containing $N \setminus i$ without macro-elements.

- $\mathcal{F}^{-i} \ni N \setminus i$: clear because $N \in \mathcal{F}$.
- $\mathcal{F}^{-i}$ is closed under intersection: take $S, S' \in \mathcal{F}^{-i}$. Then three cases arise. If $S, S' \in \mathcal{F}$, then $S \cap S' \in \mathcal{F}$ and $i \notin S \cap S'$, hence $S \cap S' \in \mathcal{F}^{-i}$. If $S \in \mathcal{F}$ and $S' \cup i \in \mathcal{F}$, then $i \notin S \cap (S' \cup i) \in \mathcal{F}$, and therefore $S \cap (S' \cup i) = S \cap S' \in \mathcal{F}^{-i}$. Lastly, if $S \cup i, S' \cup i \in \mathcal{F}$, then $i \in (S \cup i) \cap (S' \cup i) \in \mathcal{F}$, therefore $((S \cup i) \cap (S' \cup i)) \setminus i = S \cap S' \in \mathcal{F}^{-i}$.
- $\mathcal{F}^{-i}$ has no macro-element: suppose $K \subseteq N \setminus i$ is a macro-element in $\mathcal{F}^{-i}$. Take $S \in \mathcal{F}^{-i}$. Then either $S \cap K = \emptyset$ or $S \supseteq K$. If $S \in \mathcal{F}$, then $S \cap K = \emptyset$ or $S \supseteq K$ remains true. If $S \cup i \in \mathcal{F}$, then $(S \cup i) \cap K = \emptyset$ or $S \cup i \supseteq K$ is true because $K \not\ni i$. Hence K is a macro-element in $\mathcal{F}$, a contradiction.

Then $\mathcal{F}^{-i}$ is a set system containing $N \setminus i$ closed under intersection without macro-element on $N \setminus i$, and by the induction hypothesis, $\mathcal{F}^{-i}$ is nondegenerate; i.e., the system of equations $x(S) = 0, S \in \mathcal{F}^{-i}$ has a unique solution $x = 0$. Finally, observe that the system $x(S) = 0, S \in \mathcal{F}$ differs from the previous one only by the adjunction of x_i in some lines. Since $\{i\} \in \mathcal{F}$, the line $x_i = 0$ makes the two systems equivalent. Therefore, $\mathcal{F}$ is nondegenerate. □

A useful result is the following.

Lemma 3.22 (Derks and Reijnierse [87]) *The recession cone* **core**(0) *of a game on* $(N, \mathcal{F})$ *with* $N \in \mathcal{F}$ *is a linear subspace if and only if* $\mathcal{F} \setminus \{\varnothing, N\}$ *is a balanced collection or is empty.*

Proof We set $\mathcal{F}^\circ = \mathcal{F} \setminus \{\varnothing, N\}$ for simplicity.

$\Leftarrow$) Suppose $\mathcal{F}^\circ$ is balanced, and take $(\lambda_S)_{S \in \mathcal{F}^\circ}$ a system of balancing weights. Take any $x \in \mathbf{core}(0)$ and any $T \in \mathcal{F}^\circ$. We have

$$-x(T) = x^\top \cdot \frac{1}{\lambda_T}\Big(\sum_{\substack{S \in \mathcal{F}^\circ \\ S \neq T}} \lambda_S 1_S - 1_N\Big) = \frac{1}{\lambda_T} \sum_{\substack{S \in \mathcal{F}^\circ \\ S \neq T}} \lambda_S x(S) \geqslant 0,$$

proving that $-x$ is an element of **core**(0). Since **core**(0) is a cone, it is therefore a linear subspace.

⇒) Suppose that **core**(0) is a linear subspace. Then $x \in \mathbf{core}(0)$ implies that $-x \in \mathbf{core}(0)$, and it follows that inequalities $x(S) \geqslant 0, S \in \mathcal{F}^\circ, x(N) = 0$ imply the inequality $x(T) \leqslant 0$ for any $T \in \mathcal{F}^\circ$. From Farkas' Lemma II (Theorem 1.7), it follows that, for any $T \in \mathcal{F}^\circ$, there exist nonnegative constants λ_S^T for all $S \in \mathcal{F}^\circ$ and a real constant λ_N^T such that

$$\sum_{S \in \mathcal{F}^\circ} \lambda_S^T 1_S + \lambda_N^T 1_N = -1_T. \tag{3.13}$$

Summing for all T's yields

$$\lambda_N 1_N = -\sum_{S \in \mathcal{F}^\circ} 1_S\Big(1 + \sum_{T \in \mathcal{F}^\circ} \lambda_S^T\Big), \tag{3.14}$$

with $\lambda_N = \sum_{T \in \mathcal{F}^\circ} \lambda_N^T$. Observe from (3.13) that $\lambda_N^T < 0$ for all T, hence $\lambda_N < 0$. Dividing both members of Eq. (3.14) by $-\lambda_N$ yields

$$\sum_{S \in \mathcal{F}^\circ} 1_S \lambda'_S = 1_N$$

with $\lambda'_S = \frac{1+\sum_{T \in \mathcal{F}^\circ} \lambda_S^T}{-\lambda_N} > 0$. Hence, $\mathcal{F}^\circ$ is balanced. □

From the theory of polyhedra again, we know that the core is bounded if and only if its recession cone reduces to $\{\mathbf{0}\}$. Combining Lemma 3.22 and previous facts, we find:

Theorem 3.23 (Boundedness of the core) *Let v be a balanced game on a set system $\mathcal{F}$ containing N. Then* **core**(v) *is bounded if and only if $\mathcal{F}$ is nondegenerate and $\mathcal{F} \setminus \{\varnothing, N\}$ is balanced.*

This result was proved by Derks and Reijnierse [87].

3.3.3 Extremal Rays

When the core is unbounded, it is possible to find its extremal rays in the case where $\mathcal{F}$ is closed under union and intersection; i.e., basically $\mathcal{F} = \mathcal{O}(N, \preceq)$, which is a distributive lattice of height n (see Sect. 2.19.2). The next theorem was shown by Tomizawa [326] (see also Fujishige [149, Theorem 3.26]), and by Derks and Gilles [84]. We recall that for $i, j \in N$, $j \prec\!\cdot\, i$ means $j \prec i$ and $j \preceq k \preceq i$ implies $k = i$ or $k = j$.

Theorem 3.24 (Extremal rays of the core) *Let $\mathcal{F} = \mathcal{O}(N, \preceq)$ be a set system. The recession cone of the core reads*

$$\mathbf{core}(0) = \mathbf{cone}(1_{\{j\}} - 1_{\{i\}} \;:\; i, j \in N \text{ such that } j \prec\!\cdot\, i).$$

Proof Take $i,j \in N$ such that $j \prec\!\cdot\, i$. First, we prove that $r = 1_{\{j\}} - 1_{\{i\}}$ is a ray; i.e., $r \in \mathbf{core}(0)$. Since any $S \in \mathcal{F}$ is a downset, $i \in S$ implies $j \in S$. Hence in any case $r(S) \geqslant 0$, and $r(N) = 0$.

Second, we show that $\mathbf{core}(0)$ is included into the cone generated by the above rays. We use cone duality for this purpose (see Sect. 1.3.6). We claim that for $w \in \mathbb{R}^N$, $w^\top x$ is bounded for all $x \in \mathbf{core}(0)$ if $w^\top(1_{\{j\}} - 1_{\{i\}}) \leqslant 0$ for all $i,j \in N$ such that $j \prec\!\cdot\, i$. Then, by Lemma 1.10, this is equivalent to say that $w^\top x \leqslant 0$ for all $x \in \mathbf{core}(0)$ if $w^\top(1_{\{j\}} - 1_{\{i\}}) \leqslant 0$ for all $i,j \in N$ such that $j \prec\!\cdot\, i$, which means that the rays $1_{\{j\}} - 1_{\{i\}}$ generate $\mathbf{core}(0)$.

Finally, extremality of these rays is immediate because they generate $\mathbf{core}(0)$ and none of them can be written as a conic combination of the others.

PROOF OF THE CLAIM: The conditions in the claim for w read $w_j \leqslant w_i$ for $j \prec\!\cdot\, i$, hence there exists some linear extension of $\preceq$; i.e., a reordering $\sigma(1), \ldots, \sigma(n)$ of the elements of N such that $i \prec j$ implies $\sigma(i)$ is ranked before $\sigma(j)$, such that

$$w_{\sigma(1)} \leqslant w_{\sigma(2)} \leqslant \cdots \leqslant w_{\sigma(n)}.$$

Note that by construction $F_i = \{\sigma(1), \ldots, \sigma(i)\} \in \mathcal{F}$ for $i = 1, \ldots, n$. Define now the vector $y \in \mathbb{R}^{\mathcal{F}}$ by

$$y_{F_n} = -w_{\sigma(n)}, y_{F_{n-j}} = -w_{\sigma(n-j)} + w_{\sigma(n-j+1)} \qquad (j = 1, \ldots, n-1)$$

and $y_F = 0$ if $F \neq F_i$ for all i. By construction we find that $y_F \geq 0$ for all $F \in \mathcal{F}$, $F \neq N$, and $\sum_{F \ni i} y_F = -w_i$, for all $i \in N$. Hence, y is a feasible solution for the linear program

$$\min \mathbf{0}^\top z \text{ s.t. } -A^\top z = w, \quad z_F \geqslant 0 \text{ for all } F \neq N,$$

with A the matrix defining the recession cone; i.e., $\mathbf{core}(0) = \{x \in \mathbb{R}^N : -Ax \leqslant \mathbf{0}\}$. Observe that this linear program is the dual of $\max w^\top x$ s.t. $x \in \mathbf{core}(0)$. Hence, by weak duality, we deduce that $w^\top x$ is bounded.[4] □

Note that this shows that when $\mathcal{F}$ is a distributive lattice, the core of a balanced game is bounded if and only if $\mathcal{F} = 2^N$.

3.3.4 Extreme Points

Extreme points are known in the case where $\mathcal{F} = \mathcal{O}(N)$ and v is supermodular. In other cases, little can be said.

We recall from Sect. 3.2.2 that marginal vectors are defined w.r.t. permutations on N or equivalently maximal chains on 2^N. Therefore, the definition of marginal

---

[4]Thanks are due to Ulrich Faigle for providing this proof.

vectors remains unchanged provided $\mathcal{F}$ is regular: all maximal chains have length n (this is in particular the case when $\mathcal{F} = \mathcal{O}(N, \preceq)$). If $\mathcal{F}$ is not regular, some alternative notion has to be found (Remark 3.33). Supposing $\mathcal{F}$ to be regular, we denote by $\mathfrak{S}(\mathcal{F})$ the set of *feasible* permutations, i.e., those corresponding to the maximal chains of $\mathcal{F}$, and for any $\sigma \in \mathfrak{S}(\mathcal{F})$, the corresponding marginal vector is denoted by $x^{\sigma,v}$. We observe that (3.9) still holds; i.e.,

$$x^{\sigma,v}(A_i) = v(A_i) \tag{3.15}$$

for any permutation $\sigma \in \mathfrak{S}(\mathcal{F})$ and the sets $A_i = \{\sigma(1), \ldots, \sigma(i)\}$ induced by the permutation.

The first result is a generalization of Theorem 3.14. To this end, we define the Weber set as the convex hull of marginal vectors induced by the feasible permutations:

$$\mathbf{Web}(v, \mathcal{F}) = \mathbf{conv}(\{x^{\sigma,v} : \sigma \in \mathfrak{S}(\mathcal{F})\}).$$

Theorem 3.25 *(Derks and Gilles [84]) Let v be a game on $(N, \mathcal{F})$ with $\mathcal{F} = \mathcal{O}(N, \preceq)$. Then*

$$\mathbf{core}(v) \subseteq \mathbf{Web}(v, \mathcal{F}) + \mathbf{core}(0).$$

Proof Suppose there exists $x \in \mathbf{core}(v)$ such that $x \notin \mathbf{Web}(v, \mathcal{F}) + \mathbf{core}(0)$. Since $\mathbf{Web}(v, \mathcal{F}) + \mathbf{core}(0)$ is a convex polyhedron, by the separating hyperplane Theorem 1.5, there exists $y \in \mathbb{R}^N$ such that

$$\langle y, x\rangle < \langle y, w + r\rangle \qquad (w \in \mathbf{Web}(v, \mathcal{F}), r \in \mathbf{core}(0)).$$

By Theorem 3.24, this implies

$$\langle y, x\rangle < \langle y, w\rangle + \alpha\langle y, 1_{\{j\}} - 1_{\{i\}}\rangle = \langle y, w\rangle + \alpha(y_j - y_i) \tag{3.16}$$

for all $w \in \mathbf{Web}(v, \mathcal{F})$, all $\alpha \geqslant 0$, and all $i, j \in N$ such that $j \prec\!\cdot\, i$ in $(N, \preceq)$. It follows that $y_j \geqslant y_i$ for all $j \prec\!\cdot\, i$, and by transitivity

$$y_j \geqslant y_i \qquad (j \preceq i). \tag{3.17}$$

Suppose for ease of notation that $y_1 \geqslant y_2 \geqslant \cdots \geqslant y_n$ (otherwise apply some permutation), and define $A_k = \{1, \ldots, k\}$ for $k = 1, \ldots, n$. Under these conditions, $1, 2, \ldots, n$ is a linear extension of $(N, \preceq)$. Therefore $A_k \in \mathcal{F}$ for $k = 1, \ldots, n$, and these sets form a maximal chain in $\mathcal{F}$, inducing the marginal vector $x^{\mathrm{Id},v}$ w.r.t. the permutation $\sigma = \mathrm{Id}$. Letting $w = x^{\mathrm{Id},v}$ and $\alpha = 0$ in (3.16) yields

$$\begin{aligned}\langle y, x\rangle < \langle y, x^{\mathrm{Id},v}\rangle &= \sum_{i=1}^{n}\Big(v(A_i) - v(A_{i-1})\Big)y_i \\ &= v(N)y_n + \sum_{i=1}^{n-1} v(A_i)\underbrace{(y_i - y_{i+1})}_{\geqslant 0} \\ &\leqslant x(N)y_n + \sum_{i=1}^{n-1} x(A_i)(y_i - y_{i+1}) = \langle x, y\rangle,\end{aligned}$$

a contradiction. □

Remark 3.26 Though $\mathbf{core}(v) = \mathbf{conv}(\mathbf{ext}(\mathbf{core}(v)) + \mathbf{core}(0)$, the readers should not deduce that the convex part of the core is included in the Weber set: $\mathbf{conv}(\mathbf{ext}(\mathbf{core}(v)) \subseteq \mathbf{Web}(v, \mathcal{F})$ does *not* hold in general, as shown by the following simple example.

Take $N = \{1, 2, 3\}$, $\mathcal{F} = \{\varnothing, 1, 3, 13, 123\}$ induced by the poset $1 \prec 2$, $3 \prec 2$, and consider the game v defined by $v(1) = v(3) = v(13) = 1$, and $v(123) = 10$. Remark that v is not supermodular. There are two marginal vectors $(1, 9, 0)$ and $(0, 9, 1)$. Now, these two marginal vectors are vertices, but there is a third vertex, $x^3 = (1, 8, 1)$, so it is not in the convex hull of the marginal vectors.

However, observe that $x^3 = (1, 8, 1) = (1, 9, 0) + (0, -1, 1)$, where $(0, -1, 1)$ is an extremal ray, in accordance with Theorem 3.25. ◇

The next result generalizes Theorem 3.15.

Theorem 3.27 (Extreme points of the core for supermodular games) *Let v be a game on $(N, \mathcal{F})$ with $\mathcal{F} = \mathcal{O}(N, \preceq)$. The following propositions are equivalent.*

(i) *v is supermodular;*
(ii) *$x^{\sigma,v} \in \mathbf{core}(v)$ for all $\sigma \in \mathfrak{S}(\mathcal{F})$;*
(iii) *$\mathbf{core}(v) = \mathbf{Web}(v, \mathcal{F}) + \mathbf{core}(0)$;*
(iv) *$\mathbf{ext}(\mathbf{core}(v)) = \{x^{\sigma,v} : \sigma \in \mathfrak{S}(\mathcal{F})\}$.*

Proof ((i)⇔(ii): see Grabisch and Sudhölter [182])

(i)⇒(ii) Assume that v is supermodular, let $S \in \mathcal{F} \setminus \{\emptyset\}$, and $\sigma \in \mathfrak{S}(\mathcal{F})$. We have to show that

$$x^{\sigma,v}(S) \geqslant v(S). \tag{3.18}$$

Let $i_1, \ldots, i_s \in S$, $s = |S|$, be chosen so that $\sigma^{-1}(i_1) < \cdots < \sigma^{-1}(i_s)$. Then $T_k = \{i_1, \ldots, i_k\} = A_{\sigma^{-1}(i_k)} \cap S \in \mathcal{F}$ for any $k = 1, \ldots, s$, using the notation of (3.15). By supermodularity, we get:

$$v(A_{\sigma^{-1}(i_k)}) - v(A_{\sigma^{-1}(i_k)} \setminus i_k) \geqslant v(T_k) - v(T_{k-1}) \text{ for all } k = 1, \ldots, s,$$

where $T_0 = \varnothing$. Summing up all these inequalities and using (3.15) yield (3.18).

(ii)⇒(i) Let v be a game and assume that $x^{\sigma,v} \in \mathbf{core}(v)$ for all $\sigma \in \mathfrak{S}(\mathcal{F})$. Let $S, T \in \mathcal{F}$ so that $S \setminus T \neq \varnothing \neq T \setminus S$, and rank the elements in N such that

$$N = \{\underbrace{i_1, \ldots, i_k}_{S \cap T}, \underbrace{i_{k+1}, \ldots, i_l}_{T \setminus S}, \underbrace{i_{l+1}, \ldots, i_s}_{S \setminus T}, \underbrace{i_{s+1}, \ldots, i_n}_{N \setminus (S \cup T)}\}$$

and for any $j \in N$, $\{i_1, \ldots, i_j\} \in \mathcal{F}$. Then the permutation σ defined by $\sigma(j) = i_j$ for any $j \in N$ is a linear extension. Hence

$$\begin{aligned} v(S) \leqslant & \sum_{i \in S} x_i^{\sigma,v} = \sum_{i \in S} \left(v(A_{\sigma^{-1}(i)}) - v(A_{\sigma^{-1}(i)} \setminus i)\right) \\ = & \sum_{j=1}^{r} \left(v(\{i_1, \ldots, i_j\}) - v(\{i_1, \ldots, i_{j-1}\})\right) + \sum_{j=t+1}^{q} \left(v(T \cup \{i_{t+1}, \ldots, i_j\})\right. \\ & \left. - v(T \cup \{i_{t+1}, \ldots, i_{j-1}\})\right) \\ = & \; v(S \cap T) + v(S \cup T) - v(T), \end{aligned}$$

so that the proof is complete.

(ii)⇒(iii) By (ii) and the decomposition of polyhedra, $\mathbf{Web}(v, \mathcal{F}) + \mathbf{core}(0) \subseteq \mathbf{core}(v)$ holds. The reverse inclusion holds by Theorem 3.25.

(iii)⇒(iv) Identical to (iii)⇒(iv) in Theorem 3.15, thanks to (3.15).

(iv)⇒(ii) is obvious. □

Remark 3.28 These results have been shown independently by several authors. Theorem 3.25 can be found in Faigle and Kern [135, Sect. 5], and it is also mentioned in Derks and Gilles [84], where the result is shown for acyclic permission structures (which correspond to distributive lattices of the form $\mathcal{O}(N)$; see Algaba et al. [6]). Now, Theorem 3.27 (i)⇒(iv) was shown by Fujishige and Tomizawa [150] (see also Fijishige [149, Theorem 3.22]), and also by Derks and Gilles [84]. ◇

In the case of the more general regular set systems, the above properties do not hold in general. The following example shows that the statement of Theorem 3.25 is no longer true in this case.

Example 3.29 (Bilbao et al. [26]) Consider the following set system (regular set lattice but not distributive, since it contains a pentagon, figured by the red circles), with the values of the game v given in parentheses.

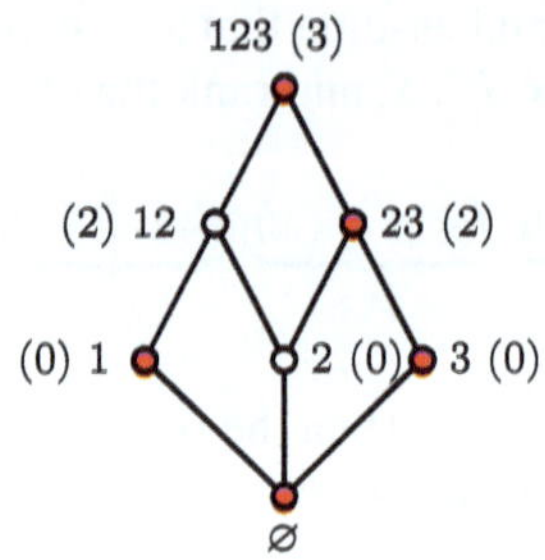

The core is determined by

$$\begin{aligned} x_1 + x_2 &\geqslant 2 \\ x_2 + x_3 &\geqslant 2 \\ x_1 + x_2 + x_3 &= 3 \\ x_1, x_2, x_3 &\geqslant 0. \end{aligned}$$

Observe that $\mathcal{F}$ is nondegenerate and balanced, hence the core is bounded. Its vertices are $(0, 2, 1), (1, 2, 0), (0, 3, 0), (1, 1, 1)$. The marginal vectors associated with the 4 maximal chains are (0,2,1), (2,0,1), (1,0,2) and (1,2,0). Clearly, $\mathbf{core}(v) \nsubseteq \mathbf{Web}(v, \mathcal{F})$ and $\mathbf{Web}(v, \mathcal{F}) \nsubseteq \mathbf{core}(v)$. ◇

Theorem 3.30 *Let $\mathcal{F}$ be a regular set system that is a set lattice,*[5] *with infimum and supremum $\wedge, \vee$, and consider a game v on $(N, \mathcal{F})$ that is monotone and supermodular on $\mathcal{F}$ w.r.t $\vee, \wedge$. Then $\mathbf{Web}(v, \mathcal{F}) \subseteq \mathbf{core}(v)$, and all marginal vectors $x^{\sigma,v}$ are extreme points of the core.*

Proof Take some $\sigma \in \mathfrak{S}(\mathcal{F})$, and consider the marginal vector $x^{\sigma,v}$ and the associated maximal chain $C = \{S_0, S_1, \ldots, S_n\}$ where $S_0 = \varnothing \subset S_1 \cdots \subset S_n = N$. It suffices to show that $x^{\sigma,v} \in \mathbf{core}(v)$; i.e, $x^{\sigma,v}(T) \geqslant v(T)$ for all $T \in \mathcal{F}$, and $x^{\sigma,v}(N) = v(N)$. The latter property is immediate from the definition of marginal vectors.

Denote by $\vee, \wedge$ the supremum and infimum in the lattice $\mathcal{F}$. If $T \in C$, then $x^{\sigma,v}(T) = v(T)$ by (3.15). Suppose then that $T \notin C$. Call S_p the smallest set in C such that $T \subset S_p$, and put $\{i\} = S_p \setminus S_{p-1}$. Note that $T \vee S_{p-1} = S_p$ and $T \ni i$.

[5] See Remark 2.113(iii).

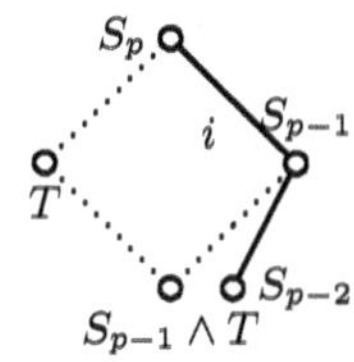

We prove $x^{\sigma,v}(T) \geqslant v(T)$ by induction on $p \geqslant 2$.

If $p = 2$, then $T = \{i\}$, $S_1 = \{j\}, j \neq i$. We have $v(\{i,j\}) + v(\varnothing) \geqslant v(\{j\}) + v(\{i\})$. By (3.15) and supermodularity,

$$x_i^{\sigma,v} = x^{\sigma,v}(\{i,j\}) - x_j^{\sigma,v} = v(\{i,j\}) - v(\{j\}) \geqslant v(\{i\}).$$

Let us suppose the property true for $p \leqslant q$ and prove it for $p = q + 1 \leqslant n$. By supermodularity and (3.15) again,

$$x^{\sigma,v}(S_p) - x^{\sigma,v}(S_{p-1}) + v(S_{p-1} \wedge T) \geqslant v(T),$$

hence $x_i^{\sigma,v} + v(S_{p-1} \wedge T) \geqslant v(T)$. By induction hypothesis, $x^{\sigma,v}(S_{p-1} \wedge T) \geqslant v(S_{p-1} \wedge T)$, hence

$$x_i^{\sigma,v} + x^{\sigma,v}(S_{p-1} \wedge T) \geqslant v(T).$$

Observe that $(S_p \wedge T) \cup i \subseteq T$. Now v monotone implies $x^{\sigma,v} \in \mathbb{R}^n_+$, hence

$$x^{\sigma,v}(T) \geqslant x_i^{\sigma,v} + x^{\sigma,v}(S_{p-1} \wedge T) \geqslant v(T),$$

which is the desired result.

Finally, since (3.15) holds, each marginal vector is an extreme point because it is the unique solution of a triangular system of n linear equalities. □

This result was shown by Xie and Grabisch [349]. The next two examples show that (1) Monotonicity cannot be left out, and (2) The conditions of the theorem are not strong enough to ensure that all extreme points are marginal vectors.

Example 3.31 Consider $N = \{1, 2, 3, 4, 5\}$ and the regular set lattice $\mathcal{F}$ depicted in Fig. 3.1. Define v on $\mathcal{F}$ as follows: $v(S) = 0$ for all $S \in \mathcal{F}$ except $v(124) = -1$. Then v is supermodular but not monotonic. Consider the maximal chain $\{\varnothing, 1, 14, 124, 1245, 12345\}$. Its associated marginal vector is $x = (0, -1, 0, 0, 1)$. Observe that $x(123) = -1 < v(123)$, hence $x \notin$ **core**(v). ◇

Example 3.32 (Example 3.29 cont'd) Change the values of the game as follows: $v(12) = v(23) = 1$. Then v is supermodular and the four marginal vectors are $(0, 1, 2), (1, 0, 2), (2, 0, 1)$ and $(2, 1, 0)$. However, observe that $(0, 3, 0)$ is an extreme point of the core, and it is not a marginal vector. ◇

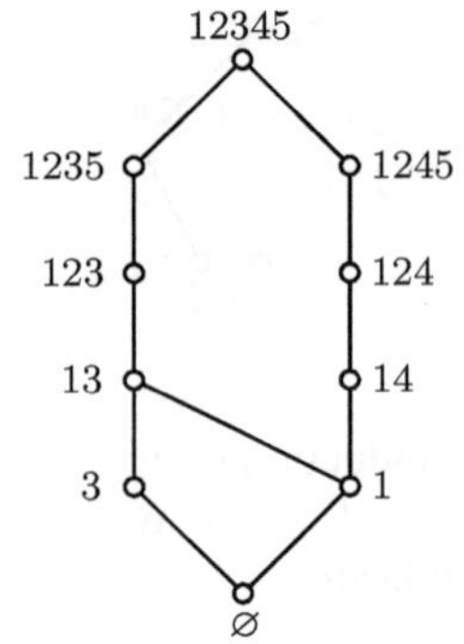

Fig. 3.1 Example of a regular set lattice (nondistributive)

Remark 3.33 Another setting was proposed by Faigle et al. [132] where the results valid in the classical case ($\mathcal{F} = 2^X$) remain true: the core is included in the Weber set and equality holds if and only if supermodularity holds. To achieve this, the authors consider the positive core $\mathbf{core}^+(v)$ of a game v on an arbitrary set system $\mathcal{F}$, which is the intersection of the core with the positive orthant. A Monge-type algorithm (see Remark 4.99) with input vector $c \in \mathbb{R}^N$ produces a list $\mu(c)$ of elements of $\mathcal{F}$ (playing the rôle of a maximal chain in $\mathcal{F}$), a list $\pi(c)$ of elements in N (which are not necessarily permutations on N because some elements in N may be missing), and a vector $y(c) \in \mathbb{R}^{|\mathcal{F}|}$. Letting $\hat{v}(c) = \langle v, y(c)\rangle$, it is proved that $\hat{v}(c)$ is an extension of v in the sense that $\hat{v}(1_S) = v(S)$ for any $S \in \mathcal{F}$, and it is the support function of $\mathbf{core}^+(v)$:

$$\mathbf{core}^+(v) = \{x \in \mathbb{R}^N : \langle c, x\rangle \geqslant \hat{v}(c), \forall c \in \mathbb{R}^N\}.$$

Next, marginal vectors $x^{\pi,v}$ are defined for each list π, μ produced by the Monge algorithm as follows: $x_i^{\pi,v} = 0$ for each $i \notin \pi$, and for the other coordinates, $x^{\pi,v}$ is the unique solution of the system

$$x(S) = v(S) \qquad (S \in \mu).$$

The Weber set is defined as the convex hull of marginal vectors, for all lists π produced by the Monge algorithm. Defining v to be convex as $\hat{v}$ being concave, the main result asserts that the positive core is included into the Weber set, with equality if and only if v is convex.

◇

3.3.5 Faces

Faces of the core have been deeply studied in combinatorial optimization when $\mathcal{F} = \mathcal{O}(N, \preceq)$ (see Fujishige [149, Chap. 2, Sect. 3.3 (d)] for a detailed account). We restrict here to basic facts.

Assuming v is a balanced game, take any $x \in \mathbf{core}(v)$ and define $\mathcal{F}(x) = \{S \in \mathcal{F} : x(S) = v(S)\}$. Then $\mathcal{F}(x)$ is a sublattice of $\mathcal{F}$ if v is supermodular. Indeed, first remark that $\varnothing, N \in \mathcal{F}(x)$. Now, take $S, T \in \mathcal{F}(x)$ and let us prove that $S \cup T$ and $S \cap T$ belong to $\mathcal{F}(x)$. We have

$$x(S) + x(T) = x(S \cup T) + x(S \cap T) \geqslant \\ v(S \cup T) + v(S \cap T) \geqslant v(S) + v(T) = x(S) + x(T),$$

which forces $x(S \cup T) = v(S \cup T)$ and $x(S \cap T) = v(S \cap T)$ because $x \in \mathbf{core}(v)$.

Assuming throughout this section that v is supermodular, define for any subsystem $\mathcal{D} \subseteq \mathcal{F}$

$$F(\mathcal{D}) = \{x : x(S) = v(S), \forall S \in \mathcal{D}, \quad x(S) \geq v(S) \text{ otherwise}\}$$
$$F^\circ(\mathcal{D}) = \{x : x(S) = v(S), \forall S \in \mathcal{D}, \quad x(S) > v(S) \text{ otherwise}\}.$$

Note that $F(\mathcal{D})$ is either empty or a face of the core provided $\mathcal{D} \ni N$, and that $F^\circ(\mathcal{D})$ is an "open" face in the sense that it does not contain any other face of lower dimension. Define

$$\mathbf{D} = \{\mathcal{D} \subseteq \mathcal{F} : \mathcal{D} \text{ is a sublattice of } \mathcal{F} \text{ containing } \varnothing, N, F^\circ(\mathcal{D}) \neq \varnothing\}.$$

Observe that any $\mathcal{D} \in \mathbf{D}$ is necessarily distributive because $\mathcal{F}$ is, and therefore is generated by a poset. It is easy to see that $\mathbf{D} = \{\mathcal{F}(x) : x \in \mathbf{core}(v)\}$. It follows that any face of the core is defined by a distributive sublattice of $\mathcal{F}$. Moreover, the dimension of a face $F(\mathcal{D})$ is $|N| - |h(\mathcal{D})|$, where $h(\mathcal{D})$ is the height of the lattice $\mathcal{D}$.

3.3.6 Bounded Faces

This section is based on papers by the author [169, 182], where the readers can find more details. We suppose throughout that $\mathcal{F} \ni N$.

Basically, finding bounded faces of the core amounts to selecting sets $T_1, \ldots, T_q$ in $\mathcal{F}$ such that replacing inequalities $x(T_i) \geqslant v(T_i)$ in $\mathbf{core}(v)$ by equalities $x(T_i) = v(T_i)$, $i = 1, \ldots, q$, results in a bounded set. Let us call *normal collection* any collection $\mathcal{N} = \{T_1, \ldots, T_q\}$ of $q < n$ nonempty subsets in $\mathcal{F} \setminus \{N\}$ such that

$$\mathbf{core}_{\mathcal{N}}(0) = \{x \in \mathbf{core}(0) : x(S) = 0 \ \forall S \in \mathcal{N}\} = \{\mathbf{0}\}.$$

Note that normal collections are determined by $\mathcal{F}$, not by v.

Supposing $\mathbf{core}(v) \neq \varnothing$, a normal collection $\mathcal{N}$ determines a bounded face of the core, denoted by $\mathbf{core}_{\mathcal{N}}(v)$:

$$\mathbf{core}_{\mathcal{N}}(v) = \{x \in \mathbf{core}(v) \,:\, x(T) = v(T), \forall T \in \mathcal{N}\}.$$

Note that $\mathbf{core}_{\mathcal{N}}(v)$ may be empty even if $\mathbf{core}(v)$ is not. The necessary and sufficient conditions for nonemptiness are given below and are similar to the Bondareva-Shapley theorem (Theorem 3.7). Considering a normal collection $\mathcal{N}$, we say that a collection $\mathcal{B} \subseteq \mathcal{F}$ is *$\mathcal{N}$-balanced* if there exists $y_S > 0$, $S \in \mathcal{B}$, such that $\sum_{S\in\mathcal{B}} y_S 1_S = \sum_{S\in\mathcal{N}\cup N} 1_S$. We call $\{y_S\}_{S\in\mathcal{B}}$ a *system of $\mathcal{N}$-balancing weights.*

Theorem 3.34 *Let $\mathcal{N}$ be a normal collection. $\mathbf{core}_{\mathcal{N}}(v) \neq \varnothing$ if and only if for every $\mathcal{N}$-balanced collection $\mathcal{B}$ with $\mathcal{N}$-balancing weights $\{y_S\}_{S\in\mathcal{B}}$, it holds*

$$\sum_{S\in\mathcal{B}} y_S v(S) \leqslant \sum_{S\in\mathcal{N}\cup N} v(S).$$

Proof We consider the following linear program with $x \in \mathbb{R}^N$:

$$\begin{aligned} \min \quad & z = \textstyle\sum_{S\in\mathcal{N}\cup N} x(S) \\ \text{s.t.} \quad & x(S) \geqslant v(S), \quad S \in \mathcal{F}. \end{aligned}$$

The optimal value z^* of z is $\sum_{S\in\mathcal{N}\cup N} v(S)$ if and only if $\mathbf{core}_{\mathcal{N}}(v) \neq \emptyset$. The dual problem reads

$$\begin{aligned} \max \quad & w = \textstyle\sum_{S\in\mathcal{F}} y_S v(S) \\ \text{s.t} \quad & \textstyle\sum_{S\ni i, S\in\mathcal{F}} y_S = \sum_{S\ni i, S\in\mathcal{N}\cup N} y_S, \quad i \in N \\ & y_S \geqslant 0, \quad S \in \mathcal{F}. \end{aligned}$$

By the duality theorem, $w^* = z^*$, which implies that any feasible solution satisfies $\sum_{S\in\mathcal{F}} y_S v(S) \leqslant \sum_{S\in\mathcal{N}\cup N} v(S)$. □

Take any balanced game $v \in \mathcal{G}(N, \mathcal{F})$, with $\mathbf{core}(0) \neq \{\mathbf{0}\}$, and consider an extremal ray r of $\mathbf{core}(0)$. We say that r is *deleted* by $T \in \mathcal{F}$ if the cone $\{x \in \mathbf{core}(0) \,:\, x(T) = 0\}$ does not contain r any more. In the case where $\mathcal{F} = \mathcal{O}(N, \preceq)$, there is a simple way to find sets deleting extremal rays. We recall from Theorem 3.24 that extremal rays are of the form $1_{\{j\}} - 1_{\{i\}}$, with $j \prec\cdot\, i$.

Lemma 3.35 *Suppose $\mathcal{F} = \mathcal{O}(N, \preceq)$, $\mathcal{F} \neq 2^N$, and take $i, j \in N$ such that $j \prec\cdot\, i$. Then the extremal ray $1_{\{j\}} - 1_{\{i\}}$ is deleted by T if and only if $T \ni j$ and $T \not\ni i$.*

Proof $\Leftarrow$) Take $T \in \mathcal{F}$ such that $T \ni j$ and $T \not\ni i$. Then $(1_{\{j\}} - 1_{\{i\}})(T) = 1$, hence $1_{\{j\}} - 1_{\{i\}}$ does not belong to $\mathbf{core}(0) \cap \{x(T) = 0\}$.

$\Rightarrow$) Suppose $1_{\{j\}} - 1_{\{i\}} \notin \mathbf{core}(0) \cap \{x(T) = 0\}$. Since $1_{\{j\}} - 1_{\{i\}}$ is a ray of $\mathbf{core}(0)$, it implies that $(1_{\{j\}} - 1_{\{i\}})(T) \neq 0$; i.e., $1_{\{j\}}(T) \neq 1_{\{i\}}(T)$. Therefore either

i or j belongs to T, but not both. Because $j \prec\cdot\, i$ and T is a downset, it must be $j \in T$ and $i \notin T$. □

This result gives in principle a systematic way of building normal collections, however the search of normal collections is highly combinatorial, as illustrated by the following example.

Example 3.36 Consider the set system $\mathcal{F}$ of Fig. 2.3 built on $N = \{1,2,3,4\}$ endowed with the partial order $1 \prec 2, 3 \prec 2, 3 \prec 4$. There are three extremal rays $(1,-1,0,0)$, $(0,-1,1,0)$ and $(0,0,1,-1)$. The simplest normal collection is $\mathcal{N}_1 = \{\{1,3\}\}$, but $\mathcal{N}_2 = \{\{1\},\{3\}\}$, $\mathcal{N}_3 = \{\{1\},\{\{1,3\}\}$, $\mathcal{N}_4 = \{\{3\},\{1,3\}\}$, $\mathcal{N}_5 = \{\{1\},\{3\},\{1,3\}\}$, $\mathcal{N}_6 = \{\{1\},\{1,2,3\},\{3,4\}\}$, $\mathcal{N}_7 = \{\{1,2,3\},\{1,3,4\}\}$, etc., are also possible. ◇

The following fact is noteworthy.

Lemma 3.37 *Suppose $\mathcal{F} = \mathcal{O}(N,\preceq)$, $\mathcal{F} \neq 2^N$. Any normal collection contains at least $h(N,\preceq)$ sets, where $h(N,\preceq)$ is the height of $(N,\preceq)$.*

Proof Take $\mathcal{N}$ a normal collection and denote $h(N,\preceq)$ by $h(N)$ for simplicity. By definition of the height, there exists a maximal chain in $(N,\preceq)$ of length $h(N)$ going from a minimal element to a maximal element, say $i_0, i_1, \ldots, i_{h(N)}$. Then by Theorem 3.24, $1_{\{i_0\}} - 1_{\{i_1\}}, \ldots, 1_{\{i_{h(N)-1}\}} - 1_{\{i_{h(N)}\}}$ are extremal rays. Because $1_{\{i_0\}} - 1_{\{i_1\}}$ is deleted, by Lemma 3.35 $\mathcal{N}$ must contain a set K_1 such that $i_0 \in K_1$ and $i_1 \notin K_1$. Moreover, since K_1 must be a downset, $i_2, \ldots, i_{h(N)}$ cannot belong to K_1. Similarly, there must exist a set K_2 deleting ray $1_{\{i_1\}} - 1_{\{i_2\}}$ such that $i_1 \in K_2$ and $i_2, \ldots, i_{h(N)} \notin K_2$. Therefore, $K_1 \neq K_2$. Continuing this process we construct a sequence of distinct $h(N)$ subsets $K_1, K_2, \ldots, K_{h(N)}$, the last one deleting ray $1_{\{i_{h(N)-1}\}} - 1_{\{i_{h(N)}\}}$. Therefore, at least $h(N)$ equalities are needed. □

We introduce the following terminology for normal collections: a normal collection is *minimal* if no subcollection is normal; a normal collection is *short* if it has cardinality $h(N,\preceq)$; a normal collection is *nested* if it is a chain in $\mathcal{F}$.

Example 3.38 (Example 3.36 cont'd) $\mathcal{N}_1, \mathcal{N}_2, \mathcal{N}_6$ are minimal, but only $\mathcal{N}_1$ is short, and only $\mathcal{N}_1$ is nested. ◇

Nested normal collections have interesting properties. In particular, they permit to define special marginal vectors. Suppose that $\mathcal{F} = \mathcal{O}(N,\preceq)$ and consider $\mathcal{N}$ a nested normal collection, and a feasible permutation $\sigma \in \mathfrak{S}(\mathcal{F})$ (linear extension of $\preceq$). We say that σ is *compatible* with $\mathcal{N}$ if the induced maximal chain $\{\varnothing, A_1, A_2, \ldots, A_n\}$ with $A_i = \{\sigma(1), \ldots, \sigma(i)\}$, contains $\mathcal{N}$. We may denote by $\mathfrak{S}(\mathcal{F},\mathcal{N})$ the set of feasible permutations compatible with $\mathcal{N}$. The following fundamental fact is immediate.

Lemma 3.39 *Suppose $\mathcal{F} = \mathcal{O}(N,\preceq)$, $\mathcal{N}$ is a nested normal collection, and take $\sigma \in \mathfrak{S}(\mathcal{F},\mathcal{N})$. Then the marginal vector $x^{\sigma,v}$ is an extreme point of the bounded face* $\mathbf{core}_{\mathcal{N}}(v)$ *if v is supermodular.*

Proof By Theorem 3.27, we know that $x^{\sigma,v} \in \mathbf{core}(v)$ and is moreover an extreme point of it. By (3.15), $x^{\sigma,v}$ satisfies $x^{\sigma,v}(T) = v(T)$ for every $T \in \mathcal{N}$, hence it belongs to $\mathbf{core}_{\mathcal{N}}(v)$. □

The following fact is also noteworthy.

Lemma 3.40 *Let $\mathcal{F} = \mathcal{O}(N, \preceq)$ and $\mathcal{N} = \{T_1, \ldots, T_q\}$ be a normal nested collection for game v on $(N, \mathcal{F})$. Then $T_k \setminus T_{k-1}$ is an antichain in $(N, \preceq)$ for $k = 1, \ldots, q$, with $T_0 = \varnothing$.*

Proof If $T_k \setminus T_{k-1}$ is not an antichain, then there exist $i, j \in T_k \setminus T_{k-1}$ such that $j \prec\cdot\, i$. Because $\mathcal{N}$ is nested, no set in $\mathcal{N}$ contains j and not i. Then by Lemma 3.35, the ray $1_{\{j\}} - 1_{\{i\}}$ is not deleted by $\mathcal{N}$. □

We introduce the union of all bounded faces of the core, called the *bounded core*:

$$\mathbf{core}^{\mathrm{b}}(v) = \bigcup_{\mathcal{N}} \mathbf{core}_{\mathcal{N}}(v).$$

In the above expression, it could happen that many bounded faces $\mathbf{core}_{\mathcal{N}}(v)$ are empty, and that many of them are redundant because included into other bounded faces. When the game is supermodular, it is possible to get an irredundant expression where each face is nonempty, by simply considering minimal nested normal collections. We call $\mathcal{MNNC}$ the set of minimal nested normal collections.

Theorem 3.41 (Maximal bounded faces of the core) *Let v be a game on $\mathcal{F} = \mathcal{O}(N, \preceq)$, and assume it is supermodular. The following holds.*

(i) For any nested normal collection $\mathcal{N}$, $\mathbf{core}_{\mathcal{N}}(v) \neq \varnothing$. Moreover, if v is strictly supermodular,[6] then $\mathbf{core}_{\mathcal{N}}(v) \neq \varnothing$ if and only if $\mathcal{N}$ is nested, and $\dim \mathbf{core}_{\mathcal{N}}(v) = n - |\mathcal{N}| - 1$.

(ii)

$$\mathbf{core}^{\mathrm{b}}(v) = \bigcup_{\mathcal{N} \in \mathcal{MNNC}(\mathcal{F})} \mathbf{core}_{\mathcal{N}}(v).$$

Moreover, no term in the union is redundant if v is strictly supermodular.

Proof (i) The first statement comes from Lemma 3.39.

Suppose now strict supermodularity of v, and let $\mathcal{N}$ be a normal collection that is not nested. Hence, there exist $S, T \in \mathcal{N}$ such that $S \setminus T \neq \varnothing \neq T \setminus S$. By strict convexity, $v(S) + v(T) < v(S \cup T) + v(S \cap T)$, so that any $y \in \mathbb{R}^N$ with $y(S) = v(S)$ and $y(T) = v(T)$ either satisfies $y(S \cap T) < v(S \cap T)$ or $y(S \cup T) < v(S \cup T)$. We conclude that $\mathbf{core}_{\mathcal{N}}(v) = \varnothing$.

[6]That is, $v(S \cup T) + v(S \cap T) > v(S) + v(T)$ whenever $S \setminus T$ and $T \setminus S$ are nonempty.

It remains to prove the assertion on the dimension. Let

$$x = \frac{1}{|\mathfrak{S}(\mathcal{F}, \mathcal{N}|} \sum_{\sigma \in \mathfrak{S}(\mathcal{F}, \mathcal{N})} x^{\sigma, v}. \tag{3.19}$$

Since v is strictly supermodular, it suffices to prove that $x(S) > v(S)$ for all $S \in \mathcal{F} \setminus (\mathcal{N} \cup \{\varnothing, N\})$. Let $\mathcal{N} \cup \{\varnothing, N\} = \{T_0, \ldots, T_r\}$ such that $T_0 = \varnothing \subset T_1 \subset \cdots \subset T_r = N$. Suppose there exists $j \in \{1, \ldots, r-1\}$ such that $T_j \setminus S \neq \varnothing \neq S \setminus T_j$, then

$$\begin{aligned} v(S) + v(T_j) < v(S \cap T_j) + v(S \cup T_j) \leqslant x(S \cap T_j) + x(S \cup T_j) \\ = x(S) + x(T_j) = x(S) + v(T_j) \end{aligned}$$

by strict convexity and because $x \in \mathbf{core}(v)$. Otherwise there exists $\ell \in \{0, \ldots, r-1\}$ such that $T_\ell \subset S \subset T_{\ell+1}$. Let $S' = T_\ell \cup (T_{\ell+1} \setminus S)$. Note that because $T_{\ell+1} \setminus T_\ell$ is an antichain by Lemma 3.40, $S' \in \mathcal{F}$. Then there exists $\widetilde{\pi} \in \mathfrak{S}(\mathcal{F}, \mathcal{N})$ such that $S' = \{\widetilde{\pi}(1), \ldots, \widetilde{\pi}(|S'|)\}$; i.e., $x^{\widetilde{\pi}, v}(S') = v(S')$. By strict convexity we conclude that $x^{\widetilde{\pi}, v}(S) > v(S)$. For any $\pi \in \mathfrak{S}(\mathcal{F}, \mathcal{N})$, $x^{\pi, v}(S) \geq v(S)$, hence $x(S) > v(S)$.

(ii) We claim that under supermodularity, for any normal collection $\mathcal{N}$, there exists a nested normal collection $\mathcal{N}'$ such that $\mathbf{core}_{\mathcal{N}}(v) \subseteq \mathbf{core}_{\mathcal{N}'}(v)$.

PROOF OF THE CLAIM: consider $\mathcal{N} \neq \varnothing$, and take $T, T' \in \mathcal{N}$. Let us show that $\mathcal{N}' = (\mathcal{N} \setminus \{T, T'\}) \cup \{T \cup T', T \cap T'\}$ is (a) normal and (b) $\mathbf{core}_{\mathcal{N}}(v) \subseteq \mathbf{core}_{\mathcal{N}'}(v)$. In view of Lemma 3.35, in order to show (a) it suffices to prove that for any $i, j \in N$, $j \prec\cdot\, i$, and $j \in T \not\ni i$, either $i \notin T \cup T'$ or $j \in T \cap T'$. Now, if $i \in T \cup T'$, then $j \in T'$ since it is a downset. Hence $j \in T \cap T'$ in this case. In order to show (b) let $x \in \mathbf{core}_{\mathcal{N}}(v)$, and show that $x(T \cup T') = v(T \cup T')$ and $x(T \cap T') = v(T \cap T')$. As v is supermodular,

$$\begin{aligned} v(T \cup T') + v(T \cap T') \leqslant x(T \cup T') + x(T \cap T') \\ = x(T) + x(T') = v(T) + v(T') \leqslant v(T \cup T') + v(T \cap T') \end{aligned}$$

so that the desired equalities follow immediately. Finally, an inductive argument that we omit shows that applying to $\mathcal{N}'$ the same transformation (that is, removing T, T' and replacing them by their union and intersection) at most $\frac{f(f-1)}{2}$ times, where $f = |\mathcal{F}|$, we get a collection containing a nested collection. □

Because minimal normal collections give largest bounded faces, the first statement is proved. In order to show the final statement, let v be strictly supermodular and let x be defined by (3.19). We have seen that $x(S) = v(S)$ if and only if $S \in \mathcal{N} \cup \{\varnothing, N\}$ so that there is no other minimal normal collection $\mathcal{N}'$ with $x \in \mathbf{core}_{\mathcal{N}'}(v)$. □

Remark 3.42 The bounded core was proposed by Grabisch and Sudhölter [181]. Equivalently, the bounded core is the set of all elements in **core**(v) that satisfy the following condition for any $i,j \in N$ such that $i \prec\cdot j$: There is no $\epsilon > 0$ such that $x + \epsilon\left(1_{\{j\}} - 1_{\{i\}}\right) \in \mathbf{core}(v)$.

3.4 Exact Games, Totally Balanced Games, Large Cores and Stable Sets

In this section we consider games defined on $(N, 2^N)$ again. Let v be such a game, we introduce the set of its *acceptable vectors* $A(v)$:

$$A(v) = \{x \in \mathbb{R}^N : x(S) \geqslant v(S), \forall S \in 2^N\}.$$

Clearly, $A(v)$ is a nonempty convex closed polyhedron bounded from below. Observe that it has minimal elements; i.e., elements $y \in A(v)$ such that $x \leqslant y$ and $x \in A(v)$ imply $x = y$. Indeed, any core element is a minimal element: suppose $\mathbf{core}(v) \neq \varnothing$ and $x \in \mathbf{core}(v)$ is not minimal in $A(v)$. Then there exists $y \in A(v)$ with $y_i \leqslant x_i$ for all $i \in N$, with at least one strict inequality. Then $y(N) < v(N)$, a contradiction with $y \in A(v)$. Now, if $\mathbf{core}(v) = \varnothing$, it suffices to increase $v(N)$ so as to make $\mathbf{core}(v) \neq \varnothing$.

Let us denote by $\underline{A}(v)$ the set of minimal elements of $A(v)$. As established just above for any game v on 2^N,

$$\mathbf{core}(v) \subseteq \underline{A}(v). \tag{3.20}$$

For any $S \in 2^N \setminus \{\varnothing\}$, the *subgame* v_S is the restriction of v to 2^S: $v_S : 2^S \to \mathbb{R}$, $T \subseteq S \mapsto v_S(T) = v(T)$.

We introduce the following notions.

Definition 3.43 Let v be a balanced game on $(N, 2^N)$; i.e., $\mathbf{core}(v) \neq \varnothing$.

(i) The *lower envelope* of v is the game v_* on $(N, 2^N)$ defined by

$$v_*(S) = \min_{x \in \mathbf{core}(v)} x(S) \qquad (S \subseteq N);$$

(ii) v is *exact* if for every $S \in 2^N \setminus \{\varnothing\}$, there exists a core element $x \in \mathbf{core}(v)$ such that $x(S) = v(S)$; i.e., $v_* = v$;
(iii) v is *totally balanced* if every subgame v_S, $S \in 2^N \setminus \{\varnothing\}$, is balanced;
(iv) v has a *large core* if for every $y \in A(v)$, there exists a core element $x \in \mathbf{core}(v)$ such that $x \leqslant y$.

Several easy connections can be established.

Lemma 3.44 *Let v be a balanced game on $(N, 2^N)$. The following holds.*

(i) $v_* \geqslant v$, $v_*(N) = v(N)$, *and* $\mathbf{core}(v_*) = \mathbf{core}(v)$*;*
(ii) If v is exact, then v is totally balanced;
(iii) If v is supermodular, then v is exact;
(iv) v has a large core if and only if $\underline{A}(v) = \mathbf{core}(v)$.

Proof

(i) $v_* \geqslant v$ and $v_*(N) = v(N)$ are obvious and imply that $\mathbf{core}(v_*) \subseteq \mathbf{core}(v)$. Now, any $z \in \mathbf{core}(v)$ satisfies $z(S) \geqslant \min_{x \in \mathbf{core}(v)} x(S) = v_*(S)$ for all $\varnothing \neq S \subseteq N$, which in view of $v_*(N) = v(N) = z(N)$, proves the reverse inclusion.
(ii) Take $S \in 2^S \setminus \{\varnothing\}$ and consider $x \in \mathbf{core}(v)$ such that $x(S) = v(S)$. Clearly, $x_{|S} \in \mathbf{core}(v_S)$, hence v is totally balanced.
(iii) Take $S \in 2^N \setminus \{\varnothing\}$ and a permutation σ on N such that $S = \{\sigma(1), \dots, \sigma(|S|)\}$. By (3.9) we know that the marginal vector $x^{\sigma,v}$ satisfies $x^{\sigma,v}(S) = v(S)$. Since v is supermodular, by Theorem 3.15(ii), $x^{\sigma,v} \in \mathbf{core}(v)$.
(iv) Suppose v has a large core. By (3.20), it remains to show that any minimal element x of A belongs to $\mathbf{core}(v)$. Suppose *per contra* that $x \notin \mathbf{core}(v)$. Then $x(N) > v(N)$. Since v has a large core, there exists a core element z such that $z \leqslant x$, $z \neq x$, a contradiction with the minimality of x.

Suppose $\underline{A}(v) = \mathbf{core}(v)$ and take $y \in A(v)$. Then there exists $x \in \underline{A}(v)$ such that $x \leqslant y$, and by assumption x is a core element. □

The *totally balanced cover* of a game v on $(N, 2^N)$ is the game $\mathbf{tbc}(v)$ on $(N, 2^N)$ defined by

$$\mathbf{tbc}(v)(S) = \max\Big\{ \sum_{T \in \mathcal{B}_S} \lambda_T v(T) \; : \; \mathcal{B}_S \text{ is a balanced collection of } S$$
$$\text{and } (\lambda_T)_{T \in \mathcal{B}_S} \text{ is a system of balancing weights}\Big\},$$

for $S \in 2^N \setminus \{\varnothing\}$, and $\mathbf{tbc}(v)(\varnothing) = 0$. We show some properties of the totally balanced cover.

Lemma 3.45 *Let v be a game on $(N, 2^N)$. The following holds.*

(i) $\mathbf{tbc}(v) \geqslant v$*;*
(ii) $\mathbf{tbc}(v)$ *is totally balanced;*
(iii) v is balanced if and only if $\mathbf{tbc}(v)(N) = v(N)$*;*
(iv) v' totally balanced and $v' \geqslant v$ imply $v' \geqslant \mathbf{tbc}(v)$*;*
(v) v is totally balanced if and only if $v = \mathbf{tbc}(v)$*. Hence* $\mathbf{tbc}(\mathbf{tbc}(v)) = \mathbf{tbc}(v)$*;*
(vi) If v is balanced, then $\mathbf{core}(v) = \mathbf{core}(\mathbf{tbc}(v))$*, and* $A(v) = A(\mathbf{tbc}(v))$*. Hence* $\mathbf{core}(v)$ *is large if and only if* $\mathbf{core}(\mathbf{tbc}(v))$ *is large.*

Proof

(i) Obvious because $\{S\}$ is a balanced collection of S with weight 1.

(ii) Take $S \in 2^N\backslash\{\varnothing\}$ and $\mathcal{B}_S^0$ a balanced collection on S with a system of balancing weights $(\lambda_T)_{T\in\mathcal{B}_S^0}$. Denote by $\mathfrak{B}(T)$ the set of balanced collections on $\varnothing \neq T \subseteq N$. We have

$$\sum_{T\in\mathcal{B}_S^0} \lambda_T \mathbf{tbc}(v)(T) = \sum_{T\in\mathcal{B}_S^0} \lambda_T \max_{\mathcal{B}_T\in\mathfrak{B}(T)} \sum_{K\in\mathcal{B}_T} \delta_K v(K) = \sum_{T\in\mathcal{B}_S^0} \lambda_T \sum_{K\in\mathcal{B}_T^*} \delta_K^T v(K)$$

$$= \sum_{K\in\bigcup_{T\in\mathcal{B}_S^0}\mathcal{B}_T^*} \lambda_T \delta_K^T v(K) \leqslant \max_{\mathcal{B}_S\in\mathfrak{B}(S)} \sum_{T\in\mathcal{B}_S} \lambda_T' v(T) = \mathbf{tbc}(v)(S).$$

The inequality holds because $\bigcup_{T\in\mathcal{B}_S^0}\mathcal{B}_T^*$ is a balanced collection with balancing weights $\lambda_T\delta_K^T$, as it can be checked.

(iii) Suppose v is balanced. Then by Definition 3.6 and (i),

$$v(N) \geqslant \mathbf{tbc}(v)(N) \geqslant v(N).$$

Now, suppose $v(N) = \mathbf{tbc}(v)(N)$. Then $v(N) \geqslant \sum_{S\in\mathcal{B}} \lambda_S v(S)$ for all balanced collection $\mathcal{B}$ of N with balancing weights λ_S; i.e., v is balanced.

(iv) Take $S \subseteq N, S \neq \varnothing$. Since v' is totally balanced, $\mathbf{core}(v'_S) \neq \varnothing$, hence for any balanced collection $\mathcal{B}_S$ on S with balancing weights $(\lambda_T)_{T\in\mathcal{B}_S}$, we have

$$v'(S) \geqslant \sum_{T\in\mathcal{B}_S} \lambda_T v'(T) \geqslant \sum_{T\in\mathcal{B}_S} \lambda_T v(T).$$

This proves $v'(S) \geqslant \mathbf{tbc}(v)(S)$.

(v) Let $\varnothing \neq S \subseteq N$. Then $\mathbf{tbc}(v_S) = \mathbf{tbc}(v)_{|S}$. Hence the equivalence holds by (iii), and the last assertion follows from (ii).

(vi) Suppose v is balanced. Take $x \in \mathbf{core}(v)$. By (iii), $x(N) = v(N) = \mathbf{tbc}(v)(N)$. It remains to show $x(S) \geqslant \mathbf{tbc}(v)(S)$ for every $S \subset N$. Take any such S. We have for every balanced collection $\mathcal{B}_S$ of S with balancing weights λ_T:

$$x(S) = \sum_{T\in\mathcal{B}_S} \lambda_T x(T) \geqslant \sum_{T\in\mathcal{B}_S} \lambda_T v(T),$$

hence $x(S) \geqslant \mathbf{tbc}(v)(S)$, and we have shown $\mathbf{core}(v) \subseteq \mathbf{core}(\mathbf{tbc}(v))$. Now, the converse inclusion is obvious by (i) and the fact that $v(N) = \mathbf{tbc}(v)(N)$. The assertions on $A(v)$ and largeness follow trivially.

□

Theorem 3.46 *If v is totally balanced and has a large core, then v is exact.*

Proof Take any $S \in 2^N \setminus \{\varnothing\}$ and $z_S \in \mathbf{core}(v_S)$. We can extend z_S to a vector $z = (z_S, z_{N\setminus S})$ in $\mathbb{R}^N$ so that $z \in A(v)$. Since v has a large core, there exists $x \in \mathbf{core}(v)$ such that $x \leqslant z$. Then $v(S) \leqslant x(S) \leqslant z(S) = v(S)$. Hence x is a core element satisfying $x(S) = v(S)$. □

From the above result and Lemma 3.45 (ii) and (vi), we obtain:

Corollary 3.47 *(Sharkey [302]) If v has large core,* $\mathbf{tbc}(v)$ *is exact.*

Remark 3.48

(i) Exact games have been introduced by Schmeidler [285], while lower envelopes play an important rôle in decision theory (see, e.g., Walley [341] and Sect. 5.3.5).

(ii) Totally balanced games and totally balanced covers have been introduced by Shapley [300] in order to characterize market games, which model exchange economies.

(iii) The notion of large core was introduced by Sharkey [302]. Van Gellekom et al. [330] have shown that largeness of the core is a *prosperity property*: let v be a game on $(N, 2^N)$ and call v^0 its restriction to $2^N \setminus \{N\}$. A property P on games is a prosperity property if for each v^0 there exists a real number $\alpha(v^0) \geqslant \sum_{i\in N} v(\{i\})$ such that v has property P if and only if $v(N) \geqslant \alpha(v^0)$. In words, the property (like largeness of the core) can be made true if $v(N)$ is large enough. Many properties of games are prosperity properties: balancedness, supermodularity for partitions (Definition 3.49), stability of the core (see Definition 3.53), extendability (see Definition 3.55), etc. On the other hand, exactness, total balancedness, and supermodularity are not prosperity properties.

(iv) As shown by Biswas et al. [31], for $n = 3$ and 4, every exact game has a large core. However for $n \geqslant 5$, this property does not hold any more.

(v) Biswas et al. [31] have shown further properties for the particular case of symmetric games. If v is totally balanced and symmetric, exactness, largeness of the core, and stability of the core are all equivalent properties.

◇

We introduce now a weaker notion of supermodularity.

Definition 3.49 (Sharkey [302]) Consider a partition $\pi = (P_1, \dots, P_k)$ of N and a family $\mathcal{Q} = \{Q_1, \dots, Q_k\}$ of subsets of N satisfying

$$Q_i \subseteq \bigcup_{j=0}^{i-1} P_j, \text{ and } Q_k \cup P_k \neq N, \text{ where } P_o = \varnothing.$$

A game v is said to be *supermodular for partitions* if for all such collections π and $\mathcal{Q}$, it holds

$$\sum_{i=1}^{k} \big(v(P_i \cup Q_i) - v(Q_i)\big) \leqslant v(N). \tag{3.21}$$

Remark 3.50 This notion was introduced by Sharkey under the name of *subconvexity*.

Lemma 3.51 *If a game v on $(N, 2^N)$ is supermodular, then it is supermodular for partitions.*

Proof Let us show by induction that $\sum_{i=1}^{\ell} \big(v(P_i \cup Q_i) - v(Q_i)\big) \leqslant v(P_1 \cup \cdots \cup P_\ell)$, for $\ell = 1, \ldots, k$. The desired result follows with $\ell = k$.

The claim is trivially true for $\ell = 1$ because $Q_1 = \varnothing$. Assume it is true for $\ell < k$ and let us show it for $\ell + 1$.

$$\begin{aligned}\sum_{i=1}^{\ell+1} \big(v(P_i \cup Q_i) - v(Q_i)\big) &= \sum_{i=1}^{\ell} \big(v(P_i \cup Q_i) - v(Q_i)\big) + v(P_{\ell+1} \cup Q_{\ell+1}) - v(Q_{\ell+1}) \\ &\leqslant v(P_1 \cup \cdots \cup P_\ell) + v(P_{\ell+1} \cup Q_{\ell+1}) - v(Q_{\ell+1}) \\ &\leqslant v(P_1 \cup \cdots \cup P_{\ell+1})\end{aligned}$$

where the last inequality comes from supermodularity of v and $Q_{\ell+1} \subseteq P_1 \cup \cdots \cup P_\ell$. □

Theorem 3.52 *(Sharkey [302]) If v is supermodular for partitions, then it has a large core.*

Proof Let $y \in A(v)$ and choose $x \in \underline{A}(v)$ such that $x \leqslant y$. For each $i \in N$, choose $S_i \ni i$ such that $x(S_i) = v(S_i)$ (always possible because x is a minimal element of $A(v)$). Define

$$P_i = S_i \setminus \bigcup_{j=1}^{i-1} S_j, \quad Q_i = S_i \cap \bigcup_{j=1}^{i-1} S_j,$$

for $i = 1, \ldots, n$, where $S_0 = \varnothing$. By construction, $P_i \cup Q_i = S_i$, $\bigcup_{i=1}^{n} P_i = N$, $Q_i \subseteq \bigcup_{j=0}^{i-1} P_i$, and $P_i \cap P_j = \varnothing$ if $i \neq j$. Hence $P_1, \ldots, P_n$ form a covering of N with disjoint sets, and discarding empty sets in this family we get a partition $\pi = \{P_1, \ldots, P_k\}$ (after renumbering) and the family $\mathcal{Q} = \{Q_1, \ldots, Q_k\}$ of corresponding sets. If $P_i \cup Q_i = N$ for some i, then $x(N) = v(N)$, and the desired result follows. Otherwise, supermodularity for partitions yields

$$\begin{aligned}x(N) &= \sum_{i=1}^{k} (x(P'_i \cup Q'_i) - x(Q'_i)) \\ &= \sum_{i=1}^{k} (v(P'_i \cup Q'_i) - x(Q'_i)) \\ &\leqslant \sum_{i=1}^{k} (v(P'_i \cup Q'_i) - v(Q'_i)) \leqslant v(N).\end{aligned}$$

Therefore the core of v is large. □

Consequently, due to Lemma 3.51, supermodular games have a large core. Figure 3.2 summarizes the relative position of the classes of games introduced so far.

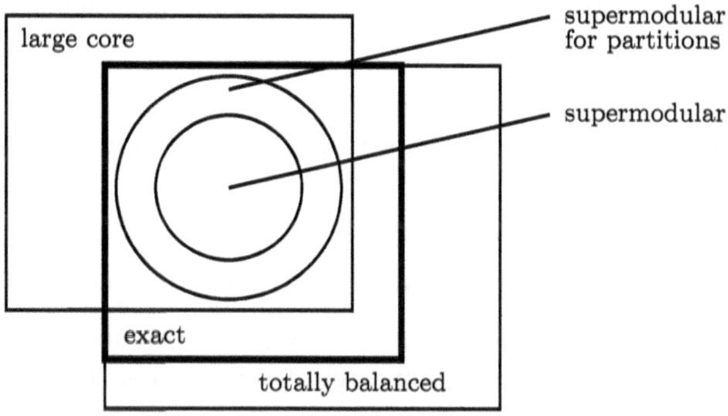

Fig. 3.2 Relations between the different classes of games

Largeness of the core is also related to its stability.

Definition 3.53 Let v be a game on $(N, 2^N)$, and consider the set $I(v) = \{x \in \mathbb{R}^N : x_i \geqslant v(\{i\}), i \in N, \quad x(N) = v(N)\}$ (set of *imputations*).

(i) Let $x, y \in I(v)$ and $\varnothing \neq S \subset N$. We say that x *dominates* y *via* S if $x_i > y_i$ for all $i \in S$ and $x(S) \leqslant v(S)$. Also, x *dominates* y if it dominates y via some set $S \subset N$.
(ii) A subset $C(v) \subseteq I(v)$ of imputations is *stable* if every imputation outside $C(v)$ is dominated by an element of $C(v)$, and if no element of $C(v)$ dominates another element of $C(v)$.

Obviously, when **core**(v) is nonempty, no core element can be dominated by another imputation, hence every stable set contains the core. Moreover, no core element can dominate another core element. However, it is not always true that the core is itself a stable set because some imputations outside the core may be undominated. If the core is stable, then it is the unique stable set.

Theorem 3.54 *(Sharkey [302]) If a game v on $(N, 2^N)$ has a large core, then the core is stable, and is therefore the unique stable set.*

Proof Assume **core**(v) is large and take an imputation x not in the core. Let S be a minimal subset such that $x(S) < v(S)$. Choose $y \in \underline{A}(v)$ such that

$$y_i = x_i + \frac{v(S) - x(S)}{|S|} \qquad (i \in S).$$

Since the core is large, there exists $z \in \mathbf{core}(v)$ such that $z \leqslant y$. Then $z(S) = v(S)$, $z_i = y_i > x_i$ for all $i \in S$; i.e., z dominates x via S. □

The notion of extendability introduced by Kikuta and Shapley [208] is also related to stability and largeness.

Definition 3.55 A balanced game v on $(N, 2^N)$ is *extendable* if for every nonempty $S \subset N$ such that $\mathbf{core}(v_S) \neq \varnothing$, and every $y \in \mathbf{core}(v_S)$, there exists $x \in \mathbf{core}(v)$ such that $x_{|S} = y$.

Theorem 3.56 *(Kikuta and Shapley [208]) Let v be a balanced game on $(N, 2^N)$. If* $\mathbf{core}(v)$ *is large, then v is extendable. If v is extendable, then* $\mathbf{core}(v)$ *is stable.*

Proof Suppose v has a large core, take $S \subset N$ such that $\mathbf{core}(v_S) \neq \varnothing$, and take $x_S \in \mathbf{core}(v_S)$. Consider the (nonempty) set of $y \in A(v)$ such that $y_{|S} = x_{|S}$, and take a minimal element y' of this set. Observe that $y' \in \underline{A}(v)$ because there is no $y'' \in \underline{A}(v)$ with $y'' \leqslant y$ with at least one strict inequality for $i \in S$. Therefore $y' \in \mathbf{core}(v)$ because v has a large core (Lemma 3.44).

Suppose now that v is extendable. Take an imputation $x \notin \mathbf{core}(v)$, and choose a minimal $S \subseteq N$ such that $x(S) < v(S)$. Define $x' \in \mathbb{R}^S$ by

$$x'_i = x_i + \frac{v(S) - x(S)}{|S|} \qquad (i \in S).$$

Then x' satisfies $x'(S) = v(S)$ and $x'(T) \geqslant v(T)$ for all $T \subset S$. Hence $x' \in \mathbf{core}(v_S)$, and it can be extended to a vector $x'' \in \mathbb{R}^N$ in the core of v. Now, x'' dominates x via S. □

Remark 3.57

(i) The notion of stable set was proposed by von Neumann[7] and Morgenstern[8] [333]. Van Gellekom et al. [330, Sect. 4.1, Example 1] have proved that the converse of Theorem 3.54 is false in general, by providing a counterexample with $n = 6$. However, the converse holds for $n \leqslant 5$, as proved by Estévez-Fernández [125]. Biswas et al. [32] have improved the result of Kikuta and Shapley by providing a stronger notion of extendability, which happens to be equivalent to largeness of the core. They also proved that extendability in the ordinary sense is equivalent to largeness when $n \leqslant 5$.

(ii) Extendability is far from being a necessary condition for core stability. Shellshear and Sudhölter have proposed a weaker notion, called *vital-exact extendability*, which implies core stability [303, Theorem 3.3].

[7]John von Neumann (Budapest, 1903 – Washington, 1957) is a Hungarian and American mathematician and physicist. He made important contributions in logic and set theory, in quantum physics, computer sciences, economics, etc. He is also considered to be the father of game theory and general equilibrium with Oskar Morgenstern.

[8]Oskar Morgenstern (Görlitz, 1902 – Princeton, 1977) is a German and American mathematician and economist. He is the founder of game theory, with J. von Neumann.

(iii) Other characterizations of largeness have been provided by Estévez-Fernández [125]. One is based on linear programming by relating largeness to basic solutions (i.e., extreme points of the core). The second one is based on minimal covers, a kind of generalization of balanced collections.

◇

3.5 The Selectope

In the whole section, we consider games defined on $(N, 2^N)$. We define formally a *value* as a mapping $\phi : \mathcal{G}(N) \rightarrow \mathbb{R}^N$, which to each game assigns a n-dim vector (interpreted as a payoff vector in game theory). We have already seen in Chap. 2 two well-known values, namely the Shapley value and the Banzhaf value (see Remark 2.43).

A *selector* is a mapping $\alpha : 2^N \setminus \{\varnothing\} \rightarrow N$ such that $S \mapsto \alpha(S) \in S$. In words, a selector selects an element in each nonempty subset of N. We denote by $\mathcal{S}(N)$ the set of selectors defined on $(N, 2^N)$. Given a game v on $(N, 2^N)$ and a selector α, we define the associated *selector value* $\phi^\alpha(v) \in \mathbb{R}^N$ as

$$\phi_i^\alpha(v) = \sum_{S : \alpha(S) = i} m^v(S) \qquad (i \in N), \tag{3.22}$$

where m^v is the Möbius transform of v. Referring to the interpretations given in Sect. 3.1, $\phi^\alpha(v)$ represents either a payoff vector for all players in N, or a probability distribution compatible with v viewed as an uncertainty measure (capacity).

Definition 3.58 Let v be a game on $(N, 2^N)$. The *selectope* of v is the convex closure of all selector values:

$$\mathbf{sel}(v) = \mathbf{conv}(\phi^\alpha(v) : \alpha \in \mathcal{S}(N)).$$

We now identify an important subfamily of selectors, whose selector values coincide with marginal vectors. Let σ be a permutation on N, v a game on $(N, 2^N)$ and consider the marginal vector $x^{\sigma,v}$ [see (3.8)]. Expressing $x^{\sigma,v}$ in terms of m^v we find

$$x_{\sigma(i)}^{\sigma,v} = \sum_{S \subseteq A_i, S \ni \sigma(i)} m^v(S) \qquad (i \in N),$$

with $A_i = \{\sigma(1), \ldots, \sigma(i)\}$. Since $\sigma(i)$ is the element of highest rank in A_i, it follows that $x^{\sigma,v}$ corresponds to the selector value ϕ^α where α is the selector selecting the element of highest rank (according to σ) in each subset S. It follows that any marginal vector of v belongs to the selectope of v.

All selectors corresponding to marginal vectors can be characterized by a simple property: a selector α is *consistent* if for any $S, T \in 2^N \setminus \{\varnothing\}$ such that $S \subset T$ and $\alpha(T) \in S$, we have $\alpha(S) = \alpha(T)$.

Lemma 3.59 *A selector α is consistent if and only if its selector value $\phi^\alpha(v)$ is a marginal vector of v, for any game v on $(N, 2^N)$.*

Proof $\Rightarrow$) Supposing α consistent, build a permutation σ as follows: put $\sigma(n) = \alpha(N)$, then $\sigma(n-1) = \alpha(N \setminus \{\sigma(n)\})$, $\sigma(n-2) = \alpha(N \setminus \{\sigma(n), \sigma(n-1)\})$, etc. Since α is consistent, the selector α choose $\sigma(i)$ in any subset containing $\sigma(i)$, hence the corresponding selector value is the marginal vector $x^{\sigma,v}$.

$\Leftarrow$) By the above property, if $\alpha(T) = i = \sigma(j) \in S \subseteq T$, then $\alpha(S) = \sigma(j)$ too. □

We give now an explicit expression of the elements of the selectope through the notion of sharing system. A *sharing system* on N is a function $\lambda : 2^N \times N \to [0, 1]$ satisfying the following conditions:

(i) $\sum_{i \in B} \lambda(B, i) = 1$ for all $\varnothing \neq B \subseteq N$;
(ii) $\lambda(B, i) = 0$ whenever $i \notin B$.

In words, a sharing system indicates which proportion should be given to each element of a subset of N. We denote by $\Lambda(N)$ the set of all sharing systems on N. A given sharing system λ induces a value, called *sharing value*, and defined by

$$\phi_i^\lambda(v) = \sum_{S \ni i} \lambda(S, i) m^v(S) \qquad (i \in N).$$

Any selector α corresponds to some particular sharing system λ_α, by the following relation

$$\lambda_\alpha(S, i) = \begin{cases} 1, & \text{if } \alpha(S) = i \\ 0, & \text{otherwise.} \end{cases}$$

Evidently, we have the identity $\phi^\alpha = \phi^{\lambda_\alpha}$, which permits to show the following relation.

Lemma 3.60 *For any game v on N,*

$$\mathbf{sel}(v) = \{\phi^\lambda(v) : \lambda \in \Lambda(N)\}.$$

Proof Consider for any v a convex combination $\beta_1 \phi^{\alpha_1}(v) + \cdots + \beta_q \phi^{\alpha_q}(v)$ of selector values. Define the following sharing system λ:

$$\lambda(S, i) = \sum_{k \in \{1, \ldots, q\} : \alpha_k(S) = i} \beta_k \qquad (\varnothing \neq S \subseteq N, i \in S),$$

and $\lambda(S,i) = 0$ if $i \notin S$. Then λ is indeed a sharing system because for every nonempty S

$$\sum_{i\in S}\lambda(S,i) = \sum_{i\in S}\sum_{k\in\{1,\dots,q\}\,:\,\alpha_k(S)=i}\beta_k = \sum_{k=1}^{q}\beta_k = 1.$$

Then by construction we have $\beta_1\phi^{\alpha_1}(v) + \cdots + \beta_q\phi^{\alpha_q}(v) = \phi^{\lambda}(v)$.

To prove the converse, because $\phi^{\alpha}, \phi^{\lambda}$ are values that are linear on $\mathcal{G}(N)$, it suffices to prove the result for any unanimity game u_T, $\varnothing \neq T \subseteq N$. Take some sharing system λ and consider $\phi^{\lambda}(u_T)$ for some nonempty $T \subseteq N$. Since m^{u_T} is nonzero only for T, we simply put

$$\beta_1 = \lambda(T, i_1), \dots, \beta_{|T|} = \lambda(T, i_{|T|})$$

with $T = \{i_1, \dots, i_{|T|}\}$. Then the coefficients $\beta_1, \dots, \beta_{|T|}$ define a convex combination, and letting $\alpha_j(T) = i_j$ for $j = 1, \dots, |T|$, we have

$$\beta_1\phi^{\alpha_1}(u_T) + \cdots + \beta_{|T|}\phi^{\alpha_{|T|}}(u_T) = \phi^{\lambda}(u_T).$$

□

Example 3.61 (The Shapley value is the uniform sharing value) Apart from selectors that choose in each subset S only one element and are thus very specific sharing systems, another remarkable instance of a sharing system is the *uniform sharing system*, which allocates to every element in S the same quantity $\frac{m^v(S)}{|S|}$. The corresponding sharing value is then

$$\phi_i(v) = \sum_{S\ni i}\frac{m^v(S)}{|S|} \qquad (i \in N).$$

Comparing with (2.41) taking $A = \{i\}$ and remembering that the Shapley value corresponds to the interaction transform for singletons, one realizes that the uniform sharing value is nothing but the Shapley value.

By contrast, the Banzhaf value is not a sharing value as one can see from (2.47) with $A = \{i\}$. Indeed, the underlying sharing system would be given by $\lambda(S,i) = \frac{1}{2}^{|S|-1}$, so that $\sum_{i\in S}\lambda(S,i) \neq 1$, unless $|S| = 2$. ◇

The next theorem gathers the main properties of the selectope. Beforehand, for any game v on N, we define the game $v^* = v^+ - \overline{v^-}$, where v^+, v^- are defined by (2.53), (2.54), and $\bar{\cdot}$ indicates conjugation; see (2.1) (recall that $v = v^+ - v^-$ and that v^+, v^- are totally monotone, hence supermodular).

Theorem 3.62 (Main properties of the selectope) *For any game v on N, the following holds:*

(i) v^* *is supermodular and* $\mathbf{sel}(v) = \mathbf{core}(v^+) - \mathbf{core}(v^-) = \mathbf{core}(v^*)$*;*
(ii) $\mathbf{core}(v) \subseteq \mathbf{Web}(v) \subseteq \mathbf{sel}(v)$*;*
(iii) $\mathbf{core}(v) = \mathbf{sel}(v)$ *if and only if* $m^v(S) \geqslant 0$ *for all* $S \subseteq N$, $|S| > 1$*;*
(iv) $\mathbf{sel}(v) = \mathbf{Web}(v)$ *if and only if all permutations* σ *are consistent*[9] *in* v*;*
(v) Let $D^+(v) = \bigcup\{S : |S| > 1, m^v(S) > 0\}$ *be the set of elements in* N *that belong to non-singleton coalitions with positive Möbius transform, and define* $D^-(v)$ *similarly. Then, if* $|D^+(v) \cap D^-(v)| \leqslant 1$, $\mathbf{sel}(v) = \mathbf{Web}(v)$.

Proof

(i) Step 1: We first prove that $\mathbf{sel}(v^+) = \mathbf{core}(v^+)$ and $\mathbf{sel}(v^-) = \mathbf{core}(v^-)$. Since v^+ is supermodular, by Theorem 3.15, it is the convex hull of all marginal vectors, hence $\mathbf{core}(v^+) \subseteq \mathbf{sel}(v^+)$. Consider now any selector value ϕ^α. We have, for any nonempty coalition S,

$$\begin{aligned}\phi^\alpha(v^+)(S) &= \sum_{i\in S}\sum_{T:\alpha(T)=i} m^{v^+}(T)\\ &\geqslant \sum_{T\subseteq S} m^{v^+}(T) = v^+(S),\end{aligned}$$

where the inequality follows by nonnegativity of m^{v^+}. This proves $\mathbf{sel}(v^+) \subseteq \mathbf{core}(v^+)$, and thus the desired equality holds. The equality for v^- is obtained in the same way.

Step 2: We prove the first equality $\mathbf{sel}(v) = \mathbf{core}(v^+) - \mathbf{core}(v^-)$. By linearity of the sharing value, we have, for any selector α, $\phi^\alpha(v) = \phi^\alpha(v^+) - \phi^\alpha(v^-)$, hence by Step 1, $\phi^\alpha(v) \in \mathbf{core}(v^+) - \mathbf{core}(v^-)$, which proves $\mathbf{sel}(v) \subseteq \mathbf{core}(v^+) - \mathbf{core}(v^-)$. For the reverse inclusion, it is enough by Step 1 to prove that $\phi^\alpha(v^+) - \phi^\beta(v^-) \in \mathbf{sel}(v)$ for arbitrary selectors α, β. Define the selector γ by $\gamma(S) = \alpha(S)$ if $m^v(S) \geqslant 0$ and $\gamma(S) = \beta(S)$ otherwise. Then $\mathbf{sel}(v) \ni \phi^\gamma(v) = \phi^\alpha(v^+) - \phi^\beta(v^-)$.

Step 3: We prove the second equality $\mathbf{core}(v^+) - \mathbf{core}(v^-) = \mathbf{core}(v^*)$. We have noted that v^+ is supermodular. Since v^- is supermodular too, $-v^-$ is submodular and by Theorem 2.20(ii), $\overline{-v^-} = -\overline{v^-}$ is supermodular, and so is v^*. It follows from Theorem 3.17(iii) that

$$\mathbf{core}(v^*) = \mathbf{core}(v^+ - \overline{v^-}) = \mathbf{core}(v^+) + \mathbf{core}(-\overline{v^-}).$$

The desired equality follows by the relation $-\mathbf{core}(-\overline{v}) = \mathbf{core}(v)$ [see (3.2)].

[9] A permutation σ is consistent in v if it is possible to find a consistent selector α satisfying $\alpha(T) = \max_\sigma(T)$ for all T such that $m^v(T) > 0$ and $\alpha(T) = \min_\sigma(T)$ for all T such that $m^v(T) < 0$, where $\max_\sigma(T)$ is the last element (highest rank) in T according to σ, and similarly for $\min_\sigma(T)$.

(ii) The first inclusion is proved in Theorem 3.14. As for the second, we have already observed that any marginal vector of v belongs to the selectope of v.

(iii) Suppose equality between the core and the selectope. In particular, this implies that every element x of the selectope satisfies $x_i \geqslant v(\{i\})$ for all $i \in N$. Assume that there exists $S \subseteq N$, $|S| > 1$ such that $m^v(S) < 0$. Choose $i \in S$ and let α be a selector satisfying $\alpha(S) = i$ and $\alpha(T) \neq i$ for every $T \neq S, \{i\}$. Then

$$\phi_i^{\alpha}(v) = v(\{i\}) + m^v(S) < v(\{i\}),$$

a contradiction.

Conversely, suppose that $m^v(S) \geqslant 0$ for all $S \subseteq N$, $|S| > 1$. Then $\mathbf{core}(v) = \mathbf{core}(v^+ - v^-) = \mathbf{core}(v^+) - \mathbf{core}(v^-) = \mathbf{sel}(v)$, where the second equality holds by Theorem 3.17(iii) and (3.2), because $-v^-$ is an additive game, and the third one follows by (i).

(iv) $\Leftarrow$) By (i), any marginal vector x^{σ,v^*} of v^* is an extreme point of $\mathbf{sel}(v)$. Moreover,

$$\begin{aligned} x^{\sigma,v^*}_{\sigma(i)} &= v^*(\{\sigma(1),\ldots,\sigma(i)\}) - v^*(\{\sigma(1),\ldots,\sigma(i-1)\}) \\ &= v^+(\{\sigma(1),\ldots,\sigma(i)\}) - v^+(\{\sigma(1),\ldots,\sigma(i-1)\}) \\ &\quad + v^-(\{\sigma(i+1),\ldots,\sigma(n)\}) - v^-(\{\sigma(i),\ldots,\sigma(n)\}) \\ &= \sum_{\substack{T\subseteq\{\sigma(1),\ldots,\sigma(i)\}:m^v(T)>0 \\ T\ni\sigma(i)}} m^v(T) + \sum_{\substack{T\subseteq\{\sigma(i),\ldots,\sigma(n)\}:m^v(T)<0 \\ T\ni\sigma(i)}} m^v(T). \end{aligned} \tag{3.23}$$

Denoting by $\max_\sigma(T)$ the last element in T according to σ, and similarly for $\min_\sigma(T)$, it follows that for any selector α satisfying $\alpha(T) = \max_\sigma(T)$ for all T such that $m^v(T) > 0$ and $\alpha(T) = \min_\sigma(T)$ for all T such that $m^v(T) < 0$, we have $x^{\sigma,v^*} = \phi^\alpha(v)$. Since σ is consistent in v, it is always possible to choose such an α that is consistent. By Lemma 3.59, ϕ^α is a marginal value of v; i.e., $\phi^\alpha = x^{\pi,v}$ for some permutation π. By assumption, any σ is consistent in v, so that $\mathbf{Web}(v) \supseteq \mathbf{sel}(v)$, implying the desired equality by (ii).

$\Rightarrow$) Suppose that there exists a permutation σ that is not consistent in v (w.l.o.g. we may assume that this is the identity permutation). Take $p \in \mathbb{R}^N$ with $p_1 < p_2 < \cdots < p_n$. Let α be a consistent selector such that $p \cdot \phi^\alpha(v)$ is maximal subject to all consistent selectors. Because σ is not consistent in v, we can take a coalition S with either $\alpha(S) \neq \max_\sigma(S)$ and $m^v(S) > 0$, or $\alpha(S) \neq \min_\sigma(S)$ and $m^v(S) < 0$. Define α' by $\alpha'(S) = \max_\sigma(S)$ if $m^v(S) > 0$ or $\alpha'(S) = \min_\sigma(S)$ if $m^v(S) < 0$, and $\alpha'(T) = \alpha(T)$ for any $T \neq S$. If $m^v(S) > 0$, we have

$$p \cdot \phi^{\alpha'}(v) = p \cdot \phi^\alpha(v) + (p_{\max_\sigma(S)} - p_{\alpha(S)})m^v(S) > p \cdot \phi^\alpha(v),$$

and if $m^v(S) < 0$, then

$$p \cdot \phi^{\alpha'}(v) = p \cdot \phi^{\alpha}(v) + (p_{\min_\sigma(S)} - p_{\alpha(S)})m^v(S) > p \cdot \phi^{\alpha}(v).$$

In both cases we have $p \cdot \phi^{\alpha'}(v) > p \cdot \phi^{\alpha}(v)$, which proves that $\mathbf{sel}(v) \neq \mathbf{Web}(v)$.

(v) In view of (i), it is sufficient to show that under the given condition, any marginal vector x^{σ,v^*} is also a marginal vector of v; i.e., $x^{\sigma,v^*} = x^{\pi,v}$ for some appropriate π. From (3.23), we see that if there is no S s.t. $|S| > 1$ and $m^v(S) < 0$, then it suffices to take $\pi = \sigma$. Therefore, assume on the contrary that such an S exists, and $D^+(v) \cap D^-(v) = \{i_0\}$. W.l.o.g, fix σ to be the identity permutation. Recall that $x^{\pi,v}_{\pi(i)} = \sum_{T \subseteq \{\pi(1),\dots,\pi(i)\}, T \ni \pi(i)} m^v(T)$.

Step 1. Suppose only one such S exists, and denote by i_f, i_l the first and last element of S according to σ. By (3.23), observe that the term $m^v(S)$ can be present in x^{σ,v^*}_i only if $i = i_f$, while for $x^{\sigma,v}$ this can happen only for $i = i_l$. It follows that $x^{\sigma,v^*}_i = x^{\sigma,v}_i$ for all $i \neq i_f, i_l$. Suppose $i_0 = i_f$. Then

$$\begin{aligned} x^{\sigma,v^*}_{i_f} &= \sum_{T \subseteq \{1,\dots,i_f\}, T \ni i_f} m^v(T) + m^v(S) \\ x^{\sigma,v^*}_{i_l} &= 0. \end{aligned}$$

If $i_0 = i_l$, we find

$$\begin{aligned} x^{\sigma,v^*}_{i_f} &= m^v(S) \\ x^{\sigma,v^*}_{i_l} &= \sum_{T \subseteq \{1,\dots,i_l\}, T \ni i_l} m^v(T). \end{aligned}$$

Lastly, if $i_0 \neq i_f, i_l$, we find $x^{\sigma,v^*}_{i_f} = m^v(S)$ and $x^{\sigma,v^*}_{i_l} = 0$. Consider the permutation π that exchanges i_l and i_f in σ. Then it can be checked that $x^{\sigma,v^*} = x^{\pi,v}$ in any case.

Step 2. Suppose several such S exists, say S_1, S_2 (proceed similarly if more than two subsets), and denote the "endpoints" of these sets by $i^1_f, i^1_l, i^2_f, i^2_l$. As in Step 1, the term $m^v(S_1)$ appears only for $i = i^1_f$, similarly for $m^v(S_2)$, and the position of i_0 on one of the endpoints determine where is the term of the form $\sum_{T \subseteq \{1,\dots,j\}, T \ni j} m^v(T)$. Therefore, the adequate permutation π consists in taking the reverse order on $S_1 \cup S_2$, and being equal to σ otherwise (e.g., with $n = 7$, if $S_1 = \{2, 5\}$ and $S_2 = \{4, 5, 6\}$, the order induced by π would be 1,6,3,5,4,2,7).

□

Remark 3.63

(i) The selectope was first proposed by Hammer et al. in 1977 [189] and rediscovered several times, most notably by Chateauneuf and Jaffray [49], and studied in depth by Derks et al. [85]. All results from this section are from the latter reference, while (ii) and (iii) in Theorem 3.62 were already proved in [49, 189], using flow methods. Sharing values were also rediscovered by Billot and Thisse [28], under the name *Möbius values*, while sharing values corresponding to selectors appear in Dubois and Prade [104].

Although never explicitly defined, sharing values and the selectope underlie many works in the theory of belief functions/measures (Chap. 7).

(ii) The necessary and sufficient condition (iv) in Theorem 3.62 is quite technical and difficult to use in practice, but condition (v) is much more handy. It is easy to see that games with nonnegative (or nonpositive) Möbius transform on non-singletons coalitions satisfy this condition.

Chapter 4
Integrals

It is well known that in the case of classical (additive) measures, the Lebesgue integral is the usual definition of an integral with respect to a measure, and it allows the computation of the expected value of random variables. The question which is addressed in this chapter is: How to define the integral of a function with respect to a nonadditive measure, i.e., a capacity or a game? As we will see, the answer is not unique, and there exist many definitions in the literature. Nevertheless, two concepts of integrals emerge: the one proposed by Choquet in 1953, and the one proposed by Sugeno in 1974. Both are based on the decumulative distribution function of the integrand w.r.t. the capacity, the Choquet integral being the area below the decumulative function, and the Sugeno integral being the value at the intersection with the diagonal. Most of the other concepts of integral are also based on the decumulative function, like the Shilkret integral, but other approaches are possible. For example, the concave integral proposed by Lehrer is defined as the lower envelope of a class of concave and positively homogeneous functionals.

Integrals being naturally defined for functions on an arbitrary (infinite) universe, we suppose in this chapter that X is an arbitrary nonempty set, in contradiction with the general philosophy of the book, which is to work on finite sets. As far as heavy topological and measure-theoretic notions are not needed, we give definitions and establish results in the general (infinite) case, before specializing to the discrete case. More detailed expositions in a fully measure-theoretic framework can be found in Denneberg [80], Marinacci and Montrucchio [235], Wang and Klir [343], see also Murofushi and Sugeno [250, 252, 255].

The chapter mainly studies in parallel the Choquet integral and the Sugeno integral. Their definition are first given for nonnegative functions (Sect. 4.2) and then extended to real-valued functions (Sect. 4.3), which lead to two kinds of integrals, the symmetric and the asymmetric one. In Sect. 4.4, the case of simple functions is studied, which leads naturally to the discrete case (Sect. 4.5). The properties of both integrals are studied in depth in Sect. 4.6, followed by

M. Grabisch, *Set Functions, Games and Capacities in Decision Making*,
Theory and Decision Library C 46, DOI 10.1007/978-3-319-30690-2_4

results on characterization (Sect. 4.8). Other minor topics are studied (expression w.r.t. various transforms, particular cases, integrands defined on the real line, etc.) before introducing other integrals (Sect. 4.11): the Shilkret integral, the concave integral, the decomposition integral and various pseudo-integrals. The chapter ends with an extension of the Choquet integral to nonmeasurable functions (Sect. 4.12).

Throughout the chapter, all capacities and games are finite; i.e., $\mu(X) < \infty$.

4.1 Simple Functions

Let X be arbitrary. A function $f : X \to \mathbb{R}$ is *simple* if its range $\mathbf{ran}f$ is a finite set. We give several ways of decomposing a simple nonnegative function f using characteristic functions. We assume $\mathbf{ran}f = \{a_1, \ldots, a_n\}$, supposing $0 \leqslant a_1 < a_2 < \cdots < a_n$. One can easily check that

$$\begin{aligned} f &= \sum_{i=1}^{n} a_i 1_{\{x\in X : f(x)=a_i\}} \\ &= \sum_{i=1}^{n} (a_i - a_{i-1}) 1_{\{x\in X : f(x)\geqslant a_i\}} \end{aligned} \tag{4.1}$$

letting $a_0 = 0$. These decompositions are respectively called the *vertical* and the *horizontal* decompositions. These names should be clear from Fig. 4.1 illustrating them.

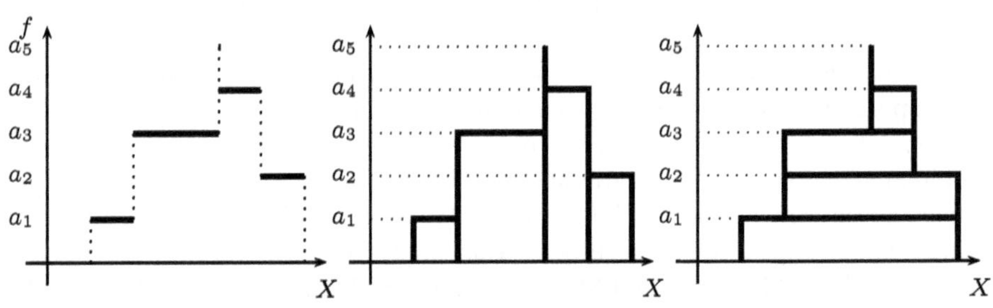

Fig. 4.1 Decompositions of a simple function f on X (function on *left*): vertical (*middle*), and horizontal decomposition (*right*)

4.2 The Choquet and Sugeno Integrals for Nonnegative Functions

We consider an arbitray set X, together with an algebra $\mathcal{F}$ (Definition 2.102). A function $f : X \to \mathbb{R}$ is *$\mathcal{F}$-measurable* if the sets $\{x : f(x) > t\}$ and $\{x : f(x) \geqslant t\}$ belong to $\mathcal{F}$ for all $t \in \mathbb{R}$.

We denote by $B(\mathcal{F})$ the set of bounded $\mathcal{F}$-measurable functions, and by $B^+(\mathcal{F})$ the set of bounded $\mathcal{F}$-measurable nonnegative functions.

Lemma 4.1 *The set $B(\mathcal{F})$ endowed with the usual order on functions is a lattice; i.e., $f, g \in B(\mathcal{F})$ imply that $f \vee g$ and $f \wedge g$ belong to $B(\mathcal{F})$.*

(The proof is left to the readers.) Evidently, the same holds for $B^+(\mathcal{F})$.

For any function $f \in B(\mathcal{F})$ and a capacity μ, we introduce the *decumulative distribution function* or *survival function* $G_{\mu,f} : \mathbb{R} \to \mathbb{R}$, which is defined by

$$G_{\mu,f}(t) = \mu(\{x \in X : f(x) \geqslant t\}) \quad (t \in \mathbb{R}). \tag{4.2}$$

We notice that $G_{\mu,f}$ is well-defined because f is $\mathcal{F}$-measurable. Some authors replace "$\geqslant$" by "$>$," but as we will see by Lemma 4.5, this is unimportant.

For convenience, we often use the shorthands $\mu(f \geqslant t)$ and $\mu(f > t)$ for $\mu(\{x \in X : f(x) \geqslant t\})$ and $\mu(\{x \in X : f(x) > t\})$.

We establish basic properties of the decumulative distribution function. Before that, we introduce the notions of essential supremum and infimum.

Definition 4.2 For any $f \in B(\mathcal{F})$ and any capacity μ on $(X, \mathcal{F})$, the *essential supremum* and *essential infimum* of f w.r.t. μ are defined by

$$\text{ess sup}_\mu f = \inf\{t : \{x \in X : f(x) > t\} \text{ is null w.r.t. } \mu\}$$

$$\text{ess inf}_\mu f = \sup\{t : \{x \in X : f(x) < t\} \text{ is null w.r.t. } \mu\}$$

respectively (see Definition 2.107 for the definition of a null set).

Lemma 4.3 *Let $f \in B^+(\mathcal{F})$ and μ be a capacity on $(X, \mathcal{F})$. Then $G_{\mu,f} : \mathbb{R} \to \mathbb{R}$*

(i) is a nonnegative nonincreasing function, with $G_{\mu,f}(0) = \mu(X)$;
(ii) $G_{\mu,f}(t) = \mu(X)$ on the interval $[0, \text{ess inf}_\mu f]$;
(iii) has a compact support, namely $[0, \text{ess sup}_\mu f]$.

Proof (i) Obvious by monotonicity of μ and the fact that $t > t'$ implies $\{x : f(x) \geqslant t\} \subseteq \{x : f(x) \geqslant t'\}$.

(ii) By definition, $N = \{f < \text{ess inf}_\mu f\}$ is a null set, hence $G_{\mu,f}(\text{ess inf}_\mu f) = \mu(X \setminus N) = \mu(X)$ by Theorem 2.108(vi).

(iii) Since f is bounded, so is its essential supremum. Now, by definition $\{x : f(x) > \text{ess sup}_\mu f\}$ is a null set, therefore $G_{\mu,f}(t) = 0$ if $t > \text{ess sup}_\mu f$. □

Figure 4.2 illustrates these definitions and properties.

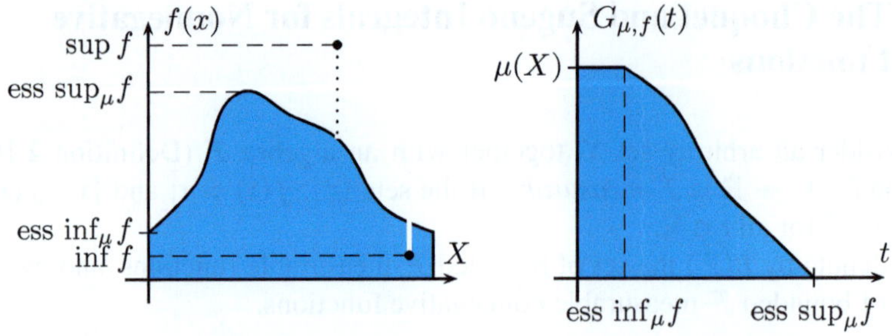

Fig. 4.2 A bounded nonnegative measurable function (*left*) and its decumulative distribution (*right*), supposing that singletons are null sets

Definition 4.4 Let $f \in B^+(\mathcal{F})$ and μ be a capacity on $(X, \mathcal{F})$. The *Choquet integral* of f w.r.t. μ is defined by

$$\int f \, \mathrm{d}\mu = \int_0^\infty G_{\mu,f}(t) \, \mathrm{d}t, \tag{4.3}$$

where the right hand-side integral is the Riemann integral.

Let us check if the Choquet integral is well-defined. As shown in Lemma 4.3, the decumulative function is a decreasing function bounded by $\mu(X) < \infty$, with compact support. Hence it is Riemann-integrable, so the Choquet integral is well-defined.

We prove now that it is equivalent to put a strict inequality in the definition of the decumulative function.

Lemma 4.5 *Let $f \in B^+(\mathcal{F})$ and μ be a capacity. Then*

$$\int_0^\infty \mu(f \geqslant t) \, \mathrm{d}t = \int_0^\infty \mu(f > t) \, \mathrm{d}t.$$

Proof (We follow Marinacci and Montrucchio [235].) Set for simplicity $G'_{\mu,f}(t) = \mu(\{x : f(x) > t\})$ for each $t \in \mathbb{R}$. We have for each $t \in \mathbb{R}$ and each $n \in \mathbb{N}$

$$\Big\{x : f(x) \geqslant t + \frac{1}{n}\Big\} \subseteq \{x : f(x) > t\} \subseteq \{x : f(x) \geqslant t\}$$

which yields

$$G_{\mu,f}\Big(t + \frac{1}{n}\Big) \leqslant G'_{\mu,f}(t) \leqslant G_{\mu,f}(t).$$

If $G_{\mu,f}$ is continuous at t, we have

$$G_{\mu,f}(t) = \lim_{n\to\infty} G_{\mu,f}\Big(t + \frac{1}{n}\Big) \leqslant G'_{\mu,f}(t) \leqslant G_{\mu,f}(t)$$

hence equality holds throughout. Otherwise, as $G_{\mu,f}$ is a nonincreasing function, it is discontinuous on an at most countable set $T \subseteq \mathbb{R}$. Hence both functions are equal for all $t \notin T$, which in turn implies that $\int_0^\infty G'_{\mu,f}(t)\,\mathrm{d}t = \int_0^\infty G_{\mu,f}(t)\,\mathrm{d}t$ by standard results on Riemann integration. □

We turn now to the Sugeno integral.

Definition 4.6 Let $f \in B^+(\mathcal{F})$ be a function and μ be a capacity on $(X, \mathcal{F})$. The *Sugeno integral* of f w.r.t. μ is defined by

$$⨍ f\,\mathrm{d}\mu = \bigvee_{t\geqslant 0}(G_{\mu,f}(t) \wedge t) = \bigwedge_{t\geqslant 0}(G_{\mu,f}(t) \vee t). \tag{4.4}$$

In words, the Sugeno integral is the abscissa of the intersection point between the diagonal and the decumulative function, while the Choquet integral is the area below the decumulative function (Fig. 4.3).

One can easily check that the second equality holds in (4.4).

Remark 4.7 As for the Choquet integral, $G_{\mu,f}(t)$ can be replaced by $G'_{\mu,f}(t) = \mu\{x : f(x) > t\}$ without change. Indeed, we have proved that $G_{\mu,f}$ and $G'_{\mu,f}$ only differ at discontinuity points, and for those points, it can be checked that the two definitions lead to the same result. ◇

We immediately give a useful alternative formula for the Sugeno integral.

Lemma 4.8 *Let $f \in B^+(\mathcal{F})$ be a function and μ be a capacity on $(X, \mathcal{F})$. The Sugeno integral can be written as follows:*

$$⨍ f\,\mathrm{d}\mu = \bigvee_{A\in\mathcal{F}}\Big(\bigwedge_{x\in A} f(x) \wedge \mu(A)\Big). \tag{4.5}$$

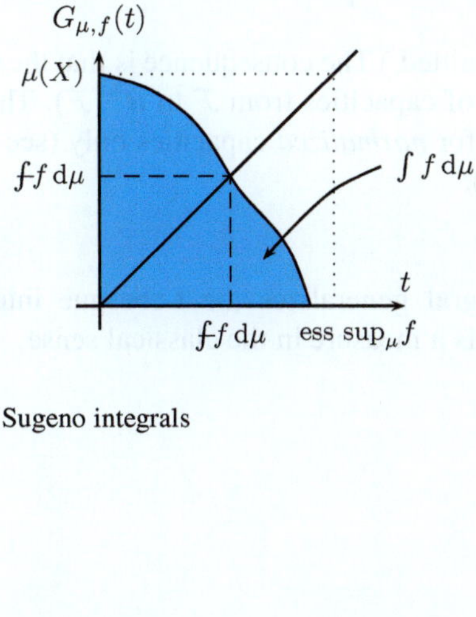

Fig. 4.3 The Choquet and Sugeno integrals

Proof For any $t \geqslant 0$, because $\bigwedge_{x:f(x)\geqslant t} f(x) \geqslant t$ and $\{f \geqslant t\} \in \mathcal{F}$ we have

$$t \wedge \mu(f \geqslant t) \leqslant \bigvee_{A\in\mathcal{F}} \Big(\bigwedge_{x\in A} f(x) \wedge \mu(A)\Big),$$

which yields

$$⨍ f\,\mathrm{d}\mu = \bigvee_{t\geqslant 0} \Big(t \wedge \mu(f \geqslant t)\Big) \leqslant \bigvee_{A\in\mathcal{F}} \Big(\bigwedge_{x\in A} f(x) \wedge \mu(A)\Big). \tag{4.6}$$

Now, for any given $A \in \mathcal{F}$, taking $t' = \bigwedge_{x\in A} f(x)$, we get $A \subseteq \{f \geqslant t'\}$. Applying monotonicity of μ, we obtain

$$\bigwedge_{x\in A} f(x) \wedge \mu(A) \leqslant t' \wedge \mu(f \geqslant t') \leqslant \bigvee_{t\geqslant 0} \Big(t \wedge \mu(f \geqslant t)\Big) = ⨍ f\,\mathrm{d}\mu$$

for any $A \in \mathcal{F}$. Consequently,

$$\bigvee_{A\in\mathcal{F}} \Big(\bigwedge_{x\in A} f(x) \wedge \mu(A)\Big) \leqslant ⨍ f\,\mathrm{d}\mu,$$

which, with (4.6), permits to conclude. □

This result is already in the original work of Sugeno [319] (see also Wang and Klir [343, Theorem 9.1]).

A fundamental fact is the following.

Lemma 4.9 *Let $A \in \mathcal{F}$ (i.e., 1_A is measurable). Then for every capacity μ*

$$\int 1_A\,\mathrm{d}\mu = \mu(A). \tag{4.7}$$

(Proof is obvious and omitted.) The consequence is that the Choquet integral can be viewed as an extension of capacities from $\mathcal{F}$ to $B^+(\mathcal{F})$. The same statement holds for the Sugeno integral for *normalized* capacities only (see Theorem 4.43(iii) for a more general statement).

Remark 4.10

(i) The Choquet integral generalizes the Lebesgue integral, and the latter is recovered when μ is a measure in the classical sense.

(ii) As the name indicates, the Choquet integral was introduced by Gustave Choquet[1] [53], although this reference does not mention explicitly any notion of integral. As many great ideas, the Choquet integral was rediscovered many times. The first appearance seems to be due to Vitali[2] [332], whose integral for inner and outer Lebesgue measures is exactly the Choquet integral for these measures. We mention also Šipoš [334], who introduced the symmetric version of the Choquet integral (Sect. 4.3.1) as the limit of finite sums computed over finite subsets of $\mathbb{R}$ containing 0. Also, the expression of the Choquet integral in the discrete case can be found in the 1967 paper of Dempster on upper and lower probabilities [77, Eq. (2.10)], as well as in the works of Lovász [226] (known under the name of Lovász extension; see Sect. 2.16.4). Up to the knowledge of the author, the first appearance of the name "Choquet integral" is due to Schmeidler [286] in 1986, followed independently by Murofushi and Sugeno in 1989 [250]. As mentioned by Chateauneuf and Cohen [48, Footnote 10], Schmeidler in fact rediscovered the Choquet integral, and became aware that is was previously introduced by Choquet through private discussions with Jean-François Mertens, who drew his attention to the 1971 paper by Dellacherie [76], showing that the Choquet integral is comonotonically additive and monotone (a fact, by the way, duly acknowledged by Schmeidler himself in [286]).

(iii) The Sugeno integral was introduced by Michio Sugeno[3] in 1972 [318–320] under the name of *fuzzy integral.*[4] As for the Choquet integral, this functional was in fact known as early as 1944, under the name of Ky Fan[5] distance [137]. This distance is defined as

$$\|f - g\|_0 = \bigvee\{x : x > 0, G_{\nu,|f-g|}(x)/x < 1\}$$

with ν a σ-additive probability. Hence, the Sugeno integral of f corresponds to the Ky Fan distance of f to the null function; i.e., $\|f - 0\|_0$.

[1]Gustave Choquet (Solesmes, 1915 – Lyon, 2006) is a French mathematician. He was professor at Université Pierre et Marie Curie in Paris and at École Polytechnique, and his main contributions include functional analysis, potential and capacity theory, topology and measure theory.

[2]Giuseppe Vitali (Ravenna, 1875 – Bologna, 1932), Italian mathematician. His contributions concern measure theory.

[3]Michio Sugeno (Yokohama, 1940–), Japanese computer scientist and mathematician. He has been professor at Tokyo Institute of Technology. Apart his contribution to measure theory, he mainly works in the field of artificial intelligence.

[4]As mentioned on p.28, Sugeno used instead of "capacity" the term "fuzzy measure," which he introduced, in the idea of representing human subjectivity.

[5]Ky Fan (Hangzhou, 1914 – Santa Barbara CA, 2010) Chinese mathematician. He received his D.Sc. degree in Paris under the supervision of M. Fréchet, and then did all his career in the United States, mainly at UCSB. He worked in convex analysis and topology. The "Fan inequality" is famous and generalizes Cauchy-Schwarz inequality.

(iv) An important observation is that the Sugeno integral can live on very poor structures for the range of the integrands and capacities: the richness of the real line is not needed, contrarily to the Choquet integral, and the definition of the Sugeno integral works on any totally ordered set L, like $\mathbb{N}$ or any finite subset of it, provided L is considered to be a set of "positive" values (see Sect. 4.3.2 for the general case where negative values are allowed). In particular, L can be taken as a *qualitative scale*; i.e., a finite totally ordered set of qualitative degrees of evaluation, like {*very bad, bad, medium, good, very good*}. The only requirement is that the range of μ and f should be included in L.

(v) The Choquet and Sugeno integrals are defined for (finite) capacities, but their definitions still work for games, provided they are of bounded variation norm (Sect. 2.19.1) in the case of the Choquet integral. However, note that in this case, the decumulative function is no longer nonincreasing in general, which causes the second equality in (4.4) not to hold any more! Therefore, it is better to consider that the Sugeno integral is not well defined in the case of nonmonotonic games. Alternatively, one may decide to define the Sugeno integral by, e.g., the expression with the supremum.

Lastly, note that the Choquet integral is not defined for set functions ξ such that $\xi(\varnothing) \neq 0$. Indeed, the area under the decumulative function may become infinite in this case.

(vi) We may define these integrals on a restricted domain $A \subseteq X$: in this case, we write

$$\int_A f \, \mathrm{d}\mu = \int_0^\infty \mu(\{f \geqslant t\} \cap A) \, \mathrm{d}t, \quad ⨍_A f \, \mathrm{d}\mu = \bigvee_{t \geqslant 0} (\mu(\{f \geqslant t\} \cap A) \wedge t). \tag{4.8}$$

◇

4.3 The Case of Real-Valued Functions

We suppose now that f is a bounded measurable real-valued function. We decompose f into its positive and negative parts f^+, f^-:

$$f = f^+ - f^-, \text{ with } f^+ = 0 \vee f, \quad f^- = (-f)^+. \tag{4.9}$$

Note that both f^+, f^- are nonnegative functions in $B^+(\mathcal{F})$ (bounded and measurable).

4.3.1 The Choquet Integral

There are basically two ways for extending the Choquet integral to functions in $B(\mathcal{F})$. The simplest one is to consider that the integral is additive with respect to the above decomposition:

$$\check{\int} f\,\mathrm{d}\mu = \int f^+\,\mathrm{d}\mu - \int f^-\,\mathrm{d}\mu. \tag{4.10}$$

Observe that since $f^+, f^- \in B^+(\mathcal{F})$, the integral is well-defined, and that it satisfies

$$\check{\int} (-f)\,\mathrm{d}\mu = -\check{\int} f\,\mathrm{d}\mu, \tag{4.11}$$

because $(-f)^+ = f^-$ and $(-f)^- = f^+$. For this reason, this integral is called the *symmetric Choquet integral.*

Example 4.11 Consider $X = \mathbb{R}$ and the function f defined by $f(x) = \text{sign}\,(x)$. Then its symmetric integral is

$$\check{\int} \text{sign}\,\mathrm{d}\mu = \int 1_{\mathbb{R}_+}\,\mathrm{d}\mu - \int 1_{\mathbb{R}_-}\,\mathrm{d}\mu = \mu(\mathbb{R}_+) - \mu(\mathbb{R}_-),$$

using Lemma 4.9. Now, consider the nonnegative function $1_{\mathbb{R}} + \text{sign}$; i.e., the function defined by $f(x) = 1 + \text{sign}\,(x)$, $x \in \mathbb{R}$. We can easily check that

$$\check{\int} (1_{\mathbb{R}} + \text{sign}\,)\,\mathrm{d}\mu = \int (1_{\mathbb{R}} + \text{sign}\,)\,\mathrm{d}\mu = 2\mu(\mathbb{R}_+).$$

◇

The use of real-valued functions permits to consider translations of functions; i.e., to consider $f + \alpha 1_X$ with $\alpha \in \mathbb{R}$. We say that a functional I on $B(\mathcal{F})$ is *translation invariant* if for every $f \in B(\mathcal{F})$ and $\alpha \in \mathbb{R}$,

$$I(f + \alpha 1_X) = I(f) + \alpha I(1_X).$$

By Lemma 4.9, for any extension $\tilde{\int}$ of the Choquet integral, this definition reduces to

$$\tilde{\int} (f + \alpha 1_X)\,\mathrm{d}\mu = \tilde{\int} f\,\mathrm{d}\mu + \alpha\mu(X).$$

Example 4.11 shows that the symmetric integral is not invariant to translation, because $\check{\int}(f + 1_X)\,\mathrm{d}\mu \neq \check{\int} f\,\mathrm{d}\mu + \mu(X)$. There exists an extension of the Choquet integral that is translation invariant, given by

$$\int f\,\mathrm{d}\mu = \int f^+\,\mathrm{d}\mu - \int f^-\,\mathrm{d}\overline{\mu} \tag{4.12}$$

where $\overline{\mu}$ is the conjugate capacity. Observe that this integral is well defined for any $\mathcal{F}$-measurable function [see (4.13)], and is clearly an extension of the Choquet integral. We use the same symbol for the Choquet integral of nonnegative functions and its extension. This is because this extension is the usual one, sometimes called *asymmetric Choquet integral* (Denneberg [80]), because in general $\int(-f)\,\mathrm{d}\mu \neq -\int f\,\mathrm{d}\mu$. Without further indication, by "Choquet integral" we always mean its asymmetric extension. An equivalent and more explicit expression is

$$\int f\,\mathrm{d}\mu = \int_0^\infty \mu(f \geqslant t)\,\mathrm{d}t + \int_{-\infty}^0 \big(\mu(f \geqslant t) - \mu(X)\big)\,\mathrm{d}t. \tag{4.13}$$

Indeed,

$$\begin{aligned} -\int f^-\,\mathrm{d}\overline{\mu} = -\int_0^\infty \overline{\mu}(-f \geqslant t)\,\mathrm{d}t &= -\int_0^\infty \big(\mu(X) - \mu(\{-f \geqslant t\}^c)\big)\,\mathrm{d}t \\ &= \int_{-\infty}^0 \big(\mu(f \geqslant t) - \mu(X)\big)\,\mathrm{d}t. \end{aligned}$$

This formula has an easy graphical interpretation (Fig. 4.4). Note that $G_{\mu,f}(\text{ess inf}_\mu f) = \mu(X)$, similarly to the case of nonnegative functions (Lemma 4.3). Let us check translation invariance using (4.13). For any $\alpha \in \mathbb{R}$,

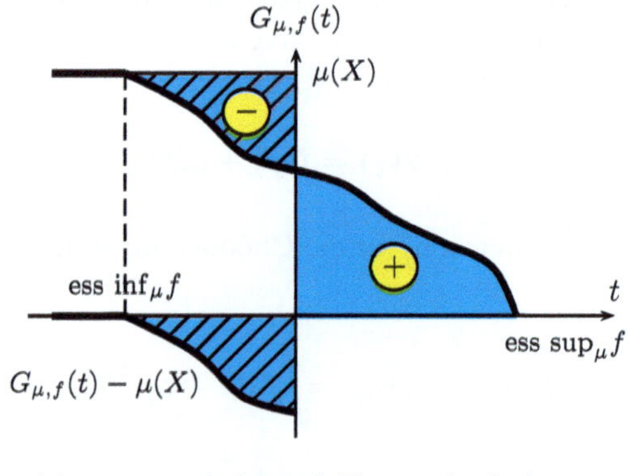

Fig. 4.4 Computation of the asymmetric integral. The two hatched areas are equal, and the value of the integral is the difference of the nonhatched area and one hatched area

$$\begin{aligned}\int (f + \alpha 1_X)\,\mathrm{d}\mu &= \int_0^\infty \mu(f + \alpha 1_X \geqslant t)\,\mathrm{d}t + \int_{-\infty}^0 \big(\mu(f + \alpha 1_X \geqslant t) - \mu(X)\big)\,\mathrm{d}t \\ &= \int_0^\infty \mu(f \geqslant t - \alpha)\,\mathrm{d}t + \int_{-\infty}^0 \big(\mu(f \geqslant t - \alpha) - \mu(X)\big)\,\mathrm{d}t \\ &= \int_{-\alpha}^\infty \mu(f \geqslant t')\,\mathrm{d}t' + \int_{-\infty}^{-\alpha} \big(\mu(f \geqslant t') - \mu(X)\big)\,\mathrm{d}t' \\ &= \int f\,\mathrm{d}\mu + \alpha\mu(X),\end{aligned}$$

letting $t' = t - \alpha$. We now prove that this extension is the only one satisfying translation invariance.

Theorem 4.12 *The unique extension of the Choquet integral on $B^+(\mathcal{F})$ to $B(\mathcal{F})$ that is translation invariant is the asymmetric integral given by (4.13).*

Proof (We follow Marinacci and Montrucchio [235].) Suppose that I is a functional on $B(\mathcal{F})$ that is translation invariant and coincides with the Choquet integral on $B^+(\mathcal{F})$. It suffices to show that I has the form given by (4.13). Take $f \in B(\mathcal{F})$ and suppose that $\inf f = \gamma < 0$. By translation invariance, $I(f - \gamma 1_X) = I(f) - \gamma I(1_X)$. As $f - \gamma 1_X$ belongs to $B^+(\mathcal{F})$, we can write:

$$\begin{aligned}I(f) &= I(f - \gamma 1_X) + \gamma I(1_X) \\ &= \int (f - \gamma 1_X)\,\mathrm{d}\mu + \gamma\mu(X) \\ &= \int_0^\infty \mu(f - \gamma 1_X \geqslant t)\,\mathrm{d}t + \gamma\mu(X) \\ &= \int_\gamma^\infty \mu(f \geqslant t')\,\mathrm{d}t' + \gamma\mu(X) \\ &= \int_\gamma^0 \mu(f \geqslant t')\,\mathrm{d}t' + \int_0^\infty \mu(f \geqslant t')\,\mathrm{d}t' - \int_\gamma^0 \mu(X)\,\mathrm{d}t'\end{aligned}$$

with $t' = t + \gamma$. As $\mu(f \geqslant t') - \mu(X) = 0$ for all $t' \leqslant \gamma$, it follows that

$$I(f) = \int_0^\infty \mu(f \geqslant t)\,\mathrm{d}t + \int_{-\infty}^0 \big(\mu(f \geqslant t) - \mu(X)\big)\,\mathrm{d}t,$$

the desired result. □

Remark 4.13 The integral introduced by Šipoš in 1979 [334] coincides with the symmetric Choquet integral. For this reason, the latter is sometimes called the *Šipoš integral*. Note also that Tversky and Kahneman [329] proposed with prospect theory a generalization of the symmetric Choquet integral (Sect. 5.2.7). ◇

4.3.2 The Sugeno Integral

The case of the extension of the Sugeno integral on functions taking negative values cause unexpected problems, because of the ordinal nature of this integral, defined solely with minimum and maximum. We begin by explaining the origin of the problem.

Suppose we want to define a symmetric Sugeno integral, in the spirit of (4.10). Simply replacing $\int$ by $⨍$ is tempting, however, one should be aware that doing so, two different algebraic systems are mixed: the algebra based on minimum and maximum, on which the Sugeno integral is built, and the usual algebra based on arithmetic operations $+, -, \times$, on which is based the Choquet integral. Mixing the two would forbid the Sugeno integral to work on finite chains, i.e., totally ordered sets [see Remark 4.10(iv)], which is one of the main interest of this integral. Hence, one has to find a substitute for the "$-$" operation.

Taking an algebraic point of view, we may see the quantity $a - b$ as $a + (-b)$, hence, considering that $\vee, \wedge$ should play the rôle of "$+$" and "$\cdot$", our task amounts to extending the minimum and maximum on negative numbers, while keeping as much as possible properties of the ring of real numbers. To this purpose, let us consider that integrands f and capacities μ take values in L, which is a set endowed with a total order $\geqslant$, and symmetric in the sense that for any $a \in L$, there exists an element $-a \in L$, and there is a unique element called "zero," denoted by 0, such that $-0 = 0$. Elements a such that $a > 0$ are called "positive," while those satisfying $a < 0$ are "negative." It is assumed that μ takes nonnegative values, and that $\mu(\varnothing) = 0$. Also, denote by L^+, L^- the set of positive and negative elements of L, respectively.

Considering that the maximum operator is defined on L^+, we seek an extension of it on L, denoted by $\varovee$, so that the symmetric integral, defined by

$$\check{⨍} f \, d\mu = ⨍ f^+ \, d\mu \varovee \left(- ⨍ f^- \, d\mu \right) \tag{4.14}$$

has suitable properties. Hence, our first requirement is:

R1. $\varovee$ coincides with $\vee$ on $(L^+)^2$.

Next, we require that a and $-a$ are symmetric w.r.t. $\varovee$:

R2. For every $a \in L$, $a \varovee (-a) = 0$.

This property would yield an integral of value 0 for a function f that has equal integrals of f^+ and f^-. Observe that surprisingly, these two simple requirements yield an operator that is not associative. Indeed, letting $L = \mathbb{Z}$, consider the aggregation of the values $-3, 3, 2$. We have

$$((-3) \otimes 3) \otimes 2) = 0 \otimes 2 = 0 \vee 2 = 2$$
$$(-3) \otimes (3 \otimes 2) = (-3) \otimes (3 \vee 2) = (-3) \otimes 3 = 0.$$

Hence, any extension of the maximum satisfying **R2** is by essence nonassociative (note however that associativity holds on L^+, and that (4.14) does not require associativity).

The lack of associativity forbids to infer the *rule of sign*, hence we are forced to put it as an axiom:

R3. For every $a, b \in L$, $(-a) \otimes (-b) = -(a \otimes b)$.

This rule is necessary for the symmetry property of the integral to hold:

$$\check{\int}(-f)\,\mathrm{d}\mu = \int f^-\,\mathrm{d}\mu \otimes \left(-\int f^+\,\mathrm{d}\mu\right) = -\left(\left(-\int f^-\,\mathrm{d}\mu\right) \otimes \int f^+\,\mathrm{d}\mu\right)$$
$$= -\check{\int} f\,\mathrm{d}\mu, \qquad (4.15)$$

and is therefore mandatory.

It can be shown that under the three requirements **R1**, **R2** and **R3**, the "best" definition for the symmetric maximum is given below, in the sense that no operator satisfying the three requirements is associative on a larger domain (for a detailed discussion including a proof of this result, see Grabisch et al. [177, Sect. 9.3] and the original paper [166]).

Definition 4.14 Let L be a symmetric totally ordered set. The *symmetric maximum* $\otimes : L^2 \to L$ is defined by:

$$a \otimes b = \begin{cases} -(|a| \vee |b|), & \text{if } b \neq -a \text{ and either } |a| \vee |b| = -a \text{ or } = -b \\ 0, & \text{if } b = -a \\ |a| \vee |b|, & \text{otherwise.} \end{cases} \qquad (4.16)$$

Observe that, except for the case $b = -a$, $a \otimes b$ equals the absolutely larger one of the two values a and b.[6]

[6] A simple image to understand the result of $a \otimes b$ when a and b have different sign is the image of a pair of scales. On one pan one puts the negative number, say a, and on the other pan one puts the positive number b, and the weight of a number is its absolute value. Then, if $a = -b$, the two pans

In summary, the symmetric Sugeno integral is defined by (4.14), with $\otimes$ given in Definition 4.14.

Remark 4.15

(i) The symmetric maximum on a symmetric real interval $[-\alpha, \alpha]$ can be obtained as the limit for $n \to \infty$ of a family of binary operators $\oplus_n$ defined on $[-\alpha, \alpha]$ by

$$a \oplus_n b = \left(a^{2n-1} + b^{2n-1}\right)^{\frac{1}{2n-1}}$$

as it can be checked (Mesiar and Komornikova [241]). Note that $\oplus_n$ is associative for all $n \in \mathbb{N}$.

(ii) What about the asymmetric Sugeno integral? It seems that yet no adequate definition exists. Mimicking (4.12) is meaningless because this definition is taylored to get a translation-invariant integral, and anyway this would lead to an integral with bad properties (see a discussion in Grabisch [164] and also [177, p. 215]). Since the Sugeno integral for nonnegative functions is *not* translation invariant, the methodology used for the Choquet integral cannot be applied. As a consequence, unless symmetry of the integral is desired, as expressed by (4.15), the most reasonable seems to restrict the usage of the Sugeno integral to nonnegative functions. ◇

4.4 The Choquet and Sugeno Integrals for Simple Functions

We deal here with simple and measurable functions (see Sect. 4.1). Our aim is to derive explicit formulas for the Choquet and Sugeno integrals.

4.4.1 *The Choquet Integral of Nonnegative Functions*

Let f be a simple, measurable and nonnegative function, with $\mathbf{ran} f = \{a_1, \dots, a_n\}$, supposing as before $0 \leqslant a_1 < a_2 < \cdots < a_n$, and let us introduce the sets $A_i = \{x \in X : f(x) \geqslant a_i\}$, for $i = 1, \dots, n$. Note that $A_1 = X$. The decumulative distribution function w.r.t. a capacity μ is now a staircase function (Fig. 4.5). The Choquet integral is the area below the decumulative function (in blue on Fig. 4.5). Dividing this area in vertical rectangles (delimited by solid lines in Fig. 4.5), we find

are balanced, otherwise, the pan with the "heavier" number goes down, and the result of $a \otimes b$ is the number contained in the latter pan.

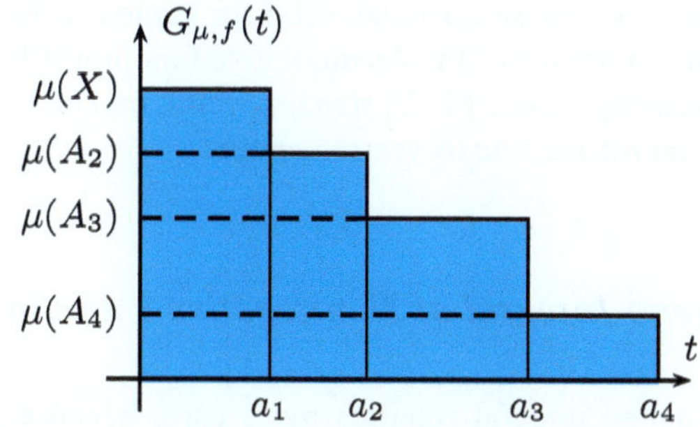

Fig. 4.5 A decumulative distribution function of a simple function taking values $a_1 < a_2 < a_3 < a_4$

the following formula:

$$\int f \, \mathrm{d}\mu = \sum_{i=1}^{n} (a_i - a_{i-1})\mu(A_i), \tag{4.17}$$

letting $a_0 = 0$.

Remark 4.16 Observe that this formula comes immediately from the horizontal (!) decomposition of a simple function [Eq. (4.1)]. Indeed, assuming that the integral is additive for the step functions $1_{\{x \in X : f(x) \geqslant a_i\}}$ and positively homogeneous, we have[7]

$$\int f \, \mathrm{d}\mu = \int \sum_{i=1}^{n} (a_i - a_{i-1}) 1_{\{x \in X : f(x) \geqslant a_i\}} \, \mathrm{d}\mu = \sum_{i=1}^{n} (a_i - a_{i-1}) \int 1_{\{x \in X : f(x) \geqslant a_i\}} \, \mathrm{d}\mu,$$

which yields the above formula by using Lemma 4.9. ◇

It is equivalent to compute the area under the decumulative function by cutting the surface into horizontal rectangles (delimited by the dashed lines on Fig. 4.5). Doing so, we obtain the equivalent formula

$$\int f \, \mathrm{d}\mu = \sum_{i=1}^{n} a_i\big(\mu(A_i) - \mu(A_{i+1})\big), \tag{4.18}$$

with the convention $A_{n+1} = \varnothing$. Note also that this formula readily obtains from a rearrangement of terms in (4.17).

[7]As it will be seen in Sect. 4.6.1, this is true because the step functions are pairwise comonotonic, hence by Theorem 4.28, additivity holds. Also, the Choquet integral is always positively homogeneous [Theorem 4.24(i)].

Extension to Games Are these formulas valid for games? It is easy to check that the vertical division of the area of the decumulative function still works even if this function is nondecreasing, hence (4.17) remains valid, and (4.18) as well, because it is obtained from the former one by rearrangement.

4.4.2 *The Sugeno Integral of Nonnegative Functions*

The case of the Sugeno integral requires more care, because the 45° line may intersect the decumulative distribution function on a horizontal or a vertical segment (Fig. 4.6). Let us compute $\bigvee_{t\geqslant 0}(G_{\mu,f}(t) \wedge t)$ for any function f and any capacity μ,

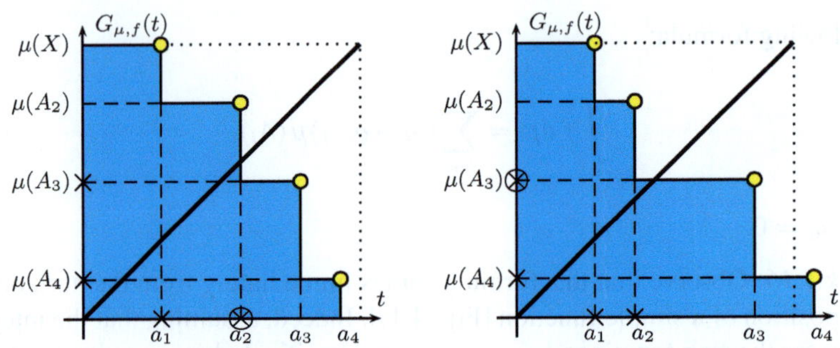

Fig. 4.6 Computation of the Sugeno integral: two cases

decomposing it as follows:

$$\bigvee_{t\geqslant 0}(G_{\mu,f}(t)\wedge t) = \bigvee_{t\in[0,a_1]}(G_{\mu,f}(t)\wedge t) \vee \bigvee_{t\in]a_1,a_2]}(G_{\mu,f}(t)\wedge t) \vee \cdots$$

$$\cdots \vee \bigvee_{t\in]a_{n-1},a_n]}(G_{\mu,f}(t)\wedge t) \vee \bigvee_{t>a_n}(\underbrace{G_{\mu,f}(t)}_{0}\wedge t),$$

where the last term could be ignored. Observe from Fig. 4.6 that for $i = 1, \ldots, n$

$$\bigvee_{t\in]a_{i-1},a_i]}(G_{\mu,f}(t)\wedge t) = \mu(A_i)\wedge a_i,$$

with $a_0 = 0$ and $A_i = \{x \in X : f(x) \geqslant a_i\}$, which yields the formula

$$⨍ f\,\mathrm{d}\mu = \bigvee_{i=1}^{n}(a_i \wedge \mu(A_i)). \tag{4.19}$$

Let us apply the same procedure to decompose $\bigwedge_{t\geqslant 0}(G_{\mu,f}(t) \vee t)$:

$$\bigwedge_{t\geqslant 0}(G_{\mu,f}(t) \vee t) = \bigwedge_{t\in[0,a_1]} (G_{\mu,f}(t) \vee t) \wedge \bigwedge_{t\in]a_1,a_2]} (G_{\mu,f}(t) \vee t) \wedge \cdots$$

$$\cdots \wedge \bigwedge_{t\in]a_{n-1},a_n]} (G_{\mu,f}(t) \vee t) \wedge \bigwedge_{t>a_n}(\underbrace{G_{\mu,f}(t)}_{0} \vee t)$$

$$= \mu(A_1) \wedge (\mu(A_2) \vee a_1) \wedge \cdots \wedge (\mu(A_n) \vee a_{n-1}) \wedge a_n,$$

which can be summarized as

$$⨍ f \, \mathrm{d}\mu = \bigwedge_{i=0}^{n} \big(a_i \vee \mu(A_{i+1})\big) \tag{4.20}$$

with the convention $A_{n+1} = \varnothing$ and $a_0 = 0$.

It is important to note that the result of the integral is either a value a_k of the integrand or the value of the capacity for a set A_k. Specifically (see Fig. 4.6), for all $a_1, \ldots, a_i$ prior the intersection of $G_{\mu,f}$ with the diagonal (circles above the diagonal), we have

$$\bigvee_{t\in]a_{k-1},a_k]} (G_{\mu,f}(t) \wedge t) = a_k \qquad (k = 1, \ldots, i).$$

This is indicated by crosses on the abscissa axis ($\times$). Similarly, for all points $a_{i+1}, \ldots, a_n$ after the intersection with the diagonal (circles below the diagonal), we have

$$\bigvee_{t\in]a_{k-1},a_k]} (G_{\mu,f}(t) \wedge t) = \mu(A_k) \qquad (k = i+1, \ldots, n).$$

Again this is indicated by crosses on the Y-axis. If the diagonal intersects a vertical segment, then the maximum over all crosses is situated on the abscissa axis (indicated by $\otimes$ on the figure); i.e., the value of the Sugeno integral is one of the values taken by the integrand f. If on the contrary, the diagonal intersects a horizontal segment, the maximum is situated on the Y-axis. In this case, the value of the Sugeno integral is one of the values taken by the capacity μ. These observations yield a third formulation, due to Kandell and Byatt [204].

Lemma 4.17 *For any simple measurable nonnegative function f on X and any capacity μ, we have*

$$⨍ f \, \mathrm{d}\mu = \mathrm{med}(a_1, \ldots, a_n, \mu(A_2), \ldots, \mu(A_n)), \tag{4.21}$$

using the above notation, and where med *is the median of a set of an odd number of real numbers.*

Proof As remarked above, the Sugeno integral is either equal to a_k or to $\mu(A_k)$ for some $k \in \{1,\dots,n\}$. Let us suppose that $⨍ f \,\mathrm{d}\mu = a_k$. Then $a_k \leqslant \mu(A_k)$, and $a_1,\dots,a_{k-1},\mu(A_{k+1}),\dots,\mu(A_n)$ are smaller or equal to a_k (hence $n-1$ values). Similarly, $a_{k+1},\dots,a_n,\mu(A_2),\dots,\mu(A_k)$ are greater or equal to a_k (again $n-1$ values). Consequently, a_k is the median of the values $a_1,\dots,a_n,\mu(A_2),\dots,\mu(A_n)$. When $⨍ f \,\mathrm{d}\mu = \mu(A_k)$, the proof is much the same. □

Extension to Games As explained in Remark 4.10(v), it is better not to consider the Sugeno integral w.r.t. games. Formula (4.19) with the supremum is not affected if $G_{v,f}$ is not monotone, however it is no longer equivalent with formula (4.20) with the infimum, nor with the expression with the median. To illustrate this, consider a simple function f with range $\{\frac{1}{4},\frac{1}{2},\frac{3}{4}\}$, and $v(A_1)=1, v(A_2)=0, v(A_3)=1$. Then

$$⨍ f \,\mathrm{d}v = \frac{1}{4} \vee 0 \vee \frac{3}{4} = \frac{3}{4} \neq \mathrm{med}\Big(\frac{1}{4},\frac{1}{2},\frac{3}{4},0,1\Big) = \frac{1}{2}.$$

4.4.3 *The Case of Real-Valued Functions*

We begin with the Choquet integral. Consider a simple measurable function $f : X \to \mathbb{R}$, with $\mathbf{ran} f = \{a_1,\dots,a_n\}$, supposing $a_1 < \cdots < a_p < 0 \leqslant a_{p+1} < \cdots < a_n$, and a capacity μ. A direct application of (4.10) and (4.17) leads to

$$\check{\int} f \,\mathrm{d}\mu = \sum_{i=1}^{p-1}(a_i - a_{i+1})\mu(1_{\{x\in X: f(x)\leqslant a_i\}}) + a_p\mu(1_{\{x\in X: f(x)\leqslant a_p\}})$$

$$+ a_{p+1}\mu(1_{\{x\in X: f(x)\geqslant a_{p+1}\}}) + \sum_{i=p+2}^{n}(a_i - a_{i-1})\mu(1_{\{x\in X: f(x)\geqslant a_i\}}). \tag{4.22}$$

For the (asymmetric) Choquet integral, simply observe that this integral is invariant by translation, and because f is bounded, $f' = f + (-a_1)1_X$ is nonnegative. By translation invariance, this yields $\int f' \,\mathrm{d}\mu = \int f \,\mathrm{d}\mu - a_1\mu(X)$, with $\mathbf{ran} f' = \{a'_1, a'_2,\dots,a'_n\}$ and $a'_i = a_i - a_1$, $i = 1,\dots,n$. Using (4.17) for f', we get, letting $a_0 = 0$,

$$\int f \,\mathrm{d}\mu = \int f' \,\mathrm{d}\mu + a_1\mu(X) = \sum_{i=1}^{n}(a'_i - a'_{i-1})\mu(A_i) + a_1\mu(X) = \sum_{i=1}^{n}(a_i - a_{i-1})\mu(A_i) \tag{4.23}$$

because $a'_1 = 0$. We recognize the formula (4.17) for nonnegative functions, hence (4.17) is valid for any real-valued function.

The formula for the symmetric Sugeno integral can be simply obtained in the same way, from (4.14) and (4.19). Defining the *symmetric minimum* as the binary operator $\otimes : L^2 \to L$ given by

$$a \otimes b = \begin{cases} -(|a| \wedge |b|), & \text{if sign } (a) \neq \text{sign } (b) \\ |a| \wedge |b|, & \text{otherwise,} \end{cases} \tag{4.24}$$

one obtains:

$$\begin{aligned} \check{\int} f \, \mathrm{d}\mu =& \Big(\mathop{\otimes}_{i=1}^{p} \big(a_i \otimes \mu(1_{\{x \in X : f(x) \leqslant a_i\}})\big) \Big) \\ & \otimes \Big(\mathop{\otimes}_{i=p+1}^{n} \big(a_i \otimes \mu(1_{\{x \in X : f(x) \geqslant a_i\}})\big) \Big), \end{aligned} \tag{4.25}$$

with the same notation as above.

Remark 4.18 Since we have only used formulas valid for games, the formulas given here are also valid for games. ◇

4.5 The Choquet and Sugeno Integrals on Finite Sets

We suppose in this section that $X = \{x_1, \ldots, x_n\}$ is a finite set of n elements and that $\mathcal{F} = 2^X$, hence all functions are measurable and simple. Thus, the expressions of the Choquet and Sugeno integrals directly obtain from the formulas for simple functions.

4.5.1 The Case of Nonnegative Functions

Take a nonnegative function f, and identify it with the vector $(f_1, \ldots, f_n)$, letting $f_i = f(x_i)$. Choose σ a permutation on X such that $f_{\sigma(1)} \leqslant f_{\sigma(2)} \leqslant \cdots \leqslant f_{\sigma(n)}$. Observe that the whole set of nonnegative functions on X can be partitioned into $n!$ subsets, depending on which permutation orders the values of f. We set

$$(\mathbb{R}_+^X)_\sigma = \{f : X \to \mathbb{R}_+ : f_{\sigma(1)} \leqslant \cdots \leqslant f_{\sigma(n)}\},$$

and similarly we define $\mathbb{R}_\sigma^X$, I_σ^X where I is an interval in $\mathbb{R}$, in particular $[0, 1]_\sigma^X$, the *canonical simplices of the unit hypercube* (Remark 2.89).

Setting

$$A_\sigma^\uparrow(i) = \{x_{\sigma(i)}, x_{\sigma(i+1)}, \ldots, x_{\sigma(n)}\} \qquad (i = 1, \ldots, n),$$

we obtain directly from (4.17) and (4.18) the following expressions, for any capacity μ:

$$\int f \,\mathrm{d}\mu = \sum_{i=1}^{n} (f_{\sigma(i)} - f_{\sigma(i-1)})\mu(A_{\sigma}^{\uparrow}(i)). \tag{4.26}$$

$$\int f \,\mathrm{d}\mu = \sum_{i=1}^{n} f_{\sigma(i)}\big(\mu(A_{\sigma}^{\uparrow}(i)) - \mu(A_{\sigma}^{\uparrow}(i+1))\big), \tag{4.27}$$

with the conventions $f_{\sigma(0)} = 0$ and $A_{\sigma}^{\uparrow}(n+1) = \varnothing$.

For the Sugeno integral, we obtain from (4.19) and (4.20):

$$⨍ f \,\mathrm{d}\mu = \bigvee_{i=1}^{n} \big(f_{\sigma(i)} \wedge \mu(A_{\sigma}^{\uparrow}(i))\big). \tag{4.28}$$

$$⨍ f \,\mathrm{d}\mu = \bigwedge_{i=0}^{n} \big(f_{\sigma(i)} \vee \mu(A_{\sigma}^{\uparrow}(i+1))\big) \tag{4.29}$$

with the same conventions.

As it is sometimes convenient, we give also the formulas where the permutation orders the function in descending order. Given a function f, denote by τ a permutation such that $f_{\tau(1)} \geqslant f_{\tau(2)} \geqslant \cdots \geqslant f_{\tau(n)}$ and define the sets

$$A_{\tau}^{\downarrow}(i) = \{x_{\tau(1)}, \ldots, x_{\tau(i)}\} \qquad (i = 1, \ldots, n).$$

Then formulas (4.26)–(4.29) become

$$\int f \,\mathrm{d}\mu = \sum_{i=1}^{n} (f_{\tau(i)} - f_{\tau(i+1)})\mu(A_{\tau}^{\downarrow}(i)) \tag{4.30}$$

$$\int f \,\mathrm{d}\mu = \sum_{i=1}^{n} f_{\tau(i)}\big(\mu(A_{\tau}^{\downarrow}(i)) - \mu(A_{\tau}^{\downarrow}(i-1))\big) \tag{4.31}$$

$$⨍ f \,\mathrm{d}\mu = \bigvee_{i=1}^{n} \big(f_{\tau(i)} \wedge \mu(A_{\tau}^{\downarrow}(i))\big) \tag{4.32}$$

$$⨍ f \,\mathrm{d}\mu = \bigwedge_{i=1}^{n+1} \big(f_{\tau(i)} \vee \mu(A_{\tau}^{\downarrow}(i-1))\big) \tag{4.33}$$

with the convention $f_{\tau(n+1)} = 0$ and $A_{\tau}^{\downarrow}(0) = \varnothing$.

Remark 4.19 Based on considerations given in Sect. 4.4, all formulas remain valid for games, except formulas (4.29), and (4.33). ◇

Example 4.20 We give a practical example of computation of Choquet and Sugeno integrals. Take $X = \{x_1, \ldots, x_4\}$ and the function $f = (0.2, 0.1, 0.5, 0.3)$. There is a unique permutation σ arranging f in increasing order: $\sigma(1) = 2$, $\sigma(2) = 1$, $\sigma(3) = 4$ and $\sigma(4) = 3$. Indeed, $f_{\sigma(1)} < f_{\sigma(2)} < f_{\sigma(3)} < f_{\sigma(4)}$. It follows that the sets $A^{\uparrow}_{\sigma}(i)$ are

$$A^{\uparrow}_{\sigma}(1) = X, \quad A^{\uparrow}_{\sigma}(2) = \{x_1, x_3, x_4\}, \quad A^{\uparrow}_{\sigma}(3) = \{x_3, x_4\}, \quad A^{\uparrow}_{\sigma}(4) = \{x_3\}.$$

Let us specify the values of the capacity μ for these sets:

$$\mu(X) = 1, \quad \mu(\{x_1, x_3, x_4\}) = 0.8, \quad \mu(\{x_3, x_4\}) = 0.6, \quad \mu(\{x_3\}) = 0.2.$$

We obtain for the integrals

$$\begin{aligned}\int f \, \mathrm{d}\mu &= f_2\mu(X) + (f_1 - f_2)\mu(\{x_1, x_3, x_4\}) + (f_4 - f_1)\mu(\{x_3, x_4\}) + (f_3 - f_4)\mu(\{x_3\}) \\ &= 0.1 + 0.08 + 0.06 + 0.04 = 0.28 \\ ⨍ f \, \mathrm{d}\mu &= (f_2 \wedge \mu(X)) \vee (f_1 \wedge \mu(\{x_1, x_3, x_4\})) \vee (f_4 \wedge \mu(\{x_3, x_4\})) \vee (f_3 \wedge \mu(\{x_3\})) \\ &= 0.1 \vee 0.2 \vee 0.3 \vee 0.2 = 0.3.\end{aligned}$$

◇

Example 4.21 (Example 2.8 continued) (Murofushi and Sugeno [250, 254]) Let $X = \{x_1, \ldots, x_n\}$ be the set of workers. We suppose that each day they all start working at 8:00, work continuously, but they leave at different times. Let us denote by $f(x_i)$ the number of worked hours for worker x_i, and suppose that $f(x_1) \leqslant f(x_2) \leqslant \cdots \leqslant f(x_n)$; i.e., x_1 leaves first and so on.

Consider that for any group of workers $A \subseteq X$, its productivity per hour is $\mu(A)$, and $\mu : 2^X \to \mathbb{R}_+$ is a capacity (or a game). Let us compute the total number of goods produced per day by the set of all workers. By definition of f, we have

- The entire group X has worked $f(x_1)$ hours;
- Then x_1 leaves and the group $X \setminus \{x_1\} = \{x_2, \ldots, x_n\}$ works in addition $f(x_2) - f(x_1)$ hours;
- Then x_2 leaves and the group $X \setminus \{x_1, x_2\} = \{x_3, \ldots, x_n\}$ works in addition $f(x_3) - f(x_2)$, etc.,
- Finally only x_n remains and he works still $f(x_n) - f(x_{n-1})$ hours alone.

Supposing linearity in production, the total production is:

$$f(x_1)\mu(X) + (f(x_2) - f(x_1))\mu(\{x_2,\dots,x_n\}) + (f(x_3) - f(x_2))\mu(\{x_3,\dots,x_n\}) \\ + \cdots + (f(x_n) - f(x_{n-1}))\mu(\{x_n\}).$$

We recognize here the Choquet integral $\int f \, d\mu$. ◇

Example 4.22 (The Hirsch index) The well-known h-index introduced by Hirsch [195] to quantify the research output of a researcher (how is this conceivable?!), is defined as follows: your h-index is α if you published α papers which are cited at least α times, and any other of your published papers has no more than α citations. This can be expressed formally through the function f defined on $X = \{x_1,\dots,x_n\}$, the set of papers of a given researcher r, by

$$\text{paper } x_i \mapsto f(x_i) = \text{ number of citations of } x_i \qquad (i = 1,\dots,n).$$

Then, taking a permutation σ on X such that $f(x_{\sigma(1)}) \geqslant \cdots \geqslant f(x_{\sigma(n)})$, the h-index of r is given by

$$h_r = \bigvee_{i=1}^{n} (f(x_{\sigma(i)}) \wedge i). \tag{4.34}$$

As noticed by Torra and Narukawa [328], this has a Sugeno integral form, taking μ to be the counting measure m_c (Example 2.4). Indeed,

$$⨍ f \, dm_c = \bigvee_{i=1}^{n} (f(x_{\sigma(i)}) \wedge m_c(\{\sigma(1),\dots,\sigma(i)\}) = \bigvee_{i=1}^{n} (f(x_{\sigma(i)}) \wedge i) = h_r.$$

◇

4.5.2 The Case of Real-Valued Integrands

First we deal with the Choquet integral. Suppose $f : X \to \mathbb{R}$, with at least one value f_i being negative, and take σ any permutation on X such that $f_{\sigma(1)} \leqslant \cdots \leqslant f_{\sigma(p)} < 0 \leqslant f_{\sigma(p+1)} \leqslant \cdots \leqslant f_{\sigma(n)}$. We obtain directly from (4.22)

$$\check{\int} f \, d\mu = \sum_{i=1}^{p-1} (f_{\sigma(i)} - f_{\sigma(i+1)})\mu(\{x_{\sigma(1)},\dots,x_{\sigma(i)}\}) + f_{\sigma(p)}\mu(\{x_{\sigma(1)},\dots,x_{\sigma(p)}\}) \\ + f_{\sigma(p+1)}\mu(\{x_{\sigma(p+1)},\dots,x_{\sigma(n)}\}) \\ + \sum_{i=p+2}^{n} (f_{\sigma(i)} - f_{\sigma(i-1)})\mu(\{x_{\sigma(i)},\dots,x_{\sigma(n)}\}). \tag{4.35}$$

For the (asymmetric) Choquet integral, referring to the case of simple functions, we find that formulas (4.26) and (4.27) remain valid.

Lastly, the symmetric Sugeno integral obtains directly from (4.25):

$$\check{\int} f \, \mathrm{d}\mu = \Big(\operatorname*{\circledvee}_{i=1}^{p} \big(f_{\sigma(i)} \circledwedge \mu(\{x_{\sigma(1)}, \ldots, x_{\sigma(i)}\})\big)\Big) \\ \circledvee \Big(\operatorname*{\circledvee}_{i=p+1}^{n} \big(f_{\sigma(i)} \circledwedge \mu(\{x_{\sigma(i)}, \ldots, x_{\sigma(n)}\})\big)\Big). \tag{4.36}$$

Note that both formulas are still valid for games.

4.5.3 *The Case of Additive Capacities*

If the capacity is additive, the formulas simplify for the Choquet integral. Indeed, additivity of μ yields $\mu(A_\sigma^\uparrow(i)) - \mu(A_\sigma^\uparrow(i+1)) = \mu(\{\sigma(i)\})$, hence we obtain from (4.27):

$$\int f \, \mathrm{d}\mu = \sum_{i=1}^{n} f_{\sigma(i)} \mu(\{\sigma(i)\}) = \sum_{i=1}^{n} f_i \mu(\{i\}). \tag{4.37}$$

(See also Corollary 4.37 and Theorem 4.63.)

4.6 Properties

Beforehand, we introduce the important property of stochastic dominance, related to decumulative functions, which is a central notion in decision under risk (Sect. 5.2).

Definition 4.23 We consider functions in $B(\mathcal{F})$ (or $B^+(\mathcal{F})$ according to the context). For two such functions f, g and a capacity μ, f is said to *stochastically dominates (at first order)* g w.r.t μ, denoted by $f \geqslant_{\mathrm{SD}}^{\mu} g$, if $G_{\mu,f}(t) \geqslant G_{\mu,g}(t)$ for every $t \in \mathbb{R}$.

In this book, because only first order stochastic dominance is considered, we drop the term "first order."

4.6.1 The Choquet Integral

Elementary Properties

We recall that $\mathcal{BV}(\mathcal{F})$ is the set of games of bounded variation norm (Sect. 2.19.1).

Theorem 4.24 *Let $f : X \to \mathbb{R}$ be a function in $B(\mathcal{F})$ and $v \in \mathcal{BV}(\mathcal{F})$. The following properties hold.*

(i) Positive homogeneity:

$$\int \alpha f \, \mathrm{d}v = \alpha \int f \, \mathrm{d}v \qquad (\alpha \geqslant 0)$$

(ii) Homogeneity of the symmetric Choquet integral:

$$\check{\int} \alpha f \, \mathrm{d}v = \alpha \check{\int} f \, \mathrm{d}v \qquad (\alpha \in \mathbb{R})$$

(iii) Translation invariance:

$$\int (f + \alpha 1_X) \, \mathrm{d}v = \int f \, \mathrm{d}v + \alpha v(X) \qquad (\alpha \in \mathbb{R})$$

(iv) Asymmetry:

$$\int (-f) \, \mathrm{d}v = -\int f \, \mathrm{d}\overline{v}$$

where $\overline{v}$ is the conjugate game;

(v) Scale inversion:

$$\int (\alpha 1_X - f) \, \mathrm{d}v = \alpha v(X) - \int f \, \mathrm{d}\overline{v} \qquad (\alpha \in \mathbb{R})$$

(vi) Monotonicity (or nondecreasingness) w.r.t. the integrand: for any capacity μ,

$$f \leqslant f' \Rightarrow \int f \, \mathrm{d}\mu \leqslant \int f' \, \mathrm{d}\mu \qquad (f, f' \in B(\mathcal{F}))$$

(vii) Monotonicity w.r.t. the game for nonnegative integrands: if $f \geqslant 0$,

$$v \leqslant v' \Rightarrow \int f \, \mathrm{d}v \leqslant \int f \, \mathrm{d}v' \qquad (v, v' \in \mathcal{BV}(\mathcal{F}))$$

(viii) *Monotonicity w.r.t. stochastic dominance: for any capacity* μ,

$$f \geqslant^{\mu}_{\mathrm{SD}} f' \Rightarrow \int f \,\mathrm{d}\mu \geqslant \int f' \,\mathrm{d}\mu \qquad (f, f' \in B(\mathcal{F}))$$

(ix) *Linearity w.r.t. the game:*

$$\int f \,\mathrm{d}(v + \alpha v') = \int f \,\mathrm{d}v + \alpha \int f \,\mathrm{d}v' \qquad (v, v' \in \mathcal{BV}(\mathcal{F}), \alpha \in \mathbb{R})$$

(x) *Boundaries:* $\inf f$ *and* $\sup f$ *are attained:*

$$\inf f = \int f \,\mathrm{d}\mu_{\min}, \quad \sup f = \int f \,\mathrm{d}\mu_{\max},$$

with $\mu_{\min}(A) = 0$ *for all* $A \subset X$, $A \in \mathcal{F}$, $\mu_{\min}(X) = 1$, *and* $\mu_{\max}(A) = 1$ *for all nonempty* $A \in \mathcal{F}$ *(Sect. 2.8.1);*

(xi) *Boundaries: for any normalized capacity* μ,

$$\text{ess}\inf_{\mu} f \leqslant \int f \,\mathrm{d}\mu \leqslant \text{ess}\sup_{\mu} f$$

(xii) *Lipschitz continuity:*

$$\left| \int f \,\mathrm{d}v - \int g \,\mathrm{d}v \right| \leqslant \|v\| \|f - g\| \qquad (f, g \in B(\mathcal{F})),$$

where $\|v\|$ *is the variation norm of* v *(Sect. 2.19.1) and* $\|f\| = \sup_{x \in X} |f(x)|$.

Proof

(i) Observe that $\alpha f(x) \geqslant t$ is equivalent to $f(x) \geqslant \frac{t}{\alpha}$. By performing the change of variable $t \to t' = \frac{t}{\alpha}$ in (4.13), the result is established.

(ii) If $\alpha \geqslant 0$, using (i) in (4.10) gives the desired result. Suppose then $\alpha < 0$. By using (4.11) (still valid for games), we have $\check{\int} \alpha f \,\mathrm{d}v = -\check{\int} |\alpha| f \,\mathrm{d}v = -|\alpha| \check{\int} f \,\mathrm{d}v = \alpha \check{\int} f \,\mathrm{d}v$.

(iii) see Theorem 4.12.

(iv) We have

$$\begin{aligned}
\int (-f) \,\mathrm{d}v &= \int_0^{\infty} v(-f \geqslant t) \,\mathrm{d}t + \int_{-\infty}^{0} \big(v(-f \geqslant t) - v(X)\big) \,\mathrm{d}t \\
&= \int_{-\infty}^{0} v(-f \geqslant -t) \,\mathrm{d}t - \int_0^{\infty} \big(v(X) - v(-f \geqslant -t)\big) \,\mathrm{d}t \\
&= \int_{-\infty}^{0} v(f \leqslant t) \,\mathrm{d}t - \int_0^{\infty} \big(v(X) - v(f \leqslant t)\big) \,\mathrm{d}t
\end{aligned}$$

$$= -\int_{-\infty}^{0} -v(f < t)\,\mathrm{d}t - \int_{0}^{\infty} \big(v(X) - v(f < t)\big)\,\mathrm{d}t$$

$$= -\int_{-\infty}^{0} \big(\overline{v}(f \geqslant t) - \overline{v}(X)\big)\,\mathrm{d}t - \int_{0}^{\infty} \overline{v}(f \geqslant t)\,\mathrm{d}t = -\int f\,\mathrm{d}\overline{v}.$$

(v) Immediate from (iii) and (iv).

(vi) $f \leqslant f'$ implies $\{f \geqslant t\} \subseteq \{f' \geqslant t\}$, and by monotonicity of μ, we get $\mu(f \geqslant t) \leqslant \mu(f' \geqslant t)$ for each $t \in \mathbb{R}$, which yields the result by (4.13).

(vii) and (viii) are obvious (see Fig. 4.4), as well as (ix).

(x) By translation invariance (iii), it suffices to show the result for nonnegative functions. Since $\mu_{\min}$ and $\mu_{\max}$ are 0-1-capacities, by properties of the decumulative function (see Lemma 4.3), it is plain that $\int f\,\mathrm{d}\mu_{\min} = \operatorname{ess\,inf}_{\mu_{\min}} f$ and $\int f\,\mathrm{d}\mu_{\max} = \operatorname{ess\,sup}_{\mu_{\max}} f$. Finally, observe that $\operatorname{ess\,inf}_{\mu_{\min}} f = \inf f$ and $\operatorname{ess\,sup}_{\mu_{\max}} f = \sup f$.

(xi) Since $\mu_{\min} \leqslant \mu \leqslant \mu_{\max}$ for any normalized capacity μ, monotonicity w.r.t. capacities (vii) and (x) permit to conclude.

(xii) (We follow Marinacci and Montrucchio [235].) Suppose first that v is a capacity. Assume $\int f\,\mathrm{d}v \geqslant \int g\,\mathrm{d}v$ (the other case is similar). As $f \leqslant g + \|f - g\|$, we have by (iii) and (vi) $\int f\,\mathrm{d}v \leqslant \int g\,\mathrm{d}v + \|f - g\| v(X)$. This implies

$$\left|\int f\,\mathrm{d}v - \int g\,\mathrm{d}v\right| \leqslant v(X)\|f - g\|, \tag{4.38}$$

which is (xii) for v monotone because in this case $\|v\| = v(X)$. Suppose now that $v \in \mathcal{BV}(\mathcal{F})$. By Aumann and Shapley [11], we know that v can be written as $v = \mu_1 - \mu_2$, where μ_1, μ_2 are capacities such that $\|v\| = \mu_1(X) + \mu_2(X)$ (see Sect. 2.19.1). By (4.38), we have then

$$\begin{aligned}\left|\int f\,\mathrm{d}v - \int g\,\mathrm{d}v\right| &\leqslant \left|\int f\,\mathrm{d}\mu_1 - \int g\,\mathrm{d}\mu_1\right| + \left|\int f\,\mathrm{d}\mu_2 - \int g\,\mathrm{d}\mu_2\right| \\ &\leqslant (\mu_1(X) + \mu_2(X))\|f - g\|,\end{aligned}$$

as desired. □

Remark 4.25

(i) It is easy to check that the symmetric Choquet integral satisfies properties (vi), (vii) and (ix) as well. However, (viii) is not satisfied in general, as one can see on the following example: take $X = \{x_1, x_2, x_3\}$ and consider the two functions $f = (-2, -1, 1)$ and $g = (-1, -2, 1)$. Observe that $f >_{\mathrm{SD}}^{\mu} g$ as soon as $\mu(\{x_1, x_3\}) < \mu(\{x_2, x_3\})$. However, the integrals on the negative parts are:

$$\int f^-\,\mathrm{d}\mu = \mu(\{x_1, x_2\}) + \mu(\{x_1\}), \quad \int g^-\,\mathrm{d}\mu = \mu(\{x_1, x_2\}) + \mu(\{x_2\}).$$

Hence $\check{\int} f \, d\mu < \check{\int} g \, d\mu$ as soon as $\mu(\{x_1\}) > \mu(\{x_2\})$, which is not in contradiction with the former inequality.

(ii) The meaning of scale inversion becomes clear if the upper bound of the range of $|f|$ is α and if v is normalized. Then $\int (\alpha 1_X - f) \, dv = \alpha - \int f \, d\overline{v}$.

(iii) Lipschitz continuity implies uniform continuity and therefore ordinary continuity. It follows by (4.10) that the symmetric integral is continuous. ◇

Comonotonic Additivity

As it is easy to see, the Choquet integral is not additive, in the sense that $\int (f+g) \, dv$ is in general different from $\int f \, dv + \int g \, dv$. Indeed, taking simply $f = 1_A, g = 1_B$ with A, B disjoint subsets of X, we find by using Lemma 4.9 that $\int (1_A + 1_B) \, dv = v(A \cup B)$, and unless v is itself additive, it differs from $v(A) + v(B)$.

We introduce now the notion of comonotonicity, which will reveal to be sufficient and necessary to ensure the additivity of the Choquet integral.

Definition 4.26 Two functions $f, g : X \to \mathbb{R}$ are *comonotonic* if

$$\big(f(x) - f(x')\big)\big(g(x) - g(x')\big) \geqslant 0 \qquad (x, x' \in X).$$

Equivalently, there is no $x, x' \in X$ such that $f(x) < f(x')$ and $g(x) > g(x')$. Roughly speaking, two comonotonic functions have a similar pattern of variation, however one should be careful that comonotonicity is in fact more demanding than simply to be increasing and decreasing on the same domains (see Fig. 4.7). Note also that a constant function is comonotonic with any other function. From this, one deduces that the binary relation "is comonotonic with" is reflexive, symmetric, but not transitive. We start by showing some equivalent conditions.

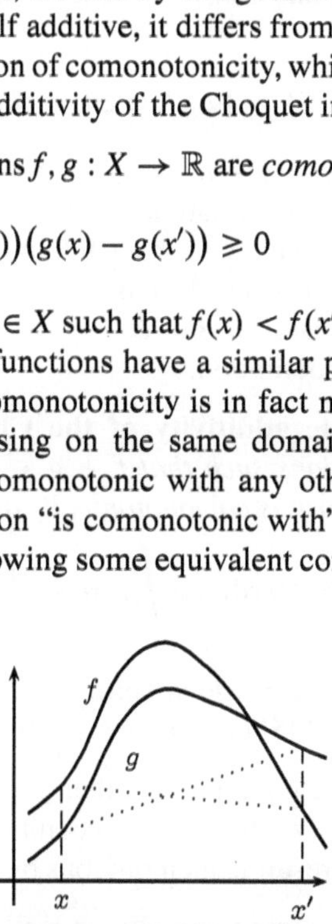

Fig. 4.7 Although f and g are increasing and decreasing on the same domain, they are not comonotonic because $f(x) > f(x')$ and $g(x) < g(x')$

Lemma 4.27 *Let $f, g : X \to \mathbb{R}$. The following propositions are equivalent.*

(i) f and g are comonotonic;

(ii) There exist nondecreasing functions $u, v : \mathbb{R} \to \mathbb{R}$ and a function $h : X \to \mathbb{R}$ such that $f = u \circ h$ and $g = v \circ h$.

(iii) The collection $\{(f \geqslant t)\}_{t\in\mathbb{R}} \cup \{(g \geqslant t)\}_{t\in\mathbb{R}}$ *is a chain;*

Moreover, suppose now that X *is finite with* $|X| = n$. *Then*

(iv) $f = (f_1, \ldots, f_n)$ *and* $g = (g_1, \ldots, g_n)$ *are comonotonic if and only if there exists a permutation* σ *on* X *such that* $f_{\sigma(1)} \leqslant \cdots \leqslant f_{\sigma(n)}$ *and* $g_{\sigma(1)} \leqslant \cdots \leqslant g_{\sigma(n)}$.

Proof (i)⇒(ii): Assume that f is not constant, otherwise if both f, g are constant, the result holds trivially. Choose any increasing function u on $\mathbb{R}$ and define $h = u^{-1} \circ f$. Now, define $v : \mathbf{ran}\, h \to \mathbb{R}$ as follows: for any $x \in X$, put $v(h(x)) = g(x)$. Let us show that v is nondecreasing on $\mathbf{ran}\, h$. Take $x, x' \in X$ such that $h(x) > h(x')$, which is possible because f is not constant and u is increasing. This is equivalent to $u^{-1}(f(x)) > u^{-1}(f(x'))$, which is in turn equivalent to $f(x) > f(x')$. Since f, g are comonotonic, it follows that $g(x) \geqslant g(x')$.

(ii)⇒(iii): Take $\alpha, \alpha' \in \mathbb{R}$ and consider the level sets $\{f \geqslant \alpha\}$ and $\{g \geqslant \alpha'\}$. We have

$$\{f \geqslant \alpha\} = \{u \circ h \geqslant \alpha\} = \{h \geqslant \beta\}$$

with $\beta = \inf u^{-1}(\alpha)$. Similarly, $\{g \geqslant \alpha'\} = \{h \geqslant \beta'\}$, with $\beta' = \inf v^{-1}(\alpha')$. These two level sets are then in inclusion relation.

(iii)⇒(i): Suppose there exist $x, x' \in X$ such that $f(x) < f(x')$ and $g(x) > g(x')$. Consider the level sets $A = \{f \geqslant f(x')\}$ and $B = \{g \geqslant g(x)\}$. Then $x \in B \setminus A$ and $x' \in A \setminus B$; i.e., it cannot be that $A \subseteq B$ or $B \subseteq A$.

(iv) This is clear from the definition. □

Theorem 4.28 (Comonotonic additivity of the Choquet integral) *Let* $f, g \in B(\mathcal{F})$ *be comonotonic functions such that* $f + g \in B(\mathcal{F})$. *Then for any game* v *in* $\mathcal{BV}(\mathcal{F})$, *the Choquet integral is comonotonically additive, that is,*

$$\int (f + g)\,\mathrm{d}v = \int f\,\mathrm{d}v + \int g\,\mathrm{d}v.$$

We give the proof for the discrete case, assuming $\mathcal{F} = 2^X$. For a general proof, see Marinacci and Montrucchio [235] or Denneberg [80].

Proof Observe that if f, g are comonotone, then $f + g$ is comonotone with f and with g. Then by Lemma 4.27(iv), there exists a common permutation σ increasingly ordering f, g and $f + g$. The result is then obvious by (4.27). □

Remark 4.29 The notion of comonotonicity as well as the above result is due to Dellacherie [76]. ◇

Horizontal Additivity

There is another way—and as we will see, equivalent to comonotonic additivity—to characterize the additivity of the Choquet integral, called horizontal min- (or max-) additivity.

Given a function $f : X \to \mathbb{R}$ and a constant $c \in \mathbb{R}$, the *horizontal min-additive decomposition* of f is:

$$f = (f \wedge c1_X) + (f - (f \wedge c1_X)). \tag{4.39}$$

This amounts to "cut" horizontally the function at level c. Similarly, the *horizontal max-additive decomposition* of f is:

$$f = (f \vee c1_X) + (f - (f \vee c1_X)). \tag{4.40}$$

A functional $I : \mathbb{R}^X \to \mathbb{R}$ is *horizontally min-additive* if for every $f : X \to \mathbb{R}$ and $c \in \mathbb{R}$,

$$I(f) = I(f \wedge c1_X) + I(f - (f \wedge c1_X)). \tag{4.41}$$

Horizontal max-additivity is defined similarly.

We note that $f \wedge c1_X$ and $f - (f \wedge c1_X)$ are comonotonic functions, as well as $(f \vee c1_X)$ and $(f - (f \vee c1_X))$. It follows that if a functional satisfies comonotonic additivity, then it also satisfies horizontal min- and max-additivity. The converse is also true, as shown in the next theorem.

Theorem 4.30 *Suppose $|X| = n$ and $\mathcal{F} = 2^X$. A functional $I : \mathbb{R}^n \to \mathbb{R}$ is comonotonically additive if and only if it is horizontally min-additive (equivalently, horizontally max-additive).*

Proof Due to the above remark, we only have to prove that horizontal min-additivity implies comonotonic additivity (the proof for horizontal max-additivity is much the same).

0. We claim that I satisfies $I(-\alpha 1_X) = -I(\alpha 1_X)$ for all $\alpha \in \mathbb{R}$. First we prove that $I(\mathbf{0}) = 0$. Taking $\alpha > 0$ and applying horizontal min-additivity with $c = 0$, we find

$$I(\alpha 1_X) = I(\mathbf{0}) + I(\alpha 1_X - 0 \cdot 1_X) = I(\mathbf{0}) + I(\alpha 1_X)$$

which forces $I(\mathbf{0}) = 0$. Next, considering $\alpha > 0$, we have by horizontal min-additivity with $c = -\alpha$

$$0 = I(\mathbf{0}) = I(-\alpha 1_X) + I(\alpha 1_X),$$

and the claim is proved.

1. We claim that I satisfies

$$I((x+y)1_X) = I(x1_X) + I(y1_X) \qquad (x, y \in \mathbb{R}). \tag{4.42}$$

Take $x, y \in \mathbb{R}$, with at most one negative number, and suppose $x \leqslant y$. Choose $c = x$. Horizontal min-additivity implies

$$I((x+y)1_X) = I(((x+y)\wedge x)1_X) + I((x+y)1_X - (((x+y)\wedge x)1_X)) = I(x1_X) + I(y1_X).$$

Now, if both x, y are negative, by Claim 0 and the above result we have

$$I((x+y)1_X) = -I((-x-y)1_X) = -(I(-x1_X) + I(-y1_X)) = I(x1_X) + I(y1_X).$$

2. We claim that I satisfies

$$I((x+y)1_A) = I(x1_A) + I(y1_A) \qquad (x, y \geqslant 0, A \subset X). \tag{4.43}$$

Take $x, y \geqslant 0$, suppose $x \leqslant y$, and choose $c = x$. Proceeding as above, the claim is shown (note that it is not true if x or y is negative).

3. We claim that I satisfies for any function $f \in \mathbb{R}^X$

$$I(f) = I(f_{\sigma(1)}1_X) + \sum_{i=2}^{n} I((f_{\sigma(i)} - f_{\sigma(i-1)})1_{A_\sigma^\uparrow(i)}), \tag{4.44}$$

where σ is a permutation on X such that $f_{\sigma(1)} \leqslant \cdots \leqslant f_{\sigma(n)}$. By repeatedly applying horizontal min-additivity on f with the successive level sets $f_{\sigma(1)}, f_{\sigma(2)} - f_{\sigma(1)}, \ldots, f_{\sigma(n)} - f_{\sigma(n-1)}$, we obtain

$$\begin{aligned} I(f) &= I(f_{\sigma(1)}1_X) + I(0, f_{\sigma(2)} - f_{\sigma(1)}, \ldots, f_{\sigma(n)} - f_{\sigma(1)}) \\ &= I(f_{\sigma(1)}1_X) + I((f_{\sigma(2)} - f_{\sigma(1)})1_{A_\sigma^\uparrow(2)}) + I(0, 0, f_{\sigma(3)} - f_{\sigma(2)}, \ldots, f_{\sigma(n)} - f_{\sigma(2)}) \\ &= \cdots = \\ &= I(f_{\sigma(1)}1_X) + \sum_{i=2}^{n} I((f_{\sigma(i)} - f_{\sigma(i-1)})1_{A_\sigma^\uparrow(i)}). \end{aligned}$$

4. Consider any two functions $f, f' \in \mathbb{R}^X$ that are comonotonic. By Lemma 4.27(iv), we have, using Claims 1, 2 and 3 above:

$$\begin{aligned} I(f+f') &= I((f+f')_{\sigma(1)}1_X) + \sum_{i=2}^{n} I(((f+f')_{\sigma(i)} - (f+f')_{\sigma(i-1)})1_{A_\sigma^\uparrow(i)}) \\ &= I((f_{\sigma(1)} + f'_{\sigma(1)})1_X) + \sum_{i=2}^{n} I((f_{\sigma(i)} + f'_{\sigma(i)} - f_{\sigma(i-1)} - f'_{\sigma(i-1)})1_{A_\sigma^\uparrow(i)}) \end{aligned}$$

$$= I(f_{\sigma(1)}1_X) + I(f'_{\sigma(1)}1_X) + \sum_{i=2}^{n} I((f_{\sigma(i)} - f_{\sigma(i-1)})1_{A_\sigma^\uparrow(i)})$$

$$+ \sum_{i=2}^{n} I((f'_{\sigma(i)} - f'_{\sigma(i-1)})1_{A_\sigma^\uparrow(i)})$$

$$= I(f) + I(f').$$

□

By application of Theorems 4.28 and 4.30, we obtain immediately:

Corollary 4.31 *Let $|X| = n$, $\mathcal{F} = 2^X$. The Choquet integral is horizontally min- and max-additive.*

Remark 4.32 Horizontal min- and max additivity was introduced by Šipoš [334], and also considered by Benvenuti et al. [19]. Theorem 4.30 was shown by Couceiro and Marichal [58]. Also, the two functions $f \vee c1_X$ and $f - (f \vee c1_X)$ are *maxmin-related* in the sense of Wakker [337]. Two functions f, g are maxmin-related if one of them plays the rôle of the max-function (say f) and the other one the rôle of the min-function, that is, for every $x \in X$, either the max-function assigns its maximal value or the min-function assigns its minimal value. This property is used in [337] to characterize the Choquet integral. ◇

Lastly, we introduce a third type a horizontal additivity, which is a mix of the two preceding ones, called horizontal median additivity (Couceiro and Marichal [58]). Given a function $f : X \to \mathbb{R}$ and a constant $c > 0$, the *horizontal median-additive decomposition* of f is:

$$f = \text{med}(-c, f, c) + (f - (f \wedge c1_X)) + (f - (f \vee -c1_X)), \tag{4.45}$$

where $\text{med}(-c, f, c) \in \mathbb{R}^n$, with ith coordinate being the middle value of $-c, f_i$, and c (Fig. 4.8).

A functional $I : \mathbb{R}^X \to \mathbb{R}$ is *horizontally median-additive* if for every $f \in \mathbb{R}^X$ and $c > 0$,

$$I(f) = I(\text{med}(-c, f, c)) + I(f - (f \wedge c1_X)) + I(f - (f \vee -c1_X)). \tag{4.46}$$

As it can be checked, $\text{med}(-c, f, c)$, $(f - (f \wedge c1_X))$ and $(f - (f \vee -c1_X))$ are pairwise comonotonic functions, so that if I is comonotonic additive it is also horizontally median-additive. We note also that if I is horizontally median-additive, we have that $I(\mathbf{0}) = 0$ (apply (4.46) with $f = \mathbf{0}$ and $c = 0$), and for any function f,

$$I(f) = I(f^+) + I(-f^-) \tag{4.47}$$

(apply (4.46) with $c = 0$).

The following result sheds light on the connection between the three types of horizontal additivity.

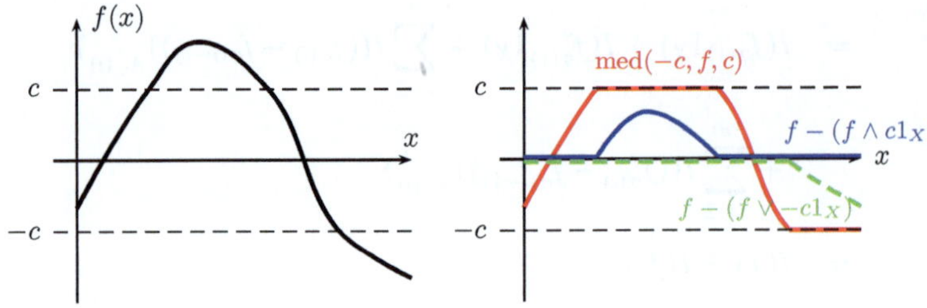

Fig. 4.8 Horizontal median-additive decomposition: mapping f (*left*) and its decomposition (*right*)

Lemma 4.33 (Couceiro and Marichal [58]) *Let $I : \mathbb{R}^X \to \mathbb{R}$ be a functional. The following propositions are equivalent:*

(i) I is horizontally median-additive;
(ii) I is horizontally min-additive on $\mathbb{R}^X_+$ with $c \geqslant 0$, horizontally max-additive on $\mathbb{R}^X_-$ with $c \leqslant 0$, and satisfies (4.47).

Proof (i)⇒(ii): we show horizontal min-additivity for nonnegative functions only. For any $f : X \to \mathbb{R}_+$, we have by horizontal median-additivity

$$I(f) = I(f \wedge c1_X) + I(f - (f \wedge c1_X)) + \underbrace{I(\mathbf{0})}_{0},$$

which is horizontal min-additivity.

(ii)⇒(i): Applying (4.47) to the function $\mathrm{med}(-c, f, c)$, we obtain:

$$I(\mathrm{med}(-c,f,c)) = I(f^+ \wedge c1_X) + I((-f^-) \vee (-c1_X)). \tag{4.48}$$

On the other hand, we have

$$\begin{aligned} I(f) = & \; I(f^+) + I(-f^-) && \text{(by (4.47))} \\ = & \; \big(I(f^+ \wedge c1_X) + I(f^+ - (f^+ \wedge c1_X))\big) + \\ & \; \big(I((-f^-) \vee (-c1_X)) + I(-f^- - ((-f^-) \vee (-c1_X)))\big) && \text{(by hor. min/max-add.)} \\ = & \; I(\mathrm{med}(-c,f,c)) + I(f - (f \wedge c1_X)) \\ & \; + I(f - (f \vee (-c1_X))), && \text{(by (4.48))} \end{aligned}$$

which is horizontal median-additivity. □

Comonotonic Modularity

A functional $I : \mathbb{R}^X \to \mathbb{R}$ is *modular* if for every $f, g : X \to \mathbb{R}$,

$$I(f \vee g) + I(f \wedge g) = I(f) + I(g). \tag{4.49}$$

It is easy to see that an additive functional is always modular:

$$I(f) + I(g) = I(f + g) = I((f \vee g) + (f \wedge g)) = I(f \vee g) + I(f \wedge g),$$

but the converse is false (take $|X| = 2$ and $I(f) = f_1^2 + f_2^2$: it is modular but not additive). When X is finite, Topkis [327] has shown that a modular functional has necessarily the form $I = \sum_{i=1}^n \varphi_i$, with $\varphi_i : \mathbb{R} \to \mathbb{R}$ and $n = |X|$.

A functional $I : \mathbb{R}^X \to \mathbb{R}$ is *maxitive* if $I(f \vee g) = I(f) \vee I(g)$ for all $f, g \in \mathbb{R}^X$, and *minitive* if $I(f \wedge g) = I(f) \wedge I(g)$. If a functional is both maxitive and minitive, then it is modular:

$$I(f \vee g) + I(f \wedge g) = (I(f) \vee I(g)) + (I(f) \wedge I(g)) = I(f) + I(g).$$

Again, the converse is false (take the same example as before).

The above notions can be turned into their comonotonic version: I is *comonotonically modular* (respectively, *comonotonically maxitive, minitive*) if it is modular (respectively, maxitive, minitive) for comonotonic functions. It follows that

(i) Comonotonic additivity implies comonotonic modularity;
(ii) Comonotonic maxitivity and comonotonic minitivity imply comonotonic modularity.

Corollary 4.34 *Let $f, g \in B(\mathcal{F})$ be comonotonic functions such that $f + g \in B(\mathcal{F})$. Then for any game in $\mathcal{BV}(\mathcal{F})$, the Choquet integral is comonotonically modular:*

$$\int (f \vee g)\, dv + \int (f \wedge g)\, dv = \int f\, dv + \int g\, dv.$$

Concavity

We continue with properties related to additivity and concavity. The next important result shows that the Choquet integral is superadditive (equivalently, concave) if and only if the game is supermodular.

Theorem 4.35 (Superadditivity of the Choquet integral) *For any game $v \in \mathcal{BV}(\mathcal{F})$, the following conditions are equivalent:*

(i) v is supermodular;

(ii) *The Choquet integral is superadditive on $B(\mathcal{F})$, that is,*

$$\int (f+g)\,\mathrm{d}v \geqslant \int f\,\mathrm{d}v + \int g\,\mathrm{d}v$$

for all $f, g \in B(\mathcal{F})$ such that $f+g \in B(\mathcal{F})$;

(iii) *The Choquet integral is supermodular on $B(\mathcal{F})$, that is,*

$$\int (f \vee g)\,\mathrm{d}v + \int (f \wedge g)\,\mathrm{d}v \geqslant \int f\,\mathrm{d}v + \int g\,\mathrm{d}v$$

for all $f, g \in B(\mathcal{F})$;

(iv) *The Choquet integral is concave on $B(\mathcal{F})$, that is,*

$$\int (\lambda f + (1-\lambda)g)\,\mathrm{d}v \geqslant \int \lambda f\,\mathrm{d}v + (1-\lambda)\int g\,\mathrm{d}v$$

for all $\lambda \in [0,1]$, $f, g \in B(\mathcal{F})$ such that $\lambda f + (1-\lambda)g \in B(\mathcal{F})$.

Proof (Marinacci and Montrucchio [235])

(i)⇒(ii) Consider $f \in B^+(\mathcal{F})$ and $S \in \mathcal{F}$. We have

$$\{f + 1_S \geqslant t\} = \{f \geqslant t\} \cup \{S \cap \{f \geqslant t-1\}\},$$

and so $f + 1_S \in B^+(\mathcal{F})$. This implies that $f + g \in B^+(\mathcal{F})$ for any simple function. Moreover, as v is supermodular, we get

$$v(f + 1_S \geqslant t) \geqslant v(f \geqslant t) + v(S \cap (f \geqslant t-1)) - v(S \cap (f \geqslant t)).$$

Consequently,

$$\begin{aligned}\int (f+1_S)\,\mathrm{d}v &= \int_0^\infty v(f + 1_S \geqslant t)\,\mathrm{d}t \\ &\geqslant \int_0^\infty v(f \geqslant t)\,\mathrm{d}t + \int_0^\infty v(S \cap (f \geqslant t-1))\,\mathrm{d}t - \int_0^\infty v(S \cap (f \geqslant t))\,\mathrm{d}t \\ &= \int f\,\mathrm{d}v + \int_{-1}^0 v(S \cap (f \geqslant t))\,\mathrm{d}t = \int f\,\mathrm{d}v + v(S).\end{aligned}$$

As the Choquet integral is positively homogeneous, we have for any $\lambda \geqslant 0$

$$\begin{aligned}\int (f + \lambda 1_S)\,\mathrm{d}v = \lambda \int \Big(\frac{f}{\lambda} + 1_S\Big)\,\mathrm{d}v &\geqslant \lambda\Big(\int \frac{f}{\lambda}\,\mathrm{d}v + v(S)\Big) \\ &= \int f\,\mathrm{d}v + \lambda v(S). \qquad (4.50)\end{aligned}$$

Let $g \in B^+(\mathcal{F})$ be a simple function. By using its horizontal decomposition (4.1) $g = \sum_{i=1}^n \lambda_i 1_{D_i}$ with $\lambda_i \geqslant 0$ for $i = 1, \ldots, n$ and $D_1 \supseteq D_2 \cdots \supseteq D_n$, we can write, using (4.50):

$$\int (f+g)\,\mathrm{d}v = \int \Big(f + \sum_{i=1}^n \lambda_i 1_{D_i}\Big)\,\mathrm{d}v \geqslant \int \Big(f + \sum_{i=2}^n \lambda_i 1_{D_i}\Big)\,\mathrm{d}v + \lambda_1 v(D_1)$$

$$\geqslant \cdots \geqslant \int f\,\mathrm{d}v + \sum_{i=1}^n \lambda_i v(D_i) = \int f\,\mathrm{d}v + \int g\,\mathrm{d}v,$$

as desired, where the last equality obtains from (4.17). Now, the inequality $\int (f+g)\,\mathrm{d}v \geqslant \int f\,\mathrm{d}v + \int g\,\mathrm{d}v$ holds for any $f, g \in B(\mathcal{F})$ because of translation invariance and continuity of the integral [Theorem 4.24(xii)].

(ii)⇒(i) Given any sets $A, B \in \mathcal{F}$, we have

$$1_{A\cup B} + 1_{A\cap B} = 1_A + 1_B.$$

Observe that $1_{A\cup B}$ and $1_{A\cap B}$ are comonotonic functions, therefore by Theorem 4.28, we get

$$v(A\cup B) + v(A\cap B) = \int 1_{A\cup B}\,\mathrm{d}v + \int 1_{A\cap B}\,\mathrm{d}v = \int (1_{A\cup B} + 1_{A\cap B})\,\mathrm{d}v$$

$$= \int (1_A + 1_B)\,\mathrm{d}v \geqslant \int 1_A\,\mathrm{d}v + \int 1_B\,\mathrm{d}v = v(A) + v(B),$$

hence v is supermodular.

(i)⇒(iii) As the Choquet integral is translation invariant, it is enough to prove the result for nonnegative f, g. The following holds for each $t \geqslant 0$:

$$\{f \vee g \geqslant t\} = \{f \geqslant t\} \cup \{g \geqslant t\}$$
$$\{f \wedge g \geqslant t\} = \{f \geqslant t\} \cap \{g \geqslant t\}.$$

Supermodularity of v yields

$$v(f \vee g \geqslant t) + v(f \wedge g \geqslant t) \geqslant v(f \geqslant t) + v(g \geqslant t).$$

It follows that

$$\int (f \vee g)\,\mathrm{d}v + \int (f \wedge g)\,\mathrm{d}v = \int_0^\infty v(f \vee g \geqslant t)\,\mathrm{d}t + \int_0^\infty v(f \wedge g \geqslant t)\,\mathrm{d}t$$
$$= \int_0^\infty \big(v(f \vee g \geqslant t) + v(f \wedge g \geqslant t)\big)\,\mathrm{d}t$$
$$\geqslant \int_0^\infty \big(v(f \geqslant t) + v(g \geqslant t)\big)\,\mathrm{d}t = \int f\,\mathrm{d}v + \int g\,\mathrm{d}v,$$

as desired.

(iii)⇒(i) We have $1_A \vee 1_B = 1_{A\cup B}$ and $1_A \wedge 1_B = 1_{A\cap B}$. Hence, putting $f = 1_A$ and $g = 1_B$, supermodularity of the integral yields supermodularity of v.

(ii)⇔(iv) holds because of positive homogeneity (see Sect. 1.3.7). □

Remark 4.36

(i) The equivalence (i)⇔(iii) was remarked by Lovász [226].
(ii) By asymmetry of the integral [Theorem 4.24(iv)] and the fact that the conjugate of supermodular games are submodular [Theorem 2.20(ii)], a similar theorem holds for submodular games, with all inequalities inverted. ◇

A consequence of the above remark is that the Choquet integral is additive if and only if v is modular; i.e., additive, because $\mathcal{F}$ is assumed to be an algebra (Theorem 2.117).

Corollary 4.37 (Additivity of the Choquet integral) *For any game $v \in \mathcal{BV}(\mathcal{F})$, the Choquet integral is additive on $B(\mathcal{F})$ if and only if v is additive.*

The next result is established in the discrete case. We recall that for any game v on $X = \{x_1, \ldots, x_n\}$ and any permutation σ on X, the marginal vector $\phi^{\sigma,v}$ is defined by

$$\phi^{\sigma,v}_{\sigma(i)} = v(\{x_{\sigma(1)}, \ldots, x_{\sigma(i)}\}) - v(\{x_{\sigma(1)}, \ldots, x_{\sigma(i-1)}\})$$

(see Sect. 3.2.2 with a different notation). Identifying the vector $\phi^{\sigma,v}$ with an additive measure, we get the following result.

Lemma 4.38 *Suppose $|X| = n$ and take $\mathcal{F} = 2^N$. If v is supermodular, then for any f on X and any permutation π on X,*

$$\int f\,\mathrm{d}v \leqslant \int f\,\mathrm{d}\phi^{\pi,v}$$

with equality if $f_{\pi(1)} \geqslant \cdots \geqslant f_{\pi(n)}$.

Proof Suppose $f_{\pi(1)} \geqslant \cdots \geqslant f_{\pi(n)}$. Then $\int f\,\mathrm{d}v$ expressed with (4.31) corresponds to $\int f\,\mathrm{d}\phi^{\pi,v}$ given by (4.37).

Suppose on the contrary that $\pi \neq \text{Id}$ does not order f in decreasing order. Without loss of generality, consider that $f_1 \geqslant f_2 \geqslant \cdots \geqslant f_n$. It is a standard result from combinatorics that one can go from the identity permutation to π by elementary switches exchanging only 2 neighbor elements; i.e., we have the sequence

$$\sigma = \text{Id} \to \cdots \pi' \to \pi'' \to \cdots \to \pi$$

with in each step $\pi'(j) = \pi''(j)$ except for $j = i, i+1$ for some $1 \leqslant i < n$, where $\pi'(i) = \pi''(i+1)$ and $\pi'(i+1) = \pi''(i)$. Consider two consecutive π', π'' in the sequence differing on $i, i+1$; we have by (4.37),

$$\begin{aligned}
\int f \, d\phi^{\pi',v} &= \sum_{j=1}^{n} f_{\pi'(j)}(v(A^{\downarrow}_{\pi'}(j)) - v(A^{\downarrow}_{\pi'}(j-1))) = \sum_{j=1}^{n}(f_{\pi'(j)} - f_{\pi'(j+1)})v(A^{\downarrow}_{\pi'}(j)) \\
&= \sum_{j=1}^{i-2}(f_{\pi'(j)} - f_{\pi'(j+1)})v(A^{\downarrow}_{\pi'}(j)) + (f_{\pi'(i-1)} - f_{\pi'(i)})v(A^{\downarrow}_{\pi'}(i-1)) \\
&\quad + (f_{\pi'(i)} - f_{\pi'(i+1)})v(A^{\downarrow}_{\pi'}(i)) + (f_{\pi'(i+1)} - f_{\pi'(i+2)})v(A^{\downarrow}_{\pi'}(i+1)) \\
&\quad + \sum_{j=i+2}^{n}(f_{\pi'(j)} - f_{\pi'(j+1)})v(A^{\downarrow}_{\pi'}(j)) \\
&\leqslant \sum_{j=1}^{i-2}(f_{\pi''(j)} - f_{\pi''(j+1)})v(A^{\downarrow}_{\pi''}(j)) \\
&\quad + (f_{\pi''(i-1)} - f_{\pi''(i+1)})v(A^{\downarrow}_{\pi''}(i-1)) \\
&\quad + (f_{\pi''(i+1)} - f_{\pi''(i)})(v(A^{\downarrow}_{\pi''}(i+1)) + v(A^{\downarrow}_{\pi''}(i-1)) - v(A^{\downarrow}_{\pi''}(i))) \\
&\quad + (f_{\pi''(i)} - f_{\pi''(i+2)})v(A^{\downarrow}_{\pi''}(i+1)) + \sum_{j=i+2}^{n}(f_{\pi''(j)} - f_{\pi''(j+1)})v(A^{\downarrow}_{\pi''}(j)) \\
&= \sum_{j=1}^{n} f_{\pi''(j)}(v(A^{\downarrow}_{\pi''}(j)) - v(A^{\downarrow}_{\pi''}(j-1))) = \int f \, d\phi^{\pi'',v},
\end{aligned}$$

where in the inequality we have used supermodularity of v, and the fact that $f_{\pi'(i)} - f_{\pi'(i+1)} \geqslant 0$, because $\pi'(i) < \pi'(i+1)$ (by construction, i and $i+1$ have not been switched before). It follows that $\int f \, dv \leqslant \int f \, d\phi^{\pi,v}$. □

From the above lemma, the following fundamental result is immediate (see Definition 3.1 for a definition of **core**(v)).

Theorem 4.39 (The Choquet integral as a lower expected value) *Suppose* $|X| = n$ *and* $\mathcal{F} = 2^X$. *Then for any function* f *on* X, *the game* v *is supermodular if and* *only if*

$$\int f\,\mathrm{d}v = \min_{\phi\in\mathbf{core}(v)} \int f\,\mathrm{d}\phi, \tag{4.51}$$

where $\phi \in \mathbf{core}(v)$ is identified with an additive measure.

Proof Suppose that v is supermodular. By Theorem 3.15, we know that any core element ϕ is a convex combination of all marginal vectors: $\phi = \sum_\pi \lambda_\pi \phi^{\pi,v}$ with $\lambda_\pi \geqslant 0$ and $\sum_\pi \lambda_\pi = 1$. Using Lemma 4.38, we have by linearity of the integral [Theorem 4.24(ix)]

$$\int f\,\mathrm{d}v = \sum_\pi \Big(\lambda_\pi \int f\,\mathrm{d}v\Big) \leqslant \sum_\pi \Big(\lambda_\pi \int f\,\mathrm{d}\phi^{\pi,v}\Big) = \int f\,\mathrm{d}\Big(\sum_\pi \lambda_\pi \phi^{\pi,v}\Big) = \int f\,\mathrm{d}\phi$$

for any core element ϕ. Since by Lemma 4.38, equality is satisfied for at least one $\phi^{\pi,v}$, (4.51) holds.

Conversely, suppose that (4.51) holds for any f. Take any function f such that $f_{\sigma(1)} \geqslant \cdots \geqslant f_{\sigma(n)}$. Letting $A^{\downarrow}_{\sigma}(i) = \{x_{\sigma(1)}, \ldots, x_{\sigma(i)}\}$, there exists a core element ϕ such that

$$\int f\,\mathrm{d}v = \sum_{i=1}^{n} (f_{\sigma(i)} - f_{\sigma(i+1)}) v(A^{\downarrow}_{\sigma}(i)) = \sum_{i=1}^{n} (f_{\sigma(i)} - f_{\sigma(i+1)}) \phi(A^{\downarrow}_{\sigma}(i)) = \int f\,\mathrm{d}\phi. \tag{4.52}$$

Since $\phi \in \mathbf{core}(v)$, we have $\phi(A^{\downarrow}_{\sigma}(i)) \geqslant v(A^{\downarrow}_{\sigma}(i))$, hence nonnegativity of $f_{\sigma(i)} - f_{\sigma(i+1)}$ and (4.52) force $\phi(A^{\downarrow}_{\sigma}(i)) = v(A^{\downarrow}_{\sigma}(i))$; i.e., ϕ is the marginal vector $\phi^{\sigma,v}$ [see (3.8) and (3.9)]. This being true for any f on X, it follows that for any permutation σ on X, the marginal vector $\phi^{\sigma,v}$ belongs to the core, a condition that is equivalent to supermodularity of v (see Theorem 3.15). □

Remark 4.40

- Again, as explained in Remark 4.36, results similar to Lemma 4.38 and Theorem 4.39 hold for submodular games, with inequalities inverted, min changed to max and the core changed to the anticore (that is, the set of efficient vectors ϕ satisfying $\phi(S) \leqslant v(S)$ for all $S \in 2^X$; see Sect. 3.1): the Choquet integral for submodular games is an upper expected value on the anticore.
- Dempster [77, Sect. 2] has shown that (4.51) holds for belief measures, a particular case of supermodular capacities.
- A result similar to Theorem 4.39 holds for the Sugeno integral; see Sect. 7.7.4.

◇

Recalling from Sect. 1.3.7 the notion of support function of a convex set, Theorem 4.39 merely says that for supermodular games, the Choquet integral is the support function of the core, because the right-hand of (4.51) can be rewritten as $\min_{x\in\mathbf{core}(v)}\langle f, x\rangle$, considering f, x as vectors in $\mathbb{R}^n$. A simple application of Theorem 1.12 leads to the following corollary.

Corollary 4.41 (The core as the superdifferential of the Choquet integral) *(Danilov and Koshevoy [66]) Suppose $|X| = n$ and $\mathcal{F} = 2^X$. Then for any supermodular*[8] *game v on X,*

$$\mathbf{core}(v) = \partial\Big(\int \cdot \, \mathrm{d}v\Big)(\mathbf{0}).$$

Remark 4.42 By Theorem 1.12 again, the support function is positively homogeneous and concave (or equivalently, superadditive). Hence, Lemma 4.38 and Theorem 4.39 constitute another proof of the equivalence of (i) and (ii) in Theorem 4.35. ◇

4.6.2 The Sugeno Integral

Theorem 4.43 *Let f be a function in $B^+(\mathcal{F})$, and μ a capacity on $(X, \mathcal{F})$. The following properties hold.*

(i) Positive $\wedge$-homogeneity:

$$⨍(\alpha 1_X \wedge f)\,\mathrm{d}\mu = \alpha \wedge ⨍ f\,\mathrm{d}\mu \qquad (\alpha \geqslant 0)$$

(ii) Positive $\vee$-homogeneity if $\operatorname{ess\,sup}_\mu f \leqslant \mu(X)$*:*

$$⨍(\alpha 1_X \vee f)\,\mathrm{d}\mu = \alpha \vee ⨍ f\,\mathrm{d}\mu \qquad (\alpha \in [0, \operatorname{ess\,sup}_\mu f]).$$

(iii) Hat function: for every $\alpha \geqslant 0$ and for every $A \in \mathcal{F}$,

$$⨍ \alpha 1_A\,\mathrm{d}\mu = \alpha \wedge \mu(A)$$

(iv) Scale inversion: if $\operatorname{ess\,sup}_\mu f \leqslant \mu(X)$,

$$⨍(\mu(X)1_X - f)\,\mathrm{d}\mu = \mu(X) - ⨍ f\,\mathrm{d}\overline{\mu},$$

where $\overline{\mu}$ is the conjugate capacity;

[8] In Danilov and Koshevoy [66], the condition of supermodularity was overlooked.

(v) *Scale translation:*

$$\fint (f + \alpha 1_X)\,\mathrm{d}\mu \leqslant \fint f\,\mathrm{d}\mu + \fint \alpha\,\mathrm{d}\mu = \fint f\,\mathrm{d}\mu + \alpha \wedge \mu(X) \qquad (\alpha \geqslant 0)$$

(vi) *Monotonicity (or nondecreasingness) w.r.t. the integrand:*

$$f \leqslant f' \Rightarrow \fint f\,\mathrm{d}\mu \leqslant \fint f'\,\mathrm{d}\mu \qquad (f, f' \in B^+(\mathcal{F}))$$

(vii) *Monotonicity w.r.t. the capacity:*

$$\mu \leqslant \mu' \Rightarrow \fint f\,\mathrm{d}\mu \leqslant \fint f\,\mathrm{d}\mu' \qquad (\mu, \mu' \text{ on } (X, \mathcal{F}))$$

(viii) *Monotonicity w.r.t. stochastic dominance:*

$$f \geqslant^{\mu}_{\mathrm{SD}} f' \Rightarrow \fint f\,\mathrm{d}\mu \geqslant \fint f'\,\mathrm{d}\mu \qquad (f, f' \in B^+(\mathcal{F}))$$

(ix) *Max-min linearity w.r.t. the capacity:*

$$\fint f\,\mathrm{d}(\mu \vee (\alpha \wedge \mu')) = \fint f\,\mathrm{d}\mu \vee \left(\alpha \wedge \fint f\,\mathrm{d}\mu'\right) \qquad (\mu, \mu' \text{ capacities on } (X, \mathcal{F}), \alpha \geqslant 0)$$

(x) *Boundaries:* $\inf f$ *and* $\sup f$ *are attained:*

$$\inf f = \fint f\,\mathrm{d}\mu_{\min}, \quad \sup f = \fint f\,\mathrm{d}\mu_{\max},$$

with $\mu_{\min}(A) = 0$ *for all* $A \subset X$, $A \in \mathcal{F}$, *and* $\mu_{\max}(A) = 1$ *for all nonempty* $A \in \mathcal{F}$;

(xi) *Boundaries:*

$$\text{ess inf}_{\mu} f \leqslant \fint f\,\mathrm{d}\mu \leqslant (\text{ess sup}_{\mu} f) \wedge \mu(X)$$

(xii) *Lipschitz continuity:*

$$\left| \fint f\,\mathrm{d}\mu - \fint g\,\mathrm{d}\mu \right| \leqslant \mu(X) \wedge \|f - g\| \qquad (f, g \in B^+(\mathcal{F}))$$

with $\|f\| = \sup_{x\in X} |f(x)|$ (Chebyshev norm). Hence, if μ is normalized and f, g are valued on $[0, 1]$, we obtain that the Sugeno integral is 1-Lipschitzian for the Chebyshev norm.

Proof (i) Note that

$$\{f \wedge \alpha 1_X \geqslant t\} = \begin{cases} \{f \geqslant t\}, & \text{if } t \leqslant \alpha \\ \varnothing, & \text{otherwise,} \end{cases}$$

which yields

$$G_{\mu, f\wedge\alpha 1_X}(t) = \begin{cases} G_{\mu,f}(t), & \text{if } t \leqslant \alpha \\ 0, & \text{otherwise.} \end{cases}$$

It follows that

$$⨍(\alpha 1_X \wedge f)\, \mathrm{d}\mu = \bigvee_{t\geqslant 0} (G_{\mu, f\wedge\alpha 1_X}(t) \wedge t) = \bigvee_{t\in[0,\alpha]} (G_{\mu,f}(t) \wedge t) \vee \underbrace{\bigvee_{t>\alpha} (0 \wedge t)}_{0}. \qquad (4.53)$$

Observe that if $⨍ f\, \mathrm{d}\mu \leqslant \alpha$, then $\bigvee_{t\in[0,\alpha]} (G_{\mu,f}(t) \wedge t) = ⨍ f\, \mathrm{d}\mu$, otherwise $\bigvee_{t\in[0,\alpha]} (G_{\mu,f}(t) \wedge t) = \alpha$. We conclude that

$$⨍(f \wedge \alpha 1_X)\, \mathrm{d}\mu = \alpha \wedge ⨍ f\, \mathrm{d}\mu.$$

(ii) For max-homogeneity, we have for any $\alpha \in [0, \operatorname{ess\,sup}_\mu f]$

$$\{f \vee \alpha 1_X \geqslant t\} = \begin{cases} \{f \geqslant t\}, & \text{if } t \geqslant \alpha \\ X, & \text{otherwise,} \end{cases}$$

which yields

$$G_{\mu, f\vee\alpha 1_X}(t) = \begin{cases} G_{\mu,f}(t), & \text{if } t \geqslant \alpha \\ \mu(X), & \text{otherwise.} \end{cases}$$

Hence

$$\fint (\alpha 1_X \wedge f)\,\mathrm{d}\mu = \bigwedge_{t\in[0,\text{ess sup}_\mu f]} (G_{\mu,f\vee\alpha 1_X}(t) \vee t)$$

$$= \underbrace{\bigwedge_{t\in[0,\alpha[} (\mu(X)\vee t)}_{\mu(X)} \wedge \bigwedge_{t\in[\alpha,\text{ess sup}_\mu f]} (G_{\mu,f}(t)\vee t).$$

As it can be checked, $\bigwedge_{t\in[\alpha,\text{ess sup}_\mu f]}(G_{\mu,f}(t) \vee t) = \alpha \vee \fint f\,\mathrm{d}\mu$. Since by assumption, ess $\sup_\mu f \leqslant \mu(X)$, by (x) we deduce $\fint f\,\mathrm{d}\mu \leqslant \mu(X)$. In summary, $\bigwedge_{t\in[0,\text{ess sup}_\mu f]}(G_{\mu,f\vee\alpha 1_X}(t) \vee t) = \alpha \vee \fint f\,\mathrm{d}\mu$ as desired.

(iii) We have for any $A \in \mathcal{F}$

$$\fint \alpha 1_A\,\mathrm{d}v = \bigvee_{t\geqslant 0}(G_{v,\alpha 1_A}(t)\wedge t) = \bigvee_{t\in[0,\alpha]} (v(A)\wedge t) \vee \underbrace{\bigvee_{t>\alpha}(v(\varnothing)\wedge t)}_{0} = v(A)\wedge\alpha.$$

(iv) We use the fact that for any numbers $x_i \in [0,a]$, $i \in J$, we have $\vee_{i\in J}(a-x_i) = a - \wedge_{i\in J}x_i$, and $\wedge_{i\in J}(a-x_i) = a - \vee_{i\in J}x_i$. We get:

$$\fint (\mu(X)1_X - f)\,\mathrm{d}\mu = \bigwedge_{t\in[0,\mu(X)]} \big(t \vee \mu(\{\mu(X)1_X - f \geqslant t\})\big)$$

$$= \bigwedge_{t'\in[0,\mu(X)]} \big((\mu(X)-t') \vee \mu(\{f \leqslant t'\})\big)$$

with $t' = \mu(X) - t$. Now, by Remark 4.7

$$\mu(X) - \fint f\,\mathrm{d}\overline{\mu} = \mu(X) - \bigvee_{t\in[0,\mu(X)]} \big(t \wedge \overline{\mu}(\{f > t\})\big)$$

$$= \mu(X) - \bigvee_{t\in[0,\mu(X)]} \big(t\wedge(\mu(X) - \mu(\{f\leqslant t\}))\big)$$

$$= \bigwedge_{t\in[0,\mu(X)]} \big(\mu(X) - (t\wedge(\mu(X)-\mu(\{f\leqslant t\})))\big)$$

$$= \bigwedge_{t\in[0,\mu(X)]} \big((\mu(X)-t)\vee\mu(\{f\leqslant t\})\big),$$

which completes the proof.

(v) Using (4.5), we obtain:

$$\begin{aligned}
\fint (f+\alpha 1_X)\,\mathrm{d}\mu &= \bigvee_{A\in\mathcal{F}} \Big(\bigwedge_{x\in A}(f(x)+\alpha)\wedge\mu(A)\Big)\\
&\leqslant \bigvee_{A\in\mathcal{F}} \Big(\Big(\bigwedge_{x\in A} f(x)\wedge\mu(A)\Big)+(\alpha\wedge\mu(A))\Big)\\
&\leqslant \bigvee_{A\in\mathcal{F}} \Big(\Big(\bigwedge_{x\in A} f(x)\wedge\mu(A)\Big)+(\alpha\wedge\mu(X))\Big)\\
&= \bigvee_{A\in\mathcal{F}} \Big(\bigwedge_{x\in A} f(x)\wedge\mu(A)\Big)+(\alpha\wedge\mu(X)) = \fint f\,\mathrm{d}\mu + \fint \alpha 1_X\,\mathrm{d}\mu,
\end{aligned}$$

by (iii).

(vi), (vii) and (viii) work as for the Choquet integral (Theorem 4.24) and are straightforward.

(ix) We have

$$\begin{aligned}
\fint f\,\mathrm{d}(\mu\vee(\alpha\wedge\mu')) &= \bigvee_{t\geqslant 0} (G_{\mu\vee(\alpha\wedge\mu'),f}(t)\wedge t)\\
&= \bigvee_{t\geqslant 0} ((\mu(f\geqslant t)\vee(\alpha\wedge\mu'(f\geqslant t)))\wedge t)\\
&= \bigvee_{t\geqslant 0} ((\mu(f\geqslant t)\wedge t)\vee(\alpha\wedge\mu'(f\geqslant t)\wedge t)))\\
&= \bigvee_{t\geqslant 0} (\mu(f\geqslant t)\wedge t)\vee\Big(\alpha\wedge\Big(\bigvee_{t\geqslant 0}(\mu'(f\geqslant t)\wedge t)\Big)\Big)\\
&= \fint f\,\mathrm{d}\mu \;\vee\; \Big(\alpha\wedge\fint f\,\mathrm{d}\mu'\Big).
\end{aligned}$$

(x) and (xi) The proof uses (vii) and is similar to the one for the Choquet integral [Theorem 4.24(x) and (xi)].

(xii) Put $a=\sup_{x\in X}|f(x)-g(x)|$. Then $f\leqslant g+a$, and by (vi) and (v) we obtain

$$\fint f\,\mathrm{d}\mu \leqslant \fint (g+a)\,\mathrm{d}\mu \leqslant \fint g\,\mathrm{d}\mu + (a\wedge\mu(X)).$$

Similarly, from $g\leqslant f+a$ we obtain

$$\fint g\,\mathrm{d}\mu \leqslant \fint f\,\mathrm{d}\mu + (a\wedge\mu(X)),$$

from which we deduce that

$$\left| \fint f \, d\mu - \fint g \, d\mu \right| \leqslant \mu(X) \wedge \sup_{x \in X} |f(x) - g(x)|.$$

□

(v) and a weaker version of (xii) can be found in Wang and Klir [343, Theorem 9.2 and Lemma 9.2].

We turn to properties related to comonotonic functions.

Theorem 4.44 (Comonotonic maxitivity and minitivity of the Sugeno integral) *Let $f, g \in B^+(\mathcal{F})$ be comonotonic functions. Then for any capacity μ on $(X, \mathcal{F})$, the Sugeno integral is comonotonically maxitive and comonotonically minitive, that is:*

$$\fint (f \vee g) \, d\mu = \fint f \, d\mu \vee \fint g \, d\mu$$

$$\fint (f \wedge g) \, d\mu = \fint f \, d\mu \wedge \fint g \, d\mu.$$

Proof First $f \vee g, f \wedge g \in B^+(\mathcal{F})$ by Lemma 4.1. Let us prove comonotonic maxitivity. We have

$$\fint (f \vee g) \, d\mu = \bigvee_{t \geqslant 0} \big(G_{\mu, f \vee g}(t) \wedge t \big).$$

Remark that $\{f \vee g \geqslant t\} = \{f \geqslant t\} \cup \{g \geqslant t\}$. It follows from Lemma 4.27(iii) that for any $t \geqslant 0$, either $\{f \geqslant t\} \subseteq \{g \geqslant t\}$ or the converse. By monotonicity of capacities, we deduce that

$$\mu(\{f \vee g \geqslant t\}) = \mu(\{f \geqslant t\}) \vee \mu(\{g \geqslant t\})$$

for every $t \geqslant 0$. We then obtain

$$\begin{aligned} \fint (f \vee g) \, d\mu &= \bigvee_{t \geqslant 0} \big((G_{\mu,f}(t) \vee G_{\mu,g}(t)) \wedge t \big) \\ &= \bigvee_{t \geqslant 0} \big((G_{\mu,f}(t) \wedge t) \vee (G_{\mu,g}(t) \wedge t) \big) \\ &= \fint f \, d\mu \vee \fint g \, d\mu. \end{aligned}$$

The proof for comonotonic minitivity is obtained in the same way, starting from $\fint f \, d\mu = \bigwedge_{t \geqslant 0} (G_{\mu,f}(t) \vee t)$. □

Using results of Sect. 4.6.1, we obtain immediately:

Corollary 4.45 *The Sugeno integral is comonotonically modular over $B^+(\mathcal{F})$, for every capacity on $(X, \mathcal{F})$.*

The next theorem shows under which condition the Sugeno integral is maxitive or minitive.

Theorem 4.46 (Maxitivity/minitivity of the Sugeno integral) *The following holds:*

(i) $⨍(f \vee g)\,\mathrm{d}\mu = ⨍ f\,\mathrm{d}\mu \vee ⨍ g\,\mathrm{d}\mu$ *for all* $f, g \in B^+(\mathcal{F})$ *if and only if* μ *is maxitive;*
(ii) $⨍(f \wedge g)\,\mathrm{d}\mu = ⨍ f\,\mathrm{d}\mu \wedge ⨍ g\,\mathrm{d}\mu$ *for all* $f, g \in B^+(\mathcal{F})$ *if and only if* μ *is minitive.*

Proof Suppose that μ is maxitive. Then for all $f, g \in B^+(\mathcal{F})$

$$
\begin{aligned}
⨍(f \vee g)\,\mathrm{d}\mu &= \bigvee_{t\geqslant 0}(t \wedge \mu(f \vee g \geqslant t)) \\
&= \bigvee_{t\geqslant 0}(t \wedge \mu(\{f \geqslant t\} \cup \{g \geqslant t\})) \\
&= \bigvee_{t\geqslant 0}(t \wedge (\mu(f \geqslant t) \vee \mu(g \geqslant t))) \\
&= \bigvee_{t\geqslant 0}(t \wedge \mu(f \geqslant t)) \vee \bigvee_{t\geqslant 0}(t \wedge \mu(g \geqslant t)) = ⨍ f\,\mathrm{d}\mu \vee ⨍ g\,\mathrm{d}\mu,
\end{aligned}
$$

where we have used mutual distributivity of $\vee, \wedge$. The converse statement is immediate by taking $f = 1_A$, $g = 1_B$, where A, B are any two sets in $\mathcal{F}$.

The proof for minitivity is similar, starting from $⨍ f\,\mathrm{d}\mu = \bigwedge_{t\geqslant 0}(t \vee G_{\mu,f}(t))$. □

Next, we give properties relating the two integrals.

Theorem 4.47 *Let μ be a normalized capacity on $(X, \mathcal{F})$. The following holds.*

(i) The equality $\int f\,\mathrm{d}\mu = ⨍ f\,\mathrm{d}\mu$ *holds for every function* $f \in B^+(\mathcal{F})$ *such that* $\operatorname{ess\,sup}_\mu f \leqslant 1$ *if and only if* μ *is a 0-1-capacity;*
(ii) For any $f \in B^+(\mathcal{F})$ *such that* $\operatorname{ess\,sup}_\mu f \leqslant 1$,

$$
\left|\int f\,\mathrm{d}\mu - ⨍ f\,\mathrm{d}\mu\right| \leqslant \frac{1}{4}.
$$

Proof

(i) Suppose μ is 0-1-valued. Then $G_{\mu,f}$ is a rectangle of height 1 and width a, $0 < a \leqslant 1$. Hence, the area under the decumulative function is a, while the intersection of the diagonal with $G_{\mu,f}$ has coordinates (a, a), which proves the equality of the integrals.

Conversely, take μ such that $0 < \mu(A) < 1$ for some $A \subset X$, and $0 < \alpha < 1$. Then

$$\int \alpha 1_A \, \mathrm{d}\mu = \alpha\mu(A) < \alpha \wedge \mu(A) = \fint \alpha 1_A \, \mathrm{d}\mu,$$

by Theorem 4.43(iii).

(ii) (Figure 4.9) Let us fix $\alpha \in [0, 1]$, and consider some $f \in B^+(\mathcal{F})$ such that $\fint f \, \mathrm{d}\mu = \alpha$. This means that the decumulative function intersects the diagonal exactly at α. The smallest possible Choquet integral is obtained for f giving the decumulative function given by the dashed line, whose area below is α^2. Similarly the largest one is obtained by the decumulative function given by the thick solid line, with area equal to $1 - (1 - \alpha)^2$. The difference between the Choquet integral and the Sugeno integral is in both cases $\alpha - \alpha^2$. This function attains its maximum value for $\alpha = \frac{1}{2}$, and then the difference is equal to $\frac{1}{4}$.

□

(See Sect. 4.9 for the expression of the Choquet or Sugeno integral w.r.t 0-1-capacities.)

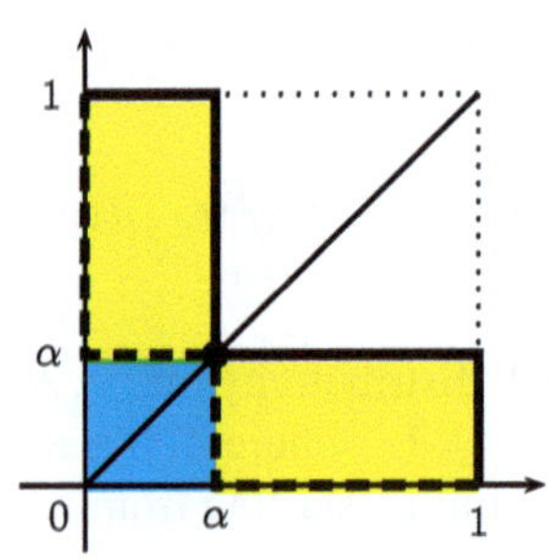

Fig. 4.9 Proof of Theorem 4.47(ii)

4.7 Expression with Respect to the Möbius Transform and Other Transforms

We consider in the whole section that X is a finite set with $|X| = n$, and that $\mathcal{F} = 2^X$.

4.7.1 The Choquet Integral

We have seen in Sect. 2.17 that transforms and bases are dual notions. From this duality and by linearity of the Choquet integral w.r.t. games, it is easy to establish a general way to express the Choquet integral in terms of a given linear and invertible transform.

Let Ψ be a linear invertible transform, $\{b_A^\Psi\}_{A\in 2^X}$ the corresponding basis of set functions given by Lemma 2.91, and $\{b_A'^\Psi\}_{A\in 2^X\setminus\{\varnothing\}}$ the corresponding basis of games given by (2.110) (see Remarks 4.10(v) and 2.92). Then for every $f \in \mathbb{R}^X$ and every game $v \in \mathcal{G}(X)$,

$$\int f\,\mathrm{d}v = \int f\,\mathrm{d}\Big(\sum_{\varnothing\neq A\subseteq X} \Psi^v(A) b_A'^\Psi\Big) = \sum_{\varnothing\neq A\subseteq X} \Psi^v(A) \int f\,\mathrm{d}b_A'^\Psi. \tag{4.54}$$

It is therefore sufficient to compute $\int f\,\mathrm{d}b_A'^\Psi$ for every $A \subseteq X, A \neq \varnothing$.

We give hereafter the expression of the Choquet integral for the main transforms.

(i) *The Möbius transform:* the associated basis being the family of unanimity games, let us compute $\int f\,\mathrm{d}u_A$ for $\varnothing \neq A \in 2^X$. We denote by σ a permutation on X ordering f in nondecreasing order. Using (4.27), let j be the leftmost index in the ordered sequence $\{\sigma(i), i \in A\}$. Then $\int f\,\mathrm{d}u_A = f_j = \wedge_{i\in A} f_i$. It follows from (4.54) that

$$\int f\,\mathrm{d}v = \sum_{A\subseteq X} m^v(A) \bigwedge_{i\in A} f_i. \tag{4.55}$$

From this and (4.10), we can derive the expression of the symmetric integral as well:

$$\check{\int} f\,\mathrm{d}v = \sum_{A\subseteq X} m^v(A)\Big(\bigwedge_{i\in A} f_i^+ - \bigwedge_{i\in A} f_i^-\Big) \tag{4.56}$$

$$= \sum_{A\subseteq X^+} m^v(A) \bigwedge_{i\in A} f_i + \sum_{A\subseteq X^-} m^v(A) \bigvee_{i\in A} f_i, \tag{4.57}$$

where $X^+ = \{i \in X : f_i \geqslant 0\}$ and $X^- = X \setminus X^+$, and f^+, f^- are defined in (4.9).

(ii) *The co-Möbius transform:* let us compute $\int f\,\mathrm{d}\check{u}_A'$ for $\varnothing \neq A \in 2^X$, where $\check{u}_A'(B) = (-1)^{|A|}$ if $A \cap B = \varnothing$ and $B \neq \varnothing$, and 0 otherwise (Table A.1). Using (4.27) with σ a permutation on X reordering f in nondecreasing order, we find easily

$$\int f\,\mathrm{d}\check{u}_A' = (-1)^{|A|+1} \bigvee_{i\in A} f_i.$$

It follows that

$$\int f\,\mathrm{d}v = \sum_{\varnothing\neq A\in 2^X} (-1)^{|A|+1} \check{m}^v(A) \bigvee_{i\in A} f_i. \tag{4.58}$$

Proceeding as above, the expression of the symmetric integral follows immediately:

$$\check{\int} f \, \mathrm{d}v = \sum_{A \cap X^+ \neq \varnothing} (-1)^{|A|+1} \check{m}^v(A) \bigvee_{i \in A} f_i + \sum_{A \cap X^- \neq \varnothing} (-1)^{|A|+1} \check{m}^v(A) \bigwedge_{i \in A} f_i. \tag{4.59}$$

(iii) *The Fourier transform:* We compute $\int f \, \mathrm{d}\chi'_A$ for $\varnothing \neq A \in 2^X$, where $\chi'_A(B) = (-1)^{|A \cap B|}$ if $B \neq \varnothing$, and 0 otherwise. Using again (4.27) with σ a permutation on X reordering f in nondecreasing order, one can check that

$$\int f \, \mathrm{d}\chi'_A = f_{\sigma(n)} + 2 \sum_{j=1}^{|A|} (-1)^j f_{i_j} \tag{4.60}$$

with $A = \{i_1, \ldots, i_{|A|}\}$ and $f_{i_1} \geqslant \cdots \geqslant f_{i_{|A|}}$. Injecting the above expression in (4.54) gives the result.

(iv) *The (Shapley) interaction transform:* instead of computing $\int f \, \mathrm{d}b'^I_A$ we proceed differently. We decompose any given $v \in \mathcal{G}(X)$ in v^+, v^-, corresponding respectively to positive and negative interaction coefficients:

$$I^{v^+}(A) = \begin{cases} I^v(A), & \text{if } I^v(A) > 0 \\ 0, & \text{otherwise} \end{cases}, \quad I^{v^-}(A) = \begin{cases} I^v(A), & \text{if } I^v(A) < 0 \\ 0, & \text{otherwise} \end{cases},$$

for every $A \in 2^X$. By linearity of I, we have $v = v^+ + v^-$, so that by linearity of the integral, $\int f \, \mathrm{d}v = \int f \, \mathrm{d}v^+ + \int f \, \mathrm{d}v^-$. Using (4.55) and (4.58), we find

$$\int f \, \mathrm{d}v^+ = \sum_{A \subseteq X} m^{v^+}(A) \bigwedge_{i \in A} f_i$$

$$\int f \, \mathrm{d}v^- = \sum_{\varnothing \neq A \in 2^X} (-1)^{|A|+1} \check{m}^{v^-}(A) \bigvee_{i \in A} f_i.$$

Using the expression of $m, \check{m}$ in terms of I (see Table A.3), we obtain:

$$\int f \, \mathrm{d}v = \sum_{A \subseteq X} \Big(\sum_{K \subseteq X \setminus A} B_{|K|} I^{v^+}(A \cup K) \Big) \bigwedge_{i \in A} f_i + \sum_{\varnothing \neq A \in 2^X} (-1)^{|A|+1} \Big(\sum_{K \subseteq X \setminus A} B_{|K|} I^{v^-}(A \cup K) \Big) \bigvee_{i \in A} f_i \tag{4.61}$$

The case of the symmetric integral can be obtained similarly:

$$
\begin{aligned}
\check{\int} f \, \mathrm{d}v = &\sum_{A \subseteq X^+} \Bigg(\sum_{K \subseteq X \setminus A} B_{|K|} I^{v^+}(A \cup K) \Bigg) \bigwedge_{i \in A} f_i \\
&+ \sum_{A \subseteq X^-} \Bigg(\sum_{K \subseteq X \setminus A} B_{|K|} I^{v^+}(A \cup K) \Bigg) \bigvee_{i \in A} f_i \\
&+ \sum_{A \cap X^+ \neq \varnothing} (-1)^{|A|+1} \Bigg(\sum_{K \subseteq X \setminus A} B_{|K|} I^{v^-}(A \cup K) \Bigg) \bigvee_{i \in A} f_i \\
&+ \sum_{A \cap X^- \neq \varnothing} (-1)^{|A|+1} \Bigg(\sum_{K \subseteq X \setminus A} B_{|K|} I^{v^-}(A \cup K) \Bigg) \bigwedge_{i \in A} f_i, \qquad (4.62)
\end{aligned}
$$

Remark 4.48 Equation (4.55) was first proved by Chateauneuf and Jaffray [49] (also by Walley [340]), extending a result of Dempster [77]. This formula makes also clear that the Lovász extension [Eq. (2.96)] is nothing but the Choquet integral.

4.7.2 The Sugeno Integral

The Sugeno integral being nonlinear w.r.t. the capacity, the methodology applied for the Choquet integral does not fit here. For the same reason, none of the invertible linear transforms introduced so far can be adequately used. Nevertheless, it is possible to define a kind of ordinal Möbius transform, which leads to a simple expression of the Sugeno integral, analogous to (4.55), and which recovers (4.5) as a particular case.

Theorem 4.49 *[166] Let $(L, \leqslant)$ be a totally ordered set, with least element 0, $(P, \preceq)$ a (finite) partially ordered set with a least element, and $f, g : P \to L$, with g being isotone. The set of solutions of the equation:*

$$
g(x) = \bigvee_{y \preceq x} f(y), \qquad (4.63)
$$

is given by the interval $[f] = [f_*, f^*]$, with, for all $x \in P$:

$$
f^*(x) = g(x)
$$

$$
f_*(x) = \begin{cases} g(x), & \text{if } g(x) > g(y), \forall y \prec\!\cdot\, x \\ 0, & \text{otherwise} \end{cases}
$$

(recall that $\prec\!\cdot$ is the covering relation of $\preceq$, see Sect. 1.3.2). Any $f \in [f_*, f^*]$ is called an ordinal Möbius transform of g.

The name "ordinal Möbius transform" is justified by the fact that (4.63) is the ordinal counterpart of Eq. (2.17), defining the Möbius transform. Since the upper bound f^* is a trivial solution, when a single value is desired, we always take the lower bound, called *canonical (ordinal) Möbius transform*.

Applied to the case of capacities, where the poset $(P, \preceq)$ is the power set $(2^X, \subseteq)$, we obtain that the ordinal Möbius transform of a capacity μ is the interval $[m] = [m_*, m^*]$, with $m^* = \mu$, and

$$m_*(A) = \begin{cases} \mu(A), & \text{if } \mu(A) > \mu(A \setminus i), \forall i \in A \\ 0, & \text{otherwise} \end{cases} \qquad (A \subseteq X). \tag{4.64}$$

We denote by m_*^μ the (canonical) ordinal Möbius transform of μ. Formula (4.64) was first proposed as the (ordinal) Möbius transform of a capacity by Marichal [234] and Mesiar [240] independently. A theory of the Möbius transform on symmetric ordered structures using operators $\circledvee, \circledwedge$ (Sect. 4.3.2), leading in particular to Theorem 4.49, was developed by the author [166].

Although many properties of the classical Möbius transform are preserved for the canonical Möbius transform, some of them are lost, in particular m_* is not maxitive as expected: $m_*^{\mu\vee\mu'} \neq m_*^\mu \vee m_*^{\mu'}$. Also, m_* is always nonnegative (see [165, 166] for details).

Theorem 4.50 *For any $f \in \mathbb{R}_+^n$ and any capacity μ on X, the Sugeno integral of f w.r.t. μ can be written as:*

$$⨍ f \, d\mu = \bigvee_{A \subseteq X} \Big(\bigwedge_{i \in A} f_i \wedge m(A) \Big) \tag{4.65}$$

where m is any function in $[m_*, m^*]$. Equivalently,

$$⨍ f \, d\mu = \bigwedge_{A \subseteq X} \Big(\bigvee_{i \in A} f_i \vee \overline{m}(A) \Big) \tag{4.66}$$

where $\overline{m}$ is any function in $[\overline{m}_*, \overline{m}^*]$, with $\overline{m}_*(A) = \mu(X \setminus A)$, and

$$\overline{m}^*(A) = \begin{cases} \mu(X \setminus A), & \text{if } \mu(X \setminus A) < \mu((X \setminus A) \cup i), \forall i \in A \\ 1, & \text{otherwise} \end{cases} \qquad (A \subseteq N).$$

In particular,

$$⨍ f \, d\mu = \bigvee_{A \subseteq X} \Big(\bigwedge_{i \in A} f_i \wedge \mu(A) \Big) = \bigwedge_{A \subseteq X} \Big(\bigvee_{i \in A} f_i \vee \mu(X \setminus A) \Big). \tag{4.67}$$

Proof In order to prove (4.65), it suffices to prove it for m^* and m_*.

(i) We have, using distributivity of $\wedge, \vee$ and monotonicity of μ:

$$\begin{aligned}
\bigvee_{A\subseteq X}\Big(\bigwedge_{i\in A} f_i \wedge \mu(A)\Big) &= \bigvee_{\substack{A\subseteq X\\ A\ni\sigma(1)}} (f_{\sigma(1)}\wedge\mu(A)) \vee \bigvee_{\substack{A\subseteq X\setminus\sigma(1)\\ A\ni\sigma(2)}} (f_{\sigma(2)}\wedge\mu(A)) \vee \\
&\quad \cdots \vee (f_{\sigma(n)}\wedge\mu(\{\sigma(n)\})) \\
&= \Big(f_{\sigma(1)}\wedge \bigvee_{\substack{A\subseteq X\\ A\ni\sigma(1)}} \mu(A)\Big) \vee \Big(f_{\sigma(2)}\wedge \bigvee_{\substack{A\subseteq X\setminus\sigma(1)\\ A\ni\sigma(2)}} \mu(A)\Big) \vee \\
&\quad \cdots \vee (f_{\sigma(n)}\wedge\mu(\{\sigma(n)\})) \\
&= (f_{\sigma(1)}\wedge\mu(X)) \vee (f_{\sigma(2)}\wedge\mu(X\setminus\sigma(1))) \vee \cdots \\
&\quad \vee(f_{\sigma(n)}\wedge\mu(\{\sigma(n)\})) \\
&= \bigvee_{i=1}^{n} \big(f_{\sigma(i)}\wedge\mu(\{\sigma(i),\ldots,\sigma(n)\})\big).
\end{aligned}$$

(ii) For a given nonempty $A \subseteq X$, if there exists some $j \in A$ such that $\mu(A)=\mu(A\setminus j)$, then $\mu(A)\wedge\bigwedge_{i\in A} f_i \leq \mu(A\setminus j)\wedge\bigwedge_{i\in A\setminus j} f_i$, hence the corresponding term in the supremum over X (in the expression with m^*) can be deleted, or equivalently, $\mu(A)$ can be replaced by 0. But $m_*(A) = 0$ if $\mu(A) = \mu(A\setminus j)$ for some j, hence the result.

The second equation can be proved in a similar way. □

We recognize Formula (4.5) (established in the general case) in the first equality of (4.67). Also, Marichal has shown the above theorem using min-max Boolean functions [228]. Note the analogy with the expression of the Choquet integral using the Möbius transform (4.55).

4.8 Characterizations

4.8.1 The Choquet Integral

We begin with the most famous characterization of the Choquet integral, shown by Schmeidler [286].

Theorem 4.51 *Let $I : B(\mathcal{F}) \to \mathbb{R}$ be a functional. Define the set function $v(A) = I(1_A)$, $A \in \mathcal{F}$. The following propositions are equivalent:*

(i) I is monotone[9] and comonotonically additive;
(ii) v is a capacity, and for all $f \in B(\mathcal{F})$, $I(f) = \int f \, \mathrm{d}v$.

We know already that (ii) implies (i) (Sect. 4.6.1). For the proof of (i)⇒(ii) in the general case, we refer the readers to Denneberg [80], Marinacci and Montrucchio [235], or the original paper [286]. We give a simple proof in the finite case.

Proof (i)⇒(ii) Assume $|X| = n$, $\mathcal{F} = 2^X$. We begin by proving that I is positively homogeneous. Comonotonic additivity implies that $I(mf) = mI(f)$ for any positive integer m and any $f \in \mathbb{R}^X$. For two positive integers k, m we have

$$\frac{m}{k}I(f) = \frac{m}{k}I\Big(k\frac{f}{k}\Big) = mI\Big(\frac{f}{k}\Big) = I\Big(\frac{m}{k}f\Big),$$

hence positive homogeneity is true for rationals. Lastly, for any positive real number r, take any increasing sequence of positive rationals r_i converging to r, and any decreasing sequence of positive rationals s_i converging to r. Then $r_iI(f) = I(r_if) \leqslant I(rf) \leqslant I(s_if) = s_iI(f)$ for every i implies $rI(f) \leqslant I(rf) \leqslant rI(f)$. On the other hand, remark that by comonotonic additivity, $I(f+\mathbf{0}) = I(f)+I(\mathbf{0})$ for any function f, hence $I(\mathbf{0}) = 0$. In summary, we have proved that

$$I(rf) = rI(f), \quad \forall r \geqslant 0. \tag{4.68}$$

Monotonicity of I entails monotonicity of v, and because $v(\varnothing) = I(\mathbf{0}) = 0$, it follows that v is a capacity.

On the other hand, $I(-1_X) = -v(X)$ because by comonotonic additivity we have

$$0 = I(\mathbf{0}) = I(1_X + (-1_X)) = I(1_X) + I(-1_X) = v(X) + I(-1_X).$$

Hence, we deduce that

$$I(r1_X) = rv(X), \quad \forall r \in \mathbb{R}. \tag{4.69}$$

Consider without loss of generality $f \in \mathbb{R}^n$ such that $f_1 \leqslant f_2 \leqslant \cdots \leqslant f_n$ (if not, apply some permutation on X). We have by (4.1)

$$f = \sum_{i=1}^{n}(f_i - f_{i-1})1_{A_i},$$

[9] In the sense of Theorem 4.24(vi) (also called nondecreasingness).

with $A_i = \{x_i, x_{i+1}, \dots, x_n\}$, and $f_0 = 0$. Define $h^i \in \mathbb{R}^X$ by $h^i = (f_i - f_{i-1})1_{A_i}$, $i = 1, \dots, n$. Then h^i and $\sum_{j=i+1}^n h^j$ are comonotonic functions for $i = 1, \dots, n-1$. Using comonotonic additivity we get

$$I(f) = I\left(h^1 + \sum_{j=2}^n h^j\right) = I(h^1) + I\left(\sum_{j=2}^n h^j\right).$$

Iterating this we finally get $I(f) = \sum_{i=1}^n I(h^i)$. Observe that $(f_i - f_{i-1})$ is nonnegative for $i \geqslant 2$, and that $1_{A_1} = 1_X$. Then by (4.68) and (4.69), we obtain that

$$I(f) = \sum_{i=1}^n (f_i - f_{i-1}) I(1_{A_i}).$$

Since $v(A) = I(1_A)$, by (4.23), we deduce that $I(f) = \int f \, dv$. □

Remark 4.52

(i) The above result in the finite case limited to nonnegative vectors was shown by de Campos and Bolaños [68], assuming in addition positive homogeneity, which can in fact be deduced from monotonicity and comonotonic additivity.
(ii) Because comonotonic additivity is equivalent to horizontal min- or max-additivity in the finite case (Theorem 4.30), the above characterization also holds with the latter properties replacing comonotonic additivity.
(iii) A more general version, where functions are allowed to be unbounded, has been shown by Wakker [338, Theorems 1.13–1.15].

◇

We give now a more general result shown by Murofushi et al. [255], where v is not limited to a capacity. We need for this the following definition.

Definition 4.53 Let $I : B(\mathcal{F}) \to \mathbb{R}$ be a functional. The *total variation* $V(I)$ of I is defined by

$$V(I) = \sup \left\{ \sum_{i=1}^n \left| I(f^i) - I(f^{i-1}) \right| \right\}, \tag{4.70}$$

where the supremum is taken over all finite chains $\mathbf{0} = f^0 \leqslant f^1 \leqslant \cdots \leqslant f^n = 1_X$ of functions in $B(\mathcal{F})$. The functional is said to be of *bounded variation* if $V(I) < \infty$.

Theorem 4.54 *Let $I : B(\mathcal{F}) \to \mathbb{R}$ be a functional. Define the set function $v(A) = I(1_A)$, $A \in \mathcal{F}$. The following propositions are equivalent:*

(i) I is comonotonically additive, positively homogeneous and of bounded variation;
(ii) I is comonotonically additive and uniformly continuous;

(iii) $v \in \mathcal{BV}(\mathcal{F})$*, i.e., it is a game of bounded variation, and for all* $f \in B(\mathcal{F})$*,* $I(f) = \int f\,\mathrm{d}v$.

(See proof in [255].)

Another characterization of the Choquet integral can be obtained in the discrete case using comonotonic modularity.

Theorem 4.55 *(Couceiro and Marichal [59, 60]) Let* $|X| = n$ *and* $\mathcal{F} = 2^X$*, and let* $I : \mathbb{R}^X \to \mathbb{R}$ *be a functional. Define the set function* $v(A) = I(1_A)$*,* $A \subseteq X$*. The following propositions are equivalent:*

(i) I *is comonotonically modular and satisfies* $I(\alpha 1_S) = |\alpha| I(\mathrm{sign}\,(\alpha) 1_S)$ *for all* $\alpha \in \mathbb{R}$ *and* $S \subseteq X$*, and* $I(1_{X \setminus S}) = I(1_X) + I(-1_S)$*;*

(ii) v *is a game and* $I(f) = \int f\,\mathrm{d}v$.

Proof (ii)⇒(i) is immediate from previous results.

(i)⇒(ii) Let us first consider $f : X \to \mathbb{R}_+$ and σ be a permutation on X such that $0 \leqslant f_{\sigma(1)} \leqslant \cdots \leqslant f_{\sigma(n)}$. By comonotonic modularity, we have for any $i = 1, \ldots, n-1$,

$$I(f_{\sigma(i)} 1_{A^{\uparrow}_{\sigma}(i)}) + I(\mathbf{0}_{|A^{\downarrow}_{\sigma}(i)}, f_{|A^{\uparrow}_{\sigma}(i+1)}) = I(f_{\sigma(i)} 1_{A^{\uparrow}_{\sigma}(i+1)}) + I(\mathbf{0}_{|A^{\downarrow}_{\sigma}(i-1)}, f_{|A^{\uparrow}_{\sigma}(i)}),$$

that is,

$$I(\mathbf{0}_{|A^{\downarrow}_{\sigma}(i-1)}, f_{|A^{\uparrow}_{\sigma}(i)}) = \Big(I(f_{\sigma(i)} 1_{A^{\uparrow}_{\sigma}(i)}) - I(f_{\sigma(i)} 1_{A^{\uparrow}_{\sigma}(i+1)})\Big) + I(\mathbf{0}_{|A^{\downarrow}_{\sigma}(i)}, f_{|A^{\uparrow}_{\sigma}(i+1)}). \tag{4.71}$$

By adding all equations (4.71) for $i = 1, \ldots, n-1$, we obtain

$$I(f) = I(\mathbf{0}) + \sum_{i=1}^{n} \Big(I(f_{\sigma(i)} 1_{A^{\uparrow}_{\sigma}(i)}) - I(f_{\sigma(i)} 1_{A^{\uparrow}_{\sigma}(i+1)})\Big). \tag{4.72}$$

Proceeding similarly for $f : X \to \mathbb{R}_-$, we find

$$I(f) = I(\mathbf{0}) + \sum_{i=1}^{n} \Big(I(f_{\sigma(i)} 1_{A^{\downarrow}_{\sigma}(i)}) - I(f_{\sigma(i)} 1_{A^{\downarrow}_{\sigma}(i-1)})\Big), \tag{4.73}$$

with $f_{\sigma(1)} \leqslant \cdots \leqslant f_{\sigma(n)} \leqslant 0$. It follows that for a real-valued function f, supposing $f_\sigma(1) \leqslant \cdots \leqslant f_{\sigma(p)} < 0 \leqslant f_{\sigma(p+1)} \leqslant \cdots \leqslant f_{\sigma(n)}$, we get from (4.72) and (4.73):

$$I(f) = I(\mathbf{0}) + \sum_{i=1}^{p} \Big(I(f_{\sigma(i)} 1_{A^{\downarrow}_{\sigma}(i)}) - I(f_{\sigma(i)} 1_{A^{\downarrow}_{\sigma}(i-1)})\Big) + \sum_{i=p+1}^{n} \Big(I(f_{\sigma(i)} 1_{A^{\uparrow}_{\sigma}(i)}) - I(f_{\sigma(i)} 1_{A^{\uparrow}_{\sigma}(i+1)})\Big). \tag{4.74}$$

Observe that $I(\mathbf{0}) = 0$ by (i), therefore $v(\varnothing) = I(\mathbf{0}) = 0$, so v is a game. Using (4.74) and (i) we get:

$$\begin{aligned}
I(f) &= \sum_{i=1}^{p} \big(I(f_{\sigma(i)} 1_{A_\sigma^\downarrow(i)}) - I(f_{\sigma(i)} 1_{A_\sigma^\downarrow(i-1)})\big) + \sum_{i=p+1}^{n} \big(I(f_{\sigma(i)} 1_{A_\sigma^\uparrow(i)}) - I(f_{\sigma(i)} 1_{A_\sigma^\uparrow(i+1)})\big) \\
&= \sum_{i=1}^{p} f_{\sigma(i)} \big(I(-1_{A_\sigma^\downarrow(i-1)}) - I(-1_{A_\sigma^\downarrow(i)})\big) + \sum_{i=p+1}^{n} f_{\sigma(i)} \big(I(1_{A_\sigma^\uparrow(i)}) - I(1_{A_\sigma^\uparrow(i+1)})\big) \\
&= \sum_{i=1}^{n} f_{\sigma(i)} \big(I(1_{A_\sigma^\uparrow(i)}) - I(1_{A_\sigma^\uparrow(i+1)})\big) \\
&= \sum_{i=1}^{n} f_{\sigma(i)} \big(v(A_\sigma^\uparrow(i)) - v(A_\sigma^\uparrow(i+1))\big).
\end{aligned}$$

We recognize the Choquet integral. □

In a similar fashion, one can obtain the following characterization of the symmetric Choquet integral (proof is similar and is therefore omitted; see [60]).

Theorem 4.56 *Let $|X| = n$ and $\mathcal{F} = 2^X$, and let $I : \mathbb{R}^X \to \mathbb{R}$ be a functional. Define the set function $v(A) = I(1_A)$, $A \subseteq X$. The following propositions are equivalent:*

(i) I is comonotonically modular and satisfies $I(\alpha 1_S) = \alpha I(1_S)$ for all $\alpha \in \mathbb{R}$ and $S \subseteq X$;
(ii) v is a game and $I(f) = \check{\int} f \, \mathrm{d}v$.

Another characterization of the symmetric Choquet integral can be obtained through horizontal median-additivity.

Theorem 4.57 *(Couceiro and Marichal [58]) Let $|X| = n$ and $\mathcal{F} = 2^X$, and let $I : \mathbb{R}^X \to \mathbb{R}$ be a functional. Define the set function $v(A) = I(1_A)$, $A \subseteq X$. The following propositions are equivalent:*

(i) I is horizontally median-additive, and satisfies $I(\alpha 1_S) = \alpha I(1_S)$ for all $\alpha \in \mathbb{R}$ and $S \subseteq X$;
(ii) v is a game and $I(f) = \check{\int} f \, \mathrm{d}v$.

Basically the proof is based on Lemma 4.33: horizontal median-additivity implies horizontal min-additivity for nonnegative functions and horizontal max-additivity for nonpositive functions, which in turn are equivalent to comonotonic additivity for nonnegative functions and nonpositive functions, respectively. Then one can proceed as in the proof of Theorem 4.51.

Other characterizations can be found in [43] and [177, Sect. 5.4.5].

4.8.2 The Sugeno Integral

We give some characterizations in the discrete case.

Theorem 4.58 *Let $|X| = n$, $\mathcal{F} = 2^X$, and let $I : (\mathbb{R}_+)^X \to \mathbb{R}_+$ be a functional. Define the set function $\mu(A) = I(1_A)$, $A \subseteq X$. The following propositions are equivalent:*

(i) I is comonotonically maxitive, satisfies $I(\alpha 1_A) = \alpha \wedge I(1_A)$ for every $\alpha \geqslant 0$ and $A \subseteq X$ (hat function property), and $I(1_X) = 1$;
(ii) μ is a normalized capacity on X and $I(f) = ⨍ f \, \mathrm{d}\mu$.

Proof (ii)⇒(i) Established in Theorem 4.43(iii) and Theorem 4.44.

(i)⇒(ii) Consider without loss of generality $f : X \to \mathbb{R}_+$ such that $f_1 \leqslant f_2 \leqslant \cdots \leqslant f_n$ (if not, apply some permutation on X). It is easy to check that

$$f = \bigvee_{i=1}^{n} f_i 1_{A_i},$$

with $A_i = \{x_i, x_{i+1}, \ldots, x_n\}$. Define $h^i \in \mathbb{R}_+^X$ by $h^i = f_i 1_{A_i}$, $i = 1, \ldots, n$. Then h^i and $\vee_{j=i+1}^n h^j$ are comonotonic vectors for $i = 1, \ldots, n-1$. By comonotonic maxitivity, we get

$$I(f) = I\left(h^1 \vee \bigvee_{j=2}^{n} h^j\right) = I(h^1) \vee I\left(\bigvee_{j=2}^{n} h^j\right).$$

Iterating this we finally get $I(f) = \bigvee_{i=1}^{n} I(h^i)$, and by the hat function property, we obtain that

$$I(f) = \bigvee_{i=1}^{n} \big(f_i \wedge I(1_{A_i})\big).$$

With $\mu(A) = I(1_A)$, then clearly $I(f) = ⨍ f \, \mathrm{d}\mu$. It remains to show that μ is a normalized capacity. We have immediately $\mu(X) = 1$ by (i). Let $\alpha \in]0, 1]$. Then by the hat function property, we have $I(\alpha 1_X) = \alpha \wedge I(1_X) = \alpha \wedge 1 = \alpha$. Now, using comonotonic maxitivity we get $\alpha = I(\alpha 1_X) = I(\alpha 1_X \vee 1_\varnothing) = I(\alpha 1_X) \vee I(1_\varnothing) = \alpha \vee I(1_\varnothing)$. Hence $I(1_\varnothing) \leqslant \alpha$ for any $\alpha \in]0, 1]$, so that $I(1_\varnothing) = \mu(\varnothing) = 0$ because I is nonnegative. Finally, let $A \subseteq B \subseteq X$. Then $1_A \leqslant 1_B$ and $1_A, 1_B$ are comonotonic. By comonotonic maxitivity, we obtain that $I(1_B) = I(1_A \vee 1_B) = I(1_A) \vee I(1_B)$, which entails $I(1_A) \leqslant I(1_B)$. Hence, μ is monotone. □

Remark 4.59 This is a generalized and simplified version of the result established by de Campos and Bolaños [68]. In the latter, functions were supposed to be defined on [0, 1] instead of $\mathbb{R}_+$, and the Sugeno integral was characterized by comonotonic maxitivity, min-homogeneity, nondecreasingness (in fact not necessary) and the normalization condition $I(1_X) = 1$. For functions defined on [0, 1], the hat function property is needed only for $\alpha \in [0, 1]$, which is equivalent to $I(\alpha \wedge 1_A) = \alpha \wedge I(1_A)$, a particular case of min-homogeneity. The latter property is therefore unnecessarily strong for the axiomatization of the Sugeno integral. ◇

The next characterization is due to Marichal [228]. Still others can be found in this reference.

Theorem 4.60 *Let $|X| = n$, $\mathcal{F} = 2^X$, and let $I : [0,1]^X \to [0,1]$ be a functional. Define the set function $\mu(A) = I(1_A)$, $A \subseteq X$. The following propositions are equivalent:*

(i) I is monotone (nondecreasing), $\vee$-homogeneous and $\wedge$-homogeneous;
(ii) μ is a normalized capacity on X and $I(f) = ⨍ f \, \mathrm{d}\mu$.

Proof (ii)⇒(i) Established in Theorem 4.43(i), (ii) and (vi).

(i)⇒(ii) $\wedge$-homogeneity yields $I(\mathbf{0}) = I(\mathbf{0} \wedge 1_X) = 0 \wedge I(1_X) = 0$ and similarly $\vee$-homogeneity implies $I(1_X) = 1$. Lastly, nondecreasingness of I implies monotonicity of μ, so that μ is a capacity.

Consider $f \in [0,1]^n$. For any $B \subseteq X$, we have by nondecreasingness and $\wedge$-homogeneity

$$I(f) \geqslant I\Big(\Big(\bigwedge_{i \in B} f_i\Big) 1_B\Big) = \mu(B) \wedge \Big(\bigwedge_{i \in B} f_i\Big)$$

This implies that

$$I(f) \geqslant \bigvee_{B \subseteq X} \Big(\mu(B) \wedge \Big(\bigwedge_{i \in B} f_i\Big)\Big).$$

Consider $B^* \subseteq X$ such that $\mu(B^*) \wedge \Big(\bigwedge_{i \in B^*} f_i\Big)$ is maximum, and define

$$A = \Big\{ j \in X \mid f_j \leqslant \mu(B^*) \wedge \Big(\bigwedge_{i \in B^*} f_i\Big)\Big\}.$$

We have $A \neq \varnothing$, for otherwise $f_j > \mu(B^*) \wedge \Big(\bigwedge_{i \in B^*} f_i\Big)$ for all $j \in X$, and since $\mu(X) = 1$, we would have

$$\mu(X) \wedge \Big(\bigwedge_{i \in X} f_i\Big) > \mu(B^*) \wedge \Big(\bigwedge_{i \in B^*} f_i\Big),$$

which contradicts the definition of B^*. Also, we have $\mu(X \setminus A) \leqslant \mu(B^*) \wedge \left(\bigwedge_{i\in B^*} f_i\right)$, for otherwise we would have, by definition of A,

$$\mu(X \setminus A) \wedge \Big(\bigwedge_{i\in X\setminus A} f_i\Big) > \mu(B^*) \wedge \Big(\bigwedge_{i\in B^*} f_i\Big),$$

which again contradicts the definition of B^*. Then, we have, by nondecreasingness and $\vee$-homogeneity

$$\begin{aligned} I(f) &\leqslant I\Big(\Big(\mu(B^*) \wedge \Big(\bigwedge_{i\in B^*} f_i\Big)\Big)1_A + 1_{X\setminus A}\Big) \\ &= \Big(\mu(B^*) \wedge \Big(\bigwedge_{i\in B^*} f_i\Big)\Big) \vee \mu(X \setminus A) \\ &= \mu(B^*) \wedge \Big(\bigwedge_{i\in B^*} f_i\Big) = \bigvee_{B\subseteq X} \Big(\mu(B) \wedge \Big(\bigwedge_{i\in B} f_i\Big)\Big). \end{aligned}$$

We recognize by (4.67) the Sugeno integral. □

4.9 Particular Cases

We give in this section the expression of the Choquet and Sugeno integrals for particular cases of capacities.

4.9.1 The Choquet Integral

0-1-Capacities

Theorem 4.61 *(Murofushi and Sugeno [253]) Let μ be a 0-1-capacity on $(X, \mathcal{F})$. For every $f \in B(\mathcal{F})$*

$$\int f \, \mathrm{d}\mu = \sup_{A\,:\,\mu(A)=1} \inf_{x\in A} f(x).$$

Proof By translation invariance [Theorem 4.24(iii)], it suffices to prove it for nonnegative functions. Let $a = \sup_{A\,:\,\mu(A)=1} \inf_{x\in A} f(x)$. By definition of the

Choquet integral, it suffices to prove that

$$\mu(f > t) = \begin{cases} 1, & \text{if } t < a \\ 0, & \text{if } t > a. \end{cases}$$

Assume $t < a$. Then there exists $A \in \mathcal{F}$ such that $\mu(A) = 1$ and $t < \inf_{x \in A} f(x)$. This inequality implies $A \subseteq \{f > t\}$, and hence that $\mu(f > t) = 1$. Assume on the contrary that $t > a$. If $\mu(f > t) = 1$, it follows that $t > a \geqslant \inf_{\{x : f(x) > t\}} f(x) \geqslant t$, a contradiction. It follows that $\mu(f > t) = 0$. □

Relation with Some Mean Operators and Statistical Estimators We suppose in this paragraph that $X = \{1, \ldots, n\}$ and $\mathcal{F} = 2^X$. Viewing the integrand as a vector in $\mathbb{R}^n$ on which can act some mean operator, the Choquet integral permits to recover two important mean operators, namely the weighted arithmetic mean and its ordered version, known in statistics as the L-estimator. We first introduce their definition.

Definition 4.62 Let $w \in [0, 1]^n$ be a weight vector; i.e., satisfying $\sum_{i=1}^n w_i = 1$. The *weighted arithmetic mean* and the *ordered weighted arithmetic mean* with weight vector w are mappings $\mathbf{WAM}_w : \mathbb{R}^n \to \mathbb{R}$, $\mathbf{OWA}_w : \mathbb{R}^n \to \mathbb{R}$ respectively, defined for any $f \in \mathbb{R}^n$ by

$$\mathbf{WAM}_w(f) = \sum_{i=1}^n w_i f_i \tag{4.75}$$

$$\mathbf{OWA}_w(f) = \sum_{i=1}^n w_i f_{\sigma(i)} \tag{4.76}$$

where σ is any permutation on X such that $f_{\sigma(1)} \leqslant \cdots \leqslant f_{\sigma(n)}$.

Observe that $\mathbf{WAM}_w$ and $\mathbf{OWA}_w$ coincide if and only if $w = (\frac{1}{n}, \ldots, \frac{1}{n})$. Interesting particular cases are obtained when $w_k = 1$ for some k in $\{1, \ldots, n\}$. Then $w_i = 0$ for all $i \neq k$, and $\mathbf{WAM}_w$ reduces to the projection on the kth coordinate, denoted by $\mathbf{P}_k$, while $\mathbf{OWA}_w$ reduces to the kth *order statistic* (or quantile), denoted by $\mathbf{OS}_k$. Note that $\mathbf{OS}_1$ and $\mathbf{OS}_n$ correspond respectively to the minimum and maximum operators, and if $n = 2k + 1$, $\mathbf{OS}_{k+1}$ is the median. Other common means used in statistics can be recovered: the *trimmed mean*, where $w_1 = \cdots = w_k = 0 = w_{n-k+1} = \cdots = w_n$ for some $1 \leqslant k < \frac{n}{2}$, and the other weights being equal, the *Winsorized mean*, where again $w_1 = \cdots = w_k = 0 = w_{n-k+1} = \cdots = w_n$ for some $1 \leqslant k < \frac{n}{2}$, but $w_{k+1} = w_{n-k} = \frac{k+1}{n}$, the rest of the weights[10] being equal to $\frac{1}{n}$, etc.

[10]This amounts to deleting the k lowest and highest values in f, and replacing them respectively by the remaining lowest and highest values; i.e., $f_{\sigma(k+1)}, f_{\sigma(n-k)}$.

Theorem 4.63 *Let w be a weight vector. Then* $\mathbf{WAM}_w = \int \cdot \,\mathrm{d}\mu$ *and* $\mathbf{OWA}_w = \int \cdot \,\mathrm{d}\mu'$, *where μ is an additive capacity determined by*

$$\mu(\{i\}) = w_i \qquad (i \in X), \tag{4.77}$$

and μ' is a symmetric capacity determined by

$$\mu'(A) = \sum_{j=0}^{|A|-1} w_{n-j} \qquad (A \subseteq X). \tag{4.78}$$

Conversely, any Choquet integral w.r.t. a normalized additive capacity is a weighted arithmetic mean, whose weight vector is given by (4.77), and any Choquet integral w.r.t. a symmetric capacity μ is an ordered weighted arithmetic mean, whose weight vector is given by

$$w_i = \mu(\{1, \ldots, n-i+1\}) - \mu(\{1, \ldots, n-i\}) \qquad (i = 2, \ldots, n)$$

and $w_1 = 1 - \sum_{i=2}^{n} w_i$.

(We recall that μ is symmetric if $|A| = |B|$ entails $\mu(A) = \mu(B)$; see Sect. 2.14.)

Proof See Sect. 4.5.3 for the result concerning **WAM**. The result for **OWA** can be checked similarly. □

From the above result we immediately get the following particular cases:

Corollary 4.64

(i) For any $f \in \mathbb{R}^n$, $\int f \,\mathrm{d}\mu = \bigwedge_{i \in X} f_i$ *if and only if* $\mu = \mu_{\min}$;
(ii) For any $f \in \mathbb{R}^n$, $\int f \,\mathrm{d}\mu = \bigvee_{i \in X} f_i$ *if and only if* $\mu = \mu_{\max}$;
(iii) $\int \cdot \,\mathrm{d}\mu = \mathbf{P}_k$ *if and only if μ is the Dirac measure* δ_k;
(iv) $\int \cdot \,\mathrm{d}\mu = \mathbf{OS}_k$ *if and only if μ is a symmetric capacity determined by* $\mu(\{1, \ldots, i\}) = 1$ *if* $i \geqslant n-k+1$, *and 0 otherwise.*

(See Sect. 2.8.1 for the definition of $\mu_{\min}$ and $\mu_{\max}$, and Example 2.4 for the definition of the Dirac measure.) Note that the two first results were already established in Theorem 4.24(xi).

Remark 4.65

(i) The ordered weighted arithmetic mean was introduced by Yager [352], under the name of ordered weighted averaging operator. In statistics however, linear combinations of order statistics are well known and used from a long time under the name of *L-estimator*, see for instance Weisberg [347].
(ii) Most of these properties were shown by Murofushi and Sugeno [253] (see also Fodor et al. [143] for the result on ordered weighted arithmetic means).

The readers can consult [253] for further results linked to some quantities in statistics.

◇

2-Additive Games We consider in this paragraph that $|X| = n$ and $\mathcal{F} = 2^X$. Suppose that v is a 2-additive game. We know from its definition (see Definition 2.50) that v needs much less coefficients than an ordinary capacity to be defined, hence one can expect a simplification of the formula of the Choquet integral. Indeed, in (4.55), the summation is over singletons and pairs instead of all subsets of X. More interestingly, one can use Formula (4.61) expressing the Choquet integral in terms of the interaction transform. A rearrangement of terms leads to the following two formulas when v is 2-additive:

$$\int f \, dv = \sum_{i,j : I_{ij} > 0} (f_i \wedge f_j) I_{ij} + \sum_{i,j : I_{ij} < 0} (f_i \vee f_j)|I_{ij}| + \sum_{i \in X} f_i \Big(\phi_i^{\mathrm{Sh}} - \frac{1}{2}\sum_{j \neq i} |I_{ij}|\Big) \tag{4.79}$$

$$= \sum_{i \in X} \phi_i^{\mathrm{Sh}} f_i - \frac{1}{2} \sum_{\{i,j\} \subseteq X} I_{ij}|f_i - f_j|, \tag{4.80}$$

with $f \in \mathbb{R}^X$, where we have used the shorthands $I_{ij} = I^v(\{i,j\})$ and $\phi_i^{\mathrm{Sh}} = I^v(\{i\})$ [Shapley value, see (2.32)]. Note that (4.79) can be obtained directly from (2.61) and the fact that (using notation of the latter formula) $I_{ij} = m_{ij}^{\mu}$ and $\phi_i^{\mathrm{Sh}} = m_i^{\mu} + \frac{1}{2}\sum_{j \neq i} m_{ij}^{\mu}$ [see (2.41)].

One can obtain in a similar fashion the expression of the symmetric Choquet integral (see [177, Sect. 5.4.3] for a proof):

$$\begin{aligned}\check{\int} f \, dv = & \sum_{i,j \in X^+ : I_{ij} > 0} (f_i \wedge f_j) I_{ij} + \sum_{i,j \in X^- : I_{ij} > 0} (f_i \vee f_j) I_{ij} \\ & + \sum_{i,j \in X^+ : I_{ij} < 0} (f_i \vee f_j)|I_{ij}| + \sum_{i,j \in X^- : I_{ij} < 0} (f_i \wedge f_j)|I_{ij}| \\ & + \sum_{i \in X^+} f_i \Big(\sum_{j \in X^- : I_{ij} < 0} |I_{ij}| \Big) + \sum_{i \in X^-} f_i \Big(\sum_{j \in X^+ : I_{ij} < 0} |I_{ij}| \Big) \\ & + \sum_{i \in X} f_i \Big(\phi_i^{\mathrm{Sh}} - \frac{1}{2} \sum_{j \neq i} |I_{ij}| \Big).\end{aligned} \tag{4.81}$$

with $X^+ = \{i \in X : f_i \geqslant 0\}$, and $X^- = X \setminus X^+$.

We postpone the interpretation of these formulas to Chap. 6 (Sect. 6.10.4). So far no nice formula was obtained for k-additive games with $k > 2$.

λ-Measures There is no special form of the Choquet integral w.r.t λ-measures (see Sect. 2.8.6). However, some properties are noteworthy, if we consider λ-measures as distorted probabilities. An obvious fact is that with $\lambda = 0$, the Choquet integral computes the usual expectation. A more interesting result is that with the extreme values of λ, one recovers the essential supremum and infimum. We recall that the distortion function is [see (2.13)]

$$s_\lambda^{-1}(u) = \frac{1}{\lambda}((1+\lambda)^u - 1) \qquad (u \in [0,1], \lambda \in]-1, \infty[).$$

Theorem 4.66 *Let P be a probability measure on $(X, \mathcal{F})$, $\lambda \in]-1, \infty[$, and consider the λ-measure (distorted probability) $\mu_\lambda = s_\lambda^{-1} \circ P$. The following holds, for every $f \in B(\mathcal{F})$:*

(i) $\lambda \leqslant \lambda'$ *implies* $\int f \, d\mu_\lambda \geqslant \int f \, d\mu_{\lambda'}$;
(ii) $\int f \, d\mu_0 = \int f \, dP$;
(iii) $\lim_{\lambda \to -1} \int f \, d\mu_\lambda = \text{ess sup}_P f$;
(iv) $\lim_{\lambda \to \infty} \int f \, d\mu_\lambda = \text{ess inf}_P f$.

Proof

(i) By Theorem 4.24(vii), it suffices to show that $\mu_\lambda \geqslant \mu_{\lambda'}$, which amounts to showing that $\frac{\partial s_\lambda^{-1}}{\partial \lambda}(u) \leqslant 0$ for $u \in [0,1]$. We have for any $u \in [0,1]$:

$$\frac{\partial s_\lambda^{-1}}{\partial \lambda}(u) = \frac{u(1+\lambda)^{u-1}\lambda - (1+\lambda)^u + 1}{\lambda^2} = \frac{(1+\lambda)^{u-1}(\lambda(u-1) - 1) + 1}{\lambda^2}.$$

We have to prove that the numerator is nonpositive; i.e., that the following inequality holds, letting $v = 1 - u$:

$$1 + \lambda v \geqslant (1+\lambda)^v \qquad (v \in [0,1], \lambda > -1). \tag{4.82}$$

Suppose first that $|\lambda| < 1$. Then the binomial expansion can be applied:

$$(1+\lambda)^v = 1 + v\lambda + \frac{v(v-1)}{2!}\lambda^2 + \cdots + \frac{v(v-1)\cdots(v-k+1)}{k!}\lambda^k + \cdots$$

Since $v \in [0,1]$ and $|\lambda| < 1$, we have

$$\left|\frac{v(v-1)}{2}\lambda^2\right| > \left|\frac{v(v-1)(v-2)}{2 \cdot 3}\lambda^3\right|$$

and similarly for all pairs of subsequent terms. Because the terms in $\lambda^2, \lambda^4, \ldots$ are negative, it follows that $(1+\lambda)^v \leqslant 1 + v\lambda$, and (4.82) is proved for $|\lambda| < 1$.

Suppose now $\lambda \geqslant 1$ and put $\lambda' = -\frac{\lambda}{\lambda+1}$. Then $\lambda' \in]-1, -\frac{1}{2}]$ and because $\lambda = -\frac{\lambda'}{\lambda'+1}$, we have to prove that

$$1 - \frac{\lambda'}{1+\lambda'}v \geqslant \left(1 - \frac{\lambda'}{1+\lambda'}\right)^v$$

which yields

$$1 + \lambda'(1-v) \geqslant (1+\lambda')^{1-v}.$$

This is identical to (4.82) since $1 - v \in [0, 1]$, and because $|\lambda'| < 1$, we have already established that the inequality holds.

(ii) Obvious

(iii) By properties of the decumulative function, it suffices to show that $\lim_{\lambda \to -1} G_{\mu_\lambda, f}(t) = 1$ if $t < \text{ess sup}_P f$. We have

$$\lim_{\lambda \to -1} \frac{1}{\lambda}((1+\lambda)^{P(f>t)} - 1) = \frac{1}{-1}(-1) = 1$$

if $P(f > t) \neq 0$.

(iv) Similarly,

$$\lim_{\lambda \to \infty} \frac{1}{\lambda}((1+\lambda)^{P(f>t)} - 1) = \lim_{\lambda \to \infty} \lambda^{P(f>t)-1} = 0$$

if $P(f > t) \neq 1$.

□

4.9.2 The Sugeno Integral

The case of 0-1-capacities was treated in Sect. 4.9.1, because the Choquet and Sugeno integrals coincide in this case. We consider for the rest of this section that $X = \{1, \ldots, n\}$ and $\mathcal{F} = 2^X$.

As for the Choquet integral, the Sugeno integral is related to some special kinds of mean operators, build with minimum and maximum. They are the counterparts of the weighted arithmetic mean and the ordered weighted arithmetic mean. To this end, we consider a different kind a weight vector, which we call *ordinal weight vector*: it is any vector $w \in [0, 1]^n$ such that $\bigvee_{i=1}^n w_i = 1$.

Definition 4.67 Let $w \in [0,1]^n$ be an ordinal weight vector. For any $f \in [0,1]^n$,

(i) The *weighted maximum* with respect to w is a mapping $\mathbf{WMax}_w : [0,1]^n \to [0,1]$ defined by

$$\mathbf{WMax}_w(f) = \bigvee_{i=1}^{n} (w_i \wedge f_i);$$

(ii) The *weighted minimum* with respect to w is a mapping $\mathbf{WMin}_w : [0,1]^n \to [0,1]$ defined by

$$\mathbf{WMin}_w(f) = \bigwedge_{i=1}^{n} ((1-w_i) \vee f_i);$$

(iii) The *ordered weighted maximum* with respect to w is a mapping $\mathbf{OWMax}_w : [0,1]^n \to [0,1]$ defined by

$$\mathbf{OWMax}_w(f) = \bigvee_{i=1}^{n} (w_i \wedge f_{\sigma(i)}),$$

with σ a permutation on X such that $f_{\sigma(1)} \leqslant \cdots \leqslant f_{\sigma(n)}$, and $w_1 \geqslant w_2 \geqslant \cdots \geqslant w_n$;

(iv) The *ordered weighted minimum* with respect to w is a mapping $\mathbf{OWMin}_w : [0,1]^n \to [0,1]$ defined by

$$\mathbf{OWMin}_w(f) = \bigwedge_{i=1}^{n} ((1-w_i) \vee f_{\sigma(i)}),$$

with σ as above and $w_1 \leqslant w_2 \leqslant \cdots \leqslant w_n$.

Remark 4.68

(i) The usual minimum and maximum are recovered from their weighted versions and weighted ordered versions when $w_i = 1$, $i = 1, \ldots, n$. The weighted minimum and weighted maximum were proposed by Dubois and Prade [105], and their ordered versions were introduced in [117] by Dubois et al.

(ii) In the case of the ordered weighted maximum, since

$$\bigvee_{i=1}^{n} (w_i \wedge f_{\sigma(i)}) = \bigvee_{i=1}^{n} \Big(\Big(\bigvee_{k=i}^{n} w_k \Big) \wedge f_{\sigma(i)} \Big)$$

the assumption $w_1 \geqslant w_2 \geqslant \cdots \geqslant w_n$ is not necessary, however it is useful in the next theorem. The same remark applies to the ordered weighted minimum.

◇

Theorem 4.69 *Let w be an ordinal weight vector. Then the following holds:*

(i) $\mathbf{WMax}_w = \fint \cdot\, \mathrm{d}\mu$, *where μ is a normalized maxitive capacity determined by* $\mu(\{i\}) = w_i$, $i \in X$;

(ii) $\mathbf{WMin}_w = \fint \cdot\, \mathrm{d}\mu$, *where μ is a normalized minitive capacity determined by* $\mu(X \setminus \{i\}) = 1 - w_i$, $i \in X$;

(iii) $\mathbf{OWMax}_w = \fint \cdot\, \mathrm{d}\mu$, *where μ is a normalized symmetric capacity determined by* $\mu(A) = w_{n-|A|+1}$, $\varnothing \neq A \subseteq X$;

(iv) $\mathbf{OWMin}_w = \fint \cdot\, \mathrm{d}\mu$, *where μ is a normalized symmetric capacity determined by* $\mu(A) = w_{n-|A|}$, $A \subset X$.

Conversely, any Sugeno integral w.r.t. a normalized maxitive (respectively, minitive, symmetric) capacity is a $\mathbf{WMax}_w$ *(respectively,* $\mathbf{WMin}_w$, $\mathbf{OWMax}_w$ *or* $\mathbf{OWMin}_w$*) operator, with w determined as above.*

Proof (i) For simplicity we write (i) instead of $\sigma(i)$. Suppose μ is normalized maxitive, and define $w_i = \mu(\{i\})$ for each $i \in X$. This defines an ordinal weight vector w. Consider the last k in the sequence $1, \ldots, n$ such that $\fint f\, \mathrm{d}\mu = f_{(k)} \wedge \mu(\{(k), \ldots, (n)\})$, and note that $\mu(\{(k), \ldots, (n)\}) = w_{(k)} \vee \cdots \vee w_{(n)}$.

First, if $k = n$ there is nothing to prove. We claim that if $k < n$, then $w_{(k)} > w_{(i)}$ for $k < i \leqslant n$. Indeed, if there exists k_0 such that $k < k_0 \leqslant n$ and $w_{(k_0)} \geqslant w_{(i)}$, $i = k, \ldots, n$ we would have

$$\fint f\, \mathrm{d}\mu = f_{(k_0)} \wedge (w_{(k_0)} \vee \cdots \vee w_{(n)})$$

because $f_{(k_0)} \geqslant f_{(k)}$, and this contradicts the definition of k. It follows that $\fint f\, \mathrm{d}\mu = f_{(k)} \wedge w_{(k)}$.

Assume that $\fint f\, \mathrm{d}\mu = f_{(k)} \leqslant \mu(\{(k), \ldots, (n)\}) = w_{(k)} \vee \cdots \vee w_{(n)}$. We have $f_{(k)} \wedge w_{(k)} = f_{(k)} \geqslant f_{(i)} \wedge w_{(i)}$ for $1 \leqslant i \leqslant k$. Also, $f_{(k)} \wedge w_{(k)} > f_{(i)} \wedge w_{(i)}$, $k < i \leqslant n$. Indeed, assume that there exists $k < k_0 \leqslant n$ such that $f_{(k)} \wedge w_{(k)} \leqslant f_{(k_0)} \wedge w_{(k_0)}$. Then

$$\fint f\, \mathrm{d}\mu = f_{(k)} \wedge w_{(k)} \leqslant f_{(k_0)} \wedge w_{(k_0)} \leqslant f_{(k_0)} \wedge \underbrace{\left(w_{(k_0)} \vee \cdots \vee w_{(n)}\right)}_{\mu(\{(k_0), \ldots, (n)\})},$$

which contradicts the definition of the Sugeno integral or the definition of k. In summary,

$$\fint f\, \mathrm{d}\mu = \bigvee_{i=1}^{n} \left(f_{(i)} \wedge w_{(i)}\right) = \bigvee_{i=1}^{n} \left(f_i \wedge w_i\right) = \mathbf{WMax}_w(f). \tag{4.83}$$

Assume on the contrary that $\fint f\, \mathrm{d}\mu = \mu(\{(k), \ldots, (n)\}) = w_{(k)} \vee \cdots \vee w_{(n)} \leqslant f_{(k)}$. We have $f_{(k)} \wedge w_{(k)} = w_{(k)} > f_{(i)} \wedge w_{(i)}$ for $k < i \leqslant n$ because $f_{(i)} \geqslant f_{(k)} \geqslant w_{(k)} > w_{(i)}$. Also, $f_{(k)} \wedge w_{(k)} > f_{(i)} \wedge w_{(i)}$ for $1 \leqslant i < k$. Indeed, suppose there exists $1 \leqslant k_0 < k$

such that $f_{(k)} \wedge w_{(k)} < f_{(k_0)} \wedge w_{(k_0)}$. Then

$$\fint f \,\mathrm{d}\mu = f_{(k)} \wedge w_{(k)} < f_{(k_0)} \wedge w_{(k_0)} \leqslant f_{(k_0)} \wedge \underbrace{\left(w_{(k_0)} \vee \cdots \vee w_{(n)}\right)}_{\mu(\{(k_0),\ldots,(n)\})},$$

which contradicts the definition of the Sugeno integral. Hence, again (4.83) holds.

Assertion (ii) holds by (i) and scale inversion [Theorem 4.43(iv)].

(iii) Take any ordinal weight vector w such that $\bigvee_i w_i = 1$ and $w_1 \geqslant \cdots \geqslant w_n$, and define a symmetric capacity by $\mu(A) = w_{n-|A|+1}$, for all $A \subseteq X, A \neq \varnothing$. Then, by (4.28), we get

$$\begin{aligned}\fint f \,\mathrm{d}\mu &= \bigvee_{i=1}^{n} \left(f_{(i)} \wedge \mu(\{(i),\ldots,(n)\})\right) \\ &= \bigvee_{i=1}^{n} \left(f_{(i)} \wedge w_i\right) = \mathbf{OWMax}_w(f).\end{aligned}$$

As above, assertion (iv) holds by (iii) and scale inversion. □

Note that the above result shows that the classes of $\mathbf{OWMax}_w$ and $\mathbf{OWMin}_w$ operators coincide.

Remark 4.70 Normalized maxitive (respectively, minitive) capacities are called possibility (respectively, necessity) measures (Sect. 2.8.3). Moreover, as it is shown in Sect. 7.7, they are generated by a possibility distribution π [Eqs. (7.40) and (7.43)]. Here the ordinal weight vector w can be seen as a possibility distribution, and denoting by Π and Nec the possibility and necessity measures generated by w, we can rewrite the weighted minimum and maximum as

$$\mathbf{WMax}_w(f) = \fint f \,\mathrm{d}\Pi; \quad \mathbf{WMin}_w(f) = \fint f \,\mathrm{dNec}.$$

◇

4.10 The Choquet Integral on the Nonnegative Real Line

4.10.1 Computation of the Choquet Integral

The properties of the Choquet integral presented so far do not permit, up to some very special cases, to perform computation of the integral similarly as one would do for classical integral calculus; i.e., using primitives, partial integration, change of variable, and so on. This is however possible if one works on the real line and considers in particular capacities that are distorted Lebesgue measures. The results presented in this section are due to Sugeno [321] (see also [322] for more results).

We consider the Lebesgue measure λ on $\mathbb{R}_+$, with $\lambda([a,b]) = b - a$ for any interval $[a,b] \subset \mathbb{R}_+$, and a *distortion function*; i.e., a function $h : \mathbb{R}_+ \to \mathbb{R}_+$ being continuous, increasing and satisfying $h(0) = 0$. The *distorted Lebesgue measure* is the mapping $\mu_h = h \circ \lambda$, and is evidently a continuous capacity on $\mathbb{R}_+$.

In the rest of this section, we deal with integrands $f : \mathbb{R}_+ \to \mathbb{R}_+$ that are nondecreasing, and with the Choquet integral on a subdomain $[0,t]$ for some $t > 0$ [Eq. (4.8)].

Theorem 4.71 *Let $f : \mathbb{R}_+ \to \mathbb{R}_+$ be nondecreasing and continuously differentiable, and let μ be a continuous capacity on $\mathbb{R}_+$, such that $\mu([\tau,t])$ is differentiable w.r.t. τ on $[0,t]$ for every $t > 0$, and $\mu(\{t\}) = 0$ for every $t \geqslant 0$. Then*

$$\int_{[0,t]} f \,\mathrm{d}\mu = -\int_0^t \frac{\partial \mu}{\partial \tau}([\tau,t]) f(\tau)\,\mathrm{d}\tau \qquad (t > 0),$$

where the right-hand side integral is the Riemann integral. In particular, for a distorted Lebesgue measure μ_h with h being continuously differentiable, we obtain

$$\int_{[0,t]} f \,\mathrm{d}\mu_h = \int_0^t \frac{\partial h}{\partial \tau}(t-\tau) f(\tau)\,\mathrm{d}\tau. \tag{4.84}$$

We give a proof when f is either constant or increasing. The general case is cumbersome and follows the same principles [321].

Proof The second equation derives immediately from the first one, on which we focus. We set for simplicity $\mu' = \frac{\partial \mu}{\partial \tau}$.

Let us consider a constant function $f(\tau) = C \ \forall \tau \geqslant 0$. Then

$$\begin{aligned}\int_0^\infty \mu(\{\tau : C \geqslant r\} \cap [0,t])\,\mathrm{d}r &= \int_0^C \mu([0,t])\,\mathrm{d}r = C\mu([0,t]) \\ &= -\int_0^t C\mu'([\tau,t])\,\mathrm{d}\tau,\end{aligned}$$

as desired.

Let us now consider an increasing function f. We have

$$\begin{aligned}\int_{[0,t]} f \,\mathrm{d}\mu &= \int_0^\infty \mu(\{f \geqslant r\} \cap [0,t])\,\mathrm{d}r \\ &= \int_0^{f(0)} \mu([0,t])\,\mathrm{d}r + \int_{f(0)}^{f(t)} \mu([f^{-1}(r),t])\,\mathrm{d}r \\ &= \mu([0,t])f(0) + \int_{f(0)}^{f(t)} \mu([f^{-1}(r),t])\,\mathrm{d}r.\end{aligned} \tag{4.85}$$

Let $r = f(\tau)$, then we have $f'(\tau)\,\mathrm{d}\tau = \mathrm{d}r$. By partial integration, we obtain

$$\begin{aligned}\int_{f(0)}^{f(t)} \mu([f^{-1}(r), t])\,\mathrm{d}r &= \int_0^t \mu([\tau, t])f'(\tau)\,\mathrm{d}\tau \\ &= \Big[\mu([\tau, t])f(\tau)\Big]_0^t - \int_0^t \mu'([\tau, t])f(\tau)\,\mathrm{d}\tau \\ &= \mu(\{t\})f(t) - \mu([0, t])f(0) - \int_0^t \mu'([\tau, t])f(\tau)\,\mathrm{d}\tau.\end{aligned}$$

Since by assumption $\mu(\{t\}) = 0$, we obtain, substituting into (4.85),

$$\int_{[0,t]} f(\tau)\,\mathrm{d}\mu = -\int_0^t \mu'([\tau, t])f(\tau)\,\mathrm{d}\tau,$$

as desired. □

Example 4.72 Taking $h(t) = t^2$ and $f(t) = e^t$, we find

$$\begin{aligned}\int_{[0,t]} e^t\,\mathrm{d}\mu_{t^2} &= \int_0^t 2(t-\tau)e^\tau\,\mathrm{d}\tau \\ &= 2t(e^t - 1) - 2\Big[\tau e^\tau\Big]_0^t + 2(e^t - 1) = 2(e^t - t - 1).\end{aligned}$$

◇

One recognizes in (4.84) a convolution product [see (1.27)], hence one can use the Laplace transform to express the result in a very simple way (see Sect. 1.3.11 for the Laplace transform and related notions). Consider the relation

$$g(t) = \int_{[0,t]} f\,\mathrm{d}\mu_h \tag{4.86}$$

under the assumptions of Theorem 4.71 for f and h. Denoting by $F(s)$, $G(s)$ and $H(s)$ the Laplace transforms of f, g, h respectively, we obtain from (4.84), (1.26), and (1.28)

$$G(s) = sH(s)F(s) \qquad (s \in \mathbb{C}). \tag{4.87}$$

Equation (4.87) can be used in several ways. Suppose one wants to compute the Choquet integral of f w.r.t. μ_h on $[0, t]$. Then the answer is the function $g(t)$ given by

$$g(t) = \mathcal{L}^{-1}(sH(s)F(s)),$$

where $\mathcal{L}^{-1}(\cdot)$ is the inverse Laplace transform.

Example 4.73 Let $h(t) = e^{at} - 1$ with $a > 0$, and $f(t) = t$. Then (see Sect. 1.3.11, Table 1.1) $H(s) = 1/(s-a) - 1/s$ and $F(s) = 1/s^2$, which yields

$$G(s) = \frac{1}{s}\left(\frac{1}{s-a} - \frac{1}{s}\right) = \frac{1}{a}\left(\frac{1}{s-a} - \frac{1}{s}\right) - \frac{1}{s^2},$$

from which we obtain $g(t) = \frac{1}{a}(e^{at} - 1) - t$. ◇

Conversely, suppose that a continuous and increasing function $g : \mathbb{R}_+ \to \mathbb{R}_+$ with $g(0) = 0$ is given. Then, the distorted Lebesgue measure being fixed, it is possible to find a continuous and increasing function f such that the relation (4.86) holds:

$$f(t) = \mathcal{L}^{-1}\left(\frac{G(s)}{sH(s)}\right). \tag{4.88}$$

Remark 4.74

(i) The latter problem, whose solution is given by (4.88), is similar to the *Radon-Nikodym*[11] *theorem*, well-known in measure theory. To introduce it, we need some additional definitions. A measure μ on $(X, \mathcal{F})$ is *σ-finite* if X can be written as $X = \bigcup_{i\in I} A_i$, with $A_i \in \mathcal{F}$, $\mu(A_i) < \infty$, and I is countable. For two measures μ, ν on $(X, \mathcal{F})$, ν is *absolutely continuous* w.r.t. μ, which is denoted by $\nu \ll \mu$, if $\mu(A) = 0$ implies $\nu(A) = 0$, $A \in \mathcal{F}$. Now, the Radon-Nikodym theorem says that, given ν, μ two measures on $(X, \mathcal{F})$ being σ-finite and such that $\nu \ll \mu$, there exists a unique (up to a null set w.r.t. μ) measurable function f such that

$$\nu(A) = \int_A f \, \mathrm{d}\mu.$$

By analogy, f is said to be the *Radon-Nikodym derivative*, denoted by $\frac{\mathrm{d}\nu}{\mathrm{d}\mu}$. Up to the knowledge of the author, there are few studies of a generalization of the Radon-Nikodym theorem for capacities and the Choquet integral, we mention Graf [184].

(ii) More applications of Theorem 4.71 can be found in [321], in particular on Abel's integral equations, and on fractional derivatives in the companion paper [322]. ◇

---

[11] Johann Radon (Děčín (Bohemia, Austria-Hungary), 1887 – Vienna, 1956), Austrian mathematician, well-known for his contribution to measure theory. He proved the above-mentioned theorem in 1913 for the special case where $X = \mathbb{R}^n$.
Otto Nikodym (Zabolotiv (Ukraine), 1887 – Utica NY, 1974), Polish mathematician. He generalized Radon's result in 1930.

4.10.2 Equimeasurable Rearrangement

We have seen in the previous section that the computation of the Choquet integral, given by Theorem 4.71, is restricted to monotone functions. There is a simple way to overcome this limitation by considering the equimeasurable (increasing or decreasing) rearrangement of a function.

Given a function $f : \mathbb{R}_+ \to \mathbb{R}_+$ and a Lebesgue measure λ, an *equimeasurable (nondecreasing) rearrangement* of f w.r.t. λ is a function $\widetilde{f} : \mathbb{R}_+ \to \mathbb{R}_+$ which is nondecreasing and satisfies

$$\lambda(f \geqslant t) = \lambda(\widetilde{f} \geqslant t) \qquad (t \in \mathbb{R}^+). \tag{4.89}$$

If such a function exists, then clearly

$$\int f \, \mathrm{d}\mu_h \doteq \int \widetilde{f} \, \mathrm{d}\mu_h$$

for every distorted Lebesgue measure $\mu_h = h \circ \lambda$, and we can extend the results from Sect. 4.10.1 to any continuously differentiable function.

Theorem 4.75 *Let $f : [0, t] \to \mathbb{R}_+$ be a continuous function, with $\max_{x \in [0,t]} f(x) = M$, and λ be the Lebesgue measure on $\mathbb{R}_+$. Then $\widetilde{f} : [0, t] \to [0, M]$ given by*

$$\widetilde{f}(\tau) = G^{-1}_{\lambda, f}(t - \tau) \tag{4.90}$$

is a nondecreasing equimeasurable rearrangement.

Proof $G_{\lambda, f}$ is decreasing, hence invertible, because f is continuous and λ is the Lebesgue measure. Let $G_{\lambda, f}(x) = \alpha$, then we have $x = \widetilde{f}(t - \alpha)$. It follows that (see Fig. 4.10)

$$G_{\lambda, \widetilde{f}}(x) = \lambda(\widetilde{f} \geqslant x) = t - (t - \alpha) = \alpha = G_{\lambda, f}(x).$$

□

Remark 4.76 The notion of equimeasurable rearrangement is known in probability theory from a long time ago, starting with the work of Hardy, Littlewood and Pólya [192], and is very useful in many areas of economic theory, in particular, insurance. The result presented here can be found in Narukawa et al. [257] and Sugeno [322, Proposition 26]. It was extended to capacities in a much more general framework by Ghossoub [151].

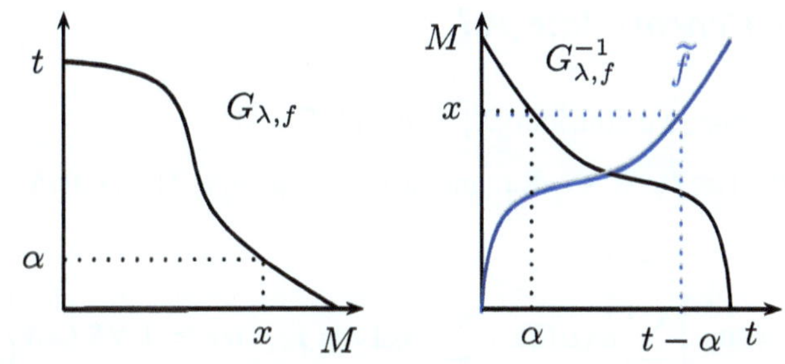

Fig. 4.10 Nondecreasing rearrangement: illustration for the proof

4.11 Other Integrals

4.11.1 The Shilkret Integral

The Shilkret integral [304] has a definition close to the Sugeno integral: the minimum is simply replaced by a multiplication.

Definition 4.77 Let $f \in B^+(\mathcal{F})$ be a function and μ be a capacity on $(X, \mathcal{F})$. The *Shilkret integral* of f w.r.t. μ is defined by

$$\int^{\mathrm{Sh}} f \,\mathrm{d}\mu = \bigvee_{t \geqslant 0} (t \cdot G_{\mu,f}(t)). \tag{4.91}$$

As for the Choquet and Sugeno integrals, the Shilkret integral has an easy graphical interpretation: it is the area of a largest rectangle that can fit below the decumulative function (Fig. 4.11). For some properties, see Theorem 4.88, because the Shilkret integral is a particular case of the decomposition integral.

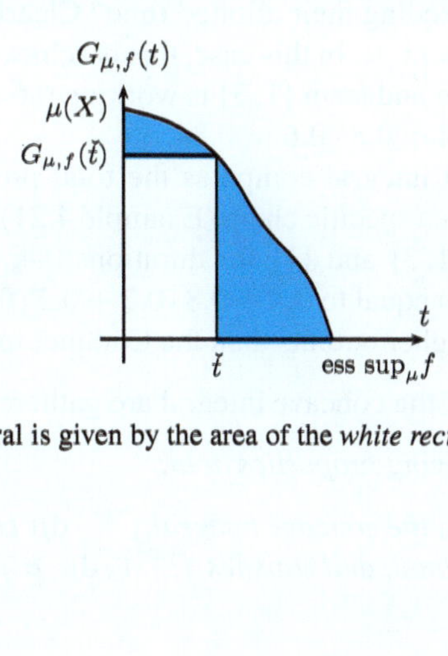

Fig. 4.11 The Shilkret integral is given by the area of the *white rectangle*

4.11.2 The Concave Integral

In the whole section, we consider $|X| = n$ and $\mathcal{F} = 2^X$.

Definition 4.78 Let $f : X \to \mathbb{R}_+$ and μ be a capacity. The *concave integral* of f w.r.t. μ is given by:

$$\int^{\mathrm{cav}} f \, \mathrm{d}\mu = \sup \left\{ \sum_{S \subseteq X} \alpha_S \mu(S) : \sum_{S \subseteq X} \alpha_S 1_S = f, \quad \alpha_S \geqslant 0, \forall S \subseteq X \right\}. \qquad (4.92)$$

We note that the supremum in the above equation can be replaced by a maximum: the concave integral is merely the optimal solution of a linear program in $\alpha_S, S \subseteq X$.

In words, the concave integral is the value achieved by the best decomposition of the integrand into hat functions.

Example 4.79 Let $X = \{1, 2, 3\}$ and μ be a capacity defined by $\mu(X) = 1, \mu(12) = \mu(23) = \frac{2}{3}, \mu(13) = \frac{1}{3}$ and $\mu(S) = 0$ otherwise. Considering the function $f = (1, 2, 1)$, possible decompositions in hat functions are for instance

$$f = 1_1 + 2 \cdot 1_2 + 1_3, \quad f = 1_{12} + 1_{23}, \quad f = 1_{123} + 1_2.$$

The second one is optimal, yielding $\int^{\mathrm{cav}} f \, \mathrm{d}\mu = \frac{2}{3} + \frac{2}{3} = \frac{4}{3}$. ◇

Example 4.80 (Example 4.21 continued) (Even and Lehrer [126]) We consider a set of three workers $X = \{1, 2, 3\}$ and a capacity μ on X representing the productivity per unit of time of a given group of workers (team). We take the following values: $\mu(1) = \mu(2) = \mu(3) = 0.2$, $\mu(12) = 0.9$, $\mu(13) = 0.8$, $\mu(23) = 0.5$ and $\mu(123) = 1$. Each worker is willing to invest a given amount of time in total, say $f_1 = 1$ for worker 1, $f_2 = 0.4$ and $f_3 = 0.6$ for workers 2 and 3. The question is: How should the workers organize themselves in teams so as to maximize the total production while not exceeding their allotted time? Clearly, the answer is given by the concave integral of f w.r.t. μ. In this case, this is achieved as follows: team $\{1, 2\}$ is working 0.4 unit of time and team $\{1, 3\}$ is working 0.6 unit of time, which yields a total production $0.9 \cdot 0.4 + 0.8 \cdot 0.6 = 0.84$.

Note that the Choquet integral computes the total productivity under the constraint that the teams form a specific chain (Example 4.21). Specifically in this case, the teams are $\{1, 2, 3\}$, $\{1, 3\}$ and $\{1\}$ for durations 0.4, 0.2 and 0.4 respectively, yielding a total production equal to $0.4 + 0.8 \cdot 0.2 + 0.2 \cdot 0.4 = 0.64$. Therefore, the concave integral yields higher outputs than the Choquet integral in general. ◇

The main properties of the concave integral are gathered below.

Theorem 4.81 *The following properties hold:*

(i) *For every capacity μ, the concave integral $\int^{\mathrm{cav}} \cdot \, \mathrm{d}\mu$ is a concave and positively homogeneous functional, and satisfies $\int^{\mathrm{cav}} 1_S \, \mathrm{d}\mu \geqslant \mu(S)$ for all $S \in 2^X$;*

(ii) For every $f \in \mathbb{R}_+^X$ and capacity μ,

$$\int^{\text{cav}} f \, \mathrm{d}\mu = \min \big\{ I(f) \; : \; I : \mathbb{R}_+^X \to \mathbb{R} \text{ concave, positively homogeneous, and such that } I(1_S) \geqslant \mu(S), \forall S \subseteq X \big\};$$

(iii) For every $f \in \mathbb{R}_+^X$ and capacity μ,

$$\int^{\text{cav}} f \, \mathrm{d}\mu = \min_{P \text{ additive}, \, P \geqslant \mu} \int f \, \mathrm{d}P;$$

(iv) For every $f \in \mathbb{R}_+^X$ and balanced capacity μ,

$$\int^{\text{cav}} f \, \mathrm{d}\mu = \min_{\phi \in \mathbf{core}(\mu)} \int f \, \mathrm{d}\phi$$

holds if and only if μ has a large core (see Definition 3.43) (core elements ϕ are identified with additive capacities);

(v) For every $f \in \mathbb{R}_+^X$ and every $c \in \mathbb{R}_+$,

$$\int^{\text{cav}} (f + c1_X) \, \mathrm{d}\mu = \int^{\text{cav}} f \, \mathrm{d}\mu + \int^{\text{cav}} c1_X \, \mathrm{d}\mu = \int^{\text{cav}} f \, \mathrm{d}\mu + c\mu(X)$$

holds if and only if μ has a large core;

(vi) For every $f \in \mathbb{R}_+^X$ and capacity μ,

$$\int f \, \mathrm{d}\mu \leqslant \int^{\text{cav}} f \, \mathrm{d}\mu,$$

and equality holds for every $f \in \mathbb{R}_+^X$ if and only if μ is supermodular.

Proof

(i) Positive homogeneity is clear. Concavity comes from the fact that for any $f, g \in \mathbb{R}_+^X, \lambda \in [0, 1]$, the decomposition of $\lambda f + (1 - \lambda)g$ into $\sum_{S \in 2^X} (\lambda \alpha_S + (1 - \lambda)\beta_S)1_S$, with α_S, β_S the coefficients of the optimal decompositions of f and g respectively, is not necessarily optimal. Now, because $\alpha_S = 1$ is a particular decomposition of 1_S, the last property follows easily.

(ii) For any $I : \mathbb{R}_+^X \to \mathbb{R}$ being positively homogeneous, concave and s.t. $I(1_S) \geqslant \mu(S), \forall S \in 2^X$, for any non-identically zero $f \in \mathbb{R}_+^X$ and for any decomposition

$\sum_{S\in 2^X} \alpha_S 1_S = f$, $\alpha_S \geqslant 0$, we have, letting $\bar{\alpha} = \sum_S \alpha_S$,

$$\begin{aligned} I(f) &= \bar{\alpha} I\left(\sum_{S\in 2^X} \frac{\alpha_S}{\bar{\alpha}} 1_S\right) \\ &\geqslant \bar{\alpha} \sum_{S\in 2^X} \frac{\alpha_S}{\bar{\alpha}} I(1_S) \\ &\geqslant \sum_{S\in 2^X} \alpha_S \mu(S). \end{aligned}$$

It follows that

$$I(f) \geqslant \max\left\{\sum_{S\subseteq X} \alpha_S \mu(S) : \sum_{S\subseteq N} \alpha_S 1_S = f, \alpha_S \geqslant 0, \forall S \subseteq X\right\} = \int^{\text{cav}} f \,\mathrm{d}\mu.$$

By (i), the concave integral satisfies all the requirements of I, therefore it is the minimum over all such functionals.

(iii) Consider the linear programming problem in the variables P_i, $i \in X$ (letting $P_i = P(\{i\})$):

$$\begin{aligned} \min \quad & z = \textstyle\sum_{i\in X} f_i P_i \\ \text{s.t.} \quad & \textstyle\sum_{i\in S} P_i \geqslant \mu(S), \quad \varnothing \neq S \subseteq X. \end{aligned}$$

It has obviously a feasible solution and because the P_i's are nonnegative (by monotonicity of μ), it is bounded. It follows by the duality theorem (Theorem 1.8) that this problem as well as the dual linear program in the variables α_S, $\varnothing \neq S \subseteq X$,

$$\begin{aligned} \max \quad & w = \textstyle\sum_{S\in 2^X\setminus\{\varnothing\}} \alpha_S \mu(S) \\ \text{s.t.} \quad & \textstyle\sum_{S\ni i} \alpha_S = f_i, \quad i \in X \\ & \alpha_S \geqslant 0, \quad \varnothing \neq S \subseteq X \end{aligned}$$

have an optimal solution, and equality of the objective functions $z^* = w^*$ holds at the optimum.

(iv) For any balanced capacity μ, we have by (iii)

$$\min_{\phi\in\mathbf{core}(\mu)} \int f \,\mathrm{d}\phi \geqslant \min_{P \text{ additive}, P\geqslant\mu} \int f \,\mathrm{d}P = \int^{\text{cav}} f \,\mathrm{d}\mu.$$

Equality for any f holds if and only if μ has a large core. Indeed, if $P \notin \mathbf{core}(\mu)$ achieves the minimum in the second term, there exists $\phi \in \mathbf{core}(\mu)$ such that $\phi \leqslant P$, so that ϕ also achieves the minimum.

(v) By (iv), largeness of the core implies that there exists an additive measure P in $\mathbf{core}(\mu)$ such that

$$\int^{\text{cav}}(f+c1_X)\,\mathrm{d}\mu = \int(f+c1_X)\,\mathrm{d}P = \int f\,\mathrm{d}P + \int c1_X\,\mathrm{d}P$$
$$= \int f\,\mathrm{d}P + cP(X) = \int^{\text{cav}} f\,\mathrm{d}\mu + c\mu(X).$$

The 4th equality comes from the fact that if P achieves the minimum for $\int(f+c1_X)\,\mathrm{d}P$, it also minimizes $\int f\,\mathrm{d}P$.

Conversely, if μ has not a large core, by Lemma 3.44(iii), there exists a minimal element in $\{P^* \geqslant \mu\}$ such that $P^*(X) > \mu(X)$. Since P^* is minimal, it must satisfy $P^*(S) = \mu(S)$ for some $S \subset X$. Take $f = 1_S$. Then $\int^{\text{cav}} f\,\mathrm{d}\mu = \min_{P\geqslant\mu}\int f\,\mathrm{d}P = P^*(S)$. Because P^* also minimizes $\int(f+c1_X)\,\mathrm{d}P$, we get for any $c > 0$:

$$\int^{\text{cav}}(f+c1_X)\,\mathrm{d}\mu = \int(f+c1_X)\,\mathrm{d}P^* = P^*(S)+cP^*(X) > P^*(S)+c\mu(X)$$
$$= \int^{\text{cav}} f\,\mathrm{d}\mu + c\mu(X).$$

(vi) Formula (4.26) shows clearly that the Choquet integral is based on a particular decomposition of the integrand, hence the inequality holds by definition of the concave integral. Now, because supermodularity of μ implies largeness of the core (see Theorem 3.52 and Lemma 3.51), it follows from (iv), (iii) and Theorem 4.39 that

$$\int f\,\mathrm{d}\mu = \min_{\phi\in\mathbf{core}(\mu)}\int f\,\mathrm{d}\phi = \min_{P\text{ additive},P\geqslant\mu}\int f\,\mathrm{d}P = \int^{\text{cav}} f\,\mathrm{d}\mu.$$

Conversely, suppose that all equalities hold. The first one implies by Theorem 4.39 that μ is supermodular.

□

Remark 4.82

(i) The concave integral was introduced by Lehrer [222] (see also Azrieli and Lehrer [12], Lehrer and Teper [223]), as an alternative to the Choquet integral for application in decision under uncertainty, where the concavity of the integral can be interpreted as uncertainty aversion (see Chap. 5). Also, Example 4.80 gives a clear motivation to the concave integral in production problems.

(ii) An essential difference with the Choquet and Sugeno integrals (as well as with their generalizations, see below Sect. 4.11.4) is that the concave integral is no longer an extension of the capacity: the equality in $\int^{\text{cav}} 1_S \,\mathrm{d}\mu = \mu(S)$ does not hold in general and is replaced by $\geqslant$ [see Theorem 4.81(i)], which can be strict: take, e.g., $X = \{1, 2\}$ and $\mu(1) = \mu(2) = \mu(12) = 1$. Then $\int^{\text{cav}} 1_{\{12\}} \,\mathrm{d}\mu = 2 > \mu(12)$.

(iii) By its definition, the concave integral can be seen as a generalization and extension of the totally balanced cover of a game (Sect. 3.4): the maximization is done over every decomposition of the function and is not restricted to balanced collections, and the domain of the game is extended to any nonnegative function.

(iv) As a consequence of (iii) and Theorem 1.12, the superdifferential at $\mathbf{0}$ of the concave integral is the set $\{P \text{ additive } : P \geqslant \mu\}$.

(v) The concave integral does not satisfy monotonicity w.r.t. stochastic dominance (Definition 4.23), as shown in the following example due to Lehrer [222]. Take $X = \{x_1, x_2, x_3\}$, $\mu(X) = \mu(\{x_2, x_3\}) = 1$, $\mu(\{x_1, x_2\}) = \mu(\{x_1, x_3\}) = 3/4$, and μ is zero for singletons. Consider $f = (1, 1, 1)$ and $f' = (0, 6/5, 6/5)$. Then $f' >^{\mu}_{\text{SD}} f$, however $\int^{\text{cav}} f \,\mathrm{d}\mu = 5/4 > 6/5 = \int^{\text{cav}} f' \,\mathrm{d}\mu$.

Moreover, Lehrer shows in [222, Example 5] that no nontrivial integral can satisfy both monotonicity w.r.t. stochastic dominance and concavity.

◇

We finish this section by giving a characterization of the concave integral. To this end, we consider functionals $I_\mu : \mathbb{R}^X_+ \to \mathbb{R}$ depending on a given capacity μ, and we introduce the following properties:

(i) *Accordance for additive measures (AAM)*: If P is an additive capacity, $I_P = \int \cdot \,\mathrm{d}P$;

(ii) *Independence of irrelevant events (IIE)*: For every $S \subseteq X$, $I_\mu(1_S) = I_{\mu_S}(1_S)$, where μ_S indicates the restriction of μ to 2^S (subgame).

Theorem 4.83 *(Lehrer [222]) Let $I_\mu : \mathbb{R}^X_+ \to \mathbb{R}$ be a functional depending on a capacity μ on X. The following propositions are equivalent:*

(i) I_μ is monotonic w.r.t. the capacity μ, concave, positively homogeneous, and satisfies (AAM) and (IIE);

(ii) $I_\mu = \int^{\text{cav}} \cdot \,\mathrm{d}\mu$ for every capacity μ.

Sketch of the Proof. (ii)⇒(i) is clear. As for the converse, monotonicity of I_μ implies that for every additive capacity satisfying $P \geqslant \mu$, we have by (AAM) $I_P = \int \cdot \,\mathrm{d}P \geqslant I_\mu$, which entails $\min_{P \geqslant \mu} \int \cdot \,\mathrm{d}P \geqslant I_\mu$. By Theorem 4.81(iii), we deduce that $\int^{\text{cav}} \cdot \,\mathrm{d}\mu \geqslant I_\mu$. On the other hand, because I_μ is positively homogeneous and concave, by Theorem 4.81(ii), it remains to prove that $I_\mu(1_S) \geqslant \mu(S)$ for all $S \in 2^X$ in order to show that $\int^{\text{cav}} \cdot \,\mathrm{d}\mu \leqslant I_\mu$, which concludes the proof. The fact that $I_\mu(1_S) \geqslant \mu(S)$ can be shown by induction on $|S|$, but the argument is somewhat involved. □

4.11.3 The Decomposition Integral

Here also we consider $|X| = n$ and $\mathcal{F} = 2^X$.

The decomposition integral, proposed by Even and Lehrer [126], is a generalization of both the Choquet and the concave integrals. The idea is simply to fix a "vocabulary" for the decompositions. If only chains are allowed for the decomposition of a function, then the Choquet integral obtains as the best achievable value for such decompositions, as will be shown in Theorem 4.86. If no restriction applies, then the concave integral is obtained. As we will see, the Shilkret integral can also be recovered. The material of this section is based on [126], to which the readers can refer for more details.

Let $f \in \mathbb{R}_+^X$ and a collection $\mathcal{D} \subseteq 2^X$ be fixed. A *$\mathcal{D}$-subdecomposition* of f is a summation $\sum_{S \in \mathcal{D}} \alpha_S 1_S$ such that

(i) $\sum_{S \in \mathcal{D}} \alpha_S 1_S \leqslant f$;
(ii) $\alpha_S \geqslant 0$ for all $S \in \mathcal{D}$.

A $\mathcal{D}$-subdecomposition is a *$\mathcal{D}$-decomposition* if equality is satisfied in (i). Considering $\mathfrak{D}$ a set of collections $\mathcal{D} \subseteq 2^X$, we say that a (sub)decomposition of f is *$\mathfrak{D}$-allowable* if it is a $\mathcal{D}$-(sub)decomposition of f with $\mathcal{D} \in \mathfrak{D}$.

Definition 4.84 Let $\mathfrak{D}$ be a set of collections $\mathcal{D} \subseteq 2^X$, $f \in \mathbb{R}_+^X$ be a function and μ a capacity on X. The *decomposition integral* of f w.r.t. μ and $\mathfrak{D}$ is given by

$$\int_{\mathfrak{D}} f \,\mathrm{d}\mu = \sup\left\{ \sum_{S \in \mathcal{D}} \alpha_S \mu(S) \,:\, \sum_{S \in \mathcal{D}} \alpha_S 1_S \text{ is a } \mathfrak{D}\text{-allowable subdecomposition of } f \right\}.$$

As for the concave integral, the supremum can be replaced by a maximum because the decomposition integral is the optimal solution over a finite number of linear programming problems (one for each $\mathcal{D} \in \mathfrak{D}$).

The following example shows why it is necessary to consider subdecompositions instead of decompositions.

Example 4.85 Suppose $X = \{1,2,3\}$, $\mu(1) = \mu(2) = \mu(3) = \frac{1}{3}$, $\mu(12) = \mu(13) = \frac{1}{2}$, $\mu(23) = \frac{11}{12}$ and $\mu(123) = 1$. Consider the function $f = (3,5,2)$ and $\mathfrak{D}$ defined as follows:

$$\mathfrak{D} = \{\{1, 23\}, \{12\}, \{2, 13\}\}.$$

The optimal subdecomposition of f is $3 \cdot 1_1 + 2 \cdot 1_{23}$, yielding the value $3 \cdot \frac{1}{3} + 2 \cdot \frac{11}{12} = 2\frac{10}{12}$ for the integral. Observe that no $\mathfrak{D}$-allowable decomposition exists. ◇

Obviously, if $\mathfrak{D} = \{2^X\}$ or $\mathfrak{D} = 2^{2^X}$, the maximum is achieved by a decomposition (because μ and the α_S's are nonnegative), and we obtain $\int_{\mathfrak{D}} = \int^{\mathrm{cav}}$, which shows that among the decomposition integrals, the concave integral yields the highest value. We introduce two additional families $\mathfrak{D}^{\text{chain}}$ and $\mathfrak{D}^{\text{sing}}$:

$$\mathfrak{D}^{\text{chain}} = \{\mathcal{D} \subseteq 2^X : \mathcal{D} \text{ is a chain in } (2^X, \subseteq)\}$$
$$\mathfrak{D}^{\text{sing}} = \{\{S\} : S \in 2^X\}$$

We show that these two families permit to recover the Choquet and Shilkret integrals.

Theorem 4.86 *For any $f \in \mathbb{R}_+^X$ and any capacity μ on X,*

$$\int f \, \mathrm{d}\mu = \int_{\mathfrak{D}^{\text{chain}}} f \, \mathrm{d}\mu$$
$$= \max\left\{\sum_{S\in\mathcal{D}} \alpha_S \mu(S) : \sum_{S\in\mathcal{D}} \alpha_S 1_S \text{ is a } \mathfrak{D}^{\text{chain}}\text{-allowable decomposition of } f\right\}.$$

Proof [126] We first prove the second equality; i.e., the subdecomposition amounts to a decomposition. Define $g = \sum_{i=1}^k \alpha_i 1_{A_i}$ with the following properties: (a) g is a $\mathfrak{D}^{\text{chain}}$-allowable subdecomposition of f, with $A_1 \supset A_2 \cdots \supset A_k$, achieving the maximum of $\sum_{i=1}^k \alpha_i \mu(A_i)$; (b) $\alpha_i > 0$ for $i = 1, \ldots, k$; (c) there is no g' satisfying (a) and (b) such that $g' \geqslant g$ with at least one strict inequality.

Let us show that g is a decomposition of f. Suppose on the contrary that there exists $j \in X$ such that $g_j < f_j$. If $j \in A_i$ for all $i = 1, \ldots, k$, then $\{A_1, \ldots, A_k, \{j\}\}$ is a chain. Letting $g' = g + (f_j - g_j)1_j$, we see that g' satisfies (a) and (b), and since $g' \geqslant g$ and $g'_j > g_j$, (c) is violated for g, a contradiction. We may therefore assume that j does not belong to all A_i. Define i_0 the smallest index such that $j \notin A_{i_0}$. Then $\{A_1, \ldots, A_{i_0} \cup \{j\}, A_{i_0}, \ldots, A_k\}$ is a chain, and let us define $g' = \sum_{i \neq i_0} \alpha_i 1_{A_i} + \beta 1_{A_{i_0} \cup \{j\}} + (\alpha_{i_0} - \beta) 1_{A_{i_0}}$ with $\beta = \min(f_j - g_j, \alpha_{i_0})$. Observe that $f \geqslant g' \geqslant g$ with in particular $g'_j > g_j$. Moreover, g' is an optimal subdecomposition because by monotonicity of μ, $\mu(A_{i_0})(\alpha_{i_0} - \beta) + \mu(A_i \cup \{j\}) \geqslant \mu(A_{i_0})\alpha_{i_0}$. Therefore g' satisfies (a) and (b), so that g does not satisfy (c), a contradiction. We conclude that $g = f$.

It remains to show that the unique (up to sets with zero coefficients) optimal decomposition of f is given by the chain $\{A_\sigma^\uparrow(1), \ldots, A_\sigma^\uparrow(n)\}$ (Sect. 4.5.1), where σ is a permutation on X such that $f_{\sigma(1)} \leqslant \cdots \leqslant f_{\sigma(n)}$. It suffices to show that if $f_m = f_\ell$, then $m \in A_\sigma^\uparrow(i) \Leftrightarrow \ell \in A_\sigma^\uparrow(i)$ for all $i = 1, \ldots, n$. Suppose on the contrary that there exists i_0 such that $\ell \in A_\sigma^\uparrow(i_0)$ and $m \notin A_\sigma^\uparrow(i_0)$. Then by (b) $f_m = g_m < g_\ell = f_\ell$, a contradiction. □

Theorem 4.87 *For any $f \in \mathbb{R}_+^X$ and any capacity μ on X,*

$$\int^{\text{Sh}} f \, \mathrm{d}\mu = \int_{\mathfrak{D}^{\text{sing}}} f \, \mathrm{d}\mu.$$

Proof

$$\int_{\mathfrak{D}^{\text{sing}}} f\,\mathrm{d}\mu = \max\Big\{\sum_{S\in\mathcal{D}} \alpha_S\mu(S)\ :$$

$$\sum_{S\in\mathcal{D}} \alpha_S 1_S \text{ is a } \mathfrak{D}^{\text{sing}}\text{-allowable subdecomposition of } f\Big\}$$

$$= \max\{\alpha\mu(S)\ :\ \alpha 1_S \leqslant f, S\subseteq X, \alpha\geqslant 0\} = \max\{\alpha\mu(f\geqslant\alpha)\ :\ \alpha\geqslant 0\}$$

$$= \int^{\text{Sh}} f\,\mathrm{d}\mu.$$

□

We summarize the main properties of the decomposition integral.

Theorem 4.88 *Let $f\in\mathbb{R}^X_+$ be a function, μ be a capacity and $\mathfrak{D}$ be a set of collections $\mathcal{D}\subseteq 2^X$. The following properties hold:*

(i) $\int_{\mathfrak{D}}\cdot\,\mathrm{d}\mu$ *is positively homogeneous;*
(ii) $\int_{\mathfrak{D}}\cdot\,\mathrm{d}\mu$ *is monotonic w.r.t. the integrand;*
(iii) $\int_{\mathfrak{D}}\cdot\,\mathrm{d}\mu$ *is monotonic w.r.t the capacity;*
(iv) $\int_{\mathfrak{D}}\cdot\,\mathrm{d}P = \int\cdot\,\mathrm{d}P$ *for every additive capacity P if and only if every function $f\in\mathbb{R}^X_+$ has a $\mathcal{D}$-decomposition with $\mathcal{D}\in\mathfrak{D}$.*

(Proof is easy and omitted.)

An interesting question is under which conditions a set of collections $\mathfrak{D}$ yields an integral smaller than the integral w.r.t another set of collections $\mathfrak{D}'$. To answer this question, we say that a collection $\mathcal{C}\subseteq 2^X$ is *independent* if the vectors $\{1_S\}_{S\in\mathcal{C}}$ are linearly independent. Clearly, if $\mathcal{D}$ is an independent collection, then for each $f\in\mathbb{R}^X_+$, there is a unique $\mathcal{D}$-decomposition.

Theorem 4.89 *Suppose that $\mathfrak{D}$ and $\mathfrak{D}'$ are two sets of collections. Then $\int_{\mathfrak{D}}\cdot\,\mathrm{d}\mu\leqslant\int_{\mathfrak{D}'}\cdot\,\mathrm{d}\mu$ for every capacity μ on X if and only if for every $\mathcal{D}\in\mathfrak{D}$ and every independent collection $\mathcal{C}\subseteq\mathcal{D}$, there exists $\mathcal{D}'\in\mathfrak{D}'$ such that $\mathcal{C}\subseteq\mathcal{D}'$.*

The proof is based on the fact that if $\mathcal{D}$ yields an optimal $\mathfrak{D}$-allowable subdecomposition of f, then there exists an independent collection $\mathcal{C}\subseteq\mathcal{D}$ that also yields an optimal $\mathfrak{D}$-allowable subdecomposition of f. Hence, if there exists $\mathcal{D}'\in\mathfrak{D}'$ such that $\mathcal{C}\subseteq\mathcal{D}'$, the highest level achieved by a $\mathcal{D}'$-subdecomposition is at least as high as $\int_{\mathfrak{D}} f\,\mathrm{d}\mu$. The converse statement is however much harder to prove [126].

The next topic of interest is related to additivity: we know from Theorem 4.28 that the Choquet integral is additive for comonotonic functions, that is, functions inducing the same chain of level sets. Now, from Theorem 4.86, we know that the "vocabulary" of the Choquet integral is precisely the set of chains. One would expect that in general, additivity of the decomposition integral holds for those functions having the same collection yielding their optimal subdecompositions. It turns out that this is not true in general, as shown by the following example.

Example 4.90 Take $X = \{1, 2\}$, $\mathfrak{D}$ be the set of partitions of X, and μ defined by $\mu(1) = \mu(2) = \frac{1}{3}$, $\mu(12) = 1$. Consider two functions $f = (\epsilon, 1)$, $g = (1, \epsilon)$, $\epsilon > 0$ small enough so that the optimal decomposition of both f and g uses $\mathcal{D} = \{\{1\}, \{2\}\}$. Then $\int_{\mathfrak{D}} f \, d\mu = \int_{\mathfrak{D}} g \, d\mu = \frac{1}{3}(1 + \epsilon)$. Observe that with $\mathcal{D}' = \{12\}$ we obtain a decomposition of $f + g$ yielding $\int_{\mathfrak{D}} (f + g) \, d\mu = 1 + \epsilon$, which is strictly greater than $\int_{\mathfrak{D}} f \, d\mu + \int_{\mathfrak{D}} g \, d\mu = \frac{2}{3}(1 + \epsilon)$. ◇

We say that f is *leaner* than g, for $f, g \in \mathbb{R}_+^X$, if there exist optimal decompositions $f = \sum_{S \in \mathcal{C}} \alpha_S 1_S$, $g = \sum_{S \in \mathcal{C}'} \beta_S 1_S$ with $\alpha_S > 0$, $\forall S \in \mathcal{C}$, $\beta_S > 0$, $\forall S \in \mathcal{C}'$, and such that $\mathcal{C} \subseteq \mathcal{C}'$ (roughly speaking, f can be decomposed with less sets than g). It is important to note that, because optimality of the decomposition depends on μ, f can be leaner than g for some capacities, but not for all of them.

Theorem 4.91 (Co-decomposition additivity of the decomposition integral) *Suppose that $\mathfrak{D}$ is a set of collections satisfying (i) Every $f \in \mathbb{R}_+^X$ has an optimal $\mathfrak{D}$-allowable decomposition for every capacity μ on X, and (ii) If $f = \sum_{S \in \mathcal{D}} \alpha_S 1_S = \sum_{S \in \mathcal{D}'} \beta_S 1_S$ with $\mathcal{D}, \mathcal{D}' \in \mathfrak{D}$, then $\mathcal{D}'' = \{S : \alpha_S > 0 \text{ or } \beta_S > 0\}$ belongs to $\mathfrak{D}$. Then, for every capacity μ on X and for every $f, g \in \mathbb{R}_+^X$ such that f is leaner than g,*

$$\int_{\mathfrak{D}} (f + g) \, d\mu = \int_{\mathfrak{D}} f \, d\mu + \int_{\mathfrak{D}} g \, d\mu. \tag{4.93}$$

Proof [126] Consider two functions $f, g \in \mathbb{R}_+^X$ such that g is leaner than f, with $\mathfrak{D}$-allowable optimal decompositions $f = \sum_{S \in \mathcal{C}} \alpha_S 1_S$, $g = \sum_{S \in \mathcal{C}} \beta_S 1_S$, with $\alpha_S > 0, \beta_S \geqslant 0$ for all $S \in \mathcal{C}$. Let $\sum_{T \in \mathcal{D}} \gamma_T 1_T$ be an optimal $\mathfrak{D}$-allowable decomposition of $f + g$. If this decomposition is equal to $\sum_{S \in \mathcal{C}} (\alpha_S + \beta_S) 1_S$, then clearly (4.93) holds. Otherwise, we have $\int_{\mathfrak{D}} (f + g) \, d\mu \geqslant \int_{\mathfrak{D}} f \, d\mu + \int_{\mathfrak{D}} g \, d\mu$.

By assumption (ii) on $\mathfrak{D}$, $\mathcal{D}' = \{S \in 2^X : \alpha_S + \beta_S > 0 \text{ or } \gamma_S > 0\}$ belongs to $\mathfrak{D}$, so that $f, g, f + g$ have optimal decompositions using $\mathcal{D}'$. Suppose that $\int_{\mathfrak{D}} (f + g) \, d\mu > \int_{\mathfrak{D}} f \, d\mu + \int_{\mathfrak{D}} g \, d\mu$. For a sufficiently small $\epsilon > 0$,

$$f = \sum_{S \in \mathcal{C}} \alpha_S 1_S - \epsilon \sum_{S \in \mathcal{C}} (\alpha_S + \beta_S) 1_S + \epsilon \sum_{T \in \mathcal{D}} \gamma_T 1_T$$

is a decomposition of f using sets in $\mathcal{D}'$. It follows that

$$\begin{aligned} \int_{\mathfrak{D}} f \, d\mu &\geqslant \sum_{S \in \mathcal{C}} \alpha_S \mu(S) - \epsilon \sum_{S \in \mathcal{C}} (\alpha_S + \beta_S) \mu(S) + \epsilon \sum_{T \in \mathcal{D}} \gamma_T \mu(T) \\ &> \int_{\mathfrak{D}} f \, d\mu - \epsilon \int_{\mathfrak{D}} f \, d\mu - \epsilon \int_{\mathfrak{D}} g \, d\mu + \epsilon \Big(\int_{\mathfrak{D}} f \, d\mu + \int_{\mathfrak{D}} g \, d\mu \Big) = \int_{\mathfrak{D}} f \, d\mu, \end{aligned}$$

a contradiction. □

The above theorem generalizes Theorem 4.28 on comonotonic additivity of the Choquet integral. This can be seen as follows. Take f, g comonotone functions and let $\mathcal{D}$ be the chain containing the sets used in their optimal decompositions.[12] Consider $h_\epsilon = \sum_{S\in\mathcal{D}} \frac{\epsilon}{n} 1_S$ for some $\epsilon > 0$, and observe that $h_\epsilon \leqslant \epsilon 1_X$; moreover, f is leaner than h_ϵ and g is leaner than $f + h_\epsilon$. By Theorem 4.91, we get

$$\begin{aligned}\int_{\mathfrak{D}^{\text{chain}}} (f + g + h_\epsilon)\,\mathrm{d}\mu &= \int_{\mathfrak{D}^{\text{chain}}} g\,\mathrm{d}\mu + \int_{\mathfrak{D}^{\text{chain}}} (f + h_\epsilon)\,\mathrm{d}\mu\\ &= \int_{\mathfrak{D}^{\text{chain}}} g\,\mathrm{d}\mu + \int_{\mathfrak{D}^{\text{chain}}} f\,\mathrm{d}\mu + \int_{\mathfrak{D}^{\text{chain}}} h_\epsilon\,\mathrm{d}\mu.\end{aligned}$$

Letting $\epsilon \to 0$, we get comonotonic additivity of the Choquet integral.

Theorem 4.91 permits also to deduce under which conditions the concave integral is additive. Since $\mathfrak{D}$ reduces to $\{2^X\}$, the conditions on $\mathfrak{D}$ in the theorem are satisfied, so that the concave integral is additive for functions having optimal decompositions such that any set involved in the decomposition of one is also used in the decomposition of the other.

As for the Shilkret integral, the theorem cannot be used because $\mathfrak{D}^{\text{sing}}$ does not fulfill the required conditions.

The next theorem investigates under which conditions the decomposition is concave, or monotonic vs. stochastic dominance (see Definition 4.23), or translation-invariant.

Theorem 4.92 *Let $\mathfrak{D}$ be a set of collections $\mathcal{D} \subseteq 2^X$. The following holds.*

(i) $\int_{\mathfrak{D}} \cdot\,\mathrm{d}\mu$ *is concave for every capacity μ on X if and only if there exists a set of collections $\mathfrak{D}'$ reduced to a singleton $\mathcal{D}$ such that* $\int_{\mathfrak{D}} \cdot\,\mathrm{d}\mu = \int_{\mathfrak{D}'} \cdot\,\mathrm{d}\mu$ *for all capacity μ;*

(ii) $\int_{\mathfrak{D}} \cdot\,\mathrm{d}\mu$ *is monotonic w.r.t. stochastic dominance if and only if there exists $k \in \mathbb{N}$ such that $\mathfrak{D}$ is a set of chains of length at most k containing all chains of length k;*

(iii) For every $f \in \mathbb{R}_+^X$, every $c \geqslant 0$ and normalized capacity μ,

$$\int_{\mathfrak{D}} (f + c1_X)\,\mathrm{d}\mu = \int_{\mathfrak{D}} f\,\mathrm{d}\mu + c$$

if and only if $\mathfrak{D}$ is a set of chains such that for every $C \in \mathfrak{D}$, there exists $C' \in \mathfrak{D}$ such that $C \subseteq C'$ and $X \in C'$.

(See [126] for a proof.)

[12] Note that these sets do *not* depend on the capacity, contrarily to the general case.

4.11.4 Pseudo-Additive Integrals, Universal Integrals

Universal integrals try to answer the following question: *What is an integral w.r.t. a capacity?*, using an axiomatic approach. They were proposed by Klement et al. [211] (see also [212]), and include as particular cases the Choquet, Sugeno and Shilkret integrals, but not the concave integral (and consequently, not the decomposition integral). The mathematical setting in [211] being different from ours (measurability, unbounded functions and measures, etc.), we do not go deeply into details and refer the readers to the above cited papers.

The name "universal" comes from the property that these integrals are defined w.r.t. any measurable space $(X, \mathcal{F})$, where $\mathcal{F}$ is a σ-algebra. Let us denote by $\mathcal{S}$ the class of all measurable spaces. We deal with functionals I whose domain is denoted by $\mathcal{D}$, defined by

$$\mathcal{D} = \bigcup_{(X,\mathcal{F})\in\mathcal{S}} \mathcal{MG}(X, \mathcal{F}) \times U(X, \mathcal{F})$$

where $\mathcal{MG}(X, \mathcal{F})$ is the set of all capacities on $(X, \mathcal{F})$, and $U(X, \mathcal{F})$ the set of all $\mathcal{F}$-measurable functions on X.

We define a *pseudo-multiplication* as an operator $\otimes : [0, \infty]^2 \to [0, \infty]$ being nondecreasing in each place, having 0 as annihilator (i.e., $a \otimes 0 = 0 \otimes a = 0$ for all $a \in [0, \infty]$), and having a neutral element e (i.e., $e \otimes a = a \otimes e = a$).

A functional $I : \mathcal{D} \to [0, \infty]$ is a *universal integral* if it satisfies the three following axioms:

(A1) For any measurable space $(X, \mathcal{F})$, the restriction of I to $\mathcal{MG}(X, \mathcal{F}) \times U(X, \mathcal{F})$ is nondecreasing in each place;

(A2) There exists a pseudo-multiplication $\otimes$ such that

$$I(\mu, c \cdot 1_A) = c \otimes \mu(A) \qquad ((\mu, c \cdot 1_A) \in \mathcal{D});$$

(A3) If $G_{\mu,f} = G_{\mu',f'}$, then $I(\mu,f) = I(\mu',f')$.

It is not difficult to prove that a functional I on $\mathcal{D}$ is a universal integral for some pseudo-multiplication $\otimes$ if and only if I is a distortion of the decumulative function: $I(\mu,f) = J(G_{\mu,f})$, where J is nondecreasing and satisfies $J(c \cdot 1_{]0,d]}) = c \otimes d$ for all $c, d \in [0, \infty]$.

It can be proved that the Sugeno and Shilkret integrals are smallest universal integrals (in the sense of the usual ordering for functions). More precisely, for a fixed $\otimes$, the smallest universal integrals have the form

$$I(\mu,f) = \sup_{t\in]0,\infty]} (t \otimes \mu(f \geqslant t)).$$

The Sugeno and Shilkret integrals are recovered with the minimum and the product, respectively.

An important class of universal integrals can be obtained as follows. For a fixed pseudo-multiplication $\otimes$, define a *pseudo-addition* as an operator $\oplus : [0,\infty]^2 \to [0,\infty]$ that is continuous, associative, nondecreasing in each place, has 0 as neutral element (from which it follows that it is also commutative), and which is left-distributive w.r.t. $\otimes$.

From $\oplus$ we define the *pseudo-difference* $\ominus$ by

$$a \ominus b = \inf\{c \in [0,\infty] : b \oplus c = a\} \qquad (0 \leqslant b \leqslant a \leqslant \infty).$$

Taking any simple function f on X, with range values $a_1 < \cdots < a_n$, defining the subsets $A_i = \{x \in X : f(x) \geqslant a_i\}$, $i = 1, \ldots, n$, we define similarly to (4.17) the quantity

$$I^{\text{simple}}_{\otimes,\oplus}(\mu,f) = \bigoplus_{i=1}^{n} \big((a_i \ominus a_{i-1}) \otimes \mu(A_i)\big) \tag{4.94}$$

with $a_0 = 0$, and for any capacity μ on some measurable space $(X, \mathcal{F})$.

Theorem 4.93 *Let $\oplus, \otimes$ be defined as above. The functional $I_{\otimes,\oplus}$ on $\mathcal{D}$ defined by*

$$I_{\otimes,\oplus}(\mu,f) = \sup\left\{I^{\text{simple}}_{\otimes,\oplus}(\mu',f') : (\mu',f') \in \mathcal{D}, f' \textit{ simple }, G_{\mu',f'} \leqslant G_{\mu,f}\right\}$$

for any $(\mu,f) \in \mathcal{D}$, is a universal integral that is an extension of $I^{\text{simple}}_{\otimes,\oplus}$.

Remark 4.94

(i) One can take $\oplus = \sup$, leading in particular to the Sugeno integral ($\otimes = \min$) and the Shilkret integral ($\otimes$ is the usual product). Moreover, the Choquet integral is recovered with $\oplus = +$ and the usual product for $\otimes$. Also, the *Choquet-like integrals* of Mesiar [239] are particular cases of such universal integrals.

(ii) A similar construction can be done also when working on the interval $[0,1]$ instead of $[0,\infty]$, considering normalized capacities and functions with range included in $[0,1]$. Then the pseudo-multiplication reduces to a *semicopula* [118] (nondecreasing, 1 is neutral element, smaller than minimum). If associativity and commutativity are also required, then one obtains triangular norms (t-norms), which are the dual operations of t-conorms (see Sect. 2.8.5). With $\oplus = \sup$ and $\otimes$ a strict t-norm (i.e., isomorphic to the product), the Sugeno-Weber integral is recovered [346].

(iii) When restricting to $[0,1]$, many other similar integrals have been defined. The *Benvenuti integral* [19] and the *Murofushi integral* [251] have exactly the same expression (4.94), however, the pseudo-multiplication and pseudo-addition have different definitions. The readers may consult the above references for more detail, as well as the survey in [177, Sect. 5.6], and the thorough analysis of Sander and Siedekum [281–283]. ◇

4.12 The Choquet Integral for Nonmeasurable Functions

So far, integrals were defined for measurable functions and games or capacities defined on an algebra $\mathcal{F}$ built on X. Usually $\mathcal{F}$ is a proper subset of the power set 2^X (especially if X is infinite), so that some functions are not measurable, and therefore their integral cannot be computed. The question we address in this section is precisely: How can we compute the integral of a nonmeasurable function? Although quite odd, the question makes sense, even in the finite case. Indeed, suppose that in some practical situation, like the one described in Chap. 6, it turns out that it is impossible to determine the value of a capacity on some subsets of X, because, e.g., of insufficient knowledge. Therefore, $\mathcal{F} \subset 2^X$ and some functions become nonmeasurable. However, if functions come from observations, any kind of function is likely to occur and we would like nevertheless to compute its integral (expected value). We will see that, with the Choquet integral and restricting to the case where X is finite, there is a natural solution, based on an approach similar to the one used for the concave integral (Sect. 4.11.2). The material of this section is based on Faigle and Grabisch [130], to which the readers can refer for more detail.

Let us consider $|X| = n$, and fix a set system $\mathcal{F}$, which is not necessarily an algebra. We know from Sect. 2.15.5 that any game v on $(X, 2^X)$ can be decomposed as $v = v^+ - v^-$, where v^+, v^- are totally monotone capacities. Here, we do the same for games on $(X, \mathcal{F})$:

$$v^+ = \sum_{A \in \mathcal{F} : m^v(A) > 0} m^v(A) u_A, \quad v^- = \sum_{A \in \mathcal{F} : m^v(A) < 0} (-m^v(A)) u_A. \tag{4.95}$$

We first define the Choquet integral w.r.t. a totally monotone capacity b on $\mathcal{F}$, by taking the smallest functional I being positively homogeneous, superadditive, and such that $I(1_A) \geqslant b(A)$ for all $A \in \mathcal{F}$. Proceeding as for the concave integral, we find that the integral of any nonnegative function f is given by

$$\int_{\mathcal{F}} f \,\mathrm{d}b = \max\left\{ \sum_{A \in \mathcal{F}} \alpha_A b(A) : \sum_{A \in \mathcal{F}} \alpha_A 1_A \leqslant f, \alpha_A \geqslant 0, \forall A \in \mathcal{F} \right\} \tag{4.96}$$

$$= \min\left\{ \sum_{i \in X} P_i f_i : \sum_{i \in A} P_i \geqslant b(A), \forall A \in \mathcal{F}, P_i \geqslant 0, \forall i \in X \right\}. \tag{4.97}$$

Note that this is the decomposition integral of f w.r.t. b with vocabulary $\mathfrak{D} = \{\mathcal{F}\}$ (Sect. 4.11.3). Now, for any game v on $(X, \mathcal{F})$, the Choquet integral of a mapping $f : X \to \mathbb{R}_+$ w.r.t. v is defined by

$$\int_{\mathcal{F}} f \,\mathrm{d}v = \int_{\mathcal{F}} f \,\mathrm{d}v^+ - \int_{\mathcal{F}} f \,\mathrm{d}v^-. \tag{4.98}$$

We summarize the main properties of this integral.

Theorem 4.95 *Let $f : X \to \mathbb{R}_+$ be a function and v be a game on $(X, \mathcal{F})$, where $\mathcal{F}$ is any set system. The following properties hold.*

(i) Positive homogeneity:

$$\int_{\mathcal{F}} \alpha f \, \mathrm{d}v = \alpha \int_{\mathcal{F}} f \, \mathrm{d}v \qquad (\alpha \geqslant 0);$$

(ii) For any $S \in \mathcal{F}$,

$$\int_{\mathcal{F}} f \, \mathrm{d}u_S = \min_{i \in S} f_i$$

where u_S is the unanimity game w.r.t. S;

(iii) If $\mathcal{F}$ is weakly union-closed,

$$\int_{\mathcal{F}} f \, \mathrm{d}v = \sum_{S \in \mathcal{F}} m^v(S) \min_{i \in S} f_i$$

where m^v is the Möbius transform of v;

(iv) If $\mathcal{F}$ is weakly union-closed,

$$\int_{\mathcal{F}} f \, \mathrm{d}v = \int f \, \mathrm{d}\hat{v}$$

where the right-hand side integral is the ordinary Choquet integral, and $\hat{v}$ is a game on $(X, 2^X)$ defined by

$$\hat{v}(S) = \int_{\mathcal{F}} 1_S \, \mathrm{d}v = \sum_{F \text{ maximum in } \mathcal{F}(S)} v(F) \qquad (S \in 2^X),$$

with $\mathcal{F}(S) = \{F \in \mathcal{F} : F \subseteq S\}$;

(v) If $\mathcal{F}$ is weakly union-closed, $\int_{\mathcal{F}} \cdot \, \mathrm{d}v$ is superadditive if and only if it is concave if and only if $\hat{v}$ is supermodular.

Proof

(i) Clear.

(ii) Let $j \in S$ such that f_j is minimum on S. Then $P^* \in \mathbb{R}^X$ defined by $P^*_j = 1$ and $P^*_i = 0$ for all $i \neq j$, is feasible for the linear program

$$\min_{P \geqslant 0} \langle f, P \rangle \text{ s.t. } P(T) \geqslant 1, \;\; T \in \mathcal{F}, T \supseteq S,$$

while $\lambda^* \in \mathbb{R}_+^{\mathcal{F}}$ with the only nonzero component $\lambda_S^* = f_j$ is feasible for the dual linear program

$$\max_{\lambda \geqslant 0} \langle u_S, \lambda \rangle \text{ s.t. } \sum_{T \supseteq S} \lambda_T 1_T \leqslant f.$$

In view of $\langle f, P^* \rangle = f_j = \langle u_S, \lambda^* \rangle$, linear programming duality guarantees optimality, which proves the result.

(iii) We write $v = v^+ - v^-$ as in (4.95). Due to the definition of the integral by (4.98), it suffices to show that

$$\int f \,\mathrm{d}\Big(\sum_{S \in \mathcal{T}} \alpha_S u_S\Big) = \sum_{S \in \mathcal{T}} \alpha_S \min_{i \in S} f_i$$

for any $\mathcal{T} \subseteq \mathcal{F}$, and $\alpha_S > 0$ for all $S \in \mathcal{T}$. Make a partition of $\mathcal{T}$ into blocks $\{S_1, \ldots, S_k\}, \{S_{k+1}, \ldots, S_{k'}\}, \ldots$, which are the smallest subcollections such that for every S in a block, one can find another set T in the same block so that $S \cap T \neq \varnothing$, and no set in a block has a nonempty intersection with a set of another block (in a sense, these are the "connected components" of $\mathcal{T}$).

The mapping f being given, for each $S \in \mathcal{T}$, pick $i_S \in S$ such that f_{i_S} realizes the minimum of f in S. In each block, number the sets S so that the values f_{i_S} are nondecreasing; i.e., $f_{i_{S_1}} \leqslant \cdots \leqslant f_{i_{S_k}}$ in the first block, and similarly for the other ones. Since $v' = \sum_S \alpha_S u_S$ is totally monotone, by (4.97), the integral is the optimal value z^* of the linear program $z = \min\langle P, f \rangle$ subject to $P(S) \geqslant v'(S)$, $S \in \mathcal{F}$, $P_i \geqslant 0, \forall i$, using the shorthand $P(S) = \sum_{i \in S} P_i$.

Let us define P^* by $P_{i_S}^* = \alpha_S$, for all $S \in \mathcal{T}$, and $P_i^* = 0$ otherwise. It is easy to check that P^* is a feasible solution of the above linear program. We remark that $z = \sum_{S \in \mathcal{T}} \alpha_S \min_{i \in S} f_i$, the desired result. Therefore, it remains to prove optimality. We use for this complementary slackness (Theorem 1.9).

The dual program is given by (4.96): maximize $w = \langle \lambda, v' \rangle$ subject to $\sum_{S \in \mathcal{F}} \lambda_S 1_S \leqslant f$, $\lambda_S \geqslant 0, \forall S \in \mathcal{F}$. We propose a solution λ^* constructed as follows. Considering the first block of the partition of $\mathcal{T}$, we define the sets

$$\mathcal{S}_j = S_j \cup S_{j+1} \cup \cdots \cup S_k \qquad (j = 1, \ldots, k),$$

and similarly for the other blocks: $\mathcal{S}_{k+1}, \ldots, \mathcal{S}_{k'}, \ldots$. Observe that because $\mathcal{F}$ is weakly union-closed, all these sets are members of $\mathcal{F}$. We put

$$\lambda_{\mathcal{S}_1}^* = f_{i_{S_1}}$$

$$\lambda_{\mathcal{S}_j}^* = f_{i_{S_j}} - f_{i_{S_{j-1}}} \qquad (j = 2, \ldots, k)$$

and similarly for the other blocks. For all other sets $S \in \mathcal{F}$, we put $\lambda^*_S = 0$. By complementary slackness, optimality of P amounts to checking that (a) λ^* is feasible, (b) is 0 for all nontight constraints in the primal program, and (c) gives tight constraints in the dual program for all i_S, $S \in \mathcal{T}$.

Proving (b) amounts to proving that $P^*(S_j) = v'(S_j)$ for all $j = 1, \dots, k$ such that $f_{i_{S_j}} > f_{i_{S_{j-1}}}$ (letting $f_{i_{S_0}} = 0$), and similarly for all other blocks. We have $P^*(S_j) = \sum_{i=j}^k \alpha_{S_j}$. On the other hand, $v'(S_j) \geqslant \sum_{i=j}^k \alpha_{S_j}$, with strict inequality if and only if S_j contains some S_ℓ with $1 \leqslant \ell < j$. But this would contradict the assumption $f_{i_{S_\ell}} < f_{i_{S_j}}$.

For every $j = 1, \dots, k$, it is easily verified that $\sum_{S \in \mathcal{F}} \lambda^*_S 1_S(i_{S_j}) = f_{i_{S_j}}$, and similarly for all blocks. This proves (c), we prove now (a). $\lambda^*_S \geqslant 0$ for all $S \in \mathcal{F}$ by construction. It remains to prove $\sum_{S \in \mathcal{F}} \lambda^*_S 1_S(i) \leqslant f_i$ for all $i \neq i_S$, $S \in \mathcal{T}$. If $i \notin \bigcup \mathcal{T}$, then clearly the left member of the inequality is 0, so that the inequality holds by nonnegativity of f. Suppose then $i \in \bigcup \mathcal{T}$, w.l.o.g., in the first block. Call j the largest index of $S_j \in \{S_1, \dots, S_k\}$ such that $i \in S_j$. We have then

$$\sum_{S \in \mathcal{F}} \lambda^*_S 1_S(i) = \sum_{\ell=1}^{j} \lambda^*_{S_\ell} = f_{i_{S_j}} \leqslant f_i,$$

where the inequality comes from the definition of $f_{i_{S_j}}$. The proof is complete.

(iv) Let $\hat{v}$ be the game on $(X, 2^X)$ defined through its Möbius transform by

$$m^{\hat{v}}(S) = \begin{cases} m^v(S), & \text{if } S \in \mathcal{F} \\ 0, & \text{otherwise.} \end{cases}$$

Then, by (iii) and (4.55), we immediately have

$$\int_{\mathcal{F}} f \, dv = \int f \, d\hat{v}.$$

Now, by (iii),

$$\int_{\mathcal{F}} 1_S \, dv = \sum_{T \in \mathcal{F}} m^v(T) \min_{i \in T} 1_S(i) = \sum_{\substack{T \subseteq S \\ T \in \mathcal{F}}} m^v(T) = \sum_{T \subseteq S} m^{\hat{v}}(T) = \hat{v}(S).$$

Lastly, if $S \in \mathcal{F}$, we have $\hat{v}(S) = \sum_{T \subseteq S} m^{\hat{v}}(T) = \sum_{\substack{T \subseteq S \\ T \in \mathcal{F}}} m^v(T) = v(S)$. Now, if $S \notin \mathcal{F}$, because $\mathcal{F}$ is weakly union-closed, the largest subsets of S belonging to $\mathcal{F}$ are necessarily disjoint, hence:

$$\begin{aligned}\hat{v}(S) = \sum_{T\subseteq S} m^{\hat{v}}(T) &= \sum_{F \text{ maximum in } \mathcal{F}(S)} \sum_{T\subseteq F} m^{\hat{v}}(T)\\ &= \sum_{F \text{ maximum in } \mathcal{F}(S)} \sum_{\substack{T\subseteq F\\ T\in\mathcal{F}}} m^{v}(T) = \sum_{F \text{ maximum in } \mathcal{F}(S)} v(F).\end{aligned}$$

(v) Since positive homogeneity holds, the first equivalence is established (Sect. 1.3.7). The second equivalence comes from (iv) and Theorem 4.35 (i) and (ii).

□

Remark 4.96

(i) Theorem 4.95(iv) shows that the Choquet integral on weakly union-closed systems essentially equals the classical Choquet integral, and therefore inherits all of its properties (in particular, comonotonic additivity).
(ii) The game $\hat{v}$ is an extension of v on $(X, 2^X)$ because it coincides with v on $\mathcal{F}$. For this reason, if f is a measurable function on $(X, \mathcal{F})$, then $\int_{\mathcal{F}} f\,\mathrm{d}v$ is the usual Choquet integral, which shows that $\int_{\mathcal{F}} \cdot\,\mathrm{d}v$ can be considered to be an extension of the Choquet integral to every function, as announced in the beginning of this Section.
(iii) The extension $\hat{v}$ is well known in cooperative game theory as *Myerson's restricted game* [256], and is used in the analysis of games on communication graphs.
(iv) If v is a capacity, then $\hat{v}$ is not necessarily a capacity, as shown by the following example: take $N = \{1, 2, 3\}$, and $\mathcal{F} = \{123, 1, 2, \varnothing\}$. Then $\hat{v}(N) = v(N)$ and $\hat{v}(12) = v(1) + v(2)$. Taking $v(N) = v(1) = v(2) = 1$ yields a nonmonotonic game. It follows that the Choquet integral w.r.t. a capacity on $\mathcal{F}$ may be not monotone w.r.t. the integrand.

◇

More results can be obtained for capacities and if $\mathcal{F}$ is a set system closed under union (note that this implies $X \in \mathcal{F}$). We recall from Sect. 2.19.2 that supermodularity is defined for weakly union-closed systems [Eq. (2.119)]. For union-closed systems, the definition simplifies as follows: a game v on $(X, \mathcal{F})$ is supermodular if for any $S, T \in \mathcal{F}$,

$$v(S \cup T) + v((S \cap T)') \geqslant v(S) + v(T), \tag{4.99}$$

where $(S \cap T)'$ is the largest subset of $S \cap T$ in $\mathcal{F}$. For the proof of the next result, it is convenient to consider $'$ as a mapping from 2^X to $\mathcal{F}$, assigning to every set its largest subset in $\mathcal{F}$.

Lemma 4.97 *Let $\mathcal{F}$ be closed under union. For any capacity μ on $\mathcal{F}$, μ is supermodular if and only if $\hat{\mu}$ is.*

Proof $\Rightarrow$) Let us take $S, T \in 2^X$, and consider S', T' the corresponding largest subsets of S, T in $\mathcal{F}$. Since $\mathcal{F}$ is union-closed, by supermodularity of μ, we have

$$\mu(S' \cup T') + \mu((S' \cap T')') \geqslant \mu(S') + \mu(T').$$

Since $S' \cup T' \subseteq (S \cup T)'$ and $(S' \cap T')' \subseteq (S \cap T)'$, monotonicity of μ implies

$$\mu((S \cup T)') + \mu((S \cap T)') \geqslant \mu(S') + \mu(T')$$

which gives

$$\hat{\mu}(S \cup T) + \hat{\mu}(S \cap T) \geqslant \hat{\mu}(S) + \hat{\mu}(T),$$

proving supermodularity of $\hat{\mu}$.

$\Leftarrow$) Take $S, T \in \mathcal{F}$. Then supermodularity of $\hat{\mu}$ reads

$$\hat{\mu}(S \cup T) + \hat{\mu}(S \cap T) \geqslant \hat{\mu}(S) + \hat{\mu}(T),$$

which can be rewritten as

$$\mu(S \cup T) + \mu((S \cap T)') \geqslant \mu(S) + \mu(T),$$

which is supermodularity of μ. □

Theorem 4.98 *Let $\mathcal{F}$ be a set system closed under union, and μ be a capacity on $(X, \mathcal{F})$. The following are equivalent:*

(i) For every function $f : X \to \mathbb{R}_+$,

$$\begin{aligned}\int_{\mathcal{F}} f \,\mathrm{d}\mu &= \max\Big\{\sum_{S\in\mathcal{F}} \lambda_S \mu(S) : \sum_{S\in\mathcal{F}} \lambda_S 1_S \leqslant f, \lambda \geqslant \mathbf{0}\Big\} \\ &= \min\Big\{\sum_{i\in X} P_i f_i : P(S) \geqslant \mu(S), \forall S \in \mathcal{F}, P \geqslant \mathbf{0}\Big\};\end{aligned}$$

(ii) $\int_{\mathcal{F}} \cdot \,\mathrm{d}\mu$ is superadditive;
(iii) μ is supermodular.

Proof (i)⇒(ii) We recognize in (i) the concave integral (see Sect. 4.11.2). Because the integral is also positively homogeneous, concavity is equivalent to superadditivity.

(ii)⇒(iii) By Theorem 4.95(iv), $\int f \, d\hat{\mu}$ is superadditive, which is equivalent to supermodularity of $\hat{\mu}$ by Theorem 4.35 (i) and (ii). We conclude using Lemma 4.97.

(iii)⇒(i) Suppose w.l.o.g. $f_1 \leqslant f_2 \leqslant \cdots \leqslant f_n$. By Theorem 4.95(iv), we know that

$$\int_{\mathcal{F}} f \, d\mu = \sum_{i=1}^{n} (f_i - f_{i-1})\hat{\mu}(\{i, \ldots, n\}) = \sum_{i=1}^{n} (f_i - f_{i-1})\mu(F_i)$$

with $f_0 = 0$, and $F_i = \{i, \ldots, n\}'$; i.e., the largest subset of $\{i, \ldots, n\}$ in $\mathcal{F}$. We claim that $\{F_1, F_2, \ldots, F_n\}$ is a chain in $\mathcal{F}$, with $F_1 \supseteq F_2 \supseteq \cdots \supseteq F_n$. Indeed, suppose that F_i and F_j, with $i < j$, are not comparable. Then $F_i \cup F_j \in \mathcal{F}$ and strictly includes F_i and F_j. By definition, F_i and F_j are subsets of $\{i, \ldots, n\}$, and therefore so is $F_i \cup F_j$. But this contradicts the choice of F_i as largest subset of $\{i, \ldots, n\}$.

Observe that several F_i's may be identical, and in this case they are necessarily consecutive. We define

$$\lambda_{F_i} = \sum_{\substack{j \geqslant i \\ F_j = F_i}} (f_j - f_{j-1}) \qquad (i = 1, \ldots, n),$$

and $\lambda_S = 0$ otherwise. We have to prove that λ is feasible and optimal. Clearly, $\lambda \geqslant \mathbf{0}$. Now, for any $i \in X$,

$$\sum_S \lambda_S 1_S(i) \leqslant \sum_{j=1}^{n} (f_j - f_{j-1}) 1_{\{j,\ldots,n\}}(i) = \sum_{j=1}^{i} (f_j - f_{j-1}) 1_{\{j,\ldots,n\}}(i) = f_i.$$

We prove optimality by complementary slackness. We propose P^* as solution of the dual defined by

$$P_i^* = \mu(F_i) - \mu(F_{i+1}) = \hat{\mu}(\{i, \ldots, n\}) - \hat{\mu}(\{i+1, \ldots, n\}) \qquad (i = 1, \ldots, n).$$

By Lemma 4.97, we know that $\hat{\mu}$ is supermodular. We recognize in P^* the marginal vector of $\hat{\mu}$ associated to permutation $n, n-1, \ldots, 1$ (Sect. 3.2.2). By (3.9), we deduce that $P^*(\{i, \ldots, n\}) = \hat{\mu}(\{i, \ldots, n\}) = \mu(F_i)$ for all i. We claim that

$P^*(\{i, \ldots, n\}) = P^*(F_i)$. Indeed, if $F_i = \{i, \ldots, n\}$ there is nothing to prove. Suppose then $j \in \{i, \ldots, n\} \setminus F_i$. Then $j \notin F_k$, $k \geqslant i$, and in particular $j \notin F_j, F_{j+1}$. It follows that $F_j = F_{j+1}$, hence

$$P_j^* = \mu(F_j) - \mu(F_{j+1}) = 0,$$

which proves the claim. Finally, P^* is feasible by Theorem 3.15(i) and (ii). As a conclusion, P^* is feasible for the dual program and has tight constraints for the nonzero variables of the primal, proving optimality. □

Remark 4.99 Our presentation of results and proofs fairly differs from the original paper [130]. In the latter, slightly more general results are proved, at the price of some more complex machinery to be introduced. In particular, a Monge-type algorithm,[13] which computes the Choquet integral, plays a central rôle in [130]. In fact, a simple version of the Monge algorithm is hidden in our proof of Theorem 4.95(iii).

See also Remark 3.33, giving a related work. There, $\hat{v}(c)$ is nothing other than the Choquet integral of c w.r.t. v in the above sense. ◇

[13]Gaspard Monge (Beaune, 1746 – Paris, 1818) is a French mathematician who is at the origin of descriptive geometry. He is also considered to be a pioneer of Operations Research, by his works on the optimal transportation of masses (*Mémoire sur la théorie des déblais et remblais, 1781*). Monge studied a geometric transportation problem in which a set of locations $s_1, \ldots, s_n$ of mass points has to be matched optimally (in the sense of minimizing the total cost) with another set of locations $t_1, \ldots, t_n$, and proved that optimality was reached if the transportation lines do not cross. This geometric fact can be expressed as follows: if the costs c_{ij} of matching objects s_i with t_j have the "uncrossing" property:

$$c_{ij} + c_{k\ell} \geq c_{\max(i,k),\max(j,\ell)} + c_{\min(i,k),\min(j,\ell)}$$

then the optimal matching is $(s_1, t_1), \ldots, (s_n, t_n)$. This is also called the "north-west corner rule." Translated into the language of set functions, the uncrossing property is in fact submodularity:

$$v(A) + v(B) \geqslant v(A \cup B) + v(A \cap B).$$

Chapter 5
Decision Under Risk and Uncertainty

This chapter opens the part of the book on applications of set functions in decision making. The foundations of decision making are mainly due to John von Neumann and Oskar Morgenstern, although the concepts of utility function and expected utility go back to Daniel Bernoulli and Blaise Pascal. The area of decision making which is addressed in this chapter is decision under risk and uncertainty. It deals with situations where the decision maker is faced with uncertainty: the consequences of his possible decisions depend on contingencies which are out of his control. The occurrence or the non-occurrence of these contingencies determine what is called the states of nature. If probabilistic information on the states of nature is available, one speaks of decision under risk. Otherwise, it is assumed that the decision maker has a personal, subjective probability measure on the states of nature in his mind, in which case one speaks of decision under uncertainty. While classical models solely rely on probability measures (expected utility), the observation of various paradoxes, unexplained by expected utility, has lead to considering capacities (viewed as nonadditive probabilities) and the Choquet integral in decision making. This chapter tries to show the emergence of these new models. It does not pretend to a full exposition of decision under risk and uncertainty, which would require a whole book. For this reason, and because already many textbooks exist on this subject, proofs of results are not given, except for some results which either are not so well known, or for which it is less easy to find comprehensive references. Recommended references for full exposition and details are Gilboa [153] and Wakker [339]; see also Wakker [335], Quiggin [270], Takemura [325] and a survey by Chateauneuf and Cohen [48]. The chapter ends with a presentation of qualitative decision making and the use of the Sugeno integral, a topic which is generally absent from monographs on decision making.

M. Grabisch, *Set Functions, Games and Capacities in Decision Making*, Theory and Decision Library C 46, DOI 10.1007/978-3-319-30690-2_5

5.1 The Framework

5.1.1 The Components of a Decision Problem

We consider a situation where an agent (called the *decision maker*, abbreviated by DM) has to make a choice between several possible actions. Each action has some *consequences* or *outcomes*, depending on what is usually called the *state of nature*. This expression means that the decision maker (DM) is unable to know everything, and contingencies, on which the DM has no control, determine the exact consequence of his action. We give in detail the components of a decision problem, fixing notation and some assumptions.

(i) We denote the set of states of nature by S. It can be finite or infinite. In the examples given below, S is most often finite, however it is easy to find examples with an infinite (and even uncountable) S: think of the unknown value of a continuous parameter, like temperature, the Dow Jones index, etc.

We make two important assumptions about S:

- S is exhaustive: the true (unknown) state of nature should belong to S. This is sometimes called the *closed world assumption*;
- The elements of S are mutually exclusive: only one of them can realize.

Subsets of S are called *events*. We say that an event $E \subseteq S$ is *true* or *realizes* if the actual state of nature belongs to E. We call S the *certain event* and $\varnothing$ the *impossible event.*

(ii) The set of consequences (outcomes) is denoted by C. For most decision problems considered in this chapter, consequences are real numbers (note that this is not the case for Example 5.3, however), which often represent amounts of money. Therefore, unless specified otherwise, we assume throughout the chapter that consequences are real numbers and identify $C = \mathbb{R}$.

(iii) We call an *act* a mapping $f : S \to C$, which represents a possible action of the decision maker, assigning to every state of nature a consequence. Acts are also called *prospects*. We make the assumption that the range of acts is finite; i.e., acts have only a finite number of outcomes. We denote the set of acts by F.

Because the range of an act f is finite, say, $\mathbf{ran} f = \{x_1, \ldots, x_n\}$, it induces a partition of S into events $E_1, \ldots, E_n$ given by $E_i = f^{-1}(x_i)$ for $i = 1, \ldots, n$. Therefore, act f can be denoted by $(E_1, x_1; \ldots; E_n, x_n)$.

Another traditional notation for acts is the following: suppose f, g are two acts, and $E \subset S$ is an event. Then $f_E g$ denotes the act equal to f on E and to g on $S \setminus E$; i.e.:

$$f_E g(s) = \begin{cases} f(s), & \text{if } s \in E \\ g(s), & \text{otherwise.} \end{cases}$$

(iv) We introduce a binary relation $\succcurlyeq$ on F, which is the *preference relation* of the decision maker over the set of acts. $f \succcurlyeq g$ reads "f is preferred to g" (with indifference allowed), or "f is at least as good as g". The preference relation is supposed to be a complete preorder, that is, complete and transitive (see Sect. 1.3.1). As usual, $\sim$ denotes the symmetric part of $\succcurlyeq$, and $f \sim g$ reads "f is indifferent to g", while $\succ$ denotes the asymmetric part, and $f \succ g$ reads "f is strictly preferred to g". To avoid trivialities, it is assumed throughout this chapter that $\succcurlyeq$ is nondegenerate; i.e., there exist two acts f, g such that $f \succ g$.

We give some examples to illustrate the previous notions.

Example 5.1 Janet is betting on a horse during a horse race. She bought a \$5 ticket, and if the horse on which she has bet wins, she will receive \$1000, otherwise nothing.

The set of states of nature is the set of horses, consequences are amounts of money, and acts are bets on horses: if horse 1 is chosen, the consequence is either \$995 if this horse wins, or −\$5 in the other cases. ◇

Example 5.2 Peter has to go to the university for an important meeting starting at 9:00. He commutes by car, but there are several possible ways to reach the university. The duration of the trip on each way depends on the state of the traffic on that road, and the possible occurrence of a car accident or any other event.

The possible actions of Peter are the different routes to reach the university. The states of nature are the possible states of the traffic on a given road (normal, accident, more or less big traffic jam, etc.), and the consequence of an action is the delay with which Peter arrives at his meeting (positive number if it is a delay, and negative if he arrives before the start of the meeting). ◇

Example 5.3 (Example 2.13 revisited) When leaving home in the morning, Leonard is wondering if he should take an umbrella or not, or even some raincoat. If the weather is sunny, it would be unnecessary to take an umbrella and quite unpleasant to wear a raincoat. If on the contrary the weather is rainy, an umbrella becomes necessary, and even a raincoat in case of heavy rain.

The states of nature are the different possibilities for the weather (sunny, some rain, heavy rain, etc.), and the acts are: to take nothing, to take an umbrella, and to take an umbrella and wear a raincoat. The consequences are, for instance, to be completely wet and catch a cold, to be safe, to sweat, to be hampered by an umbrella, etc. ◇

Remark 5.4 So far, we did not introduce any order on the set of consequences C. If C is a set of monetary values, the order is implicit, but this is not the case in general, even if $C = \mathbb{R}$. Observe that, identifying a consequence $x \in C$ with the constant act $x1_S$ yielding x for every state of nature, the preference relation $\succcurlyeq$ on F induces an ordering on C. ◇

The above components of a decision problem are supposed to be given.[1] An important aim in decision theory is to construct a numerical representation of the preference relation; i.e., to find a mapping $V : F \to \mathbb{R}$ satisfying

$$f \succcurlyeq g \Leftrightarrow V(f) \geqslant V(g) \qquad (f, g \in F). \tag{5.1}$$

The existence of such a function V is the main concern of measurement theory (see Sect. 6.2, and in particular Theorem 6.4). Instead of giving a general answer to this question, we propose several specific such functions V, and see under which conditions they exist. Since these conditions bear mainly on $\succcurlyeq$, that is, the preference relation of the DM, such results on the existence of a given function V are called *behavioral foundations* of V. A first obvious fact is that no mapping V can exist if $\succcurlyeq$ is not complete or not transitive, simply because $\geqslant$ is a complete and transitive relation on $\mathbb{R}$. This is why these conditions on $\succcurlyeq$ are primitive and are assumed henceforth.

We give a first property of V (respectively, of $\succcurlyeq$) that is considered as a basic rationality requirement. We say that V (respectively, $\succcurlyeq$) satisfies *monotonicity* if whenever $f(s) \geqslant g(s)$ for every state of nature $s \in S$, then $V(f) \geqslant V(g)$ (respectively, $f \succcurlyeq g$), and if $f(s) > g(s)$ for every $s \in S$ then $V(f) > V(g)$ (respectively, $f \succ g$). Strict monotonicity holds if in addition $f \neq g$ implies $V(f) > V(g)$ (respectively, $f \succ g$).

5.1.2 Introduction of Probabilities

The uncertainty bearing on the states of nature naturally leads to consider a probability measure P on S, endowing S with some algebra. Then, acts can be seen as real random variables, and a natural criterion is to choose an act maximizing the expected value, as far as consequences are considered as money amounts. This is called the *expected value* criterion.

[1]This is of course a kind of idealization, which hides the practical difficulties to obtain them. In particular, depending on the kind of problem considered, depending also if one has a descriptive, normative or constructive attitude, one cannot assume to know the preference relation of the DM on the entire set F. Often one tries to obtain a model from a limited knowledge of $\succcurlyeq$, with the help of some additional assumptions on the model or the preferences, which permit to determine *in fine* $\succcurlyeq$ on the whole set F. These considerations are however outside the scope of this chapter, and we refer the readers to, e.g., Wakker [339] and Takemura [325, Chap.1, Sect. 4].

There is a traditional distinction between two cases, which goes back to Frank Knight[2] (1921) [213]:

(i) An objective information on the probability distribution over the states of nature is available (e.g., based on statistics, combinatorics, etc.). Then one speaks of *decision under risk*;
(ii) No such objective information is available, only the subjective perception of uncertainty by the DM remains. This case is referred to as *decision under uncertainty*.

This distinction is well explained by John Keynes,[3] in a 1937 paper [207, pp. 213–214]:

> *By "uncertain" knowledge, let me explain, I do not mean merely to distinguish what is known for certain from what is only probable. The game of roulette is not subject, in this sense, to uncertainty; nor is the prospect of a Victory bond being drawn. Or, again, the expectation of life is only slightly uncertain. Even the weather is only moderately uncertain. The sense in which I am using the term is that in which the prospect of a European war is uncertain, or the price of copper and the rate of interest twenty years hence, or the obsolescence of a new invention, or the position of private wealth-owners in the social system in 1970. About these matters there is no scientific basis on which to form any calculable probability whatever. We simply do not know.*

The probability resulting from the subjective perception of uncertainty by the decision maker, whenever it exists, is called *subjective probability*.

5.1.3 Introduction of Utility Functions

A *utility function* is a mapping $u : \mathbb{R} \to \mathbb{R}$ (or more generally, $u : C \to \mathbb{R}$, where C is the set of consequences). Supposing that consequences are amounts of money, the utility function represents the subjective perception of wealth by the decision maker. Most often, utility functions are concave functions (like the logarithm), because the perceived difference of wealth usually diminishes as the amount of money is increasing: an increase from \$1000 to \$2000 is perceived as much more important than an increase from \$1,000,000 to \$1,001,000.

More generally, a utility function yields a numerical translation of the satisfaction of the DM regarding consequences, even nonnumerical (see Example 5.3).

---

[2]Frank Hyneman Knight (McLean County, Illinois, 1885 – Chicago, 1972) is an American economist, founder of the Chicago school. He is famous for having introduced the distinction between risk and uncertainty, but also for his debates on Pigou's social costs.

[3]John Maynard Keynes (Cambridge, 1883 – Firle, 1946) is a British economist and essayist, among the most influential theorists in economy of the twentieth century. He is the founder of Keynesian macroeconomics.

Utility functions are not given exogenously, they must be built from the preference relation of the decision maker.

5.2 Decision Under Risk

As explained above, decision under risk supposes that a probability measure is known on S, however the utility function on consequences is unknown.

To avoid intricacies, we make the following assumptions (same as in Wakker [339, Chap. 2]), repeating previous assumptions mentioned in Sect. 5.1:

(Structural assumptions for risk)

(i) Acts take finitely many values;
(ii) Preference over acts depend only on the probability distribution over the (finitely many) outcomes. Therefore, two acts $(E_1, x_1; \ldots; E_n, x_n)$ and $(E'_1, x_1; \ldots; E'_n, x_n)$ with different events but yielding the same probability distribution $(p_1, x_1; \ldots; p_n, x_n)$ are indifferent;
(iii) (richness) Every probability distribution taking finitely many values is available.

By Assumption (ii), it is enough to describe acts through their induced probability distribution on outcomes. This is why in decision under risk acts are identified with *lotteries*. Indeed, a lottery yields a finite number of possible outcomes $x_1, \ldots, x_n$, each of them occurring with a known probability $p_1, \ldots, p_n$, respectively.

The richness assumption implies that the underlying set of states of nature is infinite (since the number of outcomes n is not fixed). In summary, the set of lotteries we consider, denoted by L, and on which bears the preference relation $\succcurlyeq$, is the set of all probability distributions with finite support on $\mathbb{R}$ (they correspond to *simple probability measures* in Fishburn [141, Chap. 8]). The usual representation of a lottery is given on Fig. 5.1. A particular case of interest is the *sure lottery*, which yields some outcome α with certainty. In what follows, we write simply α instead of $(1, \alpha)$.

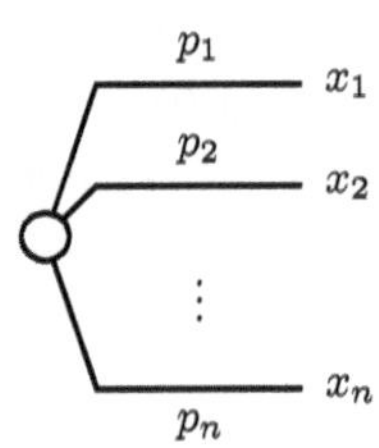

Fig. 5.1 Representation of the lottery $(p_1, x_1; p_2, x_2; \ldots; p_n, x_n)$

Given two lotteries $p = (p_1, x_1; \ldots; p_n, x_n)$, $q = (q_1, x_1; \ldots; q_n, x_n)$ on the same set of consequences,[4] one can define their convex combination or *mixture*, using some $\lambda \in [0, 1]$. The mixture $\lambda p + (1 - \lambda)q$ is a new lottery on the outcomes $x_1, \ldots, x_n$ defined by

$$\lambda p + (1 - \lambda)q = (\lambda p_1 + (1 - \lambda)q_1, x_1; \ldots; \lambda p_n + (1 - \lambda)q_n, x_n).$$

The mixture of lotteries can be considered as a lottery whose consequences are themselves lotteries (Fig. 5.2).

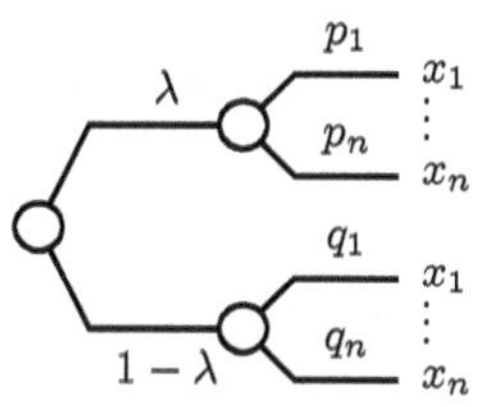

Fig. 5.2 Mixture of two lotteries p, q

5.2.1 *The Expected Utility Criterion*

Supposing that the utility function u has been determined, the *expected utility criterion* consists in choosing the lottery (act) $p = (p_1, x_1; \ldots, ; p_n, x_n)$ with the highest expected utility $\text{EU}(p)$:

$$\text{EU}(p) = \sum_{i=1}^{n} p_i u(x_i). \tag{5.2}$$

Expected utility is a simple and natural criterion, because it yields the average utility the decision maker can expect in the long run, taking into account all possible events and their probabilities. Let us give two behavioral foundations of expected utility. The first one is founded on the famous independence axiom.

Definition 5.5 A preference relation $\succcurlyeq$ on L satisfies *independence* if for all lotteries $p, q \in L$ such that $p \succ q$, the following holds

$$\lambda p + (1 - \lambda)r \succ \lambda q + (1 - \lambda)r \qquad (\lambda \in]0, 1[, r \in L).$$

[4] W.l.o.g. Indeed, any two lotteries can always be rewritten as two lotteries on the same set of consequences, by assigning zero probabilities on the missing outcomes.

A second property is needed: We say that $\succcurlyeq$ on L satisfies the *Archimedean property* if for all lotteries $p, q, r \in L$ such that $p \succ q \succ r$, there exist $\lambda, \mu \in]0, 1[$ such that

$$\lambda p + (1 - \lambda)r \succ q \succ \mu p + (1 - \mu)r. \tag{5.3}$$

The independence axiom looks rather natural[5] (a mixture with any lottery does not change the preference), but the Archimedean axiom is, however, less transparent. It is a continuity axiom, saying in substance that no matter how r is bad (respectively, good), and therefore could damage the lottery p (respectively, could improve p), it is always possible to find λ close enough to 1 (respectively, μ close enough to 0) so that the left-hand side (respectively, the right-hand side) inequality holds in (5.3).

Theorem 5.6 (Axiomatization of expected utility for risk (1st version)) *(Fishburn [141, Theorem 8.2]) Let $\succcurlyeq$ be a binary relation on L. Under the above structural assumptions for risk, the following two propositions are equivalent.*

(i) $\succcurlyeq$ is complete and transitive, and satisfies independence and the Archimedean property;
(ii) There exists a utility function u such that expected utility represents $\succcurlyeq$ (i.e., (5.1) is satisfied with $V = \mathrm{EU}$). Moreover, u is unique up to a positive affine transformation; i.e., of the form $\alpha u + \beta$, with $\alpha > 0$, $\beta \in \mathbb{R}$.

Remark 5.7 A more general version where probability measures need not be simple was proved by Fishburn [141, Theorem 8.4], completing results of von Neumann and Morgenstern [333]. Also, the above result holds for any set of outcomes, not necessarily $\mathbb{R}$. Therefore, u is not supposed to be increasing. For a proof, see the above references, or also Gilboa [153, Chap. 8]. ◇

The second axiomatization is less well-known but is more intuitive. It is based on the notion of *standard gamble*. Fix two outcomes $M, m \in \mathbb{R}$ such that $M > m$, and set $u(M) = 1$, $u(m) = 0$. Given $M > \alpha > m$, find a probability p such that

$$\alpha \sim (p, M; 1 - p, m),$$

i.e., there is indifference between the sure lottery yielding α and the lottery yielding M with probability p, and m otherwise (Fig. 5.3). Now, the *standard gamble*

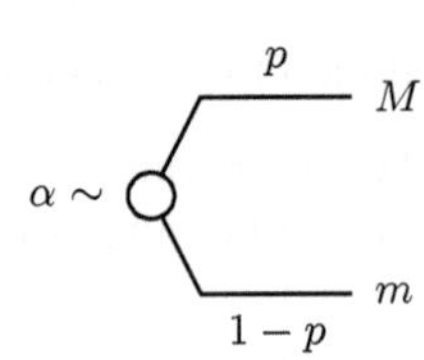

Fig. 5.3 A standard gamble

[5] Although highly controversial; see, e.g., [339, Sect. 2.7].

solvability condition stipulates that p exists for every $\alpha \in]m, M[$. Next, *standard gamble dominance* holds if for every $M > m$ and probabilities $p > q$, we have

$$(p, M; 1-p, m) \succ (q, M; 1-q, m).$$

The last property involves mixtures of lotteries. *Standard gamble consistency* says that for all outcomes α, M, m, all $\lambda \in [0, 1]$ and all lotteries $l \in L$, if $\alpha \sim (p, M; 1-p, m)$, then

$$\lambda\alpha + (1-\lambda)l \sim \lambda(p, M; 1-p, m) + (1-\lambda)l$$

should hold as well (Fig. 5.4).

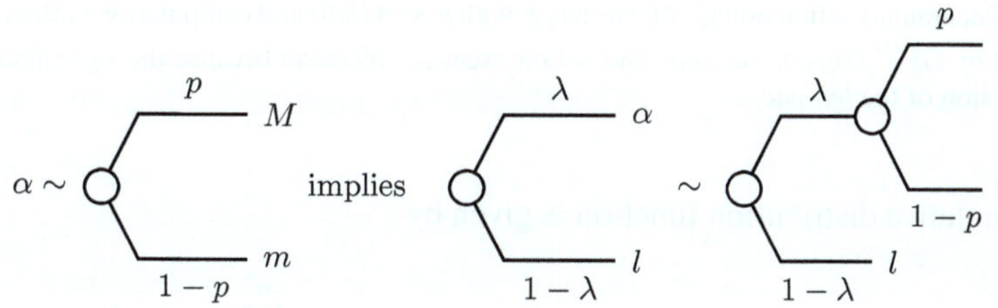

Fig. 5.4 Standard gamble consistency

Theorem 5.8 (Axiomatization of expected utility for risk (2nd version)) *(Wakker [339, Theorem 2.6.3]) Let $\succcurlyeq$ be a binary relation on L. Under the above structural assumptions for risk, the following two propositions are equivalent.*

(i) $\succcurlyeq$ is complete and transitive, and satisfies standard gamble solvability, standard gamble dominance, and standard gamble consistency;

(ii) There exists an increasing utility function u such that expected utility represents $\succcurlyeq$. Moreover, u is unique up to a positive affine transformation.

5.2.2 Stochastic Dominance

Stochastic dominance is the second example of basic rationality requirement that any preference relation or decision model should satisfy. Basically, it says that shifting probability masses to higher outcomes should improve the act.

Let us represent lotteries by their decumulative distribution function (see Sect. 4.2). For a lottery $p = (p_1, x_1; \ldots; p_n, x_n)$ with $0 < x_1 < x_2 < \cdots < x_n$,

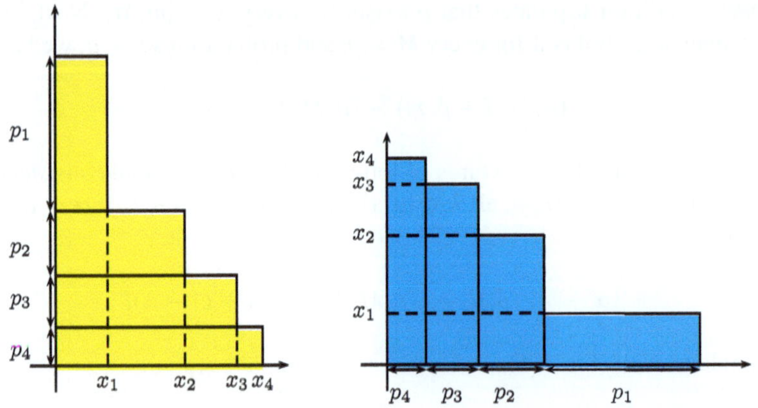

Fig. 5.5 Decumulative function G_p of a lottery p with $n = 4$ (*left*) and computation of the expected value $\mathbb{E}(p)$ by $G_p^{(-1)}$ (*right*). The *blue* and *yellow* areas are identical because the *right figure* is the flipped version of the *left one*

its decumulative distribution function is given by

$$G_p(x) = \text{Prob}(\text{lottery } p \text{ yields an outcome } \geqslant x) = \sum_{i=i(x)+1}^{n} p_i \qquad (x \in \mathbb{R}_+),$$

with $i(x) = \max\{i \in \{0, \ldots, n\} \mid x > x_i\}$ for $x > 0$, $i(0) = 0$, and $x_0 = 0$ [Fig. 5.5 (left)].

We say that a lottery p *stochastically dominates (at first order)*[6] a lottery q if $G_p(x) \geqslant G_q(x)$ for every $x \in \mathbb{R}_+$, with a strict inequality for at least one x, which is written in short by $G_p > G_q$ [see Sect. 1.1(ix)]. Then, a preference relation $\succcurlyeq$ (or its numerical representation V) satisfies *stochastic dominance* if for all lotteries p, q, $G_p > G_q$ implies $p \succcurlyeq q$. Stochastic dominance is *strict* if $p \succ q$ replaces the latter inequality.

We show that expected utility satisfies stochastic dominance if the utility function u is nondecreasing (in the strict sense if u is increasing). Let us consider the pseudo-inverse[7] of G_p denoted by $G_p^{(-1)}$; i.e., the quantile function of p. It is easy to see from Fig. 5.5 that the area under the decumulative function of a lottery $p = (p_1, x_1; \ldots; p_n, x_n)$ is equal to its expected value $\mathbb{E}(p)$, given by the area under

[6] This definition comes from the more general definition given in Chap. 4 (Definition 4.23).

[7] G_p not being a bijection because of plateaus, its inverse does not exist. It is however easy to define a "pseudo-inverse" by assigning to the inverse value of the height of a plateau some arbitrary value in the segment representing the plateau. See Denneberg [80, Chap. 1] for a formal definition.

the quantile function. Indeed,

$$\text{area} = \sum_{i=1}^{n}(x_{i+1} - x_i)\sum_{j=i}^{n} p_j = \sum_{i=1}^{n} p_i x_i = \mathbb{E}(p). \tag{5.4}$$

It follows that if p stochastically dominates q, that is, $G_p > G_q$, then we have also $G_p^{(-1)} > G_q^{(-1)}$, hence for every nondecreasing utility function u, $G_{p'}^{(-1)} \geqslant G_{q'}^{(-1)}$, with $p' = (p_1, u(x_1); \ldots; p_n, u(x_n))$ and similarly for q', which yields $\mathrm{EU}(p) \geqslant \mathrm{EU}(q)$ (strict equality is obtained when u is increasing). As a conclusion, expected utility satisfies stochastic dominance (respectively, strict stochastic dominance), as soon as the utility function is nondecreasing (respectively, increasing). The converse also holds.

5.2.3 Risk Aversion

A fundamental notion in decision under risk is risk aversion. In substance, it says that risk averse people always prefer sure lotteries to any lottery yielding the same expected value. Formally, a preference relation $\succcurlyeq$ exhibits *risk aversion* if for every consequence $\alpha \in \mathbb{R}$, $\alpha \succcurlyeq p$ for every lottery $p \in L$ such that $\mathbb{E}(p) = \alpha$. If the reverse preference holds, then $\succcurlyeq$ exhibits *risk seeking*, and if indifference always holds, then we speak of *risk neutrality*.

Under expected utility, risk aversion is characterized by concavity of the utility function.

Theorem 5.9 (Risk aversion in EU) *Under the expected utility criterion, risk aversion holds if and only if the utility function u is concave; i.e.,*

$$u(\lambda x + (1-\lambda)x') \geqslant \lambda u(x) + (1-\lambda)u(x') \qquad (x, x' \in \mathbb{R}, \lambda \in [0,1]). \tag{5.5}$$

Proof If $\succcurlyeq$ is risk averse, we must have $\mathbb{E}(p) \succcurlyeq p$ for every lottery p, i.e.,

$$\mathrm{EU}(\mathbb{E}(p)) \geqslant \mathrm{EU}(p)$$

or, letting $p = (p_1, x_1; \ldots; p_n, x_n)$,

$$u\Big(\sum_{i=1}^{n} p_i x_i\Big) \geqslant \sum_{i=1}^{n} p_i u(x_i),$$

which is equivalent to the concavity of u. □

This well-known result is quite surprising. It says that risk aversion, which is a phenomenon intimately related to the perception of probability (the underlying uncertainty), is entirely modelled by the utility function, which is supposed to model the perception of wealth! This suggests that the expected utility model is not sufficiently sophisticated in this respect. The next section shows an example where expected utility is unable to model a situation of risk aversion.

5.2.4 The Allais Paradox

In 1953, Maurice Allais[8] presented a choice problem among lotteries, where most people violate expected utility [8]. The version presented here is with two outcomes (Allais also presented a formulation with three outcomes. Many other variations can be found in [203]).

Consider the following two lotteries p, q:

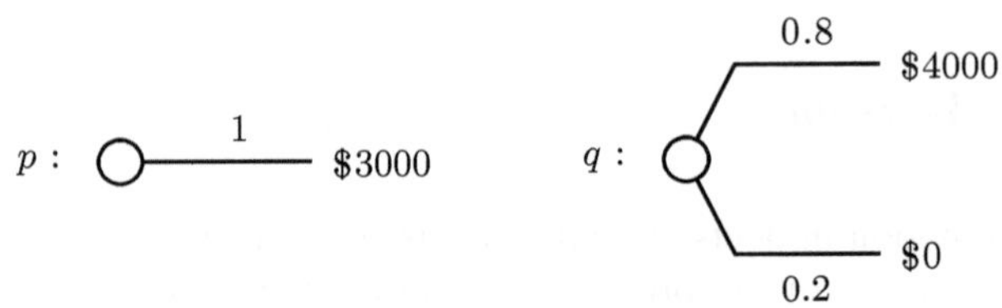

p being a sure lottery, most people have the preference $p \succ q$, because a sure substantial gain is preferable to a risky gamble, where the surplus of \$1000 is not considered as enough attractive compared to the possibility of a zero gain. In a second step, the following choice is proposed:

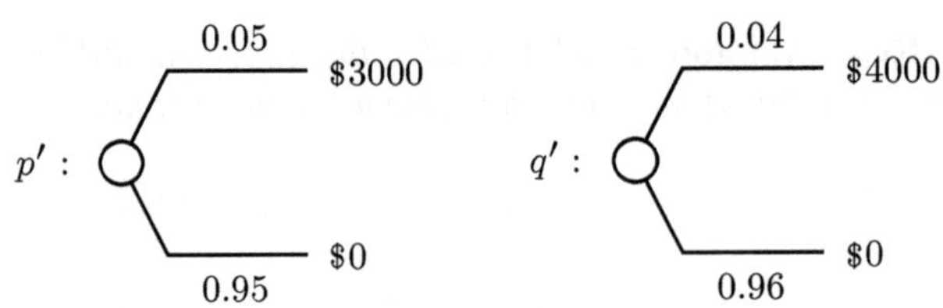

Here, most people prefer q' to p', because the probability of winning \$4000 with q' is almost the same as the probability of winning \$3000 with p'. However, a decision maker following expected utility should prefer on the contrary p' to q' if the previous preference was $p \succ q$. Indeed,

$$p \succ q \Leftrightarrow \mathrm{EU}(p) > \mathrm{EU}(q)$$

[8]Maurice Félix Charles Allais (Paris, 1911 – Saint-Cloud, 2010) is a French economist, famous for his contributions to decision theory, behavioral economics and monetary policy. He received the Nobel Prize in economics in 1988.

$$\Leftrightarrow u(3000) > 0.2u(0) + 0.8u(4000)$$
$$\Leftrightarrow 0.05u(3000) > 0.01u(0) + 0.04u(4000)$$
$$\Leftrightarrow 0.05u(3000) + 0.95u(0) > 0.96u(0) + 0.04u(4000)$$
$$\Leftrightarrow \mathrm{EU}(p') > \mathrm{EU}(q')$$
$$\Leftrightarrow p' \succ q'.$$

Hence, most people violates expected utility in this experiment.

5.2.5 Transforming Probabilities

The reason of the failure of expected utility in the Allais paradox lies in the misrepresentation of risk aversion. Expected utility can represent risk aversion only through the concavity of the utility function, but obviously risk aversion has little to do with the perception of wealth. This is why psychologists in the 50s (for example Edwards [123]) have developed the idea to transform probabilities instead of transforming wealth to represent this phenomenon. We will see that a transformation of probability done in a naive way leads to the violation of stochastic dominance, which makes it inapplicable to decision making. Our exposition follows Wakker [336] (also in [339, Sects. 5.2 and 5.3]).

Consider a lottery $p = (p_1, x_1; \ldots, ; p_n, x_n)$. The idea of Edwards [123] to represent risk aversion is to apply a *distortion function* φ to the probability distribution, i.e., an increasing bijection on [0, 1], instead of applying a utility function on the outcomes. Hence the numerical representation of preference would be the function

$$V(p) = \sum_{i=1}^{n} x_i \varphi(p_i) \qquad (p \in L). \tag{5.6}$$

If φ is not the identity function, there necessarily exist $p', p'' \in [0, 1]$ such that

$$\varphi(p' + p'') \neq \varphi(p') + \varphi(p'').$$

Suppose for example that $\varphi(p' + p'') > \varphi(p') + \varphi(p'')$ (a similar argument can be built if the reverse inequality holds). Consider a lottery $p = (p_1, x_1; \ldots ; p_n, x_n)$, with $x_1 < x_2 < \cdots < x_n$, and $p_{n-1} = p'$, $p_n = p''$. The quantity $V(p)$ is the yellow area on Fig. 5.6(a). Let us diminish the quantity x_n gradually till reaching x_{n-1}. When x_n remains strictly greater than x_{n-1} (Fig. 5.6(b), (c)), nothing special happens: the yellow area gradually diminishes, and so does $V(p)$. When x_n hits x_{n-1} (Fig. 5.6(d)), $\varphi(p')+\varphi(p'')$ is replaced by $\varphi(p'+p'')$, which creates a sudden augmentation of area, hence $V(p)$ increases although one of the outcomes of the lottery has decreased.

This is a violation of stochastic dominance, making the principle of probability transformation untenable for decision making.

5.2.6 Rank Dependent Utility

The reason of the failure is due to the fact that the probabilities p_i are transformed individually. If partial sums $p_n, p_n + p_{n-1}, p_n + p_{n-1} + p_{n-2}$, etc., are transformed instead, the problem disappears, as it becomes clear from Fig. 5.7. Note in addition that this amounts to considering the transformation of the decumulative distribution function G_p by φ instead of the transformation of p itself. The lottery is then evaluated by the area under $\varphi \circ G_p$:

$$
\begin{aligned}
V(p) &= \underbrace{\varphi(p_1 + \cdots + p_n)}_{=1} x_1 + \varphi(p_2 + \cdots + p_n)(x_2 - x_1) + \cdots \\
&\quad + \varphi(p_{n-1} + p_n)(x_{n-1} - x_{n-2}) + \varphi(p_n)(x_n - x_{n-1}) \qquad (5.7)\\
&= \big(1 - \varphi(p_2 + \cdots + p_n)\big)x_1 + \big(\varphi(p_2 + \cdots + p_n) - \varphi(p_3 + \cdots + p_n)\big)x_2 \\
&\quad + \cdots + \varphi(p_n)x_n. \qquad (5.8)
\end{aligned}
$$

Since the problem of preserving stochastic dominance with transforming probabilities has been resolved, we can reintroduce utility functions in our model and get a general expression for any lottery $(p_1, x_1; \ldots; p_n, x_n) \in L$. Recall that we assumed $x_1 < \cdots < x_n$. In the general case, let us take a permutation σ on $\{1, \ldots, n\}$ such that $u(x_{\sigma(1)}) \leqslant \cdots \leqslant u(x_{\sigma(n)})$ where u is a given increasing and continuous utility function, and denote by P the probability measure on $(\{1, \ldots, n\}, 2^{\{1,\ldots,n\}})$ associated to the distribution $p_1, \ldots, p_n$, given by $P(A) = \sum_{i \in A} p_i$ for any $A \subseteq \{1, \ldots, n\}$. A rewriting of (5.7) and (5.8) with the above notation leads to

$$
\begin{aligned}
V(p) &= \sum_{i=1}^{n} \big(u(x_{\sigma(i)}) - u(x_{\sigma(i-1)})\big)\varphi \circ P(\{\sigma(i), \ldots, \sigma(n)\}) \qquad (5.9)\\
&= \sum_{i=1}^{n} u(x_{\sigma(i)})\big(\varphi \circ P(\{\sigma(i), \ldots, \sigma(n)\}) - \varphi \circ P(\{\sigma(i+1), \ldots, \sigma(n)\})\big), \qquad (5.10)
\end{aligned}
$$

with $u(x_{\sigma(0)}) = 0$. A comparison with (4.26) and (4.27) letting $\mu = \varphi \circ P$ clearly shows that

$$
V(p) = \int (u(x_1), \ldots, u(x_n))\, \mathrm{d}(\varphi \circ P), \qquad (5.11)
$$

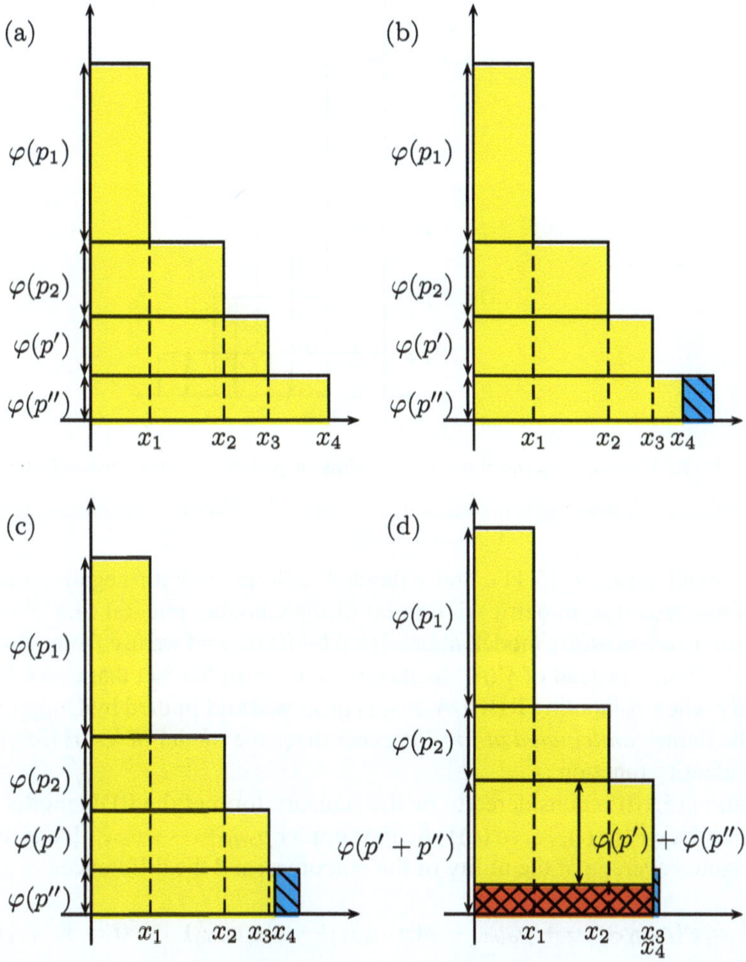

Fig. 5.6 (**a**) Computation of $V(p)$ with a distortion function φ; $n = 4$: The *yellow area* represents $V(p)$. (**b**) x_4 has diminished: the *blue hatched area* is removed from $V(p)$. (**c**) Further diminution of x_4: another *blue hatched area* is removed from $V(p)$. (**d**) x_4 hits x_3: another small *blue area* is removed but the *red crosshatched area* appears because of the difference $\varphi(p'+p'')-\varphi(p')-\varphi(p'')$

i.e., $V(p)$ is the Choquet integral of the utility of the outcomes of the lottery w.r.t. the distorted probability $\varphi \circ P$.

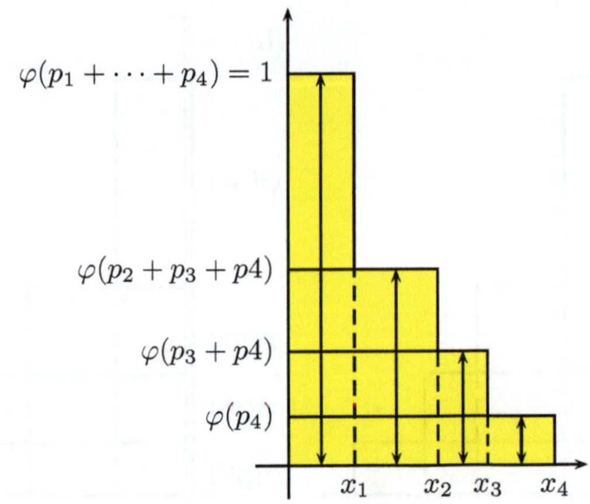

Fig. 5.7 The RDU model: transforming partial sums of probabilities (G_p) instead of probabilities (p)

The model given in (5.11), and extended to lotteries with negative outcomes through the usual (asymmetric) definition of the Choquet integral (4.13), is called the *rank dependent utility* model, abbreviated by RDU, and we use from now on the notation RDU(p) instead of $V(p)$. Its name comes from the fact that a reordering is necessary when computing RDU(p). It was proposed and studied by Quiggin [270], under the name "*anticipated utility*." It generalizes the model of Yaari [350] where u is the identity function.

Equation (5.10) is considered to be the standard form of the RDU model, where we next replace $P(\{\sigma(i),\ldots,\sigma(n)\})$ by the simpler $p_{\sigma(i)}+\cdots+p_{\sigma(n)}$. It has the form of a weighted average of the utility of the outcomes, and the differences

$$w_i^\sigma = \varphi(p_{\sigma(i)}+\cdots+p_{\sigma(n)}) - \varphi(p_{\sigma(i+1)}+\cdots+p_{\sigma(n)}) \qquad (i=1,\ldots,n)$$

are called the *decision weights*.

Example 5.10 (Allais paradox solved) Let us see how RDU can solve the Allais paradox, and take for simplicity u to be the identity function. This will show in addition that risk aversion can be accommodated without using utility functions. The preference $p \succ q$ translates into

$$3000 > 4000\varphi(0.8)$$

hence we should have $0 < \varphi(0.8) < 0.75$. Next, the preference $q' \succ p'$ imposes

$$3000\varphi(0.05) < 4000\varphi(0.04)$$

i.e., $\varphi(0.04)/\varphi(0.05) \in]0.75, 1[$. This is achieved by taking, e.g., $\varphi(0.05) = 0.05$ and $\varphi(0.04) = 0.04$. ◇

Let us examine how the distortion function φ can model optimism and pessimism (which can be seen as a form of risk aversion or risk seeking), depending on whether it is concave or convex. If φ is convex (like $\varphi(p) = p^2$), the decision weights are decreasing with i, and therefore emphasis is put on the outcomes with low values, which expresses pessimism or risk aversion. On the contrary, if φ is concave (like $\varphi(p) = \sqrt{p}$), the decision weights are increasing with i, showing optimism (Fig. 5.8).

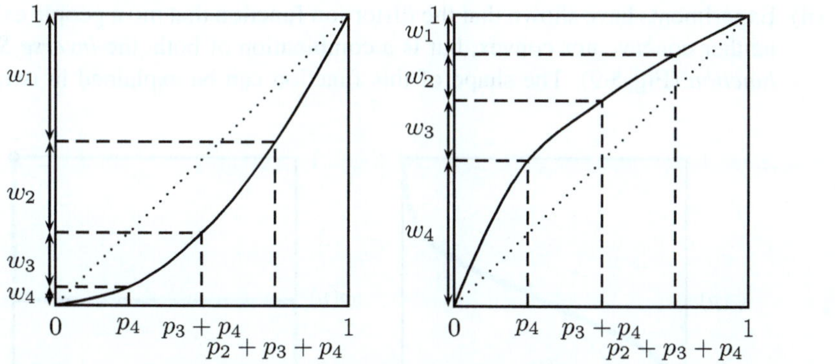

Fig. 5.8 Convex distortion function φ (*left*) and concave distortion function (*right*), with $n = 4$. The permutation σ is the identity function and is therefore omitted

Remark 5.11

(i) The question of studying risk aversion in the RDU model, and how to define it in a proper way, is in fact fairly complex. If we keep our definition of Sect. 5.2.3, that is, risk averse DMs' prefer to any lottery its expected value, while imposing u to be the identity function (i.e., Yaari's model), we immediately see that a necessary and sufficient condition for risk aversion is that, for any lottery $p = (p_1, x_1; \ldots; p_n, x_n)$ with $x_1 \leqslant \cdots \leqslant x_n$,

$$\mathbb{E}(p) \succ p \Leftrightarrow \mathrm{RDU}(\mathbb{E}(p)) > \mathrm{RDU}(p)$$

$$\Leftrightarrow \sum_{i=1}^{n} x_i p_i > \sum_{i=1}^{n} x_i\big(\varphi(p_i + \cdots + p_n) - \varphi(p_{i+1} + \cdots + p_n)\big).$$

In particular, taking binary lotteries with $n = 2$ and $x_1 = 0$, the last equivalence becomes

$$p_2 > \varphi(p_2)$$

for any $p_2 \in [0, 1]$. In other words, the distortion function should be below the diagonal, which is the case for any convex function. Chateauneuf and Cohen [47] have proved that this remains true for any differentiable u. They showed also other properties of φ and u implied by risk aversion. In particular, it is possible to have risk aversion with a convex function u (which tends to induce risk seeking behavior), provided that φ is sufficiently pessimistic in the above sense.

Chew et al. [51] proposed a stronger definition of risk aversion (being aversion to mean-preserving spreads: for a fixed mean, lotteries with narrow spread are preferred), which leads to the following result: (strong) risk aversion is equivalent to having a concave utility and a convex probability distortion.

(ii) Experiments have shown that the distortion function that most people exhibit is neither concave nor convex, but is a combination of both, the *inverse S-shape function* (Fig. 5.9). The shape of this function can be explained in two ways.

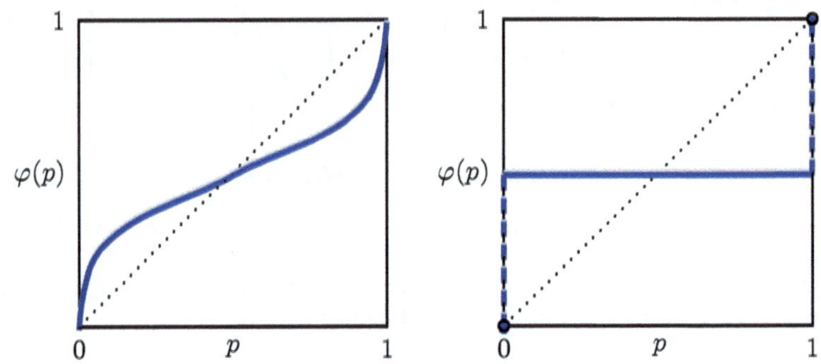

Fig. 5.9 Inverse S-shape distortion function (*left*) and its extreme case (*right*)

First, it shows the phenomenon of *likelihood insensitivity*, that is, people have only a rough perception of uncertainty, which extreme case can be seen in Fig. 5.9 (right). There, an individual is only able to distinguish between "sure to happen," "sure not to happen" and "don't know." In general, people have less sensitivity for events to be more or less certain (the flat region in the middle of the left curve), and are overly sensitive to change from impossible to possible (near $p = 0$) and from possible to certain (near $p = 1$).

Another interpretation can be done in terms of pessimism and optimism, recalling our discussion above (Fig. 5.8). Indeed, the inverse S-shape curve is a combination of a concave function (showing optimism) for low probabilities and a convex function (showing pessimism) for high probabilities. The lower part corresponds to favorable outcomes: best outcomes are overweighted (possibility effect: the probability to get the best outcome is small, but it is possible), and so are also worst outcomes in the upper part of the curve (certainty effect: the worst outcome represents what the DM is guaranteed to get at least). The inverse S-shape function then explains why an individual can gamble, hoping to win big prizes with very low probability, and at the same time can take insurance against very unfavorable outcomes, although quite improbable. We refer the readers to Wakker [339, Chap. 7] and Takemura [325, Chap. 8, Sect. 3] for a full discussion of this topic.

◇

Let us give a behavioral foundation of RDU. To this end we introduce the following property. For outcomes $\alpha, \beta, \gamma, \delta$, we say that they are related by *rank tradeoff indifference* if there exist a probability $p > 0$, consequences $x_2, \ldots, x_n$ and $y_2, \ldots, y_n$, probabilities $p_2, \ldots, p_n$ and $q_2, \ldots, q_n$ such that the lotteries $p_\alpha = (p, \alpha; p_2, x_2; \ldots; p_n, x_n)$, $q_\beta = (p, \beta; q_2, y_2; \ldots; q_n, y_n)$, $p_\gamma = (p, \gamma; p_2, x_2; \ldots; p_n, x_n)$ and $q_\delta = (p, \delta; q_2, y_2; \ldots; q_n, y_n)$ satisfy

$$p_\alpha \sim q_\beta \text{ and } p_\gamma \sim q_\delta,$$

and $\alpha, \beta, \gamma, \delta$ are such that the probability to be better are the same in their respective lotteries; i.e.,

$$\begin{aligned}\text{Prob(outcome of } p_\alpha > \alpha) &= \text{Prob(outcome of } q_\beta > \beta)\\ &= \text{Prob(outcome of } p_\gamma > \gamma) = \text{Prob(outcome of } q_\delta > \delta).\end{aligned}$$

We denote this relation as $(\alpha, \beta) \sim_{\mathrm{r}}^{\mathrm{t}} (\gamma, \delta)$, where t, r stand for "tradeoff" and "rank," respectively. It is straightforward to check that under RDU,

$$(\alpha, \beta) \sim_{\mathrm{r}}^{\mathrm{t}} (\gamma, \delta) \Rightarrow u(\alpha) - u(\beta) = u(\gamma) - u(\delta),$$

where u is the utility function of the RDU model. Now, we say that a preference relation $\succcurlyeq$ satisfies *rank tradeoff consistency* if whenever $(\alpha, \beta) \sim_{\mathrm{r}}^{\mathrm{t}} (\gamma, \delta)$, improving any outcome among $\alpha, \beta, \gamma, \delta$ breaks that relationship.

Moreover, we say that $\succcurlyeq$ is *continuous* if for any $n \in \mathbb{N}$, for any probability distribution $(p_1, \ldots, p_n)$, the sets of n-dim vectors of outcomes

$$\{(x_1, \ldots, x_n) : (p_1, x_1; \ldots; p_n, x_n) \succcurlyeq (p_1, y_1; \ldots; p_n, y_n)\} \text{ and}$$
$$\{(x_1, \ldots, x_n) : (p_1, x_1; \ldots; p_n, x_n) \preccurlyeq (p_1, y_1; \ldots; p_n, y_n)\}$$

are closed subsets of $\mathbb{R}^n$ for every $(y_1, \ldots, y_n) \in \mathbb{R}^n$.

Theorem 5.12 (Axiomatization of RDU) *(Wakker [339, Theorem 6.5.6]) Let $\succcurlyeq$ be a binary relation on L. Under the above structural assumptions for risk, the following two propositions are equivalent.*

(i) $\succcurlyeq$ is complete and transitive, is continuous, and satisfies strict stochastic dominance and rank tradeoff consistency;

(ii) There exist an increasing and continuous utility function u (unique up to a positive affine transformation) and a distortion function φ such that RDU given by (5.11) represents $\succcurlyeq$.

5.2.7 Prospect Theory

If RDU provides a fairly flexible model of decision under risk, able to represent risk aversion in a proper way, still some drawbacks exist, in particular related to what is called *reference dependence*, *loss aversion*, and *reflection effect*. It has been observed that people evaluate acts according to some reference level usually corresponding to the *status quo*; i.e., what describes their present situation. For example, the same object (like an apartment, a car, etc.) is felt differently if it is better or worse than the present apartment or car that the person possesses. In the first case, the new apartment is considered to be a gain, while it is felt as a loss in the second case. In short, the perception of outcomes is no longer absolute but relative to the present state. In addition, it has been observed that people exhibit loss aversion: they are more sensitive to losses than to gains. For example, a person earning \$1000 some day and then losing \$1000 the day after feels much more miserable than the status quo (no earning, no loss). Also, it has been observed that, by a kind of reflection phenomenon, people become risk seeking with losses.[9] This can be seen in the experiment described on Fig. 5.10.

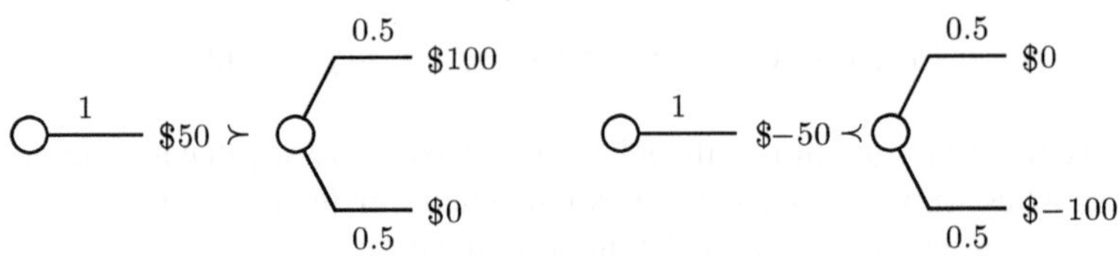

Fig. 5.10 Most people are risk averse for gains and exhibit the preference shown on the *left*. On the *right figure* where the outcomes become negative, attitude for risk is reversed

These observations have lead Kahneman and Tversky to introduce *prospect theory* [203, 329]. The original 1979 prospect theory in [203] was, roughly speaking, like expected utility where reference levels and loss aversion have been incorporated. The 1992 version (called "cumulative prospect theory") corrected the mathematical flaws of the first one and incorporated rank-dependency. We consider only the latter in this section, and call it simply "prospect theory" (abbreviated by PT), as it is the usage now.

[9] Although it seems that for losses, the behavior is closer to expected value maximization; see Wakker [339, Sect. 9.5] and Takemura [325, Chap. 8, Sect. 1] for more details.

We suppose for ease of notation that the reference point (status quo) is the outcome 0. The evaluation of lotteries under cumulative prospect theory is done by the real-valued function PT on L, defined as follows. For any lottery $p \in L$, list its outcomes in increasing order and number them as follows:

$$x_1 \leqslant \cdots \leqslant x_m < 0 \leqslant x_{m+1} \leqslant \cdots \leqslant x_{m+n}.$$

The probabilities of the outcomes are denoted accordingly: $p_1, \ldots, p_m$, $p_{m+1}, \ldots, p_{m+n}$. Let u be a strictly increasing and continuous utility function, with $u(0) = 0$, and two distortion functions φ^+, φ^-. The evaluation of p by PT is:

$$\mathrm{PT}(p) = \sum_{i=1}^{m} u(x_i)\big(\varphi^-(p_1 + \cdots + p_i) - \varphi^-(p_1 + \cdots + p_{i-1})\big) + \\ \sum_{i=m+1}^{m+n} u(x_i)\big(\varphi^+(p_i + \cdots + p_{m+n}) - \varphi^+(p_{i+1} + \cdots + p_{m+n})\big). \quad (5.12)$$

Note that RDU is recovered when only nonnegative outcomes are considered. In addition, the above expression can be expressed by the difference of two Choquet integrals:

$$\mathrm{PT}(p) = \int (\underbrace{0, \ldots, 0}_{m}, u(x_{m+1}), \ldots, u(x_{m+n}))\, \mathrm{d}(\varphi^+ \circ P) \\ - \int (-u(x_1), \ldots, -u(x_m), \underbrace{0, \ldots, 0}_{n})\, \mathrm{d}(\varphi^- \circ P), \quad (5.13)$$

where the $(m + n)$-tuples represent the integrand. Note that if $\varphi^+ = \varphi^- = \varphi$, the above expression is nothing but the symmetric Choquet integral of $(u(x_1), \ldots, u(x_m), u(x_{m+1}), \ldots, u(x_{m+n}))$ w.r.t. the capacity $\varphi \circ P$ [see (4.10); also compare (5.12) with (4.35)].

Remark 5.13 Most frequently used utility functions are concave for the positive part and convex for the negative part, with a steepest slope at the origin for the negative part, so as to represent loss aversion. A typical example is to take $u(x) = x^\alpha$ for $x > 0$ and $u(x) = -\lambda(-x)^\alpha$, with $0 < \alpha < 1$ and $\lambda > 1$ (Fig. 5.11). On the other hand, the two distortion functions are usually inverse-S-shaped and somewhat different. ◇

The following result shows that the PT model satisfies the basic rationality requirements.

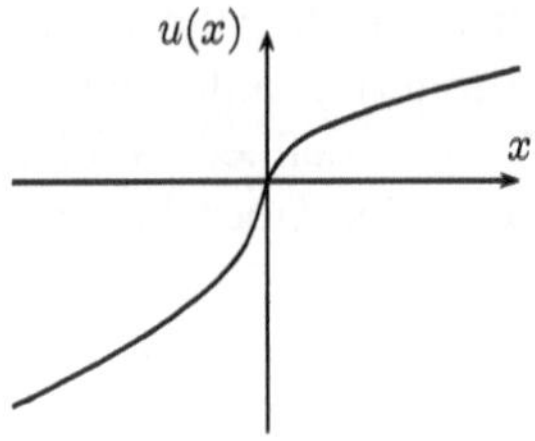

Fig. 5.11 Example of utility function for PT ($\lambda = 2, \alpha = 0.5$)

Theorem 5.14 *Under the above assumptions for risk, for any increasing continuous utility function, the PT model satisfies monotonicity and strict stochastic dominance.*

Proof Monotonicity corresponds to monotonicity w.r.t integrand for integrals, and therefore it holds because the latter property holds for the symmetric Choquet integral (see Theorem 4.24(vi) and Remark 4.25. The fact that the positive and negative parts use different capacities does not change the result).

As for stochastic dominance, consider two lotteries p, p' with $G_{p'} \geqslant G_p$. Let us separate the lotteries into their positive and negative parts as in (5.13); i.e.,

$$\mathrm{PT}(p) = \mathrm{PT}(p^+) - \mathrm{PT}(p^-).$$

Since the positive part satisfies stochastic dominance (it is merely the Choquet integral for nonnegative functions), it remains to prove that $\mathrm{PT}((p')^-) \leqslant \mathrm{PT}(p^-)$. It is enough to show that

$$\varphi^- \circ P(u^-(x) \geqslant \alpha) \geqslant \varphi^- \circ P'(u^-(x) \geqslant \alpha) \tag{5.14}$$

for every $\alpha > 0$, and where u^- is the negative part of u [see (4.9)]. We have for every $\alpha > 0$:

$$\begin{aligned}
& P'(x > -\alpha) \geqslant P(x > -\alpha) \\
\Leftrightarrow\quad & P'(x \leqslant -\alpha) \leqslant P(x \leqslant -\alpha) \\
\Leftrightarrow\quad & P'\big(u(x) \leqslant u(\alpha)\big) \leqslant P\big(u(x) \leqslant u(-\alpha)\big) \\
\Leftrightarrow\quad & P'(u^-(x) \geqslant \alpha') \leqslant P(u^-(x) \geqslant \alpha') \\
\Leftrightarrow\quad & \varphi^- \circ P'(u^-(x) \geqslant \alpha') \leqslant \varphi^- \circ P(u^-(x) \geqslant \alpha')
\end{aligned}$$

using increasingness of u, and letting $\alpha' = -u(-\alpha) > 0$. Now, the above inequality holds for any $\alpha' \in \mathbf{ran}\, u^-$, which is an interval of $\mathbb{R}^+$ by continuity of u^-, whose closure contains 0. For positive values outside the range the equality holds trivially, hence (5.14) holds for any $\alpha > 0$. □

5.3 Decision Under Uncertainty

Recall that in decision under uncertainty, the probability measure on S is unknown. Therefore, the determination of models for uncertainty requires more effort because both the utility function and the probability measure need to be determined, the latter being called "subjective" because it depends on the decision maker.

We return to our notation of acts as mappings $f, g, \ldots$ from S to the set of consequences, supposed to be $\mathbb{R}$ most of the time for simplicity. We summarize our assumptions for this section (same as in Wakker [339, Chap. 4]).

(Structural assumptions for uncertainty)

(i) Acts take finitely many values, say $x_1, \ldots, x_n$, and therefore each act f induces a partition of S into events $f^{-1}(x_i)$ for all i;
(ii) The domain of the preference relation $\succcurlyeq$ is the set of such acts, denoted by F; $\succcurlyeq$ is nondegenerate.

Before entering the main topic, we address the case where the utility function is known, or is supposed to be the identity function. Then expected utility reduces to expected value. An important concept underlying the expected value criterion is the Dutch book argument.

5.3.1 *The Expected Value Criterion and the Dutch Book Argument*

(This section is based on Wakker [339, Sects. 1.5 and 1.6].) We suppose that either the utility function is known, or is considered to be the identity function, so that we directly work with consequences, supposed to be amounts of money. Then the expected utility criterion becomes the expected value criterion, denoted by EV:

$$\mathrm{EV}(f) = \sum_{i=1}^{n} P(E_i)x_i \tag{5.15}$$

where P is a (subjective) probability measure on S, $x_1, \ldots, x_n$ are all possible outcomes of f (i.e., its range), and $E_i = \{s \in S : f(s) = x_i\}$.

We say that $\succcurlyeq$ on F is *additive* if for all acts f, g, h,

$$f \succcurlyeq g \Rightarrow f + h \succcurlyeq g + h.$$

A *certainty equivalent* to an act f is a real number denoted by $\mathrm{CE}(f)$ such that the constant act $\mathrm{CE}(f)$ is indifferent to f: $f \sim \mathrm{CE}(f)$.

A *Dutch book* consists of preferences $f^i \succcurlyeq g^i$, $i = 1, \ldots, m$, such that the preferred acts f^i, when combined, always yield less than the nonpreferred acts g^i.

That is,

$$\sum_{i=1}^{m} f^i(s) < \sum_{i=1}^{m} g^i(s) \qquad (s \in S).$$

In words, a Dutch book is a sequence of bets yielding a sure loss at the end. It is known in finance under the name *arbitrage*.

Example 5.15 (Wakker [339, Example 1.6.2]) Suppose you are a vendor on the beach. Three states of nature are considered: no rain, some rain, and all rain. Your profit depends on the product you are selling (ice creams, hot dogs, umbrellas, sun glasses, etc.) and the weather. The following table gives the profit for different acts (products) $f^1, f^2, f^3, g^1, g^2, g^3$. Suppose you have the following preference: you

Act	No rain	Some rain	All rain
f^1	0	100	100
f^2	100	0	100
f^3	100	100	0
g^1	300	0	0
g^2	0	300	0
g^3	0	0	300

prefer acts yielding a positive profit for most of the three states of nature (so that any f^i is preferred to any g^j). Then a Dutch book can be made. Indeed,

$$f^1(s) + f^2(s) + f^3(s) = 200 < 300 = g^1(s) + g^2(s) + g^3(s) \qquad (s \in S).$$

The following fundamental result shows that, essentially, a decision maker whose preference is not representable by expected value, can be made victim of a Dutch book.

Theorem 5.16 (De Finetti's Dutch book argument) *(Wakker [339, Theorem 1.6.1]) Let $\succcurlyeq$ be a binary relation on F. Under the above structural assumptions for uncertainty, the following propositions are equivalent.*

(i) $\succcurlyeq$ is complete, transitive, each act has a certainty equivalent, and no Dutch book is possible;
(ii) $\succcurlyeq$ is complete, transitive, additive, monotone, and each act has a certainty equivalent;
(iii) There exists a unique probability measure P on S such that EV given by (5.15) represents $\succcurlyeq$.

Proof (ii)⇒(i) Consider acts $f^1, \ldots, f^m, g^1, \ldots, g^m$ such that $f^i \succcurlyeq g^i$ for all i and $\sum_i f^i(s) < \sum_i g^i(s)$ for every $s \in S$. By monotonicity, we get $\sum_i f^i \prec \sum_i g^i$.

By additivity, $f^1 \succcurlyeq g^1$ implies $f^1 + f^2 \succcurlyeq g^1 + f^2$. Now, $f^2 \succcurlyeq g^2$, hence by additivity again $g^1 + f^2 \succcurlyeq g^1 + g^2$, which by transitivity yields $f^1 + f^2 \succcurlyeq g^1 + g^2$. Proceeding similarly, we eventually get

$$\sum_i f^i \succcurlyeq \sum_i g^i,$$

a contradiction.

(i)⇒(ii) Let us first prove monotonicity, considering two acts f, g. Suppose first that $f(s) > g(s)$ for every s. Then assuming $g \succcurlyeq f$ would constitute a Dutch book with $m = 1$. Suppose then that $f(s) \geqslant g(s)$ for every $s \in S$, and $g \succ f$. Then $\mathrm{CE}(f) < \mathrm{CE}(g)$, and on the other hand

$$\mathrm{CE}(f) \succcurlyeq f$$
$$g \succcurlyeq \mathrm{CE}(g).$$

Since no Dutch book is possible, there must exist a state s such that

$$\mathrm{CE}(f) + g(s) \geqslant f(s) + \mathrm{CE}(g),$$

hence $0 \geqslant g(s) - f(s) \geqslant \mathrm{CE}(g) - \mathrm{CE}(f) > 0$, a contradiction.

Let us prove that CE is additive. Suppose on the contrary that $\mathrm{CE}(f + g) > \mathrm{CE}(f) + \mathrm{CE}(g)$ for some acts f, g. We have

$$\mathrm{CE}(f + g) \preccurlyeq f + g$$
$$f \preccurlyeq \mathrm{CE}(f)$$
$$g \preccurlyeq \mathrm{CE}(g).$$

Since no Dutch book is possible, there must exist a state s such that

$$\mathrm{CE}(f + g) + f(s) + g(s) \leqslant f(s) + g(s) + \mathrm{CE}(f) + \mathrm{CE}(g),$$

a contradiction. Doing similarly by reversing inequalities, one concludes that only $\mathrm{CE}(f + g) = \mathrm{CE}(f) + \mathrm{CE}(g)$ is possible.

Based on this, we prove additivity of $\succcurlyeq$. Suppose $f \succcurlyeq g$ and $f+h \prec g+h$ for some acts f, g, h. Then $\mathrm{CE}(f+h) < \mathrm{CE}(g+h)$, hence $\mathrm{CE}(f)+\mathrm{CE}(h) < \mathrm{CE}(g)+\mathrm{CE}(h)$, a contradiction.

(iii)⇒(ii) Obvious.

(ii)⇒(iii) Since CE represents preference, we have to show that CE is the expected value w.r.t. some probability measure P on S. First, we prove that CE is additive. By $f \sim \mathrm{CE}(f)$ and $g \sim \mathrm{CE}(g)$ for any two acts f, g, we get using additivity and transitivity

$$f + g \sim \mathrm{CE}(f) + \mathrm{CE}(g),$$

hence $\mathrm{CE}(f+g) \sim \mathrm{CE}(f) + \mathrm{CE}(g)$, i.e., $\mathrm{CE}(f+g) = \mathrm{CE}(f) + \mathrm{CE}(g)$.

Let us define $P(E) = \mathrm{CE}(1_E0)$, the certainty equivalent of the act with outcome 1 on E and 0 otherwise, for every event E. Let us prove that P is a probability measure. We have $P(S) = \mathrm{CE}(1) = 1$, $P(\varnothing) = \mathrm{CE}(0) = 0$, and $P(E) \geqslant 0$ for any E by monotonicity. Now, take disjoint events A, B. Then, using additivity of CE,

$$P(A \cup B) = \mathrm{CE}(1_{A\cup B}0) = \mathrm{CE}(1_A0) + \mathrm{CE}(1_B0) = P(A) + P(B).$$

It remains to prove that $\mathrm{CE}(\lambda_E0) = \lambda P(E)$ for all event E and all $\lambda \in \mathbb{R}$. For $\lambda \geqslant 0$, it suffices to use the classical trick starting with $\lambda \in \mathbb{N}$, then $\lambda \in \mathbb{Q}_+$ and then $\lambda \in \mathbb{R}_+$ (see Proof of Theorem 4.51). It remains to prove the property for $\lambda < 0$. Observe that

$$0 = \mathrm{CE}(0) = \mathrm{CE}(f - f) = \mathrm{CE}(f) + \mathrm{CE}(-f),$$

hence $\mathrm{CE}(-f) = -\mathrm{CE}(f)$, and the desired property follows. □

5.3.2 *The Expected Utility Criterion*

Expected utility is defined as for decision under risk. Using the above notation, expected utility holds if an act f is evaluated by

$$\mathrm{EU}(f) = \sum_{i=1}^{n} u(x_i)P(E_i) \tag{5.16}$$

where u is a utility function (supposed to be increasing when outcomes are monetary values), P is a (subjective) probability measure, and $E_i = f^{-1}(x_i)$ for $i = 1, \ldots, n$. Since P is subjective, the model is often called *subjective expected utility theory* and abbreviated by SEU.

The first behavioral foundation of expected utility for decision under uncertainty was given by Leonard Savage[10] [284]. His characterization of expected utility requires S to be infinite, and is based on seven axioms. A complete description of this characterization is fairly complicated, and because it is well known and can be found in many references, we do not detail it here. We recommend to the interested readers,

[10]Leonard Jimmie Savage (1917, Detroit – 1971, New Haven) (born Leonard Ogashevitz) is an American mathematician and statistician. His most famous work is his 1954 book "Foundations of Statistics" giving the basis of subjective expected utility theory.

e.g., the excellent description in Gilboa [153, Chap. 10]. Nevertheless, there is an axiom proposed by Savage that cannot be ignored: his famous P2 axiom, referred to as the *sure-thing principle*. It says the following: for any acts $f, g, h, h' \in F$, for any event $E \subseteq S$,

$$f_E h \succcurlyeq g_E h \Leftrightarrow f_E h' \succcurlyeq g_E h'. \tag{5.17}$$

In words, the preference between two acts that are identical for some event (here $S \setminus E$) should not depend on that common part.

Let us give a simpler characterization, similar to the one of RDU given in Sect. 5.2.6. We consider compound acts that are constant on some event E, denoted by $\alpha_E f$ with $\alpha \in \mathbb{R}$ and $f \in F$ (as a simplification of the more correct $(\alpha 1_S)_E f$). We say that $(\alpha, \beta) \in \mathbb{R}^2$ and $(\gamma, \delta) \in \mathbb{R}^2$ are related by *tradeoff indifference* if there exist acts $f, g \in F$, and a nonnull event $E \subseteq S$ such that

$$\alpha_E f \sim \beta_E g \text{ and } \gamma_E f \sim \delta_E g.$$

We denote this by $(\alpha, \beta) \sim^t (\gamma, \delta)$. In words, α instead of β, when E occurs, offsets f instead of g on $S \setminus E$, and so does γ instead of δ. Now, a preference relation $\succcurlyeq$ satisfies *tradeoff consistency* if whenever $(\alpha, \beta) \sim^t (\gamma, \delta)$, improving any outcome among $\alpha, \beta, \gamma, \delta$ breaks that relationship.

Continuity of $\succcurlyeq$ is defined as for decision under risk: $\succcurlyeq$ is continuous if for every $n \in \mathbb{N}$ and every partition $(E_1, \ldots, E_n)$ of S, the sets of n-dim vectors of outcomes

$$\{(x_1, \ldots, x_n) : (E_1, x_1; \ldots; E_n, x_n) \succcurlyeq (E_1, y_1; \ldots; E_n, y_n)\} \text{ and}$$
$$\{(x_1, \ldots, x_n) : (E_1, x_1; \ldots; E_n, x_n) \preccurlyeq (E_1, y_1; \ldots; E_n, y_n)\}$$

are closed subsets of $\mathbb{R}^n$ for every $(y_1, \ldots, y_n) \in \mathbb{R}^n$.

Theorem 5.17 (Axiomatization of EU for uncertainty) *(Köbberling and Wakker [214], Wakker [339, Theorem 4.6.4]) Let $\succcurlyeq$ be a binary relation on F. Under the above structural assumptions for uncertainty, the following two propositions are equivalent.*

(i) $\succcurlyeq$ is complete and transitive, is continuous, and satisfies monotonicity and tradeoff consistency;
(ii) There exist a continuous and increasing utility function u (unique up to a positive affine transformation) and a unique probability measure P on S such that EU given by (5.16) represents $\succcurlyeq$.

5.3.3 The Ellsberg Paradox

Ellsberg [124] proposed the following betting situation. Consider an urn containing 90 balls, which could be red (R), yellow (Y) or black (B). We know that the urn contains 30 red balls, but for the remaining 60 balls, it is only known that they are yellow or black, in an unknown proportion (there could be 30 yellow and 30 black balls, as well as 0 yellow and 60 black balls, etc.).

A ball is picked at random, and the following bets are proposed: betting on red, or betting on black. If the color of the picked ball corresponds, you get \$100, otherwise nothing. Then a second choice is offered: to bet on [red or yellow] or on [black or yellow], and again you get \$100 if the color of the picked ball matches your choice, otherwise nothing.

Experimental results show that the majority of people prefer to bet on red in the first step, because they are sure that $1/3$ of the balls are red, while there is no guarantee about black balls (there could be no black ball at all). In the second step, most people prefer to bet on [black or yellow] for the same reason: $2/3$ of the balls are black or yellow, while for [red or yellow], only $1/3$ is guaranteed. Moreover, the majority of people follow the above choices in both steps. We show below that such a behavior cannot be explained by expected utility.

Let us formalize the problem in terms of states of nature and acts. The states of nature are the possible colors of the picked ball, i.e., $S = \{R, Y, B\}$, and the acts are the four possible bets. We summarize the outcomes of the different acts in the table below. We immediately see from the table that the preferences $f_1 \succ f_2$ and $f_4 \succ f_3$

Act	Red (R)	Yellow (Y)	Black (B)
f_1: bet on R	100	0	0
f_2: bet on B	0	0	100
f_3: bet on R or Y	100	100	0
f_4: bet on B or Y	0	100	100

violate the sure-thing principle. Indeed, acts f_1 and f_2 coincide on Y, hence by the sure-thing principle, only R and B matter. But since f_3, f_4 are identical to f_1, f_2 on R, B, the choice $f_1 \succ f_2$ forces $f_3 \succ f_4$, which is not the observed behavior. The sure-thing principle being implied by expected utility, it follows that most people are not expected utility maximizers.

Even worse, the RDU model cannot explain this choice either, whatever the probability measure P is chosen on S or the distortion function φ is. Indeed, on the one hand, u denoting the utility function,

$$
\begin{aligned}
f_1 \succ f_2 \Leftrightarrow \quad & u(100)\varphi(P(\{R\}) > u(100)\varphi(P(\{B\}) \\
\Leftrightarrow \quad & \varphi(P(\{R\}) > \varphi(P(\{B\})) \\
\Leftrightarrow \quad & P(\{R\}) > P(\{B\}), \qquad (5.18)
\end{aligned}
$$

by increasingness of φ. On the other hand,

$$\begin{aligned} f_4 \succ f_3 \Leftrightarrow \quad & u(100)\varphi(P(\{Y,B\}) > u(100)\varphi(P(\{R,Y\}) \\ \Leftrightarrow \quad & \varphi(P(\{Y,B\}) > \varphi(P(\{R,Y\})) \qquad (5.19) \\ \Leftrightarrow \quad & \varphi(1 - P(\{R\})) > \varphi(1 - P(\{B\})) \\ \Leftrightarrow \quad & P(\{R\}) < P(\{B\}). \qquad (5.20) \end{aligned}$$

Clearly, (5.18) and (5.20) are contradictory.

More generally, no model assuming probabilistic sophistication[11] can explain this choice. Indeed, if one prefers f_1 to f_2, this is because there might be fewer black balls than red balls [i.e., $P(B) < P(R)$]. Based on this assumption, one should infer that there might be more balls being red or yellow than balls being black or yellow ($P(R,Y) > P(B,Y)$); i.e., f_3 should be preferred to f_4. But this is precisely not the choice of the majority of people. In the rest of this chapter, we provide two explanations of this decision behavior.

A similar phenomenon happens with the *home bias* (see Wakker [339, Example 10.1.2]). An American investor has to choose between gaining \$1000 if the Dow Jones index goes up tomorrow (event denoted by DJ^+), or gaining \$1000 if the Nikkei index goes up tomorrow (NK^+). Being American, he prefers the former act. Suppose now that the choice is given instead between gaining \$1000 if the Dow Jones index goes down (DJ^-) or gaining the same amount if the Nikkei index goes down (NK^-). Again, the former is preferred (home bias). However, as in the Ellsberg paradox, these choices cannot be explained by probabilistic sophistication because the first choice entails $P(\mathrm{DJ}^+) > P(\mathrm{NK}^+)$, while the second entails $P(\mathrm{DJ}^-) > P(\mathrm{NK}^-)$, which is impossible.

5.3.4 Choquet Expected Utility

The great discovery of the Ellsberg paradox is that most people are not probabilistically sophisticated, at least in the kind of situation depicted in Sect. 5.3.3. A natural conclusion is that their representation of uncertainty is not based on probability measures (we will return to this conclusion in Sect. 5.3.5). The essence of probability measures being the additivity property, we have to look for measures of uncertainty that are not additive, and the class of (normalized) capacities provides a large class of such measures (see Chap. 2). Indeed, an examination of (5.18) and (5.20) reveals that additivity is the very cause of the contradiction in the Ellsberg

[11]*Probabilistic sophistication* says that the uncertainty of events can be quantified by a probability measure, so that acts $(E_1, x_1; \ldots; E_n, x_n)$ can be replaced by their equivalent lotteries $(p_1, x_1; \ldots; p_n, x_n)$. It *does not* say, however, that acts should be evaluated by expected utility (but could be for example by RDU, supposing a distortion of probability).

paradox. If one stops the computation at (5.19), then the contradiction disappears. Using a capacity μ for representing the uncertainty on S, we obtain

$$f_1 \succ f_2 \Leftrightarrow \mu(\{R\}) > \mu(\{B\})$$
$$f_4 \succ f_3 \Leftrightarrow \mu(\{Y,B\}) > \mu(\{R,Y\}).$$

One could take for example the capacity given in Table 5.1 to satisfy these inequalities. Note that this capacity is supermodular. We will return to this capacity

Event E	R	Y	B	R,Y	R,B	Y,B	R,Y,B
$\mu(E)$	1/3	0	0	1/3	1/3	2/3	1

Table 5.1 Definition of a capacity μ solving the Ellsberg paradox

in Sect. 5.3.5.

Replacing probability measures by capacities forces to reconsider the computation of the expected value of an act by using the Choquet integral, leading to what is usually called *Choquet expected utility*. The valuation of an act f is then given by

$$\mathrm{CEU}(f) = \int u(f)\,\mathrm{d}\mu \tag{5.21}$$

where u is a utility function (supposed to be increasing when the outcomes are monetary values) and μ a capacity on S. This can be seen as a generalization of the RDU model (with $\mu = \varphi \circ P$), and for this reason this model is also called RDU under uncertainty.

We give a behavioral foundation of CEU, which is very close to the one given for expected utility. The only condition to change is tradeoff consistency, where similarly to the axiomatization of RDU we need to introduce rank dependence in it.

Consider $\alpha, \beta, \gamma, \delta \in \mathbb{R}$ related by tradeoff indifference (Sect. 5.3.2); i.e., $(\alpha, \beta) \sim^{\mathrm{t}} (\gamma, \delta)$, meaning that there exist an event E and acts $f, g \in F$ such that $\alpha_E f \sim \beta_E g$ and $\gamma_E f \sim \delta_E g$. We say that $\alpha, \beta, \gamma, \delta \in \mathbb{R}$ are related by *rank tradeoff indifference*, denoted by $(\alpha, \beta) \sim^{\mathrm{t}}_{\mathrm{r}} (\gamma, \delta)$, if in addition, the ranks of $\alpha, \beta, \gamma, \delta$ are the same in their respective acts, that is, letting $x_1 < x_2 < \cdots < x_n$ and $y_1 < y_2 < \cdots < y_m$ be the outcomes of f and g respectively,

$$\begin{aligned} &x_1 < \cdots < x_r \leqslant \alpha \leqslant x_{r+1} < \cdots < x_n, \quad y_1 < \cdots < y_r \leqslant \beta \leqslant y_{r+1} < \cdots < y_m, \\ &x_1 < \cdots < x_r \leqslant \gamma \leqslant x_{r+1} < \cdots < x_n, \quad y_1 < \cdots < y_r \leqslant \delta \leqslant y_{r+1} < \cdots < y_m, \end{aligned} \tag{5.22}$$

for some r, and the events pertaining to x_i and y_i are the same (i.e., $f^{-1}(x_i) = g^{-1}(y_i)$) for $i = r+1, \ldots, n$ (compare with the corresponding definition for risk in Sect. 5.2.6). As with RDU, it is straightforward to check that under CEU,

$$(\alpha, \beta) \sim_{\rm r}^{\rm t} (\gamma, \delta) \Rightarrow u(\alpha) - u(\beta) = u(\gamma) - u(\delta).$$

Now, a preference relation $\succcurlyeq$ satisfies *rank tradeoff consistency* if whenever $(\alpha, \beta) \sim_{\rm r}^{\rm t} (\gamma, \delta)$, improving any outcome among $\alpha, \beta, \gamma, \delta$ breaks that relationship.

Theorem 5.18 (Axiomatization of CEU) *(Köbberling and Wakker [214], Wakker [339, Theorem 10.5.6]). Let $\succcurlyeq$ be a binary relation on F. Under the above structural assumptions for uncertainty, the following two propositions are equivalent.*

(i) $\succcurlyeq$ is complete and transitive, is continuous, and satisfies monotonicity and rank tradeoff consistency;
(ii) There exist a continuous and increasing utility function u (unique up to a positive affine transformation) and a unique capacity μ on S such that CEU given by (5.21) represents $\succcurlyeq$.

Remark 5.19

(i) Schmeidler was the first to introduce and axiomatize Choquet expected utility in his 1989 paper [287]. The way it was done is fairly different from what is presented above. Schmeidler takes as basis the framework of Anscombe and Aumann [9] for subjective expected utility, where the set of outcomes C is the set of finite lotteries (our set L). The axiomatization of Anscombe and Aumann for EU is very close to the one of Fishburn/von Neumann and Morgenstern (given in Theorem 5.6) for expected utility under risk, using independence, the Archimedean property and monotonicity. Then, in the axiomatization of Schmeidler, the independence axiom is weakened, by requiring that independence should hold only for comonotonic acts. Specifically, the axiom reads: for all comonotonic acts f, g, h and $\lambda \in]0, 1[$,

$$f \succ g \Rightarrow \lambda f + (1-\lambda)h \succ \lambda g + (1-\lambda)h \tag{5.23}$$

where acts f, g, h are *comonotonic* if there exists a permutation σ on S (supposing S to be finite) such that

$$f(s_{\sigma(1)}) \succcurlyeq \cdots \succcurlyeq f(s_{\sigma(n)})$$
$$g(s_{\sigma(1)}) \succcurlyeq \cdots \succcurlyeq g(s_{\sigma(n)})$$
$$h(s_{\sigma(1)}) \succcurlyeq \cdots \succcurlyeq h(s_{\sigma(n)}).$$

In the above, $\succcurlyeq$ on consequences is induced by $\succcurlyeq$ on acts (Remark 5.4). The comonotonicity property above is of course a direct generalization of the notion of comonotonic functions (Definition 4.26).

(ii) The Choquet integral being characterized by comonotonic additivity (see Theorem 4.51), it is not surprising that most of the axiomatizations of CEU come from axiomatizations of EU where some axioms are restricted to comonotonic acts. The axiomatization of Schmeidler is a good example, others are the axiomatization of Chew and Wakker [52] using the comonotonic sure-thing principle, and the one of Gilboa [152] using a variant of the sure-thing principle. Rank tradeoff consistency hides also some kind of comonotonicity, although in a weaker sense, because when $(\alpha,\beta) \sim_{\mathrm{r}}^{\mathrm{t}} (\gamma,\delta)$, the acts $\alpha_E f, \beta_E g, \gamma_E f, \delta_E g$ are not necessarily comonotonic.

◇

We present now a simplified model of the CEU type due to Chateauneuf [46], where the utility function in (5.21) is the identity function. The setting in [46] is more general than ours, because acts are measurable and bounded mappings from S to $\mathbb{R}$, where S is endowed with some σ-algebra $\mathcal{S}$. Let us still call F the set of such acts.

The first axioms are basic requirements ensuring the existence of a numerical representation of $\succcurlyeq$ with usual properties (monotonicity, continuity):

A1. $\succcurlyeq$ is complete and transitive;
A2. Continuity w.r.t. monotone uniform convergence:

$$[f_n, f, g \in F, f_n \succcurlyeq g, f_n \downarrow^u f] \Rightarrow f \succcurlyeq g$$

$$[f_n, f, g \in F, g \succcurlyeq f_n, f_n \uparrow^u f] \Rightarrow g \succcurlyeq f$$

A3. Monotonicity. For any $\epsilon > 0$,

$$f \geqslant g + \epsilon 1_S \Rightarrow f \succ g.$$

The key axiom forcing the representation by the Choquet integral is:

A4. For all acts f, g, h such that f and h are comonotonic, and g and h are comonotonic,

$$f \sim g \Rightarrow f + h \sim g + h.$$

Let us comment on the latter axiom. It says that the indifference between two acts f, g is kept when adding a new act h, provided the new act is comonotonic with them. Adding a common act to two others without perturbing the preference is reminiscent of the independence condition in the Fishburn/von Neumann and Morgenstern axiomatic characterization of expected utility under risk, and the comonotonicity condition that is required is as in the axiomatization of Schmeidler (Remark 5.19). Now, adding two comonotonic acts f, h yields an act $f + h$ that still has the same kind of variation pattern, but with increasing amplitude. As a result, the "uncertainty" of the act is not reduced, in the sense that the outcomes vary a lot depending on the state of nature. On the other hand, if the two acts are not comonotonic, there exists some region of S where the uncertainty is reduced, due to the fact that the acts f, h have opposite variation in this region. This is called a *hedging effect*. Since uncertainty is reduced for $f + h$, the decision maker could change his preference on $f + h, g + h$, according to his attitude towards uncertainty. The following example illustrates this point.

Example 5.20 Consider the following acts f, g, h (letters on branches indicate events).

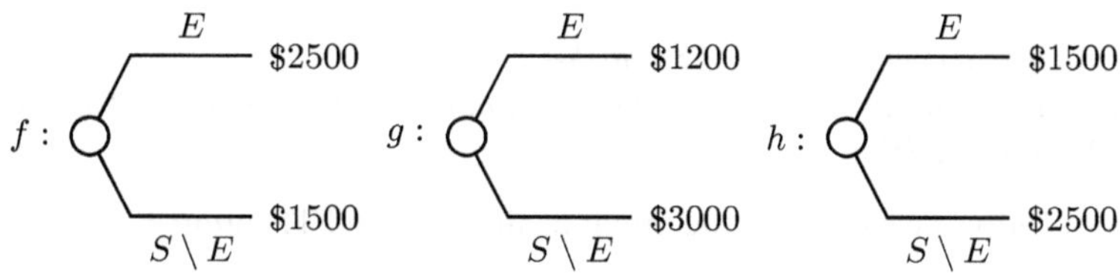

Suppose $f \sim g$. Observe that h and g are comonotonic, but not h and f. It follows that there is a reduction of uncertainty for the act $f + h$ (i.e., it is closer to a constant act; indeed it becomes a constant act), but not for $g + h$:

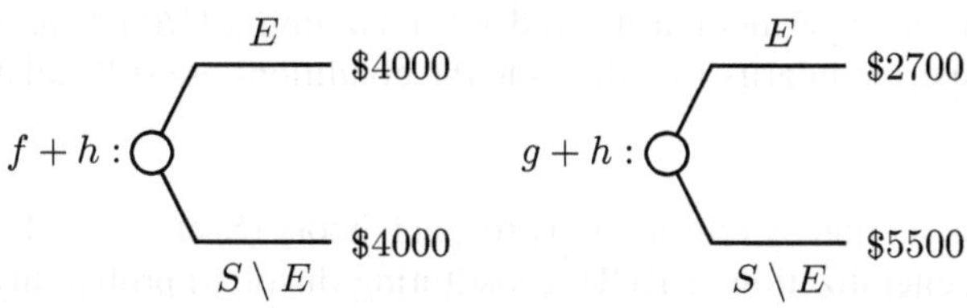

If the decision maker prefers constant acts to uncertain acts, the preference will be $f + h \succ g + h$. ◇

Theorem 5.21 (Axiomatization of CEU without utility) *(Chateauneuf [46]) Let $\succcurlyeq$ be a binary relation on F (as defined in this section). The following two propositions are equivalent.*

*(i) $\succcurlyeq$ satisfies axioms **A1, A2, A3** and **A4**;*
(ii) There exists a unique capacity μ on S such that CEU given by (5.21) represents $\succcurlyeq$.

The proof follows from the result of Schmeidler (Theorem 5.18).

By modifying Axiom **A4**, it is possible to have a characterization of uncertainty aversion by supermodularity of the capacity.

A4'. For all acts f, g, h such that g and h are comonotonic,

$$f \sim g \Rightarrow f + h \succcurlyeq g + h.$$

The axiom describes a situation where there is no hedging effect for $g + h$, but there could be one for $f + h$, inducing a reduction of uncertainty. Then uncertainty averse[12] decision makers prefer $f + h$ to $g + h$.

Theorem 5.22 (Characterization of uncertainty aversion) *(Chateauneuf [46]) Let $\succcurlyeq$ be a binary relation on F (as defined in this section). The following two propositions are equivalent.*

*(i) $\succcurlyeq$ satisfies axioms **A1**, **A2**, **A3** and **A4'**;*
(ii) There exists a unique supermodular capacity μ on S such that CEU given by (5.21) represents $\succcurlyeq$.

The proof follows easily from the previous theorem and Theorem 4.35.

Remark 5.23

(i) Obviously, uncertainty seeking is characterized by submodularity.
(ii) Recall that in our solution of the Ellsberg paradox (see Table 5.1), the capacity was convex, which tends to indicate that most people are uncertainty averse. However, note that the conditions $\mu(\{R\}) > \mu(\{B\})$ and $\mu(\{Y, B\}) > \mu(\{R, Y\})$ do not force supermodularity. Indeed, changing $\mu(\{B\})$ from 0 to, e.g., $1/6$ makes supermodularity to fail, while the conditions are still satisfied.

◇

We end this section by returning to prospect theory (Sect. 5.2.7). In the same way as CEU is a generalization of RDU, substituting distorted probability measures by capacities, one can adapt Prospect Theory to the uncertainty framework, using two capacities, one for losses and one for gains. This yields the following model:

$$\mathrm{PTU}(f) = \int u(f^+)\,\mathrm{d}\mu^+ - \int u(f^-)\,\mathrm{d}\mu^-. \tag{5.24}$$

We do not elaborate on this point and refer the readers to Wakker [339, Chap. 12].

5.3.5 *Ambiguity and Multiple Priors*

Ambiguity refers to a situation where the probability of some events is not known with precision, and for this reason one speaks also of *imprecise probabilities* (Walley [341]). The Ellsberg paradox provides a nice example of it, and shows that most

[12] So far we have not defined this term, and we suggest to take Axiom **A4'** as a definition of uncertainty aversion. By inverting the inequality, one obtains uncertainty seeking. Schmeidler [287] has defined uncertainty aversion in his Anscombe-Aumann framework as follows: if $f \sim g$, then $\alpha f + (1 - \alpha)g \succcurlyeq g$ for every $\alpha \in [0, 1]$. It turns out that, as in Theorem 5.22, uncertainty aversion is equivalent to supermodularity of the capacity.

people exhibit ambiguity aversion, because they prefer to bet on events whose probability is exactly known (like "red" and "black or yellow").

A natural modeling of a decision situation involving ambiguity is to consider a set of probability measures, which could be compatible or coherent with the available information. Each such probability measure being called a prior probability measure, this leads to the often used term *multiple priors*. It is assumed that the set of priors is a convex set, which we denote by $\mathcal{P}$. This convex set may be defined by some inequalities the probability of events should follow (e.g., interval-valued probabilities).

Given a convex set $\mathcal{P}$ of priors, the *maxmin expected utility* model consists in taking the act maximizing the minimum of its expected utility, taken over all priors. Hence, the numerical representation of such a model is

$$\mathrm{MEU}(f) = \min_{P \in \mathcal{P}} \int u(f)\, \mathrm{d}P \qquad (f \in F). \tag{5.25}$$

The MEU model yields a pessimistic evaluation of acts, because the less favorable probability measure is chosen. A dual of the MEU model is to take instead the most favorable one, replacing min by max. A convex combination of both models yields the *α-maxmin expected utility* model (Hurwicz [200]), expressed by

$$\alpha \min_{P \in \mathcal{P}} \int u(f)\, \mathrm{d}P + (1-\alpha) \max_{P \in \mathcal{P}} \int u(f)\, \mathrm{d}P \qquad (f \in F) \tag{5.26}$$

for some fixed $\alpha \in [0, 1]$.

We give some properties of the multiple prior models.

(i) A particular case of interest is when $\mathcal{P}$ is the core of some normalized capacity μ (see Chap. 3), that is, $\mathcal{P}$ is defined as the set of probabilities satisfying inequalities of the type $P(A) \geqslant \mu(A)$ for every event A. Note however that this is far from being the general case: a given convex set $\mathcal{P}$ is scarcely the core of some capacity. Take for example $S = \{s_1, s_2, s_3\}$ and the constraint $P(\{s_1\}|\{s_1, s_2\}) \geqslant 1/3$. This translates into the inequality $2P(\{s_1\}) \geqslant P(\{s_2\})$, which is not of the type $P(A) \geqslant \alpha$ (Gilboa [153, Sect. 17.2]).

(ii) On the other hand, any convex set $\mathcal{P}$ yields a capacity by taking its *lower envelope*, defined by

$$P_*(A) = \inf_{P \in \mathcal{P}} P(A)$$

for every event A. Obviously, P_* is a normalized capacity on S, and $P_* \leqslant P$ eventwise for any $P \in \mathcal{P}$. It follows that for every act f,

$$\int f\, \mathrm{d}P_* \leqslant \min_{P \in \mathcal{P}} \int f\, \mathrm{d}P.$$

Suppose now that $\mathcal{P}$ is the core of some capacity μ. Do we have $P_* = \mu$? The answer is negative in general, and a capacity is said to be *exact* if this equality holds (in which case, $\mathbf{core}(P_*) = \mathcal{P}$). Exact capacities are studied in detail in Sect. 3.4. In particular, it is known that supermodular capacities are exact.

(iii) Let us assume that $\mathcal{P} = \mathbf{core}(\mu)$ for some normalized capacity μ. The question is: when does CEU coincide with MEU? The answer is given by Theorem 4.39: the equality

$$\int f \, \mathrm{d}\mu = \min_{P \in \mathbf{core}(\mu)} \int f \, \mathrm{d}P$$

holds for any act f if and only if μ is supermodular. Hence, the two models differ in general, and moreover none is more general than the other. To see that there are some CEU models not representable as a MEU model, just consider the capacity $\mu_{\max}$ defined by $\mu_{\max}(A) = 1$ for every nonempty event A. Then, $\text{CEU}(f) = \max_{s \in S} u(f(s))$ [see Theorem 4.24(x)]. This is clearly not representable by MEU because for every probability measure P, there exists an act f such that $\int u(f) \, \mathrm{d}P < \max u(f)$. For an example of a MEU model not representable by CEU, see Wakker [339, Sol. of Exercise 11.9.1, p. 452].

(iv) MEU, as well as the α-maxmin model, may violate strict monotonicity, as shown by the following example: Let $S = \{s_1, s_2, s_3, s_4, s_5\}$, u be the identity function, and $\mathcal{P}$ be the set of all probability measures. Consider the two acts $f = (1, 1, 1, 1, 0)$ and $g = (1, 0, 0, 0, 0)$, where the ith coordinate of the vector is the outcome of the act for state s_i. The minimal value of EU for both f and g is 0, while the maximal value is 1. It follows that $f \sim g$ for MEU and any α-maxmin model, hence strict monotonicity is violated (Wakker [339, Appendix 11.9], see also Appendix 11.10).

Example 5.24 (The Ellsberg paradox revisited) We return to our solution of the Ellsberg paradox given in Sect. 5.3.4. Having found a capacity solving the paradox, we claimed that the uncertainty representation of the majority of people is not based on probability, but rather on non-additive measures, namely capacities. However, having a closer look to the capacity given in Table 5.1, it is possible to interpret it in terms of multiple priors. Indeed, the set of probability measures compatible with information on the urn is simply

$$\mathcal{P} = \{P \text{ on } \{R, Y, B\} \,:\, P(\{R\}) = 1/3\}.$$

Now, observe that the lower envelope of $\mathcal{P}$ is precisely μ given in Table 5.1. Since μ is convex, it follows from previous results that CEU w.r.t. μ coincides with MEU w.r.t. $\mathcal{P}$. Hence, the behavior of people faced with the Ellsberg urn can be interpreted as a pessimistic evaluation considering all possible compositions of the urn (i.e., proportions of black and yellow balls). ◇

Remark 5.25 The latter interpretation of the Ellsberg paradox based on the MEU model may appear more convincing and easy to explain than with the CEU model.

This is indeed the most attractive feature of the MEU and related models, because the decision behavior can be explained by a combination of ambiguity with the simple expected utility model. There is however an important drawback: it is not easy in a practical situation to determine the set of priors because of its so many degrees of freedom, and by comparison the CEU model is much more parsimonious (but maybe difficult to interpret). ◇

We end this section by briefly giving a behavioral foundation of MEU, due to Gilboa and Schmeidler [154]. It is based on the Anscombe-Aumann model, like the axiomatization of Schmeidler. The only change with respect to the latter is that the independence axiom for comonotonic acts [see (5.23)] is changed as follows: For all $f, g \in F$, for all constant act $h \in F$ and all $\lambda \in]0, 1[$, (5.23) holds. The comparison of the two axioms explains why CEU and MEU are distinct and none of them is a generalization of the other, because independence w.r.t. a constant act is neither weaker nor stronger than comonotonic independence.

We refer the readers to Wakker [339, Chap. 11] and Gilboa [153, Chap. 17] for further details and similar models.

5.4 Qualitative Decision Making

So far we have taken for granted that representations of preference should be numerical, that is, with all the power and structure of real numbers (more precisely, all of our representations so far in this chapter yield interval scales; see Sect. 6.2.2 for an explanation of this term). On the other hand, we have seen that, as far as a descriptive point of view is taken, the decision mechanisms of people are sometimes based on a very rough perception of uncertainty. A good example of this is the inverse S-shape curve of the distortion function of probabilities, exhibiting likelihood insensitivity, whose extreme case is the crude distinction between "sure to happen," "sure not to happen" and "don't know" (Sect. 5.2.6). Also, some well-known very simple decision models, like the maximin and the maximax criteria, which consist in taking the act maximizing either the worst outcome (maximin model) or the best outcome (maximax model), and thus describing an extreme pessimism or optimism, need only a comparison between the utility of the outcomes.

The question is then: *Do we need the richness of the set of real numbers and its powerful algebraic structure to build decision models that are supposed to mimic human behavior?* The crude perception of uncertainty and the minimax and maximax models evoked above certainly do not. On the other hand, everyone will agree that these examples, although having some reality, are extreme simplifications of it, and that something more involved is needed, ideally something *in between* the rich models around expected utility and the primitive models above. This is exactly where qualitative decision making enters the picture.

Qualitative decision making deals with models built on ordinal scales (see Sect. 6.2.2), that is, scales where only comparison between numbers are allowed.

In addition, models in qualitative decision making use finite scales most of the time, hence the name "qualitative," which refers to a totally ordered set of labels, like "very bad," "bad," "good," etc. The natural representation of uncertainty in this framework is not probability theory, which requires the full power of interval or ratio scales, but possibility theory, which can accommodate with a fully ordinal framework (the readers are advised to read beforehand Sect. 7.7 on possibility theory). It seems that most of the work along this line has been done by Dubois, Prade and colleagues. Our exposition follows [115, 116], where the readers can find a full exposition including proofs, as well as a detailed bibliography. In what follows, the notation already introduced before (states of nature, consequences, etc.) remains valid.

5.4.1 Decision Under Risk

In decision under risk, we assume that a possibility distribution π is known on S. We define a qualitative scale L, i.e., a totally ordered finite set, denoting its least and greatest element by 0 and 1 respectively. Then π takes values in L, as well as the utility function $u : C \to L$. A peculiarity of this setting (and certainly one of its questionable aspects) is that the same scale is used to represent uncertainty and utility, so that there is a commensurability problem which must be solved in practice.

Supposing π and u to be known, two models have been proposed, the pessimistic one (Whalen [348]) and the optimistic one (Yager [351]). For any act f, they are respectively defined as follows:

$$U_*(f) = \inf_{s \in S}(n(\pi(s)) \vee u(f(s))) \tag{5.27}$$

$$U^*(f) = \sup_{s \in S}(\pi(s) \wedge u(f(s))), \tag{5.28}$$

where n is the (unique because L is finite) negation on L; i.e., the mapping putting L upside down. We make two comments, helping in understanding the formulas.

(i) If $\pi(s) = 1$ for every s (total ignorance), then the above formulas reduce respectively to the maximin and the maximax criteria alluded to above. Hence, the above models can be seen as a generalization of these simplistic criteria. The effect of π is to penalize (by a threshold effect) consequences corresponding to states of nature that are not fully possible.
(ii) By comparison with Definition 4.67, one sees that $U_*(f)$ is nothing but the weighted minimum applied to the vector of utility of consequences of f, and π plays the rôle of the weight vector, while $U^*(f)$ corresponds to the weighted maximum with the same substitution. It follows that, using Theorem 4.69 and Remark 4.70, these models are particular Sugeno integrals. They can be

rewritten as follows:

$$U_*(f) = \fint u(f)\,\mathrm{dNec} \tag{5.29}$$

$$U^*(f) = \fint u(f)\,\mathrm{d}\Pi, \tag{5.30}$$

where Π, Nec are the possibility and necessity measures generated by π. The analogy with expected utility is now clear: instead of taking the classical expectation, one takes expectation w.r.t. a possibility or a necessity measure, through the Sugeno integral.

Remark 5.26 It should be noted that these optimistic and pessimistic models do not satisfy monotonicity although we considered this as a basic rationality requirement. Take for example three states of nature, $L = \{0, 1, 2, 3, 4\}$, the acts f, g whose utility vectors are $(1, 2, 3)$ and $(2, 4, 4)$, and $\pi = (4, 2, 2)$. Then

$$U^*(f) = (4 \wedge 1) \vee (2 \wedge 2) \vee (2 \wedge 3) = 2$$
$$U^*(g) = (4 \wedge 2) \vee (2 \wedge 4) \vee (2 \wedge 4) = 2.$$

A similar example can be found for the pessimistic model. ◇

As in the classical framework, one can identify acts with lotteries, at the difference that probability distributions are replaced by possibility distributions. We denote them by $(\pi_1, x_1; \ldots; \pi_n, x_n)$, where $x_1, \ldots, x_n$ are consequences/outcomes, and $\pi_1, \ldots, \pi_n$ their associated possibility degrees. The mixture of lotteries with two consequences is defined as follows: take $\pi = (\alpha, x; \beta, y)$ and consider the compound lottery $(\lambda, x; \mu, \pi)$. Then

$$(\lambda, x; \mu, \pi) = ((\lambda \vee (\mu \wedge \alpha)), x; \mu \wedge \beta, y).$$

In [116], based on this mixture operation, an axiomatization of the pessimistic model was proposed, similar to the one of Fishburn/von Neumann-Morgenstern.

The following example [116] illustrates the use of the pessimistic model, and shows how well it fits intuition. It is based on an example given by Savage [284, pp. 13–15].

Example 5.27 (The Savage's omelette revisited) Suppose you are preparing an omelette, and you have already broken five eggs into a bowl and mixed them. So far everything is fine. A sixth egg is remaining, but you are wondering if this sixth egg is fresh or not. Breaking this egg into the omelette would make the omelette bigger, but if the egg is rotten the whole preparation is wasted. Having read this chapter, you judiciously define two states of nature: egg is fresh (denoted by s_f) and egg is rotten (denoted by s_r). After thought, you find that there are three possible ways to act:

(i) Break the egg into the omelette (this is act f_{bo});
(ii) Break the egg into another bowl to examine it (this is act f_{bb});
(iii) Throw away the egg (this is act f_{th}).

Enumerating the consequences of each act under each state of nature yields:

(i) act f_{bo}, state s_f: you have a 6-egg omelette (consequence denoted by $6e$);
(ii) act f_{bo}, state s_r: your omelette is wasted and it is better to think of another menu (ow);
(iii) act f_{bb}, state s_f: you have a 6-egg omelette, but another bowl to wash ($6eb$);
(iv) act f_{bb}, state s_r: you have a 5-egg omelette, and another bowl to wash ($5eb$);
(v) act f_{th}, state s_f: you have a 5-egg omelette and an egg is wasted ($5ew$);
(vi) act f_{th}, state s_r: you have a 5-egg omelette ($5e$).

About preference on these consequences, you may have the following: the worst situation is to waste the omelet, then you prefer not to waste an egg. Then, if possible, you prefer not to have a bowl to wash if the egg is rotten (it is better to throw it away immediately). Finally, the best situation is a 6-egg omelet, and if possible, no bowl to wash. Defining a qualitative scale $L = \{0, 1, 2, 3, 4, 5\}$ (labels could be used instead of numbers), the previous preferences on consequences can be translated into a utility function $u : C \to L$ defined by:

$$u(6e) = 5, \quad u(6eb) = 4, \quad u(5e) = 3$$
$$u(5eb) = 2, \quad u(5ew) = 1, \quad u(ow) = 0.$$

It remains to compute the pessimistic utility U_* for each act. Recall that π is defined on L, too. Since S has only two states, we have $n(\pi(s_f)) = \text{Nec}(\{s_r\})$ and $n(\pi(s_r)) = \text{Nec}(\{s_f\})$. In addition, we have $\text{Nec}(\{s_f\}) \wedge \text{Nec}(\{s_r\}) = 0$. Hence, we obtain

$$\begin{aligned}
U_*(f_{bo}) &= \big(\text{Nec}(\{s_r\}) \vee u(6e)\big) \wedge \big(\text{Nec}(\{s_f\}) \vee u(ow)\big) \\
&= \text{Nec}(\{s_f\}) \\
U_*(f_{bb}) &= \big(\text{Nec}(\{s_r\}) \vee u(6eb)\big) \wedge \big(\text{Nec}(\{s_f\}) \vee u(5eb)\big) \\
&= 4 \wedge \big(\text{Nec}(\{s_f\}) \vee 2\big) \\
U_*(f_{th}) &= \big(\text{Nec}(\{s_r\}) \vee u(5ew)\big) \wedge \big(\text{Nec}(\{s_f\}) \vee u(5e)\big) \\
&= \begin{cases} 1, & \text{if Nec}(\{s_f\}) > 0 \\ \big(\text{Nec}(\{s_r\}) \vee 1\big) \wedge 3, & \text{otherwise.} \end{cases}
\end{aligned}$$

The best decisions are therefore:

(i) If $\text{Nec}(\{s_f\}) = 5$ (you are sure that the egg is fresh), it is better to break the egg into the omelet;
(ii) If $\text{Nec}(\{s_f\}) \in \{2, 3, 4\}$ (you are rather sure that the egg is fresh), it is indifferent between breaking the egg into the omelet or in a bowl;

(iii) If $\text{Nec}(\{s_f\}) < 2$ and $\text{Nec}(\{s_r\}) < 2$ (you are rather ignorant of the quality of the egg), it is better to break the egg in a bowl;
(iv) If $\text{Nec}(\{s_r\}) = 2$ (you have a little doubt on the egg), it is indifferent to break the egg in a bowl or to throw it away;
(v) If $\text{Nec}(\{s_r\}) > 2$ (you do not think that the egg is fresh), it is better to throw it away.

Note that the numbers play no rôle (in the cardinal sense) in the computation, only order matters. ◇

5.4.2 Decision Under Uncertainty

The foregoing pessimistic and optimistic models are the qualitative equivalents of expected utility, and could be used in decision under uncertainty as well. As expected utility is a particular case of Choquet expected utility, and because the pessimistic and optimistic models are particular Sugeno integrals, it seems natural to introduce *Sugeno expected utility* (abbreviated by SugEU) as the qualitative version of Choquet expected utility. This model is given by, using the same notation as in Sect. 5.4.1,

$$\text{SugEU}(f) = \fint u(f)\, \mathrm{d}\mu, \tag{5.31}$$

where μ is a capacity on S.

Dubois et al. [115] have proposed an axiomatization of SugEU, in a way very close to the axiomatization of Savage. Recall that the sure-thing principle is the central axiom in Savage's framework. It is easy to see that it is not satisfied even by the foregoing optimistic and pessimistic models. Indeed, a strict preference $f_E h \succ g_E h$ can easily be turned into an indifference $f_E h' \sim g_E h'$, even with U^* being the standard maximum. Take for example $S = \{s_1, s_2\}$, u the identity function, $L = \{0, 1, 2, 3\}$, $E = \{s_1\}$, $\pi(s_1) = 3, \pi(s_2) = 3$. Then, with $f = (2, 0)$, $g = (1, 0)$, $h = (0, 0)$, we get $U^*(f_E h) = 2 > 1 = U^*(g_E h)$. However, with $h' = (0, 3)$ we get $U^*(f_E h') = 3 = U^*(g_E h')$. Similar examples can be found with the minimum as well. It can be shown that these models indeed satisfy a weak version of the sure-thing principle:

$$f_E h \succ g_E h \Rightarrow f_E h' \succcurlyeq g_E h' \qquad (f, g, h, h' \in F).$$

In the following axiomatization of SugEU, axiom WP3, similar to P3 of Savage, is central and can be considered to replacing the sure-thing principle. For a consequence $\alpha \in C$, the constant act yielding α for every state of nature is denoted by α_S.

WP3 For all $E \subseteq S$, all $h \in F$, all $\alpha, \beta \in C$ such that $\alpha_S \succcurlyeq \beta_S$, we have $\alpha_E h \succcurlyeq \beta_E h$.

The next two axioms capture the ordinal nature of the model, by involving the min and max operations on acts.

RCD (Restricted Conjunctive Dominance) For all acts $f, g \in F$ and all consequences $\alpha \in C$, if $g \succ f$ and $\alpha_S \succ f$, then $g \wedge \alpha_S \succ f$;
RDD (Restricted Disjunctive Dominance) For all acts $f, g \in F$ and all consequences $\alpha \in C$, if $f \succ g$ and $f \succ \alpha_S$, then $f \succ g \vee \alpha_S$.

The fact that SugEU satisfies these two properties is immediate from the fact that the Sugeno integral satisfies positive $\wedge$- and $\vee$-homogeneity [Theorem 4.43(i) and (ii)]. In addition, these properties together with monotonicity characterize the Sugeno integral (Theorem 4.60). The following theorem can be seen as an analog of the latter in a decision framework.

Theorem 5.28 (Axiomatization of SugEU) *Let $\succcurlyeq$ be a binary relation on F, and suppose S and C to be finite. The following two propositions are equivalent.*

(i) $\succcurlyeq$ is complete and transitive, and satisfies axioms WP3, RCD and RDD;
(ii) There exist a finite qualitative scale $(L, \geqslant)$, a unique utility function $u : C \to L$, and a unique capacity μ on 2^S such that SugEU given by (5.31) represents $\succcurlyeq$.

We now give an axiomatization of the pessimistic and optimistic models U_*, U^*, which are particular cases of SugEU. The two key axioms leading to these models are:

PES (pessimism) For all acts $f, g \in F$, all events $E \subseteq S$,

$$f_E g \succ f \Rightarrow f \succcurlyeq g_E f;$$

OPT (optimism) For all acts $f, g \in F$, all events $E \subseteq S$,

$$f \succ f_E g \Rightarrow g_E f \succcurlyeq f.$$

The meaning of PES is the following. Consider an act f that can be improved by changing its consequences if E^c realizes (put differently, f is not favorable when E^c realizes). Then there is no way of improving f by changing its consequences on E. This rather strange conclusion can be explained if the decision maker is pessimist, in the sense that E and E^c are considered to be equally plausible, and the decision maker focuses on bad consequences. The following example illustrates this.

Example 5.29 (Dubois et al. [116, p. 468]) Consider Gamble 1, consisting in tossing a coin, and makes you win \$10,000 if heads comes up, and lose \$10,000 if tails comes up. This gamble is normally less preferred than Gamble 2, yielding \$10,000 if heads, and nothing if tails comes up. Consider now Gamble 3, where you can win \$20,000 if heads, and lose \$10,000 if tails. If you are pessimist in the above sense, then, although preferring Gamble 2 to Gamble 1, you are indifferent between Gamble 1 and 3. ◇

This very peculiar behavior, which could be called *extreme pessimism* (and quite questionable in a quantitative framework) is of course not representable by EU, and is the essence of the pessimistic model. A similar interpretation can be obtained for axiom OPT.

The characterization of the pessimist and optimist models are given in the next theorems.

Theorem 5.30 (Axiomatization of the pessimist qualitative model) *Let $\succcurlyeq$ be a binary relation on F, and assume S and C to be finite. The following two propositions are equivalent.*

(i) *$\succcurlyeq$ is complete and transitive, and satisfies axioms WP3, RDD and PES;*
(ii) *There exist a finite qualitative scale $(L, \geqslant)$, a unique utility function $u : C \to L$, and a unique possibility distribution on S such that U_* given by (5.27) represents $\succcurlyeq$.*

Theorem 5.31 (Axiomatization of the optimist qualitative model) *Let $\succcurlyeq$ be a binary relation on F, and assume S and C to be finite. The following two propositions are equivalent.*

(i) *$\succcurlyeq$ is complete and transitive, and satisfies axioms WP3, RCD and OPT;*
(ii) *There exist a finite qualitative scale $(L, \geqslant)$, a unique utility function $u : C \to L$, and a unique possibility distribution on S such that U^* given by (5.28) represents $\succcurlyeq$.*

Chapter 6
Decision with Multiple Criteria

This second application chapter explores decision with multiple criteria, usually called "multicriteria decision making" (MCDM). It deals with situations where the decision maker has to make decision considering together several points of view (criteria), which are often antagonistic. This covers many everyday life decision problems, like choosing a restaurant or a movie, buying a new car or renting an apartment, etc. Our presentation of the topic is unconventional, although based on classical concepts and results from multiattribute utility theory (MAUT) and measurement theory. We start from scratch, and ask ourselves under which conditions does the decomposable model (a very commonly used model consisting in assigning numerical scores on each criterion and aggregating them into a single overall score), exist, and how to build it. While the answer to the first question (conditions of existence) is well known and is a standard result of measurement theory, the answer to the second question is less obvious, and it is precisely here that we are unconventional. We show that in order to build scores on criteria, either difference measurement can be applied (under the assumption of weak difference independence), or reference points must be found on each criterion, which permits to apply the MACBETH method. With the latter method, we show that we are naturally lead to the use of the Choquet integral for aggregating the scores of the criteria. Interestingly enough, with the former approach based on the assumption of weak difference independence, it is known since the seventies through the works of Dyer and Sarin, Keeney and Raiffa, that the only possible model of aggregation is the multilinear model, also called Owen extension of a capacity. Remembering that the Choquet integral is the Lovász extension of a capacity, this shows in a striking way that capacities are firmly rooted in multicriteria decision making.

Section 6.10 introduces the notion of importance of criteria and interaction between criteria. It is shown that the interaction transform and Banzhaf interaction transform of Chap. 2 can be seen as interaction indices (defined through the average total variation of an aggregation function) of the Choquet integral and the multilinear model, respectively. The chapter ends with Sect. 6.11 on ordinal models, that is,

M. Grabisch, *Set Functions, Games and Capacities in Decision Making*,
Theory and Decision Library C 46, DOI 10.1007/978-3-319-30690-2_6

built without numbers. Here, the Sugeno integral appears to be the natural tool for aggregating scores on criteria, however, the model has a rather poor discriminative power, which can be improved using lexicographic refinement.

The readers may consult the classical textbook of Keeney and Raiffa [205] for full details on MAUT (see also Dyer [119]), as well as other monographs on MCDM, e.g., Pomerol and Barba-Romero [269], the edited book by Figueira et al. [139], and the survey by the author and Labreuche [175] on the application of the Choquet integral in multicriteria decision making.

6.1 The Framework

Decision under multiple criteria deals with situations where an agent (called the *decision maker*) has to choose between several objects, alternatives, options, etc. (called hereafter *alternatives*), considering together several points of view or *criteria* pertaining to different aspects, descriptors or *attributes*, which describe the alternatives under consideration. In contrast to decision under risk and uncertainty (Chap. 5), the decision maker is supposed to have full information on the alternatives, that is, on the values taken by the attributes, hence there is here no uncertainty nor contingencies entering the picture. The decision problem is nevertheless difficult because most of the time, there are antagonistic points of view: some alternatives may be best preferred under some point of view, but are much less attractive under another point of view. The fundamental difficulty behind is simply that there is no natural complete order on multidimensional objects.

We now introduce the main ingredients of a multicriteria decision problem. An object or alternative of interest x is represented by a vector $(x_1, \dots, x_n)$, where x_i is the value of the ith *attribute* or *criterion*[1] of x, $i = 1, \dots, n$. We denote by X_i the set of all possible values of the ith attribute, $i = 1, \dots, n$. The set of all *potential* alternatives, considering all possible combinations of the values taken by the attributes, is $X = X_1 \times \cdots \times X_n$. We set $N = \{1, \dots, n\}$ the index set of attributes/criteria.

NOTE: As all the subsequent discussion becomes void if $n = 1$, we assume throughout the chapter that $n \geqslant 2$.

The preference of the decision maker over the alternatives is represented by a binary relation $\succcurlyeq$ on X, which is supposed to be a complete preorder (complete and transitive, see Sect. 1.3.1). We denote by $\sim$ and $\succ$ the symmetric and asymmetric parts of $\succcurlyeq$. As in Chap. 5, $x \succcurlyeq y$ reads "x is preferred to y" or "x is at least as good

[1]We indifferently use the two terms, although a distinction is usually made between them, "attribute" referring to the objective description of the object (e.g., the maximum speed of this car is 200 km/h) and "criterion" to the subjective perception of the decision maker (e.g., this car is fast).

as y," $x \succ y$ reads "x is strictly preferred to y, and $x \sim y$ reads "x is indifferent to y." The relation $\sim$ is called the *indifference relation*, and is an equivalence relation.

Example 6.1 Many everyday life decision problems fall under the scope of decision with multiple criteria: buy a new car, rent a house, buy a flight ticket, choose a movie, choose a menu in a restaurant, etc. Taking the example of the flight ticket, we consider X as the set of possible flight tickets, say, from Montréal to Moscow, for a given date. The set of attributes could be: price of the ticket, duration of the flight, departure time, arrival time, number of transfers, company. We remark that attributes 1, 2 and 5 are numerical (nonnegative numbers), 3 and 4 are time indications in format hh:mm, and the sixth one is alphabetic (string of characters). ◇

One of the main aims of decision theory is to build a numerical representation of preference, in the following sense: find a function $u : X \to \mathbb{R}$ such that

$$x \succcurlyeq y \Leftrightarrow u(x) \geqslant u(y) \qquad (x, y \in X). \tag{6.1}$$

The function u is called the (overall) *utility* or *value function*.[2] A fairly general and commonly used model for representing $\succcurlyeq$ is the *monotone decomposable model*:

$$u(x) = F(u_1(x_1), \ldots, u_n(x_n)) \qquad (x \in X) \tag{6.2}$$

where $u_1, \ldots, u_n$ are (marginal) value functions from X_i to $\mathbb{R}$, and F is a function which is increasing in each variable.

The functions u_i play the same rôle as the utility function in decision under uncertainty or risk: it translates "outcomes" (here, values on attributes, which are not necessarily numerical) into real numbers. The intuition behind is that $u_i(x_i)$ reflects the satisfaction of the decision maker regarding the value x_i: the higher the satisfaction, the higher the quantity $u_i(x_i)$, and the higher the (positive) impact on $u(x)$ by means of the increasing function F. For this reason, we may call *score*[3] of x on attribute i the quantity $u_i(x_i)$. Taking the foregoing example of the flight ticket, one may define value functions on a scale from 0 (worst) to 100 (best), and writing u_6(Syldavian Airlines) = 90 would indicate that the decision maker likes a lot this company for some reason (quality of on board service, meals, movies, possibility to get bonus miles, etc.).

Function F aggregates the scores obtained on each criterion to produce the overall value (score) of the alternative x. A very common example for F is the weighted arithmetic mean, with positive weights. One may be less restrictive and

[2] In MCDM and MAUT, the usage is to call it a value function. The term "utility function" is more for decision under risk and uncertainty.

[3] Anticipating the topic of Sect. 6.2, readers should be aware that the term "score" suggests a cardinal interpretation of the numbers, that is, it quantifies the intensity of preference. As we will see, this should not be taken for granted.

allow F to be only nondecreasing in each place. Also, a more general model is the *decomposable model*, where F is one-to-one in each variable.

The (monotone) decomposable model is intuitively appealing because it corresponds to the way in which people naively do when faced with multicriteria decision making: for each alternative, put scores for each criterion and aggregate the scores to get an overall score, for example using a weighted arithmetic mean. This apparent simplicity and obviousness hides many theoretical and conceptual intricacies, which will be developed in this chapter. Basically, two fundamental questions arise:

Q1 *Under which conditions on X and $\succcurlyeq$ does the (monotone) decomposable model exist?*

Q2 *How to build $u_1, \ldots, u_n$, how to choose (and build) F?*

We will answer these questions in the sequel, and show that the Choquet integral appears as a natural candidate for the function F. The first question will be elucidated with the help of measurement theory, presented in the next section.

6.2 Measurement Theory

Measurement theory, not to be confused with measure theory, tells us the art of measuring properties of objects, either of the real world or more conceptual ones, that is, how to put numbers on these objects so that they represent physical comparisons or manipulations made on the objects. The interested readers may consult the impressive monograph of Krantz et al. [217], as well as the one of Roberts [275]. More accessible surveys including recent developments related to decision with multiple criteria and multiattribute utility theory as well as a wide literature can be found in [39, 119].

6.2.1 The Fundamental Problem of Measurement

Formally, let us consider A a set of objects, $\succcurlyeq$ a binary relation on A, called the *comparison relation*, which pertains to some property of the objects, and $* : A \times A \to A$ a binary operation on A, called *concatenation*. We call $\mathcal{A} = (A, \succcurlyeq, *)$ a *relational system*. The concatenation operation may be omitted in the relational system.

Example 6.2 We may consider for A any kind of objects, like pencils, turnips, bananas, gnus, movies, etc. Common examples of physical comparison relation are: length, weight, temperature, etc., but could be also preference: $a \succcurlyeq b$ would read "a is longer (heavier, hotter) than b," or "a is preferred to b." The concatenation relation depends on $\succcurlyeq$ and is not always defined. For length it consists in putting two pencils end to end and considering their total length, for weight, in putting two turnips in the same basket and weighing the basket, etc. There is no significant concatenation relation for temperature, nor for preference. ◇

The fundamental problem of measurement is to find a homomorphism between two relational systems $\mathcal{A} = (A, \succcurlyeq, *)$ and $\mathcal{B} = (B, \succcurlyeq', *')$, that is, a function $f : A \to B$ satisfying

$$a \succcurlyeq b \Leftrightarrow f(a) \succcurlyeq' f(b) \tag{6.3}$$

$$f(a * b) = f(a) *' f(b). \tag{6.4}$$

Most often $\mathcal{B}$ is nothing other than $(\mathbb{R}, \geqslant, +)$, the set of real numbers endowed with the usual comparison relation, and addition. In this case, we already note that our definition of numerical representation of a preference, given by (6.1), is nothing but (6.3). Such kind of measurement problem, without considering concatenation, is called *ordinal measurement*.[4]

We call *scale* the triplet $(\mathcal{A}, \mathcal{B}, f)$, and say that the scale is *numerical* if $B = \mathbb{R}$. If the context is understood, we may call f the scale. In the above example, a function f assigning to each pencil its length in centimeters is clearly a numerical scale.

6.2.2 Main Types of Scales

We know that even for a given comparison relation on a property like length, many scales have been defined in the past: meters, but also feet, inches, miles, light years, etc., and the same is true for weight, temperature, magnetic field, and all physical scales. In most cases, all is a matter of a multiplicative constant, but not always: the conversion from Celsius degrees (°C) to Fahrenheit degrees (°F) is achieved through the transformation $x \mapsto \frac{9}{5}x + 32$.

---

[4] Although in the sequel we will not refer any more to the concatenation operation, we give a few explanations on it. Measurement including a concatenation relation is called *extensive measurement*. The following important theorem can be shown [217, Theorem 3.1]: Let $\mathcal{A}=(A, \succcurlyeq, *)$ be a relational system. Then (6.3) and (6.4) both hold if and only if $\succcurlyeq$ is complete and transitive, and the concatenation operation satisfies:

- Weak associativity: $a * (b * c) \sim (a * b) * c$;
- Monotonicity: $a \succcurlyeq b$ iff $a * c \succcurlyeq b * c$ iff $c * a \succcurlyeq c * b$;
- Archimedean: If $a \succcurlyeq b$, then for any $c, d \in A$, there exists a positive integer n s.t. $na{*}c \succcurlyeq nb{*}d$, where na is defined inductively as $1a = a$, $(n + 1)a = na * a$.

Proving this theorem amounts to reducing it to its analog for strictly ordered groups, the well-known Hölder's theorem: Let $\succcurlyeq$ be a transitive, complete and antisymmetric relation on A, and suppose that $(A, *)$ is a group with neutral element e satisfying (1) $a \succcurlyeq b$ implies $a * c \succcurlyeq b * c$ and $c * a \succcurlyeq c * b$, and (2) $a \succ e$ and $b \in A$ imply $na \succ b$ for some $n \in \mathbb{N}$ (ordered group). Then $(A, \succcurlyeq, *)$ is isomorphic to a subgroup of $(\mathbb{R}, \geqslant, +)$.

More generally, if f is a scale and $\varphi : B \to B$ a mapping such that $\varphi \circ f$ is still a homomorphism between $\mathcal{A}$ and $\mathcal{B}$, φ is said to be an *admissible transformation*. Classes of admissible transformations define types of scale, whose main ones are summarized in Table 6.1. Ratio scales are the most common for

Type of scale	Admissible transformations	Examples
Absolute scale	$\varphi = \text{Id}$	Counting
Ratio scale	$\varphi(x) = \alpha x, \alpha > 0$	Mass, length, temperature in K
Interval scale	$\varphi(x) = \alpha x + \beta, \alpha > 0$	Temperature in °C, calendar
Ordinal scale	φ increasing	Mohs scale (hardness), Beaufort scale (wind)
Nominal scale	φ arbitrary	List of nouns

Table 6.1 Main types of scale

physical magnitudes. They are characterized by the presence of a "true" zero, having an absolute physical meaning (e.g., 0 K indicates total absence of molecular movement). By contrast, interval scales have a zero whose position is a matter of convention and may be shifted: 0 °C indicates the freezing temperature of water, while 0 °F (−17.8 °C) is the lowest temperature observed by Fahrenheit in Dantzig[5] during winter 1708–1709. Similarly, the first year in a calendar depends on the cultural context where it is used: the Gregorian calendar starts with the birth of Jesus-Christ, while in Japan, a new calendar starts each time a new emperor is going to reign.

Numbers on an ordinal scale do not have in general their usual cardinal meaning and should be manipulated with care, because only order matters: If objects a, b, c, d have magnitude $1, 2, 3, 4$ on an ordinal scale, any sequence of four increasing numbers would constitute another valid scale, like $-124, 0, \pi^2, e^4$. As a consequence, propositions made on these objects and involving usual arithmetic operations like $+, \times$ (for instance, a is twice as b) are in general meaningless, in the sense that they do not remain valid under an admissible transformation of the scale. Only propositions involving comparisons (using minimum, maximum) remain always valid.

6.2.3 *Ordinal Measurement*

As we said above, ordinal measurement corresponds exactly to the problem of the numerical representation of preferences; see (6.1). When A is finite, there exists a trivial numerical representation, because it suffices to consider the equivalence classes of $\sim$ and number them in increasing preference order. When A is countable, the result still holds (see, e.g., Fishburn [141, Theorem 2.2] for a proof). If A is

[5]Nowadays, Gdańsk, in Poland.

uncountable, such a representation may not exist, as the following famous example shows.

Example 6.3 (The lexicographic ordering (Debreu [74])) The *lexicographic ordering* is a binary relation on $\mathbb{R}^2$ defined as follows:

$$(a,b) \succ_{\text{lex}} (c,d) \Leftrightarrow \begin{cases} a > c, & \text{or} \\ a = c \text{ and } b > d \end{cases}.$$

There is no numerical representation of this order, for, if such a representation f would exist, by $(a,1) \succ_{\text{lex}} (a,0)$ we would have $f(a,1) > f(a,0)$. Since $\mathbb{Q}$ is dense in $\mathbb{R}$, there would exist $g(a) \in \mathbb{Q}$ such that $f(a,1) > g(a) > f(a,0)$, and this would define a mapping $g : \mathbb{R} \to \mathbb{Q}$. Now, this mapping would be injective because taking $a > b$, we have

$$g(a) > f(a,0) > f(b,1) > g(b).$$

But no injective mapping from the real numbers to the rational numbers exists. ◇

A necessary and sufficient condition for the existence of a numerical representation of a preorder is order-denseness. Let $\succcurlyeq$ be a complete, antisymmetric and transitive binary relation on A. We say that $B \subseteq A$ is *order-dense* in $(A, \succcurlyeq)$ if for all $a, b \in A \setminus B$ such that $a \succ b$, there exists $c \in B$ such that $a \succ c \succ b$. This definition is a generalization of the usual notion of denseness (e.g., $\mathbb{Q}$ is dense in $\mathbb{R}$).

Theorem 6.4 (Existence of ordinal measurement) *Let $(A, \succcurlyeq)$ be a relational system. There exists a homomorphism from $(A, \succcurlyeq)$ to $(\mathbb{R}, \geqslant)$ if and only if $\succcurlyeq$ is a complete preorder, and the quotient set $A/\sim$ contains a countable subset that is order-dense in A.*

(See Krantz et al. [217, Chap. 2, Theorem 2], or Fishburn [141, Theorem 3.1] for a proof.)

Remark 6.5 This fundamental result was proved many times, the first one by G. Cantor[6] [42] for the countable case. It seems that the representation part ("if" part) was first proved by Milgram [242], and there is an incomplete proof by Birkhoff [30, 1948 edition, pp. 31–32]. The above theorem is by the way often called the Birkhoff-Milgram theorem (Roberts [275]). The representation part was also proved by Debreu [74]. See also Herden and Mehta [194] for a generalization and a detailed study of this result. ◇

Going back to our representation problem (6.1), Theorem 6.4 tells us that there exists a mapping $u : X \to \mathbb{R}$ if and only if $\succcurlyeq$ is a complete preorder (which is

[6]Georg Ferdinand Ludwig Philipp Cantor (Saint Petersburg, 1845 – Halle, 1918) is a German mathematician, to whom we owe set theory and a deep and revolutionary study of infinity, through the notions of cardinal and ordinal numbers.

assumed), and $X/_\sim$ has a countable order-dense subset. It remains to find a condition to force the decomposability; i.e., u has the form (6.2).

Definition 6.6

(i) A preorder is *weakly independent* if for every $i \in N$, every $x, y, z, z' \in X$, we have

$$(x_i, z_{-i}) \succcurlyeq (y_i, z_{-i}) \Leftrightarrow (x_i, z'_{-i}) \succcurlyeq (y_i, z'_{-i}),$$

where we have omitted braces for singletons (see Sect. 1.1(xi) for the notation).

(ii) A preorder is *weakly separable* if for every $i \in N$, every $x, y, z \in X$, $(x_i, z_{-i}) \succ (y_i, z_{-i})$ implies that there is no $z' \in X$ such that $(y_i, z'_{-i}) \succ (x_i, z'_{-i})$.

Clearly, weak independence implies weak separability, but not the converse. In words, weak independence says that the preference between two values of an attribute, *ceteris paribus*, should not depend on the values of the remaining attributes. Taking the example of the flight ticket, where the criteria are cost, duration, departure time, arrival time, direct/non-direct, company, you may have the preference

$$(1000€, 12\text{h}, 21{:}00, 12{:}00, \text{direct, Syldavian Airlines}) \succcurlyeq (1000€, 12\text{h}, 21{:}00, 12{:}00, \text{direct, Bordurian Airlines}).$$

Weak independence implies that the values on attributes cost, duration, departure and arrival time, direct/non-direct, may be changed without reversing preference, for example:

$$(600€, 18\text{h}, 21{:}00, 18{:}00, \text{non-direct, Syldavian Airlines}) \succcurlyeq (600€, 18\text{h}, 21{:}00, 18{:}00, \text{non-direct, Bordurian Airlines}).$$

This requirement is natural in most cases, although one can find examples where this is not true, as in the following one.

Example 6.7 (Violation of weak separability) Let us consider X to be a set of cars, described with four attributes: color, type of car (sports, sedan, wagon, supermini), speed and comfort. Suppose the decision maker has the following preference:

$$(\text{metallic grey, sedan, 200 km/h, comfortable}) \succ (\text{red, sedan, 200 km/h, comfortable}),$$

which means that metallic grey is preferred over red for a sedan type car. Now, it would not be surprising if, when considering a sports car, this preference would be reversed:

$$\text{(red, sports, 300 km/h, not comfortable)} \succ \text{(metallic grey, sports, 300 km/h, not comfortable)},$$

because for a sports car, red is more appealing than metallic grey. In this case, the preference over colors is *conditional* on the type of car. ◇

Remark 6.8

(i) The name "weak independence" suggests the existence of a stronger notion of independence. Indeed, in MAUT, a nonempty subset $I \subset N$ is said to be *(preference) independent* of its complement $N \setminus I$ if $(x_I, z_{-I}) \succcurlyeq (y_I, z_{-I})$ is equivalent to $(x_I, z'_{-I}) \succcurlyeq (y_I, z'_{-I})$ (see, e.g., Keeney and Raiffa [205], Dyer and Sarin [120], Dyer [119]). Hence, weak independence means independence for every singleton. If independence holds for every nonempty subset $I \subset N$, $\succcurlyeq$ is said to satisfy *mutual preference independence*. Observe that the latter property is the counterpart of the sure-thing principle in decision under uncertainty (Sect. 5.3.2);

(ii) Since in MAUT several notions of independence exist (like preference independence, utility independence and additive independence), we should speak of weak preference independence. However, we nevertheless drop the term "preference" because we are concerned in this chapter with only this type of independence. Note also that weak independence is sometimes called "weak separability", e.g., by Wakker [335].

◇

The following can be shown and gives a complete answer to Question Q1.

Theorem 6.9 (Existence of a monotone decomposable model) *Let* $X = X_1 \times \cdots \times X_n$ *and* $\succcurlyeq$ *be a binary relation on X. The following holds.*

(i) (Krantz et al. [217, Chap. 7, Theorem 1]) A numerical representation of $\succcurlyeq$ *in the form (6.2) with F increasing in each place exists if and only if* $\succcurlyeq$ *is a complete and weakly independent preorder, and* $X/_\sim$ *has a countable order-dense subset;*

(ii) (Bouyssou and Pirlot [38, Proposition 8]) A numerical representation of $\succcurlyeq$ *in the form (6.2) with F nondecreasing in each place exists if and only if* $\succcurlyeq$ *is a complete and weakly separable preorder, and* $X/_\sim$ *has a countable order-dense subset.*

A similar result for the decomposable model (with F being one-to-one in each place) exists. It suffices to replace $\succcurlyeq$ by $\sim$ in the definition of weak independence.

Remark 6.10 Another way to interpret weak independence is the following: under this assumption, it is meaningful to define a preference relation $\succcurlyeq_i$ on a single attribute X_i, for each $i \in N$, by letting $x_i \succcurlyeq_i y_i$ if $(x_i, z_{-i}) \succcurlyeq (y_i, z_{-i})$ for some $z \in X$. Now, $\succcurlyeq_i$ inherits the properties of $\succcurlyeq$, so that by Theorem 6.4, there exists a value function u_i on X_i representing $\succcurlyeq_i$.

The answer to Question Q2 takes much more effort. We first try to build the value functions $u_1, \ldots, u_n$. To this end, still further notions of measurement theory are necessary.

6.2.4 Difference Measurement

According to our classification of scales in Sect. 6.2.2, ordinal measurement yields an ordinal scale, which means that the numerical representation conveys little information. Indeed, the function u is defined up to an increasing transformation, making the numbers $u(x), u(y)$ assigned to alternatives x, y without cardinal meaning because usual arithmetic operations on them are in general not meaningful.

Therefore, a richer type of scale is needed, like an interval scale, defined up to a positive affine transformation. In decision making, interval scales are the more commonly used. A ratio scale is an interval scale for which one can find a "zero" with an absolute meaning. To build such scales requires much more information from the decision maker, and difference measurement is one possible way to build an interval scale. We describe in Sect. 6.4 another way to achieve this. Both methods are based on the perception by the decision maker of *intensity of preference*.

We consider a *quaternary relation* $\succcurlyeq^*$ on A, that is, a subset of $A^2 \times A^2$. We write $ab \succcurlyeq^* st$ for $a, b, s, t \in A$ if the difference of intensity of preference of a over b is greater or equal to the difference of intensity of preference of s over t. We use the same notation as before and denote by $\sim^*, \succ^*$ the symmetric and asymmetric parts of $\succcurlyeq^*$.

Given such a quaternary relation, *difference measurement* amounts to finding a mapping $f : A \to \mathbb{R}$ such that

$$ab \succcurlyeq^* st \Leftrightarrow f(a) - f(b) \geqslant f(s) - f(t). \tag{6.5}$$

Sufficient conditions for such a representation are given in the next theorem. This result relies on the following notion: a *standard sequence* in $(A, \succcurlyeq^*)$ is a sequence $a_1, \ldots, a_n$ of elements in A such that $a_{i+1}a_i \sim^* a_2a_1$ for $i = 2, \ldots, n-1$, and $a_2a_1 \not\sim^* a_1a_1$. In words, a sequence is standard if its elements are "equally spaced," in terms of preference. A standard sequence $a_1, \ldots, a_n$ is *strictly bounded* if there exist $s, t \in A$ such that $st \succ^* a_ia_1 \succ^* ts$, for $i = 2, \ldots, n$.

Theorem 6.11 (Existence of difference measurement) *(Krantz et al. [217, Chap. 4, Theorem 2]) Let $(A, \succcurlyeq^*)$ where $\succcurlyeq^*$ is a quaternary relation on A. There exists a mapping $f : A \to \mathbb{R}$ such that (6.5) is satisfied if*

(i) *The binary relation* $\succcurlyeq$ *on* A *defined by* $(a,b) \succcurlyeq (s,t)$ *if and only if* $ab \succcurlyeq^* st$ *is a complete preorder;*
(ii) $ab \succcurlyeq^* st$ *implies* $ts \succcurlyeq^* ba$;
(iii) $ab \succcurlyeq^* a'b'$ *and* $bc \succcurlyeq^* b'c'$ *imply* $ac \succcurlyeq^* a'c'$;
(iv) $ab \succcurlyeq^* st$ *and* $st \succcurlyeq^* aa$ *imply that there exist* $u, v \in A$ *such that* $au \sim^* st$ *and* $vb \sim^* st$;
(v) *Every strictly bounded standard sequence is finite.*

Moreover, if such a mapping f *exists, it is defined up to a positive affine transformation, and therefore is an interval scale.*

Remark 6.12 As it is clear from (ii) in Theorem 6.11, when writing ab, it is not assumed that a is preferred to b; i.e., that the difference $f(a) - f(b)$ is nonnegative. There exists a version of Theorem 6.11 dealing only with what is called *positive differences*, which is slightly more complicated [217, Chap. 4, Theorem 1], and has the same conclusion, that is, the existence of a function f satisfying (6.5), which is unique up to a positive affine transformation. ◇

Theorem 6.11 can be seen as the counterpart of Theorem 6.4 for quaternary relations. As we did for ordinal measurement, we have now to consider the case where the set of objects A is multidimensional in order to go back to our original representation problem [Eqs. (6.1) and (6.2)]. In particular, we are interested in building the value function u_i on each attribute X_i. Here arises a particularly thorny question. Ideally, one would like to build separately these value functions, mentally isolating each attribute from the rest, asking to the decision maker questions like:

Do you prefer a flight ticket at 400€ or 600€? (ordinal measurement)

or

Is the difference of preference between 300€ and 500€ greater or equal to the difference of preference between 400€ and 600€? (difference measurement)

Strictly speaking, these questions are meaningless if one considers that the decision maker is only able to compare alternatives, that is, elements of $X = X_1 \times \cdots \times X_n$. The above questions concern a single attribute X_i, and would suppose the existence of relations $\succcurlyeq_i$ and $\succcurlyeq_i^*$ on X_i; however our assumption is that only $\succcurlyeq, \succcurlyeq^*$ on X are available. The only way to make these questions meaningful would be that, for ordinal measurement, weak independence holds so that the comparison of elements $x_i, y_i \in X_i$ does not depend on the values of the other attributes, provided they are identical for the two alternatives (see Remark 6.10). Therefore, a similar property is needed for difference measurement as well, which is called weak difference independence.

Definition 6.13 A quaternary relation $\succcurlyeq^*$ satisfies *weak difference independence* if for every $i \in N$ and every $x, y, z, w, s, t \in X$, we have

$$(x_i, t_{-i})(y_i, t_{-i}) \succcurlyeq^* (z_i, t_{-i})(w_i, t_{-i}) \Leftrightarrow$$
$$(x_i, s_{-i})(y_i, s_{-i}) \succcurlyeq^* (z_i, s_{-i})(w_i, s_{-i}).$$

If weak independence was still a reasonable condition, weak difference independence is more demanding and is not satisfied in any situation. The following example illustrates this.

Example 6.14 Suppose you intend to buy a new car, considering essentially three attributes: price, performance of engine and equipment. You would prefer a low price, high performance and good equipment, and because all these criteria are important for you, you would prefer a car with a balanced profile rather than an unbalanced one, e.g., (reasonably expensive, reasonably good performance, reasonably good equipment)$\succ$(expensive, very good performance, reasonably good equipment). In other words, you think that a good point cannot compensate a weak point.

Now, consider four cars a, b, c, d having very good performance and very good equipment, whose prices are respectively (in increasing order) "reasonably expensive", "reasonably expensive + 5000€", "expensive", "expensive + 5000€". Suppose that you are indifferent between these increments of 5000€; i.e., $ab \sim^* cd$, the reason being that all these cars have anyway unbalanced profiles.

Consider next four other cars a', b', c', d', whose prices are respectively the prices of a, b, c, d, but now all these cars have reasonable performance and equipment. Then clearly a' is your preferred car, and the perceived difference of price between a' and b' is much more important than between c' and d', because b', being less balanced than a', is definitely less preferred than a', and c', d' are anyhow out of consideration. Hence $a'b' \succ^* c'd'$, violating weak difference independence. ◇

In the case where weak difference independence does not hold, the construction of value functions must be done considering a fixed and particular level for the remaining attributes, or considering several such levels. We give in Sect. 6.4.2 a construction considering the former method.

We leave aside measurement theory and borrow some notions from psychology.

6.3 Affect, Bipolarity and Reference Levels

Psychologists think that "affect" is a fundamental element in the mechanism of decision of humans. Affective reactions to stimuli are often the very first reactions, guiding information processing and judgment. Choices made are then justified *a posteriori* by various (rational) reasons. We borrow the definition given by Slovic [308]:

> *Affect: The specific quality of "goodness" and "badness," (1) experienced as a feeling state (with or without consciousness) and (2) demarcating a positive and negative quality of a stimulus.*

Hence, affect is closely related to our ability to distinguish between good and bad alternatives. The neurologist A. Damasio [65] asserts from clinical observations that individual having brain damages destroying their ability to feel while keeping basic intelligence, memory, and capacity for logical thinking intact,[7] lose their ability to make rational decisions, making them socially dysfunctional even though they remain intellectually capable of analytical reasoning.

6.3.1 Bipolarity

The most prominent feature of affect is its *bipolar* nature: there are two opposite poles (good/bad) corresponding to positive and negative stimuli, as well as a frontier between the two, which could be called a neutral zone or level, neither good nor bad. Hence, scales (in the measurement theoretic sense) used in a model of preference representation should reflect the bipolar nature of the affect. However, classical measurement theory does not incorporate bipolarity when building numerical representations.

There are basically two ways for representing bipolarity on a scale: the *bipolar univariate model* (Osgood et al. [260]), and the *unipolar bivariate model* (Cacioppo et al. [41]). The bipolar univariate model reflects the above definition of bipolarity and simply consists in a single axis with a central 0 value representing the neutral level (see Fig. 6.1). On this scale, positive values encode intensity of stimuli

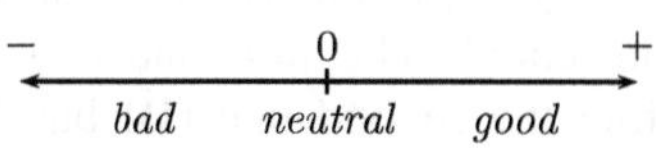

Fig. 6.1 The bipolar univariate model

corresponding to good alternatives, or good values of attributes, etc., while negative values encode the intensity of stimuli corresponding to bad alternatives. The central 0 value represents the neutral level. On the other hand, the more recent unipolar bivariate model uses two independent axes that are unipolar, bounded below by 0 (Fig. 6.2). The horizontal axis encodes positive stimuli, while the vertical axis

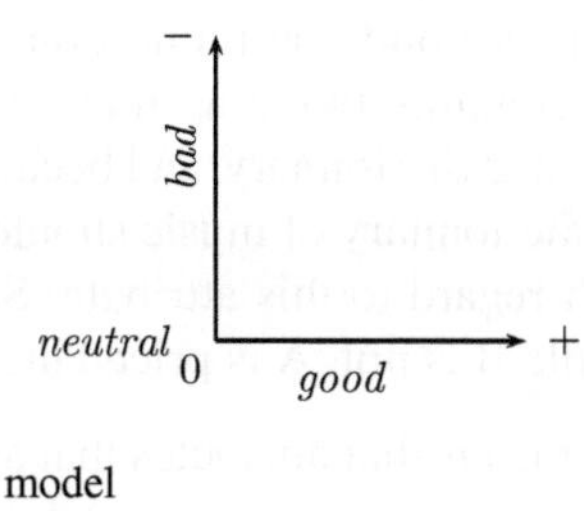

Fig. 6.2 The unipolar bivariate model

[7]See the famous case of Phineas P. Gage, in 1848, who was wounded by an iron bar traversing the front of his brain, but nevertheless survived this accident (described in Damasio [65]).

encodes negative stimuli. In this model, a given alternative may cause both a positive and a negative stimulus at the same time, and in an independent way. For example, eating chocolate may give some gustative pleasure (positive affect), while at the same time one may feel some greediness (negative affect).

Although the unipolar bivariate model may provide a richer model, we do not think it is suitable for modeling affect concerning values taken by a given attribute X_i, because an attribute is already considered to be a single "dimension" describing alternatives, which in principle cannot be subdivided into subdimensions (at least, into subdimensions of interest for the analysis). Clearly, the chocolate example is multidimensional and refers to two attributes (taste, effect on diet).

6.3.2 Reference Levels

Another important discovery in psychology concerning decision making is *evaluability* and *dominance of proportion*. Evaluability means that attributes whose perception is imprecise or without clear reference level have little importance in the final decision (or preference relation). This is well illustrated by the following experiment, due to Hsee [198].

Example 6.15 (The dictionnary experiment) At a second-hand book sale, two music dictionnaries are presented. The first one (A) has 10,000 entries and is in excellent condition, while the second one (B) has 20,000 entries, is in good condition except that its cover is torn.

In the first experiment, the two dictionnaries A and B are presented together to subjects, who are asked to estimate a price they are ready to pay for them. In general, subjects are inclined to pay more for B than for A.

In the second experiment, the question asked to subjects is the same but to each subject only one dictionnary (A or B) is shown. Surprisingly, the price estimated for A is now higher than for B.

This can be explained as follows: in the first experiment, subjects evaluate the dictionnaries mainly on the first attribute (number of entries), which is considered to be more important than the second one (condition of the dictionnary). Then B is priced higher because it contains twice as many entries as A. In the second experiment, subjects see only one dictionnary, and because they have no precise idea of how many entries a good dictionnary of music should contain, they are unable to evaluate the dictionnary with regard to this attribute. Since on the other hand A is clearly in good condition while B is not, A is priced higher than B. ◇

Dominance of proportion means that attributes that are expressed as a proportion or percentage have more impact than those expressed in an absolute way.

Again, this is a matter of reference level. If an attribute is expressed as a proportion or percentage, the implicit reference level is 100 %, and one can see how far from the reference level is the value under consideration. This explains why people prefer a small cup overflowing with ice cream rather than a half-filled big cup, even if they contain exactly the same quantity of ice cream, because the size of the cup fixes the reference level. The following experiment done by Slovic [308] illustrates this phenomenon.

Example 6.16 Two groups of people are asked if they would support an airport-safety-measure expected to save lives. The response scale ranged from 0 (would not support at all) to 20 (very strong support). In the first group, the safety measure is supposed to save 150 lives, while in the second group, the safety measure is said to save a portion of 150 lives, expressed in percentage. The results are summarized in Table 6.2 below. They show that saving a percentage of 150 lives receives more support than does saving 150 lives. Saving 150 lives is diffusely good, and therefore is not easily evaluable, whereas saving 98 % of 150 lives appears to be clearly very good (and therefore evaluable) because it is very close to the upper bound. ◇

	Potential benefit				
	Save 150 lives	Save 98 %	Save 95 %	Save 90 %	Save 85 %
Mean	10.4	13.6	12.9	11.7	10.9
Median	9.8	14.3	14.1	11.3	10.8

Table 6.2 Support on a [0-20]-scale for an airport-safety-measure

6.3.3 *Bipolar and Unipolar Scales*

The above considerations oblige us to revisit the notion of scale as established in Sect. 6.2.1, in order to incorporate bipolarity and reference levels. We consider again a set A of objects endowed with a complete preorder $\succcurlyeq$, and we suppose that a (ordinal) scale $f : (A, \succcurlyeq) \to (\mathbb{R}, \geqslant)$ exists (Theorem 6.4).

A *neutral level* is an element in A denoted by $\mathbb{O}$ such that for every $a \in A$, if $a \succ \mathbb{O}$, then object a is considered to be "good" by the decision maker (positive affect), and if $a \prec \mathbb{O}$, then the object is considered to be "bad" (negative affect). It is convenient to fix $f(\mathbb{O}) = 0$. If A has a neutral level, we say that the scale f is *bipolar*, otherwise it is *unipolar*.

Example 6.17 Binary relations stemming from pairs of opposite words in natural language and related to affect give rise to a bipolar scale. This is clearly the case for the relations "more attractive than," "better than," "preferred than," where the corresponding pairs of opposite words are "attractive/repulsive," "good/bad" and "like/dislike." On the other hand, binary relations like "more likely than," "more prioritary than" and "belongs more to category C than" do not clearly exhibit a neutral level, because even if the two first of them are related to pairs of opposite words ("likely/unlikely," "prioritary/secondary"), these words are not exactly opposite ("unlikely" means absence of likelihood rather than an opposite notion of likelihood, and similarly for "secondary," which indicates a low level of priority) and are not related to affect. As for the last example, this is also a case where we have absence of a property (an object does not belong to category C) rather than the existence of the opposite notion of membership. ◇

We say that the scale f is *bounded from above* if there exists a greatest element $\mathbb{1} \in A$; i.e., such that $\mathbb{1} \succcurlyeq a$ for all $a \in A$. We can fix for convenience $f(\mathbb{1}) = 1$. Similarly, f is a scale *bounded from below* if there exists a least element in A, that is, an element $\underline{a} \in A$ such that $a \succcurlyeq \underline{a}$ for every $a \in A$. When the scale is unipolar, the least element is denoted by $\mathbb{O}$, and we take the convention $f(\mathbb{O}) = 0$. Otherwise, for bipolar scales, we denote it by $-\mathbb{1}$, and take the convention $f(-\mathbb{1}) = -1$.

Example 6.18 (Example 6.17 continued) The above bipolar scales are neither bounded from above, nor from below in general, because it is always possible to find objects that are more (or less) attractive than, better than (or worse than), more (or less) preferred than a given object. The relations "likely/unlikely" and "belongs more to the category C than" yield scales that are bounded from below. The least elements correspond to objects which, respectively, never occur (impossible event), and do not belong to the category C (i.e., they do not possess any feature characterizing category C). By contrast, the relation "more prioritary than" does not yield a least element in general, because it is possible to find objects that are less prioritary than a given one. ◇

We have seen that in many cases, scales are not bounded, and especially when dealing with binary relations that are used in a context of decision making: "more attractive than," "better than," "preferred than." However, Sect. 6.3.2 has taught us that reference levels (indicating what the decision maker means by a "good" element) are necessary; without them the corresponding attributes are not actually used in the process of decision making. The reference level should concern an object that is considered to be satisfactory regarding the binary relation $\succcurlyeq$, in the sense that the decision maker would be quite satisfied if he could obtain it, even if better elements could exist. The existence of such a level, whose definition may appear as imprecise, is one of the fundamental theses defended/supported by Herbert Simon[8] in his theory of *bounded rationality* [271, 305, 306]. The main idea of this theory is that in a real situation, an agent is not able to find an *optimal* alternative or solution of a decision problem (e.g., chess, search for food by an animal, etc.) in the strict mathematical sense, because of limited resource (mainly time and energy, but also information, intelligence, memory, etc.), but the search for a solution will stop as

[8]Herbert Alexander Simon (Milwaukee, 1916 – Pittsburgh, 2001) is an American economist and sociologist, who studied also political science and cognitive psychology. This led him to construct a theory of human behavior in decision making (Administrative behavior, 1947). Very early convinced of the importance of computers, he was one of the pioneers of artificial intelligence. He was awarded the Nobel Prize in economics in 1978.

soon as the agent has found a *satisficing*[9] solution; i.e., bringing a sufficiently high level of satisfaction. Quoting Samuel Eilon,

Optimizing is the science of the ultimate; Satisficing is the art of the feasible.

We denote the *satisfactory level* as defined above by $\mathbb{1}$. In the case of a bipolar scale, we assume by symmetry the existence of a nonsatisfactory level denoted by $-\mathbb{1}$, and take the same convention as above: $f(\mathbb{1}) = 1, f(-\mathbb{1}) = -1$.

6.4 Building Value Functions with the MACBETH Method

6.4.1 The MACBETH Method

The MACBETH method (Measuring Attractiveness by a Categorical Based Evaluation TecHnique) is due to Bana e Costa and Vansnick (see, e.g., [13–16]), and permits to construct an interval scale from information on preference and intensity of preference given by the decision maker.

Let $\mathcal{A}$ be a finite set of objects. The decision maker is asked for each pair $(a, b) \in \mathcal{A}^2$ the following question:

Is object a more attractive than b? (yes/no)

If the answer is "yes," then we set aPb. If neither aPb nor bPa hold, a and b are considered to be indifferent, which is denoted by aC_0b. If aPb, a second question is asked to the decision maker:

Is the difference of attractiveness between a and b very weak, weak, moderate, strong, very strong, extreme? (choose only one category[10]*)*

These six categories define six binary relations C_1 (corresponding to "very weak"),..., C_6 (corresponding to "extreme"). Together with C_0, their union forms a complete binary relation on $\mathcal{A}$, which we denote by $\succcurlyeq$. The asymmetric part $\succ$ is given by $P = C_1 \cup \cdots \cup C_6$, while the symmetric part $\sim$ is C_0.

The first question pertains to ordinal measurement and yields $\succcurlyeq$. If it is transitive, we know from Theorem 6.4 that there exists $f : \mathcal{A} \to \mathbb{R}$ representing $\succcurlyeq$. The second one is related to difference measurement. More precisely, the numerical representation f should satisfy

$$aC_kb \text{ and } cC_{k'}d, k > k' \Leftrightarrow f(a) - f(b) > f(c) - f(d). \tag{6.6}$$

Bana e Costa and Vansnick [14] have shown that the existence of a function f representing $\succcurlyeq$ and satisfying (6.6) is equivalent to the existence of real numbers

[9] Word coined by H. Simon, as a contraction of "satisfying" and "sufficing."

[10] In later versions, it is allowed to choose several categories, provided they are contiguous [13].

$0 = t_1 < \cdots < t_6$ such that

$$aP_kb \Leftrightarrow f(a) > f(b) + t_k \qquad (k = 1, \ldots, 6) \tag{6.7}$$

with $P_k = C_k \cup \cdots \cup C_6$. Applying a general result of Doignon [92], Bana e Costa and Vansnick show the following result, for which we need to introduce some additional concepts. Considering binary relations $R_1, \ldots, R_m$ on a finite set $\mathcal{A}$, a *cycle* is a sequence

$$x_1Qx_2, x_2Qx_3, \ldots, x_{j-1}Qx_j, x_jQx_1$$

where $Q = R_1 \cup \cdots \cup R_m \cup R_1^{cd} \cup \cdots \cup R_m^{cd}$, and R_k^{cd} is the *codual* of R_k, defined by $xR_k^{cd}y$ if and only if not(yR_kx). A *m-cyclone* is a union of at most m cycles, and we say that a cyclone is *balanced* if for every $k = 1, \ldots, m$, there is the same number of pairs in R_j and in R_j^{cd}.

Then, a function f representing $\succcurlyeq$ and satisfying (6.6) exists if and only if no m-cyclone w.r.t. $P_1, \ldots, P_6$ is balanced, where m is the number of nonempty relations among $C_1, \ldots, C_6$, and whenever $aC_0b, a \succ c$ and $d \succ a$, we have $aC_kc \Leftrightarrow bC_kc$ and $dC_ka \Leftrightarrow dC_kb$ for every $k = 1, \ldots, m$. This condition being difficult to check in practice, a linear program can be used to check the existence of the thresholds $t_1, \ldots, t_6$ directly.

Once the existence of f is ensured, it is possible to get such a function by linear programming. However, the solution is not unique in general, and each value $f(a)$ for $a \in \mathcal{A}$ lies in an interval. The decision maker has to fix $f(a)$ for any $a \in \mathcal{A}$, and once this is done, f is an interval scale on $\mathcal{A}$ (see all details in [13]).

6.4.2 *Determination of the Value Functions*

The MACBETH method can be applied to determine the value functions u_i, $i = 1, \ldots, n$. Besides, we suppose that for each attribute X_i, we have determined beforehand its neutral level $\mathbb{O}_i$ (supposing X_i underlies a bipolar scale, otherwise take $\mathbb{O}_i$ as the lower bound of X_i), as well as its satisfactory level $\mathbb{1}_i$, and the unsatisfactory level $-\mathbb{1}_i$, in case of a bipolar scale. Recall that $\succcurlyeq_i$ is the preference relation on X_i, induced by $\succcurlyeq$ (unambiguously defined by weak independence).

All value functions being determined separately, we can focus on a single attribute X_i. We distinguish the unipolar and bipolar cases.

Unipolar Case We consider the set of alternatives

$$\tilde{X}_i = \{(x_i, \mathbb{O}_{-i}) : x_i \in X_i\} \subseteq X$$

and select a finite set $\mathcal{A}$ from it containing the alternatives $\mathbb{O}_N$ and $(\mathbb{1}_i, \mathbb{O}_{-i})$. The MACBETH method is applied to the set $\mathcal{A}$, and as a result, an interval scale f is obtained. Since f is defined up to a positive affine transformation, it suffices to arbitrarily fix the value of f on two points in order to fix f. We set

$$f(\mathbb{O}_N) = 0, \quad f(\mathbb{1}_i, \mathbb{O}_{-i}) = 1,$$

and define the value function u_i by

$$u_i(x_i) = f(x_i, \mathbb{O}_{-i}) \qquad ((x_i, \mathbb{O}_{-i}) \in \mathcal{A}),$$

which implies $u_i(\mathbb{O}_i) = 0$, $u_i(\mathbb{1}_i) = 1$.

Bipolar Case We proceed similarly but distinguish the positive and negative sides, and introduce

$$\tilde{X}_i^+ = \{(x_i, \mathbb{O}_{-i}) \,:\, x_i \in X_i, x_i \succcurlyeq_i \mathbb{O}_i\}$$
$$\tilde{X}_i^- = \{(x_i, \mathbb{O}_{-i}) \,:\, x_i \in X_i, x_i \preccurlyeq_i \mathbb{O}_i\}$$

and select finite sets $\mathcal{A}^+, \mathcal{A}^-$ from them, containing respectively $(\mathbb{1}_i, \mathbb{O}_{-i})$ and $(-\mathbb{1}_i, \mathbb{O}_{-i})$, and both containing $\mathbb{O}_N$. Applying the MACBETH method on $\mathcal{A}^+, \mathcal{A}^-$, interval scales f^+,f^- are obtained, which are fixed by setting

$$f^+(\mathbb{O}_N) = f^-(\mathbb{O}_N) = 0, \quad f^+(\mathbb{1}_i, \mathbb{O}_{-i}) = 1, \quad f^-(-\mathbb{1}_i, \mathbb{O}_{-i}) = -1,$$

and define the value function u_i by

$$u_i(x_i) = \begin{cases} f^+(x_i, \mathbb{O}_{-i}), & \text{if } x_i \succcurlyeq_i \mathbb{O}_i \\ f^-(x_i, \mathbb{O}_{-i}), & \text{if } x_i \prec_i \mathbb{O}_i. \end{cases} \tag{6.8}$$

In both cases, interpolation methods could be used for determining u_i on the entire set X_i.

Note that all value functions vanish on the corresponding neutral levels, which correspond to the same alternative $\mathbb{O}_N$, and they are equal to 1 on the satisfactory levels. Since these levels have an absolute meaning, in the sense that the decision maker feels the same intensity of satisfaction with them, the value functions are said to be *commensurate*.

Remark 6.19 Strictly speaking, in the bipolar case, the value function u_i given by (6.8) cannot be considered as an interval scale, because it has three fixed values, on $\mathbb{1}_i$, $\mathbb{O}_i$ and $-\mathbb{1}_i$ respectively. In fact, u_i comes from two independent interval scales, one being defined for levels on the positive side, and the other for levels on the negative side. Therefore, differences of preference on the positive side cannot be compared with differences of preference on the negative side through u_i. ◇

6.5 Summary of the Construction of Value Functions

At this point, we have achieved the construction of the value functions $u_1, \ldots, u_n$. We have presented two radically different methods to do this:

(i) Difference measurement: under the assumption of weak difference independence, value functions $u_1, \ldots, u_n$ can be obtained independently by difference measurement applied on the attributes $X_1, \ldots, X_n$, provided the conditions of Theorem 6.11 are fulfilled. No commensurateness is needed between the u_i's, and therefore there is no need of reference levels.

(ii) MACBETH method: value functions $u_1, \ldots, u_n$ are obtained separately as interval scales built on special sets of alternatives $\tilde{X}_1, \ldots, \tilde{X}_n$, which need the determination of two reference levels: the neutral and the satisfactory levels. Weak difference independence is not required, however, conditions (6.7) of difference measurement must hold, and can be checked by a linear program. Commensurateness is necessary, and is ensured by imposing equality of the value functions on the reference levels.

Supposing the value functions to be determined, it remains to build the aggregation function F.

6.6 The Weighted Arithmetic Mean as an Aggregation Function

Most of the methods in MCDM use for F the weighted arithmetic mean, putting weights $w_1, \ldots, w_n$ on criteria in order to represent their importance. However, in many situations, the weighted arithmetic mean gives counterintuitive results.

Consider three alternatives a, b, c evaluated on two attributes, as follows:

$$u_1(a_1) = 0.45, \quad u_1(b_1) = 0, \quad u_1(c_1) = 1$$

$$u_2(a_2) = 0.45, \quad u_2(b_2) = 1, \quad u_2(c_2) = 0.$$

If we assign the value 1 to the satisfactory level, we see that alternative b is satisfactory on the second criterion but not on the first one, while it is the opposite for alternative c. If both criteria are considered important by the decision maker, neither b nor c are acceptable, and a could appear as the best option because it is balanced on both criteria, with a value that is, although not at a satisfactory level, still acceptable. Surprisingly, no weighted arithmetic mean is able to represent the preference $a \succ b$, $a \succ c$: letting w_1, w_2 be the weights on criteria 1 and 2, with

$w_1, w_2 \geqslant 0$ and $w_1 + w_2 = 1$, we get

$$0.45(w_1 + w_2) > w_2$$
$$0.45(w_1 + w_2) > w_1,$$

which is impossible to satisfy since $w_1 + w_2 = 1$.

This example is not an isolated case, but is a well-known phenomenon in multiobjective optimization, where it is known that the weighted arithmetic mean is unable to explore the concave parts of the Pareto frontier.[11] This can be easily explained on Fig. 6.3, with two criteria.

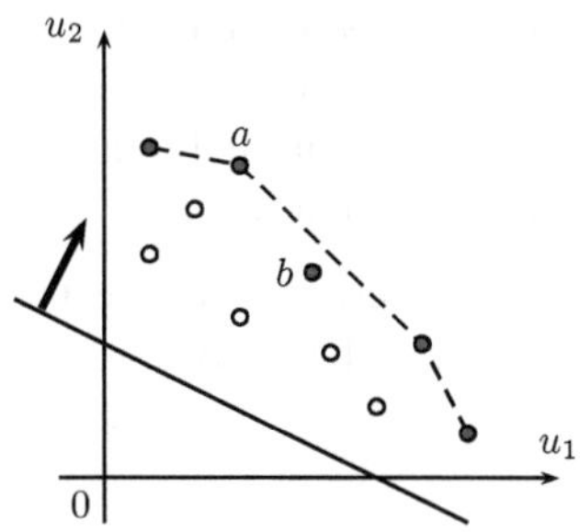

Fig. 6.3 *Red points* indicate the Pareto frontier, and the *dashed line* indicates its convex hull. Alternative b is in the concave part of the frontier, so that maximizing a weighted arithmetic mean (as represented by a *straight line* moving upwards) can only yield one of the *red points* on the convex hull as an optimal solution (in this case, a will be obtained)

Another drawback of the weighted arithmetic mean is that some simple and intuitive ways of aggregating scores cannot be handled, for example veto criteria. Roughly speaking, saying that criterion 3 is a veto means that whatever good scores an alternative obtains on other criteria, a bad score on criterion 3 cannot be compensated and the alternative receives an overall bad score. However, the weighted arithmetic mean is compensatory, in the sense that even if the score on criterion 3 is 0, any overall score can be obtained by taking a sufficiently high score on another criterion whose weight is sufficiently high.

The veto example is a particular case of *interaction* between criteria: criteria are interacting whenever their rôle and effect on the overall score are mutually dependent (see Sect. 6.10 for a rigorous definition). For example, the decision maker may desire that if criterion 3 is satisfied, then criterion 2 is not important, otherwise criterion 2 is important. Clearly, the weighted arithmetic mean is unable

[11]The *Pareto frontier* of a set of points $\mathcal{A}$ is the set of its undominated points, that is, $\{a \in \mathcal{A} : \nexists b \in \mathcal{A}, b \gg a\}$.

to represent such effects, because weights are fixed and pertain to a single criterion. The following example, borrowed from [161], illustrates this phenomenon.

Example 6.20 (Evaluation of students) Suppose that students in some faculty of sciences are evaluated in mathematics (M), physics (P) and literature (L), among other topics. Since the evaluation is done in a faculty of sciences, mathematics and physics should receive a more important weight than literature. However, we may want to take into account the two following requirements:

(i) It is observed that in general, students good in mathematics are also good in physics, and vice versa, because both subjects require skills in abstract formalization and calculus, etc. Therefore, students having good marks in both mathematics and physics should not be overevaluated.
(ii) Literature should not be considered to be negligible, because it is important for general education, and moreover a good student should be balanced in sciences and humanities. Therefore, a bad mark in literature should be penalized and not compensated by good marks in sciences.

These are examples of interacting criteria, where their weight of importance is depending somehow on the scores of the criteria. ◇

6.7 Towards a More General Model of Aggregation

If we return to our example in the beginning of Sect. 6.6, the inability of the weighted arithmetic mean to represent the preference comes from the fact that weights are put on isolated criteria. Clearly, if the decision maker has a preference for balanced alternatives rather than for unbalanced ones, this suggests that there is some weight w_{12} on the group of criteria 1 and 2, which is significantly greater than weights w_1, w_2 on individual criteria, or more exactly, greater than their sum $w_1 + w_2$.

Generalizing this idea for n criteria, one would define a weight for each group of criteria, or at least for each group where it is significant to do so (i.e., where the sum of weights of criteria in a group does not represent the overall importance of this group).

This being taken for granted, it remains to determine these weights. In the case of the classical weighted arithmetic mean with weights $w_1, \dots, w_n$ on individual criteria, we deduce immediately from the determination of the value functions by the MACBETH method (Sect. 6.4.2) that

$$u(\mathbb{1}_i, \mathbb{O}_{-i}) = F(u_i(\mathbb{1}_i), u_{-i}(\mathbb{O}_{-i})) = F(1_i, 0_{-i}) = w_i \qquad (i \in N).$$

Hence, importance weights correspond to the overall score of alternatives being satisfactory on a single criterion and neutral elsewhere. We generalize this view by considering alternatives being satisfactory on several criteria. We distinguish the unipolar and bipolar cases.

6.7.1 The Unipolar Case

We consider *binary alternatives*, i.e., of the form $(\mathbb{1}_A, \mathbb{O}_{-A})$ with $A \subseteq N$, and apply the MACBETH method to determine their overall score. Specifically, we set

$$\mathcal{A} = \{(\mathbb{1}_A, \mathbb{O}_{-A}) \, : \, A \subseteq N\}$$

and applying MACBETH on $\mathcal{A}$, as a result we get an interval scale that is nothing but the overall value function u, which we fix arbitrarily as follows:

$$u(\mathbb{O}_N) = 0, \quad u(\mathbb{1}_N) = 1. \tag{6.9}$$

This is related to the aggregation function F as follows:

$$u(\mathbb{1}_A, \mathbb{O}_{-A}) = F(u_A(\mathbb{1}_A), u_{-A}(\mathbb{O}_{-A})) = F(1_A) \qquad (A \subseteq N). \tag{6.10}$$

Observe that the above equation determines the unknown function F on all vertices of the hypercube $[0, 1]^N$. Moreover, increasingness of F leads to

$$A \subseteq B \Rightarrow u(\mathbb{1}_A, \mathbb{O}_{-A}) \leqslant u(\mathbb{1}_B, \mathbb{O}_{-B}). \tag{6.11}$$

By performing the change of notation $u(\mathbb{1}_A, \mathbb{O}_{-A}) \rightarrow \mu(A)$, where $\mu : 2^N \rightarrow \mathbb{R}$ is a set function, we realize by (6.9) and (6.11) that μ is a normalized capacity on N. Hence $F(1_A) = \mu(A)$, so that F is an extension of μ, and we know by Chap. 4 that the Choquet integral and the Sugeno integral are possible candidates for F [see Lemma 4.9 and Theorem 4.43(iii)].

In the sequel of this section, we show that the Choquet integral is in a sense the most natural candidate.

The determination of F on the whole hypercube $[0, 1]^N$ (and possibly outside) knowing its value on the vertices can be seen as an interpolation problem. Among all methods of interpolation, the linear interpolation is the simplest one. Note however that our problem of interpolation is multivariate: the function F has n variables, and unlike the usual case of univariate interpolation, it is not obvious how to choose for a given $x \in [0, 1]^n$ the interpolating vertices, the only condition being that the convex closure of the chosen vertices contains x. Figure 6.4 shows that even with two variables, there are several possibilities. Of course, at least $n + 1$ points are necessary to interpolate F at x (unless x lies on the boundary of the hypercube), and in the case of $n + 1$ points, we say that the interpolation is *parsimonious*. In this case, the convex region spanned by the vertices is a simplex. If all the 2^n vertices are used, one speaks of *multilinear interpolation*. Even if restricting to parsimonious interpolation, Fig. 6.4 shows that there are still several possibilities. It is however easy to see that only one of them can yield an interpolation, namely the left one.

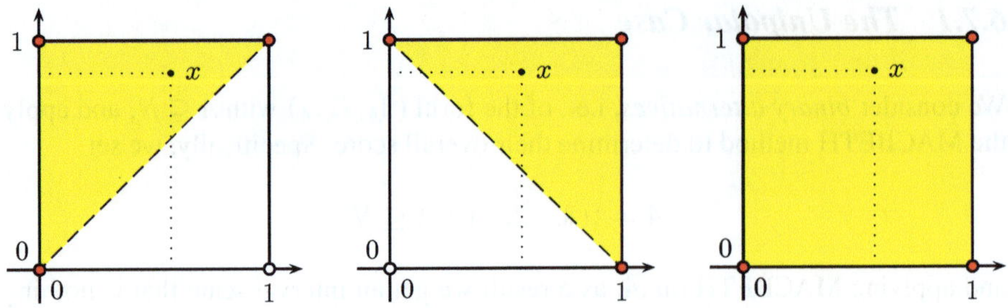

Fig. 6.4 Interpolation for x with the vertices of $[0,1]^2$: three possibilities. Vertices used for the interpolation are in *red*, and their convex closure is in *yellow*

Indeed, in this case the interpolation reads, noting that $F(0,0) = 0$:

$$F(x_1, x_2) = 0 + (\alpha_1 x_1 + \alpha_2 x_2)F(0,1) + (\beta_1 x_1 + \beta_2 x_2)F(1,1).$$

Taking $(x_1, x_2) = (0,1)$ and $(1,1)$ yields the following two triangular systems

$$\begin{aligned} \alpha_2 &= 1, & \beta_2 &= 0 \\ \alpha_1 + \alpha_2 &= 0, & \beta_1 + \beta_2 &= 1 \end{aligned}$$

whose unique solution yields the following interpolation formula:

$$F(x_1, x_2) = (x_2 - x_1)F(0,1) + x_1 F(1,1) \qquad (0 \leqslant x_1 \leqslant x_2 \leqslant 1). \tag{6.12}$$

Remembering that $F(0,1) = \mu(\{1\})$ and $F(1,1) = \mu(\{1,2\})$, one recognizes in (6.12) the Choquet integral of the vector x w.r.t. μ.

On the other hand, the interpolation in the center figure reads

$$F(x_1, x_2) = (\alpha_1 x_1 + \alpha_2 x_2)F(0,1) + (\beta_1 x_1 + \beta_2 x_2)F(1,0) + (\gamma_1 x_1 + \gamma_2 x_2)F(1,1).$$

Taking alternately $(x_1, x_2) = (0,1)$, $(1,0)$ and $(1,1)$, we find the following three systems:

$$\begin{aligned} \alpha_2 &= 1, & \beta_2 &= 0, & \gamma_2 &= 0 \\ \alpha_1 &= 0, & \beta_1 &= 1, & \gamma_1 &= 0 \\ \alpha_1 + \alpha_2 &= 0, & \beta_1 + \beta_2 &= 0, & \gamma_1 + \gamma_2 &= 1 \end{aligned}$$

which have clearly no solution. The reason is that vertex $(0,0)$ does not belong to this region, hence the linear system has one equation too many.

In the general case, it can be shown that the only valid partitioning of the hypercube into simplices is given by the $n!$ canonical simplices $[0,1]^n_\sigma$ (Sect. 4.5.1), where σ is any permutation on $[n]$. As suggested by the case $n = 2$, one finds the following result.

Theorem 6.21 (The Choquet integral as a parsimonious linear interpolator) *The parsimonious linear interpolation using the* $n!$ *canonical simplices yields the Choquet integral:*

$$F(x) = \int x \, d\mu, \qquad (x \in [0, 1]^n)$$

where $\mu(A) = F(1_A)$ *for all* $A \subseteq N$.

This result was shown by Singer [307] (Remark 2.89; see also Grabisch et al. [177, Proposition 5.25]).

6.7.2 The Bipolar Case

We consider now alternatives mixing the three reference levels $\mathbb{O}_i, \mathbb{1}_i, -\mathbb{1}_i$ on the various criteria. As a general model considering any combination of these three levels would be complicated, we may make two simplifying assumptions:

(i) There is independence between the negative and the positive parts. By this, we mean that the contribution of the positive part adds independently to the contribution of the negative part; i.e., the overall value function u would satisfy, for any disjoint $A, B \subseteq N$,

$$u(\mathbb{1}_A, -\mathbb{1}_B, \mathbb{O}_{-(A\cup B)}) = u(\mathbb{1}_A, \mathbb{O}_{-A}) + u(-\mathbb{1}_B, \mathbb{O}_{-B}).$$

(ii) There is symmetry between the positive and the negative parts, which means that the score of the binary alternative $(-\mathbb{1}_A, \mathbb{O}_{-A})$ is just the opposite of the score of $(\mathbb{1}_A, \mathbb{O}_{-A})$:

$$u(-\mathbb{1}_A, \mathbb{O}_{-A}) = -u(\mathbb{1}_A, \mathbb{O}_{-A}) \qquad (A \subseteq N).$$

If both assumptions hold, then we obtain

$$u(\mathbb{1}_A, -\mathbb{1}_B, \mathbb{O}_{-(A\cup B)}) = u(\mathbb{1}_A, \mathbb{O}_{-A}) - u(\mathbb{1}_B, \mathbb{O}_{-B}) \qquad (A, B \subseteq N, A \cap B = \varnothing).$$

Defining the capacity μ by $\mu(A) = u(\mathbb{1}_A, \mathbb{O}_{-A})$ as for the unipolar case and performing a parsimonious interpolation, the positive quadrant yields the Choquet integral, and we see that the aggregation function F obtained on $[-1, 1]^n$ is nothing but the symmetric Choquet integral (4.10):

$$F(x) = \check{\int} x \, d\mu = \int x^+ \, d\mu - \int x^- \, d\mu \qquad (x \in [-1, 1]^n),$$

with $x^+ = x \vee \mathbf{0}$ and $x^- = (-x)^+$ the positive and negative parts of x [see (4.9)].

Nothing ensures that the symmetry assumption is satisfied in practice, though. What do we obtain with the sole independence assumption? In this case, we have to apply the MACBETH method one time for the binary alternatives as for the unipolar case, and a second time for the negative binary alternatives $(-1\!\!1_A, 0\!\!\!0_{-A})$, yielding two interval scales u^+, u^-, fixed arbitrarily as follows:

$$u^+(0\!\!\!0_N) = u^-(0\!\!\!0_N) = 0, \quad u^+(1\!\!1_N) = 1, \quad u^-(-1\!\!1_N) = -1.$$

This gives two normalized capacities μ^+, μ^- on N defined by

$$\mu^+(A) = u^+(1\!\!1_A, 0\!\!\!0_{-A}), \quad \mu^-(A) = -u^-(-1\!\!1_A, 0\!\!\!0_{-A}) \qquad (A \subseteq N).$$

Applying the parsimonious interpolation on the positive and the negative quadrants, one obtains two independent Choquet integrals, and the aggregation function F reads

$$F(x) = \int x^+ \, \mathrm{d}\mu^+ - \int x^- \, \mathrm{d}\mu^- \qquad (x \in [-1, 1]^n). \tag{6.13}$$

Remark that this model has the same form as PT, the prospect theory model used in decision under uncertainty (5.24). Also, the (asymmetric) Choquet integral is recovered when $\mu^- = \overline{\mu^+}$, the conjugate of μ^+ [see (4.12)], while the symmetric Choquet integral is recovered with $\mu^- = \mu^+$.

Finally, what if none of the independence and symmetry assumptions is satisfied? This case can arise in practice, when the decision behaviors are different in the positive and negative sides. The following example illustrates this.

Example 6.22 (Evaluation of students (continued)) (Grabisch and Labreuche [173]) Consider again the problem of evaluating students on three topics, which are mathematics (M), physics (P), and literature (L), and suppose we have the following four students to evaluate (marks are given on a 0–20 scale) Since students A and

	Mathematics (M)	Physics (P)	Language (L)
Student A	14	16	7
Student B	14	15	8
Student C	9	16	7
Student D	9	15	8

B are good in mathematics and physics, more attention is given to literature. Then the one point difference between A and B in language weighs more than the one in physics, which leads to $B \succ A$. On the contrary, because students C and D are bad in mathematics, more attention is given to physics, which yields $C \succ D$.

First, this preference cannot be represented by a unipolar Choquet integral model. Assuming that the above figures can be considered as scores, the preference $B \succ A$ yields, after simplification,

$$1 > \mu(\{M,P\}) + \mu(\{P\})$$

while $C \succ D$ yields

$$\mu(\{M,P\}) + \mu(\{P\}) > 1,$$

a contradiction. Considering 10 as the neutral level and translating the scale so that the neutral level is equal to 0, we obtain the new table:

	Mathematics (M)	Physics (P)	Language (L)
Student A	4	6	−3
Student B	4	5	−2
Student C	−1	6	−3
Student D	−1	5	−2

Applying (6.13), we obtain for the four students

$$\begin{aligned} u(A) &= F(4,6,-3) = 4\mu^+(\{M,P\}) + 2\mu^+(\{P\}) - 3\mu^-(\{L\}) \\ u(B) &= F(4,5,-2) = 4\mu^+(\{M,P\}) + \mu^+(\{P\}) - 2\mu^-(\{L\}) \\ u(C) &= F(-1,6,-3) = 6\mu^+(\{P\}) - \mu^-(\{M,L\}) - 2\mu^-(\{L\}) \\ u(D) &= F(-1,5,-2) = 5\mu^+(\{P\}) - \mu^-(\{M,L\}) - \mu^-(\{L\}). \end{aligned}$$

Preferences $B \succ A$ and $C \succ D$ yield

$$\begin{aligned} \mu^-(\{L\}) &> \mu^+(\{P\}) \\ \mu^+(\{P\}) &> \mu^-(\{L\}), \end{aligned}$$

again a contradiction. ◇

Hence, in some situations, we have to give up these two assumptions and consider the general case. Then we deal with *ternary alternatives*, of the form $(1\!\!1_A, -1\!\!1_B, 0\!\!\!0_{-(A\cup B)})$ for disjoint $A, B \subseteq N$. Applying the MACBETH method on the set of ternary alternatives, we get an interval scale u, which can be fixed by letting as before $u(1\!\!1_N) = 1$ and $u(0\!\!\!0_N) = 0$. Increasingness of F yields

$$A \subseteq C, B \supseteq D \Rightarrow u(1\!\!1_A, -1\!\!1_B, 0\!\!\!0_{-(A\cup B)}) \leqslant u(1\!\!1_C, -1\!\!1_D, 0\!\!\!0_{-(C\cup D)}).$$

Letting $b(A,B) = u(1\!\!1_A, -1\!\!1_B, \mathbb{O}_{-(A\cup B)})$ for every disjoint A, B, we have created a mapping $b : \mathcal{Q}(N) \to \mathbb{R}$, where

$$\mathcal{Q}(N) = \{(A,B) \in 2^N \times 2^N \,:\, A \cap B = \varnothing\} \equiv 3^N,$$

being nondecreasing in first place and nonincreasing in second place, which is called a *bicapacity* (Grabisch and Labreuche [171]). The definition of the Choquet integral w.r.t. a bicapacity is obtained in the same way, as a parsimonious interpolation in each region where a fixed subset of attributes takes value in the positive side. Its explicit definition is (Grabisch and Labreuche [172]):

$$\int x \,\mathrm{d}b = \int |x| \,\mathrm{d}v_x \qquad (x \in [-1,1]^n)$$

where v_x is a game on 2^N defined by

$$v_x(A) = b(A \cap N^+, A \cap N^-) \qquad (A \in 2^N)$$

with $N^+ = \{i \in N \,:\, x_i \geqslant 0\}$ and $N^- = N \setminus N^+$.

Example 6.23 (Evaluation of students (continued)) Let us solve the representation of preferences among students A, B, C, and D using bicapacities. Computing the Choquet integral w.r.t. a bicapacity b yields:

$$\begin{aligned}
u(A) &= F(4,6,-3) = 3b(\{M,P\},\{L\}) + b(\{M,P\},\varnothing) + 2b(\{P\},\varnothing)\\
u(B) &= F(4,5,-2) = 2b(\{M,P\},\{L\}) + 2b(\{M,P\},\varnothing) + b(\{P\},\varnothing)\\
u(C) &= F(-1,6,-3) = b(\{P\},\{M,L\}) + 2b(\{P\},\{L\}) + 3b(\{P\},\varnothing)\\
u(D) &= F(-1,5,-2) = b(\{P\},\{M,L\}) + b(\{P\},\{L\}) + 3b(\{P\},\varnothing).
\end{aligned}$$

Preferences $B \succ A$ and $C \succ D$ yield

$$\begin{aligned}
&b(\{M,P\},\varnothing) > b(\{M,P\},\{L\}) + b(\{P\},\varnothing)\\
&b(\{P\},\{L\}) > 0,
\end{aligned}$$

and there is no contradiction any more.

The symmetries in the different models are summarized in Fig. 6.5.

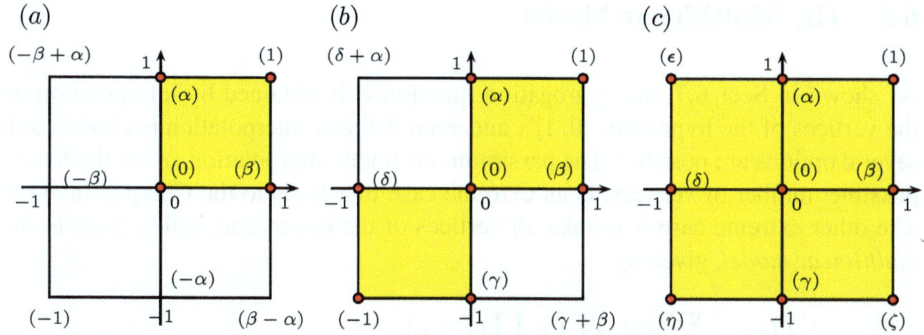

Fig. 6.5 Case of a bipolar scale with $n = 2$: (**a**) symmetric model, (**b**) prospect theory model, (**c**) general bipolar using bicapacities. *Red circles* indicate fictive alternatives determining the model, values in *parentheses* indicate overall scores; scores not in *yellow* parts are computed from scores in *yellow* parts

Remark 6.24

(i) As it can be seen, bipolarity introduces much complexity in the model and its determination. While a unipolar model has a complexity of order 2^n, the complexity of a general bipolar model jumps to 3^n, severely limiting their usage in practice, at least for the fully general version;

(ii) The idea to define mappings on $\mathcal{Q}(N)$ appeared in different domains. Let us cite for example *ternary voting games*, which are $\{-1, 1\}$-valued bicapacities, introduced by Felsenthal and Machover [138] for modeling abstention in voting games, and *bicooperative games* which are bicapacities without the monotonicity assumption, introduced by Bilbao et al. [24]. Also, real-valued mappings on $\mathcal{Q}(N)$ are called *biset functions* or *signed set functions*, and are used in combinatorial optimization (Fujishige [149, Sect. 3.5. (b)]). Bicapacities have been proposed by Labreuche and Grabisch [220], and developed in [171, 172]. There is already an important literature on the topic, in the above-mentioned domains. We do not detail further this topic, which is complex and not unified for the time being. Lastly, we mention that the concept of bipolarization can be applied to mathematical structures like semilattices, permitting the definition of very general Choquet-like functionals (see Grabisch and Labreuche [174]).

(iii) As we have claimed before, our view of bipolarity follows the bipolar univariate model of Osgood et al. The other view, namely the unipolar bivariate model, advocated by Cacioppo et al., has given rise to the notion of *bipolar capacity* proposed by Greco et al. [186], which is a two-valued function on $\mathcal{Q}(N)$.

◇

6.8 The Multilinear Model

As shown in Sect. 6.7, the aggregation function F is obtained by interpolation on the vertices of the hypercube $[0, 1]^n$, and even if linear interpolation is chosen, still several options are possible. The parsimonious linear interpolation, using the fewest possible number of vertices, is an extreme case that leads to the Choquet integral. The other extreme case is to take all vertices of the hypercube, which leads to the *multilinear model*, given by

$$\begin{aligned} F(x) &= \sum_{A\subseteq N} F(1_A)\prod_{i\in A} x_i \prod_{i\in A^c}(1 - x_i) \\ &= \sum_{A\subseteq N} \mu(A)\prod_{i\in A} x_i \prod_{i\in A^c}(1 - x_i) \qquad (x \in [0, 1]^n). \end{aligned} \tag{6.14}$$

At this point, it is interesting to view our *interpolation* problem of a function F known on the vertices of the hypercube as an *extension* problem of μ, remembering that $\mu(A) = F(1_A)$ for every $A \subseteq N$. We then return to the problem of the extension of pseudo-Boolean functions, studied in Sect. 2.16.4. This shows that F obtained by the parsimonious linear interpolation is the Lovász extension of μ, while the multilinear model is the Owen extension (2.83), denoted by f_μ^{Ow}, whose alternative expression using the Möbius transform of μ is

$$f_\mu^{\mathrm{Ow}}(x) = \sum_{A\subseteq N} m^\mu(A)\prod_{i\in A} x_i \qquad (x \in [0, 1]^n). \tag{6.15}$$

The following result shows that the multilinear model arises as the only solution for F when one considers difference measurement under the weak difference independence assumption instead of the MACBETH method with reference levels on each attribute. Before, we need the following definition.

Definition 6.25 (Dyer and Sarin [120]) The set of attributes $X_1, \dots, X_n$ is *bounded from above* (respectively, *bounded from below*) if there exists an element $\mathbb{1}_i \in X_i$ (respectively, $\mathbb{O}_i \in X_i$) such that $(\mathbb{1}_i, x_{-i}) \succcurlyeq (x_i, x_{-i})$ (respectively, $(x_i, x_{-i}) \succcurlyeq (\mathbb{O}_i, x_{-i})$) for all $x_i \in X_i$, all $x_{-i} \in X_{-i}$, and all $i \in N$. The set of attributes is *bounded* if it is bounded from above and from below.

This definition is similar to the foregoing definition of a bounded scale.[12]

Theorem 6.26 (Characterization of the Owen extension) *Assume that the conditions of Theorem 6.11 are fulfilled,*[13] *i.e., there exists a value function u on X*

[12]The readers should not confuse these bounds with the satisfactory and neutral levels introduced for the MACBETH method: we are precisely in the case where we do not assume that they exist. Moreover, no assumption of commensurateness is made here.

[13]See also Remark 6.12: the conditions for the positive difference model could be taken as well.

representing the quaternary relation $\succcurlyeq^*$ on X, and suppose in addition that the set of attributes is bounded. Then $\succcurlyeq^*$ satisfies weak difference independence if and only if there exists a unique capacity μ on N and value functions $u_1, \ldots, u_n$ unique up to a positive affine transformation such that F is the Owen extension of μ.

This result is obtained by combining Corollary 2 of Dyer and Sarin [120] and Theorem 6.3 of Keeney and Raiffa [205]. For convenience, we give a comprehensive proof of it.

Proof The "if" part is left to the readers.

For the "only if" part, we know by Theorem 6.11 that u is defined up to a positive affine transformation. Then observe that weak difference independence implies

$$u(x_i, x_{-i}) = g(x_{-i}) + h(x_{-i})u(x_i, z_{-i}) \qquad (i \in N) \tag{6.16}$$

for all $x_i \in X_i, x_{-i}, z_{-i} \in X_{-i}$, where $g(x_{-i})$ and $h(x_{-i}) > 0$ depend only on x_{-i}. Indeed, weak difference independence implies that the functions $u(x_i, x_{-i})$ and $u(x_i, z_{-i})$ rank the differences in the same way, hence by Theorem 6.11 they should be equal up to a positive affine transformation.

Furthermore, let us fix the scaling as follows:

$$u(\mathbb{1}_N) = 1, u(\mathbb{O}_N) = 0.$$

Letting $x_i = \mathbb{O}_i$ and $z_{-i} = \mathbb{O}_{-i}$ in (6.16) we get $g(x_{-i}) = u(\mathbb{O}_i, x_{-i})$, hence (6.16) becomes

$$u(x) = u(\mathbb{O}_i, x_{-i}) + h(x_{-i})u(x_i, \mathbb{O}_{-i}) \qquad (i \in N). \tag{6.17}$$

Define $u_i : X_i \to \mathbb{R}$ by $k_i u_i(x_i) = u(x_i, \mathbb{O}_{-i})$, with $k_i > 0$ such that $u_i(\mathbb{1}_i) = 1$. Note that $u_i(\mathbb{O}_i) = 0$ so that u_i is normalized, and we have $k_i = u(\mathbb{1}_i, \mathbb{O}_{-i})$.

Setting $h'(x_{-i}) = k_i h(x_{-i}) > 0$, (6.17) becomes

$$u(x) = u(\mathbb{O}_i, x_{-i}) + h'(x_{-i})u_i(x_i) \qquad (i \in N). \tag{6.18}$$

Now, letting $x_i = \mathbb{1}_i$ in (6.18) yields

$$u(\mathbb{1}_i, x_{-i}) = u(\mathbb{O}_i, x_{-i}) + h'(x_{-i})u_i(\mathbb{1}_i) \qquad (i \in N),$$

hence

$$h'(x_{-i}) = u(\mathbb{1}_i, x_{-i}) - u(\mathbb{O}_i, x_{-i}) \qquad (i \in N). \tag{6.19}$$

Substituting into (6.18) and rearranging we get

$$u(x) = u(\mathbb{1}_i, x_{-i})u_i(x_i) + u(\mathbb{O}_i, x_{-i})\big(1 - u_i(x_i)\big) \qquad (i \in N). \tag{6.20}$$

Define the set function $\mu : 2^N \to \mathbb{R}$ by $\mu(A) = u(\mathbb{1}_A, \mathbb{O}_{-A})$. Observe that $h'(x_{-i}) > 0$ for every x_{-i} and every $i \in N$ permits to deduce from (6.19) that $u(\mathbb{1}_A, \mathbb{O}_{-A}) \leqslant u(\mathbb{1}_B, \mathbb{O}_{-B})$ whenever $A \subseteq B$. Hence, by the normalization conditions, we see that μ is a normalized capacity on N.

The rest of the proof consists in repeatedly substituting (6.20) into itself for $i = 1, \ldots, n$, so that u disappears from the right hand side. In the first step, we substitute (6.20) with $i = 2$ in (6.20) with $i = 1$:

$$\begin{aligned}
u(x) = \quad & u_1(x_1)\big(u(\mathbb{1}_{12}, x_{-12})u_2(x_2) + u(\mathbb{1}_1, \mathbb{O}_2, x_{-12})(1 - u_2(x_2))\big) \\
& + (1 - u_1(x_1))\big(u(\mathbb{O}_1, \mathbb{1}_2, x_{-12})u_2(x_2) + u(\mathbb{O}_{12}, x_{-12})(1 - u_2(x_2))\big) \\
= \quad & u(\mathbb{O}_{12}, x_{-12}) + \big(u(\mathbb{1}_1, \mathbb{O}_2, x_{-12}) - u(\mathbb{O}_{12}, x_{-12})\big)u_1(x_1) \\
& + \big(u(\mathbb{O}_1, \mathbb{1}_2, x_{-12}) - u(\mathbb{O}_{12}, x_{-12})\big)u_2(x_2) \\
& + \big(u(\mathbb{1}_{12}, x_{-12}) - u(\mathbb{1}_1, \mathbb{O}_2, x_{-12}) - u(\mathbb{O}_1, \mathbb{1}_2, x_{12}) + u(\mathbb{O}_{12}, x_{-12})\big)u_1(x_1)u_2(x_2),
\end{aligned}$$

where 12 stands for $\{1, 2\}$. This ultimately yields

$$u(x) = \sum_{A \subseteq N, A \neq \varnothing} k_A \prod_{i \in A} u_i(x_i) \tag{6.21}$$

with $k_A = \sum_{B \subseteq A}(-1)^{|A \setminus B|}u(\mathbb{1}_A, \mathbb{O}_{-A})$, $\varnothing \neq A \subseteq N$, and recalling that $\mu(A) = u(\mathbb{1}_A, \mathbb{O}_{-A})$, we see that $k_A = m^\mu(A)$, the Möbius transform of μ. Then (6.21) coincides with the expression of u where F is given by (6.15). □

Remark 6.27

(i) An important consequence of the theorem is the following: the use of the Choquet integral forbids the construction of the value functions separately on each X_i. Attributes cannot be isolated, i.e., weak difference independence does not hold, and commensurateness must be ensured between the attributes. This fundamental fact can already be guessed from Formula (4.17), because a reordering of the arguments $a_1, \ldots, a_n$ in increasing order implies that they should lie on the same scale.

(ii) An axiomatization of the Choquet integral together with value functions, similar to Theorem 6.26, was done by Labreuche [219]. Up to now, we are not aware of any other attempt in this direction.

(iii) Theorem 6.26 is valid under the assumption that the set of attributes is bounded. We have seen in Sect. 6.3.3 that this assumption does not always hold, especially when preferences are involved. An examination of the proof reveals that the upper and lower bounds are merely used as particular levels to ensure normalization and the definition of μ. Hence, in the absence of lower and upper bounds, the same result holds provided one chooses $\mathbb{1}_i, \mathbb{O}_i$ so that $(\mathbb{1}_i, x_{-i}) \succ (\mathbb{O}_i, x_{-i})$ for every $x_{-i} \in X_{-i}$, where $\succcurlyeq$ is deduced from $\succcurlyeq^*$ by $x \succcurlyeq y$ if and only if $xw \succcurlyeq^* yw$ for all $w \in X$.

However, it must be noted that, contrarily to the Choquet integral, the Owen extension is *not* monotone in general when going outside $[0, 1]^n$, as shown by the following example: Take $n = 2$, with $\mu(\{1\}) = \mu(\{2\}) = 0.9$, then

$$f_\mu^{\mathrm{Ow}}\left(\frac{1}{2}, \frac{1}{2}\right) = \frac{1}{4}0.9 + \frac{1}{4}0.9 + \frac{1}{4} = 0.7$$

$$f_\mu^{\mathrm{Ow}}(3, 3) = (-6)0.9 + (-6)0.9 + 9 = -1.8 < f_\mu^{\mathrm{Ow}}\left(\frac{1}{2}, \frac{1}{2}\right).$$

See [177, Proposition 5.39] for properties of the Owen extension as an aggregation function.

(iv) The underlying assumptions of the multilinear model exclude the possibility of considering bipolar scales. However, nothing prevents from defining a bipolar version of (6.15). Indeed, the multilinear and Choquet integral are both extensions of a capacity, using a different kind of interpolation; it suffices then to replace the capacity by a bicapacity to get the bipolar multilinear model.

◇

6.9 Summary on the Construction of the Aggregation Function

(limited to the unipolar case) Our two methods to construct value functions on each criterion have naturally led to two different aggregation functions F, both depending on a capacity μ on X, defined by $\mu(A) = u(\mathbb{1}_A, \mathbb{O}_{-A})$ for all $A \subseteq N$:

(i) The Choquet integral: it is obtained as the parsimonious linear interpolation of F over the vertices of the unit hypercube, supposing that all value functions are commensurate (obtained via the MACBETH method). Here, $\mathbb{1}_i, \mathbb{O}_i$, $i \in N$, are absolute reference levels, having the meaning of "satisfactory" and "neutral," respectively. It also corresponds to the Lovász extension of μ.

(ii) The multilinear model: it is obtained as the linear interpolation using all vertices of the unit hypercube. Here, $\mathbb{1}_i, \mathbb{O}_i$, $i \in N$, need not be commensurate and should be taken as the upper and lower bounds on each attribute. However, weak difference independence must be satisfied by $\succcurlyeq^*$. It also corresponds to the Owen extension of μ.

These two aggregation functions being extensions of capacities, they should be able to represent interaction between criteria. This is the topic of the next section.

6.10 Importance and Interaction Indices

We recall from (6.10) that the value of the capacity μ for some set $A \subseteq N$ represents the overall score of the binary alternative $(\mathbb{1}_A, \mathbb{O}_{-A})$, and that F is an extension of μ. Our aim is to study how the importance of criteria and the interaction among them are captured by F, and because F is an extension of μ, in a first step we study how to define the importance of criteria and their interaction through the capacity μ.

6.10.1 *Importance and Interaction Indices for a Capacity*

One may define the importance of a criterion as its effect on the overall score of a binary alternative when this criterion becomes satisfied (i.e., its score has a satisfactory level). As this effect may depend on the situation (i.e., which other criteria are satisfied), one may consider all possible situations and make an average.

Specifically, consider a binary alternative $(\mathbb{1}_A, \mathbb{O}_{-A})$, where $\mathbb{1}_i, \mathbb{O}_i$, $i \in N$, are the satisfactory and neutral levels respectively (or the upper and lower bounds as well), and $A \subseteq N \setminus i$. Then the effect of i is measured by the difference of the scores of $(\mathbb{1}_{A\cup i}, \mathbb{O}_{-(A\cup i)})$ and $(\mathbb{1}_A, \mathbb{O}_{-A})$, hence by $\mu(A \cup i) - \mu(A)$, which we may call the marginal contribution of i to A, and is equal to $\Delta_i\mu(A)$, the derivative of μ w.r.t. i at A (see Sect. 2.5). It follows that the importance of i should be defined as the average of $\Delta_i\mu(A)$ over $A \subseteq N \setminus i$.

We have seen in Chap. 2, Remark 2.43, that two such kinds of quantities have been defined, namely the Shapley value $\phi^{\mathrm{Sh}}(\mu)$ and the Banzhaf value $\phi^{\mathrm{B}}(\mu)$ of μ (which could be called in this context *Shapley and Banzhaf importance indices*):

$$\phi_i^{\mathrm{Sh}}(\mu) = \sum_{S\subseteq N\setminus i} \frac{s!(n-s-1)!}{n!} \Delta_i\mu(S) \tag{6.22}$$

$$\phi_i^{\mathrm{B}}(\mu) = \frac{1}{2^{n-1}} \sum_{S\subseteq N\setminus i} \Delta_i\mu(S), \tag{6.23}$$

for every $i \in N$. Note that the Banzhaf value is the arithmetic mean of the first order derivative, or giving a probabilistic interpretation, it is its expected value with the uniform distribution over $2^{N\setminus i}$. The Shapley value can be interpreted in a similar way, remarking that

$$\frac{s!(n-s-1)!}{n!} = \frac{1}{n}\frac{1}{\binom{n-1}{s}}. \tag{6.24}$$

One can obtain the numbers in (6.24) by taking the expectation in two steps, first by drawing the size of the subset $S \subseteq N\setminus i$ with a uniform distribution, then by drawing a subset of the given size with a uniform distribution.

An important property of the Shapley value is *efficiency*:

$$\sum_{i\in N} \phi_i^{\mathrm{Sh}}(\mu) = \mu(N),$$

a property that is not satisfied by the Banzhaf value. Its normalized version, $\frac{\phi^{\mathrm{B}}(\mu)}{\sum_{i\in N}\phi_i^{\mathrm{B}}(\mu)}$, is of course efficient but loses linearity w.r.t. the capacity.

We turn to the modeling of interaction. Take two distinct criteria $i,j \in N$. We say that there is a *positive interaction* or *synergy* between i and j in the presence of $S \subseteq N \setminus \{i,j\}$ if the satisfaction of both criteria has a greater effect in terms of marginal contribution than the sum of their contributions taken separately; i.e.,

$$\mu(S \cup \{i,j\}) - \mu(S) \geqslant \big(\mu(S \cup i) - \mu(S)\big) + \big(\mu(S \cup j) - \mu(S)\big),$$

which can be rewritten as

$$\mu(S \cup \{i,j\}) - \mu(S \cup i) - \mu(S \cup j) + \mu(S) \geqslant 0. \tag{6.25}$$

Similarly, there is *negative interaction* or *synergy* between i and j in the presence of S if the converse inequality holds, which leads to

$$\mu(S \cup \{i,j\}) - \mu(S \cup i) - \mu(S \cup j) + \mu(S) \leqslant 0. \tag{6.26}$$

The case of equality depicts a situation where the marginal contribution of both criteria is exactly the sum of individual marginal contributions. In this case, we say that there is *independence* between i and j in the presence of S. Remark that in (6.25) and (6.26), the key quantity is nothing other than $\Delta_{\{i,j\}}\mu(S)$, the second-order derivative of μ at S.

It remains to define an index accounting for the overall interaction between i and j in all situations, as an average of the second order derivatives. Mimicking what we did for the importance index, we could consider averages corresponding to the same kind of probabilistic interpretation pertaining to the Shapley and Banzhaf values, and obtain the *Shapley* and *Banzhaf interaction indices*:

$$I_{ij}^{\mathrm{Sh}}(\mu) = \sum_{S\subseteq N\setminus\{i,j\}} \frac{s!(n-s-2)!}{(n-1)!}\Delta_{\{i,j\}}\mu(S) \tag{6.27}$$

$$I_{ij}^{\mathrm{B}}(\mu) = \frac{1}{2^{n-2}} \sum_{S\subseteq N\setminus\{i,j\}} \Delta_{\{i,j\}}\mu(S). \tag{6.28}$$

We recognize in (6.27) and (6.28) particular cases of the Shapley and Banzhaf interaction transforms (Definitions 2.41 and 2.42), precisely:

$$I_{ij}^{\mathrm{Sh}}(\mu) = I^{\mu}(\{i,j\}), \quad I_{ij}^{\mathrm{B}}(\mu) = I_{\mathrm{B}}^{\mu}(\{i,j\}).$$

Recall from Remark 2.43 that the Shapley and Banzhaf values are also particular instances of these transforms, because $\phi_i^{\mathrm{Sh}}(\mu) = I^{\mu}(\{i\})$ and $\phi_i^{\mathrm{B}}(\mu) = I_{\mathrm{B}}^{\mu}(\{i\})$. It follows that the Shapley and Banzhaf interaction transforms provide an adequate tool for the representation of interaction and importance. In particular, these transforms permit to define interaction indices for any group T of criteria, letting $I_T^{\mathrm{Sh}}(\mu) = I^{\mu}(T)$ and $I_T^{\mathrm{B}}(\mu) = I_{\mathrm{B}}^{\mu}(T)$.

A precise interpretation of the (Shapley) importance and interaction indices can be obtained from the formula of the Choquet integral w.r.t. a 2-additive capacity, which we present in Sect. 6.10.4.

Remark 6.28

(i) The Shapley and Banzhaf values are central concepts in cooperative game theory, where they serve as a means to share among the players in N the benefit $\mu(N)$ obtained from the cooperation of the members of N (see, e.g., Owen [263], Peters [268]). Apart from the classical axiomatizations of these values by their authors, numerous axiomatizations have been proposed (see Peters [268, Chap. 17] for the most representative axiomatizations of the Shapley value, including the classical one).

(ii) The first appearance of the interaction index for two criteria seems to be due to Owen in the context of cooperative game theory, under the name of *co-value* [261], but this notion seems to have gone unnoticed. It was rediscovered by Murofushi and Soneda [253], and applied to multicriteria decision making.

◇

The next section gives an approach of interaction from the point of view of the aggregation function.

6.10.2 Importance and Interaction Indices for an Aggregation Function

The material of this section is borrowed from [177, Sects. 10.3 and 10.4], to which the readers are referred for more details.

Let us consider an n-ary aggregation function defined on some real interval $[a, b]$; i.e., a mapping $F : [a, b]^n \to [a, b]$, being nondecreasing in each coordinate and satisfying $F(a, \ldots, a) = a$, $F(b, \ldots, b) = b$. The *total variation* of F w.r.t. coordinate i is the function

$$\Delta_i F(x) = F(b_i, x_{-i}) - F(a_i, x_{-i}) \qquad (x \in [a, b]^n).$$

Similarly, the *second-order total variation* of F w.r.t. coordinates i, j is the function

$$\begin{aligned}\Delta_{ij} F(x) &= \Delta_i(\Delta_j F(x)) = \Delta_j(\Delta_i F(x)) \\ &= F(b_{ij}, x_{-ij}) - F(b_i, a_j, x_{-ij}) - F(b_j, a_i, x_{-ij}) + F(a_{ij}, x_{-ij}),\end{aligned}$$

where ij stands for $\{i, j\}$.

Example 6.29 Taking basic aggregation functions with $[a, b] = [0, 1]$, we find:

$$\Delta_{ij} \min(x) = \bigwedge_{k \neq i,j} x_k \geqslant 0$$

$$\Delta_{ij} \max(x) = -1 + \bigvee_{k \neq i,j} x_k \leqslant 0$$

$$\Delta_{ij}\Big(\frac{1}{n}\sum_{i=1}^{n} x_i\Big) = 0.$$

◇

Generalizing the above definitions, the *total variation* of F w.r.t. $K \subseteq N$ is the function

$$\Delta_K F(x) = \sum_{L \subseteq K} (-1)^{|L|} F(a_L, b_{K \setminus L}, x_{-K}) \qquad (x \in [a, b]^n). \tag{6.29}$$

The *interaction index* of $K \subseteq N$ on F is defined as the average on the whole domain of the corresponding total variation:

$$I_K(F) = \frac{1}{(b-a)^n} \int_{[a,b]^n} \frac{\Delta_K F(x)}{b-a} \, \mathrm{d}x. \tag{6.30}$$

The next theorem relates the definitions of interaction indices for capacities and for aggregation functions.

Theorem 6.30 (Correspondence between interaction for capacities and for aggregation functions) *(Grabisch et al. [178]) Consider* $[a, b] = [0, 1]$ *and* μ *a normalized capacity. The following holds.*

(i) The interaction index of $K \subseteq N$ *for the Choquet integral (Lovász extension) is the interaction transform at* K*:*

$$I_K\Big(\int \cdot \, \mathrm{d}\mu\Big) = I^{\mu}(K).$$

(ii) *The interaction index of* $K \subseteq N$ *for the multilinear model (Owen extension) is the Banzhaf interaction transform at* K*:*

$$I_K(f_\mu^{\mathrm{Ow}}) = I_{\mathrm{B}}^\mu(K).$$

Proof

(i) Using (6.29) and the expression of the Lovász extension (2.96), we get

$$\begin{aligned}\Delta_K f_\mu^{\mathrm{Lo}}(x) &= \sum_{L\subseteq K}(-1)^{|L|} f_\mu^{\mathrm{Lo}}(0_L, 1_{K\setminus L}, x_{-K})\\ &= \sum_{L\subseteq K}(-1)^{|L|}\sum_{T\subseteq N\setminus L}\Big(m^\mu(T)\bigwedge_{i\in T\setminus K} x_i\Big)\\ &= \sum_{T\subseteq N}\Big(m^\mu(T)\bigwedge_{i\in T\setminus K} x_i\Big)\sum_{L\subseteq K\setminus T}(-1)^{|L|} = \sum_{T\supseteq K}\Big(m^\mu(T)\bigwedge_{i\in T\setminus K} x_i\Big),\end{aligned}$$

using Lemma 1.1(i). From (2.97), we see that $\Delta_K f_\mu^{\mathrm{Lo}}(x)$ is the derivative of the Lovász extension (the notation fortunately coincides!), so that the desired result holds by Theorem 2.90.

(ii) On the one hand, we have, using (6.29) and (2.83)

$$\begin{aligned}\Delta_K f_\mu^{\mathrm{Ow}}(x) &= \sum_{L\subseteq K}(-1)^{|L|} f_\mu^{\mathrm{Ow}}(0_L, 1_{K\setminus L}, x_{-K})\\ &= \sum_{L\subseteq K}(-1)^{|L|}\sum_{A\subseteq N\setminus K}\mu((K\setminus L)\cup A)\prod_{i\in A} x_i \prod_{i\in N\setminus(A\cup K)}(1-x_i).\end{aligned} \tag{6.31}$$

On the other hand, using Lemma 2.83 and the definition of the derivative (Sect. 2.5), we get:

$$\begin{aligned}\frac{\partial^k f_\mu^{\mathrm{Ow}}}{\partial x_{|K}}(x) &= \sum_{A\subseteq N\setminus K}\Delta_K\mu(A)\prod_{i\in A} x_i \prod_{i\in N\setminus(A\cup K)}(1-x_i)\\ &= \sum_{A\subseteq N\setminus K}\sum_{L\subseteq K}(-1)^{|K\setminus L|}\mu(A\cup L)\prod_{i\in A} x_i \prod_{i\in N\setminus(A\cup K)}(1-x_i).\end{aligned}$$

Letting $L' = K\setminus L$ in (6.31) shows that $\Delta_K f_\mu^{\mathrm{Ow}}(x)$ and $\frac{\partial^k f_\mu^{\mathrm{Ow}}}{\partial x_{|K}}(x)$ are identical, and so are their integrals on $[0, 1]^n$. We conclude by using (2.89). □

Remark 6.31 The proof of Theorem 6.30 establishes that the total variation w.r.t K of the Lovász and Owen extensions coincide with their (partial) derivatives w.r.t. $x_{|K}$. ◇

6.10.3 A Statistical Approach: The Sobol' Indices

The concept of interaction has also been studied in statistics, and is based on the analysis of variance (ANOVA) (Fisher and Mackenzie [142], Hoeffding [196]). In this context, consider n independent random variables $Z_1, \dots, Z_n$, with uniform distribution on $[0, 1]$, and our aggregation function F as a multivariate function of $Z_1, \dots, Z_n$. We put $Z = (Z_1, \dots, Z_n)$ and $Y = F(Z)$, the random output of F. Also, we adopt the same notation as above with variables; i.e., Z_S stands for $(Z_i)_{i\in S}$, etc. For the sake of clarity, we may specify on which variables the expectation is taken, writing, e.g., $\mathbb{E}_{Z_i}[Y]$, etc.

Any multivariate function can be decomposed in the following way (ANOVA decomposition):

$$Y = F(Z) = F_\varnothing + \sum_{i=1}^n F_i(Z_i) + \sum_{i<j} F_{ij}(Z_i, Z_j) + \cdots + F_N(Z) = \sum_{S\subseteq N} F_S(Z_S),$$

with

$$\begin{aligned}
F_\varnothing &= \mathbb{E}[Y]\\
F_i(Z_i) &= \mathbb{E}[Y|Z_i] - F_\varnothing\\
F_{ij}(Z_i, Z_j) &= \mathbb{E}[Y|Z_i, Z_j] - F_i(Z_i) - F_j(Z_j) - F_\varnothing\\
&= \mathbb{E}[Y|Z_i, Z_j] - \mathbb{E}[Y|Z_i] - \mathbb{E}[Y|Z_j] + E[Y]\\
\vdots &= \vdots\\
F_S(Z_S) &= \mathbb{E}_{Z_{-S}}[Y|Z_S] - \sum_{T\subset S} F_T(Z_T) = \sum_{T\subseteq S} (-1)^{|S\setminus T|}\mathbb{E}_{Z_{-T}}[Y|Z_T]\\
\vdots &= \vdots\\
F_N(Z) &= \sum_{T\subseteq N} (-1)^{|N\setminus T|}\mathbb{E}_{Z_{-T}}[Y|Z_T].
\end{aligned}$$

The property of this decomposition is that each term has zero mean, except the first one, $F_\varnothing$. It follows that the variance of Y can be decomposed as follows:

$$\mathrm{Var}[Y] = \sum_{\varnothing\neq S\subseteq N} \mathrm{Var}[F_S(Z_S)].$$

The *first-order Sobol' indices* [314, 315] are the quantities $\frac{\mathrm{Var}[F_S(Z_S)]}{\mathrm{Var}[Y]}$, although one can omit the normalization factor. The next theorem establishes the close link between Sobol' indices and the Fourier transform (and consequently the Banzhaf transform) for the multilinear model.

Theorem 6.32 (Relation between Sobol' indices and the Fourier transform) *(Grabisch and Labreuche [176]) Consider the multilinear model f_μ^{Ow} w.r.t. a capacity μ. Then the (nonnormalized) Sobol index for a subset $\varnothing \neq S \subseteq N$ is given by*

$$\mathrm{Var}[(f_\mu^{\mathrm{Ow}})_S] = \frac{1}{3^s}\left(\widehat{\mu}(S)\right)^2,$$

where $\widehat{\mu}$ is the Fourier transform of μ.

Proof We set for simplicity $f = f_\mu^{\mathrm{Ow}}$. We compute

$$f_S = \sum_{K\subseteq S}(-1)^k \mathbb{E}(f \mid Z_{S\setminus K}) \qquad (S \subseteq N, |S| > 0).$$

We have for any such S:

$$\begin{aligned}\mathbb{E}(f|Z_{S\setminus K}) &= \int_{[0,1]^{N\setminus(S\setminus K)}} f \,\mathrm{d}z_{N\setminus(S\setminus K)} = \sum_{T\subseteq[n]} m^\mu(T)\int_{[0,1]^{N\setminus(S\setminus K)}} \prod_{i\in T} z_i \,\mathrm{d}z_{N\setminus(S\setminus K)}\\
&= \sum_{T\subseteq[n]} m^\mu(T)\frac{1}{2^{|T\setminus(S\setminus K)|}}\prod_{i\in T\cap(S\setminus K)} z_i\\
&= \sum_{L\subseteq N\setminus S}\sum_{T\subseteq S} m^\mu(L\cup T)\frac{1}{2^{|L\cup(T\cap K)|}}\prod_{i\in T\setminus K} z_i. \qquad (6.32)\end{aligned}$$

It follows that

$$\begin{aligned}f_S &= \sum_{K\subseteq S}(-1)^k\sum_{L\subseteq N\setminus S}\sum_{T\subseteq S} m^\mu(L\cup T)\frac{1}{2^{|L\cup(T\cap K)|}}\prod_{i\in T\setminus K} z_i\\
&= \sum_{L\subseteq N\setminus S}\frac{1}{2^l}\sum_{T\subseteq S} m^\mu(L\cup T)\sum_{K\subseteq S}(-1)^k\frac{1}{2^{|T\cap K|}}\prod_{i\in T\setminus K} z_i.\end{aligned}$$

Letting $T' = T\cap K$, we have

$$\sum_{K\subseteq S}(-1)^k\frac{1}{2^{|T\cap K|}}\prod_{i\in T\setminus K} z_i = \sum_{T'\subseteq T}(-1)^{t'}\frac{1}{2^{t'}}\prod_{i\in T\setminus T'} z_i \underbrace{\sum_{K'\subseteq S\setminus T}(-1)^{k'}}_{=0 \text{ except if } S\setminus T=\varnothing}.$$

It follows that

$$f_S = \sum_{L \subseteq N \setminus S} \frac{1}{2^l} m^\mu(L \cup S) \sum_{T \subseteq S} (-1)^t \frac{1}{2^t} \prod_{i \in S \setminus T} z_i. \tag{6.33}$$

Observe that

$$\sum_{T \subseteq S} (-1)^t \frac{1}{2^t} \prod_{i \in S \setminus T} z_i = \frac{1}{2^s} \prod_{i \in S} (2z_i - 1),$$

hence we finally get by (2.74):

$$f_S = (-1)^s \prod_{i \in S} (2z_i - 1) \widehat{v}(S). \tag{6.34}$$

We obtain finally

$$\mathbb{E}[f_S^2] = \int_{[0,1]^S} \Big(\prod_{i \in S} (2z_i - 1) \widehat{v}(S) \Big)^2 \mathrm{d}z_S = \frac{1}{3^s} (\widehat{v}(S))^2.$$

□

6.10.4 The 2-Additive Model

Recall from Chap. 2 that a capacity μ is 2-additive if its Möbius transform (equivalently, its interaction transform) vanishes for subsets of cardinality greater than 2:

$$m^\mu(S) = 0, \quad I^\mu(S) = 0 \qquad (S \subseteq N, |S| > 2)$$

[see Definition 2.50 and Remark 2.51(ii)]. Therefore, such capacities are uniquely determined through their (Shapley) importance indices and interaction indices for pairs of criteria.

Considering F to be the Choquet integral, using Eq. (4.79), the overall score of an alternative $x \in X$ when the capacity μ is 2-additive can be written as follows:

$$u(x) = \sum_{i,j : I_{ij}^{\mathrm{Sh}} > 0} (u_i(x_i) \wedge u_j(x_j)) I_{ij}^{\mathrm{Sh}} + \sum_{i,j : I_{ij}^{\mathrm{Sh}} < 0} (u_i(x_i) \vee u_j(x_j)) |I_{ij}^{\mathrm{Sh}}|$$
$$+ \sum_{i \in N} u_i(x_i) \Big(\phi_i^{\mathrm{Sh}} - \frac{1}{2} \sum_{j \neq i} |I_{ij}^{\mathrm{Sh}}| \Big),$$

where we have dropped (μ) in the importance and interaction indices. This expression gives a very clear picture of the model:

(i) If interaction between i and j is positive, the aggregation of the scores on these criteria is purely conjunctive: criterion i *and* criterion j must be satisfied in order to get a satisfactory score. In other words, the decision maker is intolerant w.r.t. these criteria, because both must be satisfied. Another way to interpret this is to say that criteria i and j are *complementary*, which means that each of them represents a necessary ingredient in the decision process, without any possible compensation between them;

(ii) If interaction between i and j is negative, the aggregation of the scores on i and j is now purely disjunctive: criterion i *or* criterion j must be satisfied to get a satisfactory score. We may say that the decision maker is tolerant relatively to i and j, because it is enough that one of them is satisfied. Another interpretation is to say that the criteria are *substitutable*, in the sense that if i (respectively, j) is not satisfied, it can be substituted with j (respectively, i), provided this criterion is satisfied;

(iii) The computation of the overall score is obtained by summing three types of terms: conjunctive terms for any pair of positively interacting criteria, disjunctive terms for any pair of negatively interacting criteria, and a linear term which is a weighted sum of all criteria. Observe that the weight of each criterion in the latter term is equal to its importance index, diminished by half the sum of all interaction magnitudes pertaining to this criterion.

More importantly, the overall score is obtained as a convex combination of all these terms. Indeed, the coefficients of the conjunctive and disjunctive terms are clearly positive. As for the linear term, using Theorem 2.45(iii), observe that the monotonicity of μ in terms of its interaction indices reduces to, by 2-additivity:

$$\phi_i^{\mathrm{Sh}} + \frac{1}{2}\sum_{j\in L} I_{ij}^{\mathrm{Sh}} - \frac{1}{2}\sum_{j\notin L} I_{ij}^{\mathrm{Sh}} \geqslant 0 \qquad (i \in N, L \subseteq N \setminus i).$$

For a fixed $i \in N$, taking $L = \{j \in N \setminus i : I_{ij}^{\mathrm{Sh}} < 0\}$, we find that

$$\phi_i^{\mathrm{Sh}} - \frac{1}{2}\sum_{j\neq i} |I_{ij}^{\mathrm{Sh}}| \geqslant 0.$$

As a conclusion, all coefficients are nonnegative. It remains to show that they sum up to 1. We have

$$\sum_{i,j: I_{ij}^{\mathrm{Sh}}>0} I_{ij}^{\mathrm{Sh}} + \sum_{i,j: I_{ij}^{\mathrm{Sh}}<0} |I_{ij}^{\mathrm{Sh}}| + \sum_{i\in N} \Big(\phi_i^{\mathrm{Sh}} - \frac{1}{2}\sum_{j\neq i} |I_{ij}^{\mathrm{Sh}}|\Big)$$

$$= \sum_{\{i,j\}\subseteq N} |I_{ij}^{\mathrm{Sh}}| + \sum_{i\in N} \phi_i^{\mathrm{Sh}} - \frac{1}{2}\sum_{i\in N}\sum_{j\neq i} |I_{ij}^{\mathrm{Sh}}| = \sum_{i\in N} \phi_i^{\mathrm{Sh}} = 1.$$

This result can be obtained also as a simple consequence of Theorem 2.65, because by using transformation formulas between the Möbius transform and the interaction transform, one can see that Eq. (2.61), giving the convex decomposition of a 2-additive capacity, yields (4.79) just by applying the Choquet integral on it.

The fact that the expression of u is a convex combination of conjunctive, disjunctive and linear terms is particularly appealing for the explanation of the model to the decision maker, because this gives a precise idea of how much linear or nonlinear is the model, or how much conjunctive or disjunctive is the model, and what are the terms in the model which are most important.

6.11 The Case of Ordinal Measurement

A close examination of the construction of our models, either based on Choquet integral or the multilinear model, reveals that in one or the other, a kind of difference measurement is used. This is perfectly clear with the multilinear model, because it is based on the assumption of weak difference independence, a concept rooted in difference measurement. In the case of the Choquet integral, the MACBETH method forces the decision maker to think in terms of intensity of preference, which is a disguised form of difference measurement: indeed, saying that a is strongly preferred to b, while c is weakly preferred to d induces the fact that $ab \succ^* cd$.

The question is then: What happens if the conditions for difference measurement (those in Theorem 6.11) are not fulfilled? Or if the decision maker is unable to tell in a consistent way intensities of preference? In this situation we are back to ordinal measurement: nothing more than "a is preferred to b" can be said, and as a consequence, value functions should be considered to be ordinal scales, where numbers have no cardinal meaning. This amounts to saying that the range of the value functions is a totally ordered set $(L, \leqslant)$, either finite or infinite, without the algebraic structure of the field of real numbers, but with only $\vee$ (supremum) and $\wedge$ (infimum) as algebraic operations. The consequence is, if we still want to adhere to the decomposable model (6.2), that the aggregation function F, now seen as a mapping from L^n to L, must be built solely from the operations $\wedge, \vee$.[14]

6.11.1 The Emergence of the Sugeno Integral Model

A mapping $F : L^n \to L$ that is an arbitrary finite combination of $\vee, \wedge$ is called a *weighted lattice polynomial function* [231]. These functions are a generalization of *lattice polynomial functions* (see, e.g., Grätzer [185, Sect. I.4]), where some variables are considered to be constants. Formally, the class $\mathcal{WLP}(L; n)$ of weighted lattice polynomial functions from L^n to L is inductively defined as follows:

(i) For any $k \in [n]$ and any $c \in L$, the projection $x \mapsto x_k$ and the constant function $x \mapsto c$ are elements of $\mathcal{WLP}(L; n)$;
(ii) If $p, q \in \mathcal{WLP}(L; n)$, then $p \vee q, p \wedge q \in \mathcal{WLP}(L; n)$;
(iii) Every element of $\mathcal{WLP}(L; n)$ is formed by finitely many applications of rules (i) and (ii).

A function $p \in \mathcal{WLP}(L; n)$ is said to be *idempotent* if $p(a, \ldots, a) = a$ for every $a \in L$. Assuming in addition that L is a complete lattice with top and bottom elements denoted respectively by 1 and 0, p is *endpoint-preserving* if $p(0, \ldots, 0) = 0$ and $p(1, \ldots, 1) = 1$.

The fundamental fact about (weighted) lattice polynomial functions is that they can be expressed in *disjunctive* or *conjunctive normal forms* (see, e.g., Birkhoff [30, Sect. II.5]). For any $p \in \mathcal{WLP}(L; n)$,

$$p(x) = \bigvee_{j=1}^{k} \Big(a_j \wedge \bigwedge_{i \in A_j} x_i\Big) = \bigwedge_{j=1}^{\ell} \Big(b_j \vee \bigvee_{i \in B_j} x_i\Big),$$

for some appropriate $k, \ell \in [n]$, $a_j, b_j \in L$, and $A_j, B_j \subseteq [n]$. Observe from this formula that any weighted lattice polynomial function is necessarily nondecreasing w.r.t. each variable. An equivalent form is:

$$p(x) = \bigvee_{A \subseteq [n]} \Big(\alpha(A) \wedge \bigwedge_{i \in A} x_i\Big) = \bigwedge_{A \subseteq [n]} \Big(\beta(A) \vee \bigvee_{i \in A} x_i\Big), \tag{6.35}$$

where $\alpha, \beta : 2^{[n]} \to L$ are set functions. These set functions are not unique, and as shown in [231], if L is a complete lattice, one can take $\alpha^*(A) = p(1_A)$, $A \subseteq [n]$

[14] If L is finite, there is evidently no increasing function from L^n to L (unless $|L| = 1$), and few nondecreasing functions. Therefore, in the finite case, one should consider mappings from L^n to L', where L' has a cardinality sufficiently larger than $|L|$.

as a particular function α. Based on this decomposition, the following fundamental result is easily obtained:

Theorem 6.33 (Sugeno integrals are idempotent weighted lattice polynomial functions) *(Marichal [229, 231]) Suppose that the totally ordered set $(L, \leqslant)$ is a complete lattice, and consider $F : L^n \to L$. The following propositions are equivalent:*

(i) There exists a unique normalized capacity μ such that $F(\cdot) = \fint \cdot \, d\mu$;
(ii) F is an idempotent weighted lattice polynomial function;
(iii) F is an endpoint-preserving weighted lattice polynomial function.

Proof (i)⇒(ii)⇒(iii) is obvious. We prove (iii)⇒(i). Let F satisfying the conditions of (iii). We know that the disjunctive normal form of F can be written as

$$F(x) = \bigvee_{A \subseteq [n]} \Big(F(1_A) \wedge \bigwedge_{i \in A} x_i \Big).$$

Letting $\mu(A) = F(1_A)$, we observe by comparison with (4.67) that F is nothing but the Sugeno integral. □

As a consequence, the aggregation function F in the decomposable model must be the Sugeno integral w.r.t. some normalized capacity:

$$u^{\mathrm{Sug}}(x) = \fint (u_1(x_1), \ldots, u_n(x_n)) \, d\mu \qquad (x \in X). \tag{6.36}$$

It is important to note that the determination of the utility functions and capacity cannot be conducted in the same way as for the Choquet integral, that is, with the help of reference levels $\mathbb{1}_i, \mathbb{O}_i$, $i \in N$, and in two steps, the first one consisting in the determination of the u_i's and the second one in the determination of the capacity. Indeed, for an alternative of the form $(x_i, \mathbb{O}_{-i})$, we have from (6.36):

$$u^{\mathrm{Sug}}(x_i, \mathbb{O}_{-i}) = u_i(x_i) \wedge \mu(\{i\}).$$

Then, if $u_i(x_i) > \mu(\{i\})$, the value of $u_i(x_i)$ cannot be observed because it is hidden by $\mu(\{i\})$, which acts like a threshold. A full description of a method of determination of the u_i's and of the capacity based on reference levels is given in [173, Sect. 5.4].

We now give a characterization result of the model given in (6.36), which does not use reference levels. This result was given without proof by Greco et al. [187]. Later, Bouyssou et al. [37] have proposed a proof of this deep result. Greco et al. introduced the following notion: $\succcurlyeq$ on X is said to be *strongly 2-graded on attribute*

$i \in N$ if, for all $x, y, z, w \in X$ and all $a_i \in X_i$,

$$\left.\begin{array}{c} x \succcurlyeq z \\ \text{and} \\ y \succcurlyeq w \\ \text{and} \\ z \succcurlyeq w \end{array}\right\} \Rightarrow \left\{\begin{array}{c} (a_i, x_{-i}) \succcurlyeq z \\ \text{or} \\ (x_i, y_{-i}) \succcurlyeq w. \end{array}\right.$$

The relation $\succcurlyeq$ is *strongly 2-graded* if it is strongly 2-graded for every attribute $i \in N$. The condition is rather complex, but can be explained relatively simply when considering $z = w$: the strong 2-gradedness condition on attribute i induces a partition of X_i into satisfactory and unsatisfactory elements, relatively to w. Indeed, suppose that $y \succcurlyeq w$, and suppose that $(x_i, y_{-i}) \not\succcurlyeq w$. Since we have $(y_i, y_{-i}) \succcurlyeq w$, it means that level x_i is not satisfactory enough w.r.t. w, while y_i is. Now, there is no worse level w.r.t. w than x_i, because assuming in addition that $(x_i, x_{-i}) \succcurlyeq w$ implies by strong 2-gradedness that $(a_i, x_{-i}) \succcurlyeq w$ for any element $a_i \in X_i$. Hence, in a sense, x_i belongs to the category of unsatisfactory elements of X_i while y_i is a satisfactory element.

Theorem 6.34 (Characterization of the Sugeno integral model) *Let $L = [0, 1]$ and $\succcurlyeq$ be a binary relation on X. Then $\succcurlyeq$ can be represented by the Sugeno integral model (6.36) if and only if it is a complete preorder that is strongly 2-graded, and $X/_\sim$ has a countable order-dense subset.*

The sufficiency is difficult to prove (see [37]), however one can easily explain the necessity of the strong 2-gradedness condition. Suppose that there exist $x, y, z, w \in X$ and $a_i \in X_i$ such that $x \succcurlyeq z \succcurlyeq w$, $y \succcurlyeq w$, $(a_i, x_{-i}) \not\succcurlyeq z$ and $(x_i, y_{-i}) \not\succcurlyeq w$, and let us denote for simplicity $⨍(u_1(x_1), \dots, u_n(x_n))\, d\mu$ by $S_\mu(u(x))$. From $(y_i, y_{-i}) \succcurlyeq w$ and $(x_i, y_{-i}) \not\succcurlyeq w$ we obtain $u_i(x_i) < S_\mu(u(w))$ [use (4.67)]. Since $z \succcurlyeq w$, it follows that $S_\mu(u(z)) \geqslant S_\mu(u(w)) > u_i(x_i)$. Now, $x \succcurlyeq z$ and $S_\mu(u(z)) > u_i(x_i)$ imply that there exists $A \subseteq N$ such that $i \notin A$, $\mu(A) \geqslant S_\mu(u(z))$ and $u_j(x_j) \geqslant S_\mu(u(z))$ for all $j \in A$ [again use (4.67)]. This implies $S_\mu(u(a_i, x_{-i})) \geqslant S_\mu(u(z))$, which contradicts $(a_i, x_{-i}) \not\succcurlyeq z$.

6.11.2 Monotonicity Properties of the Sugeno Integral Model

Theorem 6.33 forces the use of the Sugeno integral if one wants to use the (weakly) monotone decomposable model in a purely ordinal framework. This being taken for granted, it appears nevertheless that the Sugeno integral model has a poor discriminative power. This can already be seen from the strong 2-gradedness condition, because it implies a rather coarse treatment of the attribute values. A more striking example is the following fact: it may happen that improving an alternative on each attribute does not change the value of the Sugeno integral, as illustrated by the next example.

Example 6.35 Consider $n = 3$ and an alternative $x \in X$ with vector of scores in $[0, 1]$ given by $(0.1\ \ 0.5\ \ 0.7)$. Take any normalized capacity μ satisfying $\mu(\{2, 3\}) = 0.5$. Then

$$u^{\text{Sug}}(x) = (0.1 \wedge 1) \vee (0.5 \wedge 0.5) \vee (0.7 \wedge \mu(\{3\}) = 0.5,$$

because $\mu(\{3\}) \leqslant \mu(\{2, 3\}) = 0.5$. Then for $x' \in X$ with vector of scores $(0.5\ \ 1\ \ 1)$, we obtain

$$u^{\text{Sug}}(x') = (0.5 \wedge 1) \vee (1 \wedge 0.5) \vee (1 \wedge \mu(\{3\}) = 0.5.$$

Although x' is far better than x on every attribute, its evaluation by the Sugeno integral model is identical to the evaluation of x (see also Remark 5.26). ◇

This unpleasant behavior is an example of the peculiar properties of the Sugeno integral with respect to monotonicity. It is worthwhile to study in detail this point, and to this end, we recall and introduce some notions of monotonicity.

We use the notation $a \leqslant b$, $a < b$ and $a \ll b$ for two vectors $a, b \in L^n$ given in Sect. 1.11.1. A function $F : L^n \to L$ is *nondecreasing* (or *monotone*) if it is nondecreasing w.r.t each variable; i.e., if $a \leqslant b$ then $F(a) \leqslant F(b)$, for any $a, b \in L^n$. It is *increasing* if $a < b$ implies $F(a) < F(b)$, and it is *weakly increasing* if it is nondecreasing and $a \ll b$ implies $F(a) < F(b)$. Obviously, increasingness implies weak increasingness, which implies nondecreasingness.

Example 6.35 has shown that the Sugeno integral is not weakly increasing in general, but we know from Theorem 4.43(vi) that it is always nondecreasing. The following result summarizes the monotonicity properties of the Sugeno integral.

Theorem 6.36 (Monotonicity properties of the Sugeno integral) *Consider the Sugeno integral w.r.t. a normalized capacity μ on N, both valued on a totally ordered set L that is a complete lattice, with top and bottom elements 1, 0. The following holds:*

(i) *The Sugeno integral is nondecreasing for every capacity μ;*
(ii) *The Sugeno integral is weakly increasing if and only if μ is a 0-1-capacity;*
(iii) *The Sugeno integral is never increasing.*

Proof

(i) is established in Theorem 4.43(vi).
(ii) Assume that μ is 0-1-capacity. If $a \ll b$, we have $\min_{i \in A} a_i < \min_{i \in A} b_i$ for every nonempty $A \subseteq N$, which implies that

$$\max_{A : \mu(A)=1} \Big(\min_{i \in A} a_i\Big) < \max_{A : \mu(A)=1} \Big(\min_{i \in A} b_i\Big).$$

Using the fact that the Sugeno and Choquet integrals coincide for 0-1-capacities [see Theorem 4.47(i)], it follows from Theorem 4.61 that the above equation is nothing other than $⨍ a \, \mathrm{d}\mu < ⨍ b \, \mathrm{d}\mu$.

Conversely, assume that μ is not a 0-1-capacity. Then there exists a nonempty subset $A \subset N$ such that $0 < \mu(A) < 1$. Define $a, b \in L^n$ by

$$a_i = \begin{cases} \mu(A), & \text{if } i \in A \\ 0, & \text{otherwise} \end{cases}, \quad b_i = \begin{cases} 1, & \text{if } i \in A \\ \mu(A), & \text{otherwise} \end{cases}.$$

Then $a \ll b$, and

$$⨍ a \, \mathrm{d}\mu = 0 \vee (\mu(A) \wedge \mu(A)) = \mu(A) = \mu(A) \vee (1 \wedge \mu(A)) = ⨍ b \, \mathrm{d}\mu. \tag{6.37}$$

Hence, the Sugeno integral w.r.t. μ is not weakly increasing.

(iii) Since $|N| > 1$, there exists a nonempty proper subset of N, say A. Defining $a, b \in L^n$ with A like in (ii), we have $a < b$ but equality of the integrals holds by (6.37), showing that the Sugeno integral is not increasing.

□

Remark 6.37

(i) These results were proved by Marichal [228]; see also Murofushi [249]. Our exposition follows the latter reference.

(ii) What about the Choquet integral? We know from Theorem 4.24(vi) that nondecreasingness holds. Murofushi [249] has shown that weak increasingness always holds, while increasingness holds if and only if the capacity is strictly monotone. The same results hold for the multilinear model [177, Proposition 5.39].

◇

6.11.3 Lexicographic Refinement

Theorem 6.36 shows that the Sugeno integral model has a poor discriminative power, in the sense that, for a fixed capacity μ and value functions $u_1, \ldots, u_n$, u^{Sug} remains constant over large domains of X. As a consequence, many preference relations on X cannot be represented by a Sugeno integral model. We refer the readers to Rico et al. [274] for a detailed study of the abilities of the Sugeno integral for the representation of preferences.

A possible solution for these drawbacks is to escape (at least for a while) from the decomposable model, by refining the preference relation it induces through a lexicographic procedure. Given two complete preorders $\succcurlyeq, \succcurlyeq'$, we say that $\succcurlyeq'$ *refines* $\succcurlyeq$ if whenever $a \succ b$, we have $a \succ' b$, and $\succcurlyeq, \succcurlyeq'$ are distinct. We begin by giving the well-known lexicographic refinements of minimum and maximum, called leximin and leximax [88, 248].

The *leximin* and *leximax orderings* on L^n, denoted respectively by $\succcurlyeq_{\text{lmin}}$, $\succcurlyeq_{\text{lmax}}$, are defined as follows, for every $a, b \in L^n$:

$$a \succcurlyeq_{\text{lmin}} b \Leftrightarrow (a_{(1)}, \ldots, a_{(n)}) \succcurlyeq_{\text{lex}} (b_{(1)}, \ldots, b_{(n)})$$
$$a \succcurlyeq_{\text{lmax}} b \Leftrightarrow (a_{(n)}, \ldots, a_{(1)}) \succcurlyeq_{\text{lex}} (b_{(n)}, \ldots, b_{(1)}),$$

where $(\cdot)$ indicates a permutation of the indices so that the coordinates are in nondecreasing order: $a_{(1)} \leqslant a_{(2)} \leqslant \cdots \leqslant a_{(n)}$, and $\succcurlyeq_{\text{lex}}$ is the lexicographic order (see Example 6.3). Note that $a \sim_{\text{lmin}} b \Leftrightarrow a \sim_{\text{lmax}} b \Leftrightarrow a_{(i)} = b_{(i)}$ for all $i = 1, \ldots, n$. It is easy to see that the leximin and the leximax orderings respectively refine the preorders induced by the minimum and the maximum:

$$\min_i a_i > \min_i b_i \Rightarrow a \succ_{\text{lmin}} b, \quad \max_i a_i > \max_i b_i \Rightarrow a \succ_{\text{lmax}} b.$$

It is important to note that the leximin and leximax orderings do *not* correspond to some real-valued aggregation functions (see Example 6.3), hence they cannot induce some decomposable model. However, a well-known fundamental fact is that when L is finite, it is indeed possible to represent these orderings simply by means of a sum. To see this, let us first consider the case of the ordering induced by the maximum. Letting $L = \{l_0, \ldots, l_p\}$ with $l_0 < \cdots < l_p$, one can find a mapping $\phi : L \to \mathbb{R}$ such that

$$\max_{i=1}^n a_i > \max_{i=1}^n b_i \Rightarrow \sum_{i=1}^n \phi(a_i) > \sum_i \phi(b_i).$$

Indeed, it suffices to take

$$\phi(l_i) = m^i \qquad (i = 0, \ldots, p),$$

with m any integer greater than n. Such a mapping is an example of *superincreasing* function; i.e., satisfying $\phi(l_j) > \sum_{k=0}^{j-1} \phi(l_k)$, $j = 1, \ldots, p$. It is easy to check that the same function ϕ can do the job for the leximax, and even more we have that

$$a \succ_{\text{lmax}} b \Leftrightarrow \sum_{i=1}^n \phi(a_i) > \sum_i \phi(b_i),$$

so that the real-valued function $\sum_{i=1}^n \phi(\cdot)$ is a numerical representation of $\succcurlyeq_{\text{lmax}}$. A similar construction can be done for the leximin, using the function $\psi(l_i) = 1 - m^i$.

Going one step further, we combine the leximax and leximin orderings for ordering matrices. For example, the leximin could be used to order the rows, then the leximax is used to order the matrices. Denoting this order by $\succcurlyeq_{\text{lmax}(\succcurlyeq_{\text{lmin}})}$, by the previous considerations it clearly refines the maxmin ordering $\succcurlyeq_{\text{maxmin}}$ on matrices defined by:

$$A \succcurlyeq_{\text{maxmin}} B \Leftrightarrow \max_{i=1}^{n} \min_{j=1}^{m} a_{ij} \geqslant \max_{i=1}^{n} \min_{j=1}^{m} b_{ij},$$

where A, B are $n \times m$-dimensional matrices with entries $a_{ij}, b_{ij} \in L$.

Example 6.38 Take $n = 4, m = 3$ and the matrices A, B with integer values:

$$A = \begin{bmatrix} 4 & 3 & 2 \\ 2 & 4 & 5 \\ 2 & 3 & 1 \\ 2 & 5 & 1 \end{bmatrix}, \quad B = \begin{bmatrix} 1 & 2 & 4 \\ 5 & 1 & 2 \\ 2 & 3 & 4 \\ 2 & 5 & 4 \end{bmatrix}.$$

Clearly, $A \sim_{\text{maxmin}} B$ because the maxmin of each matrix is 2. However, these matrices can be distinguished by the $\succcurlyeq_{\text{lmax}(\succcurlyeq_{\text{lmin}})}$ order. First, we order the rows in increasing order according to the leximin. We obtain the matrices

$$A' = \begin{bmatrix} 2 & 3 & 1 \\ 2 & 5 & 1 \\ 4 & 3 & 2 \\ 2 & 4 & 5 \end{bmatrix}, \quad B' = \begin{bmatrix} 1 & 2 & 4 \\ 5 & 1 & 2 \\ 2 & 3 & 4 \\ 2 & 5 & 4 \end{bmatrix}.$$

We observe that the last rows are indifferent by the leximin: $[2\ \ 4\ \ 5] \sim_{\text{lmin}} [2\ \ 5\ \ 4]$, and so are the third and second rows. For the first row, we have $[1\ \ 2\ \ 4] \succ_{\text{lmin}} [2\ \ 3\ \ 1]$, hence finally $B \succ_{\text{lmax}(\succcurlyeq_{\text{lmin}})} A$. ◇

The interest of this is that the Sugeno integral can be casted into this framework. Indeed, as shown by (4.32) or (4.67), the computation of the Sugeno integral amounts to a maxmin computation, with $m = 2$. Hence the $\succcurlyeq_{\text{lmax}(\succcurlyeq_{\text{lmin}})}$ order refines the Sugeno integral. Let us mention in addition that Dubois and Fargier [98] consider yet a different formula for the Sugeno integral, writing:

$$⨍ a \, d\mu = \bigvee_{l \in L} (l \wedge \mu(\{i : a_i \geqslant l\})) \qquad (a \in L^n). \tag{6.38}$$

However, note that the different equivalent formulas for the Sugeno integral, once refined by this procedure, lead to different orderings, making the study of the refinement of the Sugeno integral rather complex (see [98], as well as the literature cited therein).

As for the leximin and leximax orderings, when L is finite it is possible to have a numerical representation of $\succcurlyeq_{\text{lmax}(\succcurlyeq_{\text{lmin}})}$, and therefore of any refinement of the Sugeno integral. Taking for example (6.38), we find that the numerical representation is:

$$E_L^{\text{lSug}}(a) = \sum_{k=0}^{p} \chi(l_k)\chi(\mu(\{i : a_i \geqslant l_k\})) \qquad (a \in L^n),$$

with $L = \{l_0, \ldots, l_p\}$ and $\chi : L \to [0, 1]$ given by

$$\chi(l_0) = 0, \quad \chi(l_p) = 1, \quad \chi(l_k) = \frac{m}{m^{2^{p-k}}}, \quad k = 1, \ldots, p-1,$$

with m an integer satisfying $m > n$. Note that this is close, but different, to a Choquet integral. The question whether the Sugeno integral can be refined by a Choquet integral is addressed in detail in [98]. It is true that this is possible, but the $\succcurlyeq_{\text{lmax}(\succcurlyeq_{\text{lmin}})}$ ordering cannot be represented by a Choquet integral.

Chapter 7
Dempster-Shafer and Possibility Theory

The last chapter presents an application of a particular class of normalized capacities (belief and plausibility measures) to the representation of uncertainty. This class has very specific properties and can be obtained through very different approaches (upper and lower probabilities, evidence theory and random sets, at least). Moreover, there exists a subclass of particular interest, the class of possibility and necessity measures, which has lead to a whole theory, called "possibility theory". For these reasons, belief and plausibility measures (or functions, as we call them in this chapter) occupy a central position among set functions, and similarly to Chap. 2, we give here an in-depth study of their properties, providing nearly all proofs of the results. Viewed as an alternative tool of modeling uncertainty, belief and plausibility functions are close to probability measures in their usage. This is why the topic of defining a conditional belief or plausibility measure/function is of central importance. Section 7.5 is devoted to this topic, which happens to be complex since several definitions are possible, all extending classical conditional probabilities. The chapter ends with a generalization of belief functions, defined on lattices (Sect. 7.8).

As additional reading, we recommend, apart from the original book of Shafer [296], the monograph of Kramosil [216] (one of the few monographs entirely devoted to belief functions, presented in a rigorous style and complete), Chaps. 5 and 6 of Kruse et al. [218], a collection of classic papers by Yager and Liu [353], a recent paper of Dempster [78], and the monograph of Couso et al. on random sets [62].

M. Grabisch, *Set Functions, Games and Capacities in Decision Making*, Theory and Decision Library C 46, DOI 10.1007/978-3-319-30690-2_7

7.1 The Framework

Introducing what is commonly called Dempster-Shafer theory is not an easy task, because as the name may already suggest, there is not a unique way of presenting it: at least there are three different (although nearly equivalent) ways, namely, the upper and lower probabilities of Dempster, the evidence theory of Shafer, and the theory of random sets of Matheron and Kendall. Since each of them has its interest and that finally one can pass from one representation to another one, we present all three of them, and we mention in addition the probabilistic approach proposed by Kramosil. In the remaining sections however, we will take as main representation the evidence theory of Shafer, with slight modifications and generalizations.

7.1.1 Dempster's Upper and Lower Probabilities

In 1967, Arthur P. Dempster [77] proposed the following framework. Let Ω, X be two nonempty sets, and consider a correspondence $\Gamma : \Omega \rightrightarrows X$; i.e., a mapping from Ω to 2^X (also called a multivalued mapping). We suppose that Ω, the set of outcomes of some random experiment, is endowed with a probability measure P defined on an algebra $\mathcal{B}$ on Ω (set of measurable events). On the other hand, X is another set of outcomes, where for the moment no probability measure is defined. X can be viewed as the set of observations, the mapping Γ being the device for observing outcomes of Ω. The fact that Γ is multivalued can be interpreted by saying that the observation is imprecise or incomplete: to a given outcome in Ω correspond several possible outcomes in X, the true one being unknown, or it could also be the case that naturally the observation yields a subset. The case where for some $\omega \in \Omega$, $\Gamma(\omega) = X$ can also be interpreted by saying that observation was not possible, hence any outcome in X is possible. We give some examples to illustrate these different cases.

Example 7.1 (Incomplete observation) The random experiment consists in drawing a card from a deck of 52 cards (the set Ω). Suppose that $X = \{\clubsuit, \diamondsuit, \heartsuit, \spadesuit\}$, and that the observation Γ only reveals the color (black or red). Then if ω is the king of clubs, $\Gamma(\omega) = \{\clubsuit, \spadesuit\}$. ◇

Example 7.2 (Imprecise observation) An automatic radar on the side of the road measures the speed of vehicles. Due to imprecision of measurement, only an interval is obtained. ◇

Example 7.3 Let Ω be the set of patients in a hospital. Each patient has to go through a number of examinations (blood pressure measurement, body temperature, blood analysis, radiography, etc.). To each examination corresponds a mapping Γ with a specific set X. If for some reason, the examination was not possible for patient ω, we set $\Gamma(\omega) = X$. Here also, imprecision can happen for a measurement, and if the result of a test is not enough clear, $\Gamma(\omega)$ can be multivalued. ◇

The question is now if some probability measure μ could be induced on X by Γ. If Γ were an ordinary (single-valued) mapping, this would cause no difficulty, because one would have $\mu = P \circ \Gamma^{-1}$. However, Γ being multivalued, a natural solution is to derive upper and lower approximations of this probability. For any subset $T \subseteq X$, consider the sets:

$$T^* = \{\omega \in \Omega : \Gamma(\omega) \cap T \neq \varnothing\} \tag{7.1}$$

$$T_* = \{\omega \in \Omega : \Gamma(\omega) \neq \varnothing, \Gamma(\omega) \subseteq T\}. \tag{7.2}$$

Observe that $X^* = X_*$ is the domain of Γ where Γ has a nonempty image. Define $\mathcal{E}$ the family of subsets T of X such that $T^*, T_* \in \mathcal{B}$, and suppose that $X \in \mathcal{E}$ and $P(X^*) \neq 0$. We define the *upper probability* $\mu^*(T)$ and *lower probability* $\mu_*(T)$ of any $T \in \mathcal{E}$ by

$$\mu^*(T) = \frac{P(T^*)}{P(X^*)} \tag{7.3}$$

$$\mu_*(T) = \frac{P(T_*)}{P(X^*)}. \tag{7.4}$$

It is easy to interpret these notions if one keeps in mind that the true outcome that is observed lies in $\Gamma(\omega)$, but is unknown. Then T^* may be regarded as the largest event in Ω that is possibly related to T (because the true outcome in $\Gamma(\omega)$ may lie in T), hence $\mu^*(T)$ is the largest possible amount of probability that can be assigned to T. Similarly, T_* is the smallest event in Ω that is for sure related to T (because any outcome in $\Gamma(\omega)$ lies in T, and so does the true one), hence $\mu_*(T)$ is the minimal amount of probability that can be assigned to T.

7.2 Shafer's Evidence Theory

Almost 10 years later, Glenn Shafer published a book in 1976 called *A Mathematical Theory of Evidence* [296] which reformulates in a completely different language the ideas of Dempster.

Let X be a finite set representing the possible outcomes of an experiment, or the possible answers to a question.[1] Accordingly, subsets of X are called *events*. We

[1]This corresponds to our second interpretation of X in Sect. 2.4.1, and X is called *frame of discernment* by Shafer. As in decision under risk and uncertainty (Chap. 5), X can also be interpreted as the set of states of nature.

define a function $m : 2^X \to [0, 1]$ satisfying the two conditions:

$$m(\varnothing) = 0, \quad \sum_{A \subseteq X} m(A) = 1.$$

This function is originally called by Shafer *basic probability assignment*, but we prefer to it the commonly used *(belief) mass distribution*, as employed, e.g., by Kruse et al. [218]. The interpretation of $m(A)$ for some $A \subseteq X$ is that the quantity $m(A)$ represents the belief committed to the event A and *only to it*, in the sense that it could not be committed to any proper subset of A. Events such that their (belief) mass is positive are called *focal sets*.[2] We denote by $\mathbf{supp}(m)$ the set of focal sets of m (support of the function m).

Let us be more explicit on the interpretation of this framework, along the line of Sect. 2.4.1 (see in particular Examples 2.10–2.13). X being the set of possible outcomes or answers to a given question, one tries to guess the true answer by collecting pieces of evidence (hence the name "evidence theory"), ultimately represented as subsets of X endowed with some certainty represented by the mass distribution. Letting $m(A) = 1$ means that the true answer lies in A with full certainty. Subsets that do not correspond to any piece of evidence receive no mass at all. A piece of evidence could be simply the opinion of some expert (this fits particularly well the antique vase example (Example 2.12), while piece of evidence in the usual sense fits better the murder example (Example 2.11). For a given piece of evidence, the total mass of belief (which is 1) has to be distributed among the different focal sets. Taking the murder example, one could deduce from some piece of evidence that the murderer is most probably left-handed. If Alice and Charles are left-handed but not Bob, then one could represent this as

$$m(\{\text{Alice, Charles}\}) = 0.8, \quad m(\{\text{Alice, Bob, Charles}\}) = 0.2.$$

The number 0.8 translates the qualifier "most probably." Note that the remaining mass 0.2 is given to X, representing some ignorance because giving some belief to X does not bring any information (i.e., does not help to find the true answer, here the murderer; more on this later). Also, this example should make clear the meaning of $m(A)$ as the belief committed to A and only to A: indeed, the fact that the murderer is left-handed does not give any clue on Alice alone, nor on Charles alone, but only on the pair {Alice,Charles}. Two last remarks on interpretation:

(i) Each piece of evidence should be encoded by a single mass distribution. We will explain later how to combine different pieces of evidence;
(ii) Following Shafer, we have set $m(\varnothing) = 0$ although some authors allow a positive mass for the empty set. The latter means that there is some possibility

[2]The original name given by Shafer is "focal element," which is somewhat misleading because these are subsets of X.

that the true answer lies outside X (open world hypothesis). This leads to intricacies and we devote Sect. 7.2.1 to this delicate issue.

Supposing the mass distribution m to be given, we build several set functions from it:

(i) The *belief function* Bel : $2^X \to [0, 1]$, defined by

$$\text{Bel}(A) = \sum_{\substack{B \subseteq A \\ B \neq \varnothing}} m(B) \qquad (A \subseteq X); \tag{7.5}$$

(ii) The *plausibility function* Pl : $2^X \to [0, 1]$, defined by

$$\text{Pl}(A) = \sum_{B \cap A \neq \varnothing} m(B) \qquad (A \subseteq X); \tag{7.6}$$

(iii) The *commonality function* $q : 2^X \to [0, 1]$, defined by

$$q(A) = \sum_{B \supseteq A} m(B) \qquad (A \subseteq X). \tag{7.7}$$

Remark 7.4

(i) The readers may find strange that we mention $B \neq \varnothing$ in the sum defining Bel. Indeed, because $m(\varnothing) = 0$ is assumed, that makes no difference. However, this is the correct definition when $m(\varnothing) \neq 0$ (see Sect. 7.2.1).

(ii) For a given event A, Bel(A) quantifies the total mass of belief that is with certainty committed to A, while Pl(A) quantifies the total mass of belief that is possibly related to A. Both measure the uncertainty of the event A. The commonality function, which also appears in Dempster's paper [77, Eqs. (5.2) and (5.3)] (the notation q comes from this source, but not the name) is more tricky to interpret, and is not a measure of uncertainty (we will return to this later: see Sects. 7.2.3 and 7.7). It is evident from the definition that

$$\text{Bel}(A) \leqslant \text{Pl}(A) \qquad (A \in 2^X),$$

and also that they are conjugate of each other: $\overline{\text{Bel}} = \text{Pl}$ and $\overline{\text{Pl}} = \text{Bel}$. Also, they are normalized capacities. By contrast, the commonality function is not a capacity because it is anti-monotone and differently normalized:

$$A \subseteq B \Rightarrow q(A) \geqslant q(B), \qquad q(\varnothing) = 1.$$

(iii) Since $m(\varnothing) = 0$, comparing (7.5) and (2.16) reveals that the mass distribution m is nothing but the Möbius transform of Bel: see also (2.22) that is identical to (7.6). Since m is defined as a nonnegative function, it follows from Theorem 2.33(v) that the belief functions of this chapter (fortunately!) coincide with the belief measures introduced in Chap. 2; i.e., normalized totally monotone capacities. Similarly, plausibility functions and plausibility measures coincide. The equivalence between belief functions generated by a mass distribution and normalized totally monotone capacities was shown by Shafer [296, Theorem 2.1].

On the other hand, we see by comparing (7.7) and (2.39) that q is the co-Möbius transform of Bel.

(iv) We now show that the frameworks of Dempster and Shafer coincide, supposing X to be finite. As it is relatively apparent from the definitions, the upper and lower probabilities correspond respectively to the plausibility and belief functions, while the mass distribution plays the rôle of Γ and P. Specifically,

$$\begin{aligned} P(T^*) &= P(\{\omega \in \Omega \,:\, \Gamma(\omega) \cap T \neq \varnothing\}) \\ &= \sum_{\omega : \Gamma(\omega)\cap T \neq \varnothing} P(\{\omega\}). \end{aligned}$$

Defining $m^* : 2^X \to [0,1]$ by $m^*(\varnothing) = 0$ and

$$m^*(A) = P(\Gamma^{-1}(A)) \qquad (\varnothing \neq A \subseteq X),$$

we obtain $P(T^*) = \sum_{A\cap T\neq\varnothing} m^*(A)$. Similarly,

$$\begin{aligned} P(T_*) &= P(\{\omega \in \Omega \,:\, \Gamma(\omega) \subseteq T, \Gamma(\omega) \neq \varnothing\}) \\ &= \sum_{\omega : \Gamma(\omega)\subseteq T, \Gamma(\omega)\neq\varnothing} P(\{\omega\}) = \sum_{A\subseteq T} m^*(A). \end{aligned}$$

Defining $m : 2^X \to [0,1]$ by $m(A) = \frac{m^*(A)}{\sum_B m^*(B)}$ yields the mass distribution m.

◇

Since both probability measures and belief functions model uncertainty, how do they differ? The approach of Dempster clearly shows that belief and plausibility functions can be interpreted as upper and lower probabilities; i.e., probability models with insufficient knowledge. This can be directly seen through the mass distribution, even on the simplest non-trivial case; i.e., $X = \{x_1, x_2\}$. Denoting $m(\{x_1\})$ and $m(\{x_2\})$ by m_1, m_2 respectively, we have

$$\mathrm{Bel}(\{x_i\}) = m_i, i = 1,2, \quad \mathrm{Bel}(X) = m_1 + m_2 + m(X) = 1.$$

If Bel would be a probability measure, we would have additivity, hence $1 = \mathrm{Bel}(\{x_1\}) + \mathrm{Bel}(\{x_2\}) = m_1 + m_2$, forcing $m(X) = 0$. As we explained above, $m(X)$ represents the quantity of ignorance in the piece of evidence. It follows that a purely probabilistic model conveys no ignorance. The following example illustrates this.

Example 7.5 (The antique vase: Example 2.12 revisited (after Ex. 1.1 in Shafer [296])) A man enters an antique shop in Upper Lascar Row in Honk Kong. Seeing a magnificent (and expensive!) vase of the Ming dynasty, he wonders if this vase is genuine or is counterfeit. Let us define $X = \{x_1, x_2\}$ as the set of possible answers to this question, with x_1 corresponding to "genuine" and x_2 to "counterfeit."

Suppose that this person is a tourist, without any knowledge of ancient chinaware. Therefore he has no specific reason to think that x_1 or x_2 is true, and all the mass of belief is allocated to X (total ignorance):

$$m(\{x_1\}) = m(\{x_2\}) = 0, \quad m(X) = 1.$$

This yields

$$\mathrm{Bel}(\{x_1\}) = \mathrm{Bel}(\{x_2\}) = 0$$
$$\mathrm{Pl}(\{x_1\}) = \mathrm{Pl}(\{x_2\}) = 1.$$

Suppose now that this person is an expert in antique chinaware, and that after careful examination, there are as many clues in favor of authenticity than in favor of counterfeiting. This is modelled as follows:

$$m(\{x_1\}) = m(\{x_2\}) = \frac{1}{2}, \quad m(X) = 0 \text{ (no ignorance)},$$

which yields

$$\mathrm{Bel}(\{x_1\}) = \mathrm{Bel}(\{x_2\}) = \frac{1}{2}$$
$$\mathrm{Pl}(\{x_1\}) = \mathrm{Pl}(\{x_2\}) = \frac{1}{2}.$$

Observe that a modelling based on probability yields in both cases

$$P(\{x_1\}) = P(\{x_2\}) = \frac{1}{2}$$

which make them indiscernable in probability theory. ◇

Remark 7.6 In the case of an arbitrary finite X, belief functions are probability measures (i.e., additive) if and only if the focal sets of m are singletons, which is clear by (7.5). Then Bel = Pl, which in the framework of Dempster is equivalent to have Γ single-valued. ◇

We finish this section by presenting some examples of particular mass distributions and their induced belief functions. The *vacuous mass distribution* m_X on X is defined by $m_X(X) = 1$ and $m_X(A) = 0$ for all $A \subset X$. The induced *vacuous belief function* is then nothing other than the unanimity game centered on X:

$$u_X(A) = 0, \forall A \subset X, \quad u_X(X) = 1.$$

Simple mass distributions are of the form

$$m_{B,\alpha}(A) = \begin{cases} 1-\alpha, & \text{if } A = B \\ \alpha, & \text{if } A = X \\ 0, & \text{otherwise} \end{cases}, \tag{7.8}$$

with $\alpha \in [0, 1[$ and $B \in 2^X \setminus \{\varnothing, X\}$. The corresponding *simple belief function* is given by

$$\text{Bel}_{B,\alpha}(A) = \begin{cases} 1-\alpha, & \text{if } A \in [B, X[\\ 1, & \text{if } A = X \\ 0, & \text{otherwise} \end{cases}.$$

According to our interpretation, the vacuous mass distribution represents total ignorance, hence its name. It was illustrated by the antique vase example, when representing the belief of the tourist. Simple mass distributions are very common because they express that there is a clue focusing on a single subset, endowed with some ignorance. Note that when $\alpha = 0$ (no ignorance), simple belief functions are unanimity games.

7.2.1 The Case Where $m(\varnothing) > 0$

Before going to the presentation of the next approach, we study the delicate issue of allowing $m(\varnothing) > 0$. It seems that Dubois and Prade [104] have been the first to consider this possibility and to study its consequences.

The comparison with Dempster's approach has revealed that the focal sets of m correspond to the image of Γ (more precisely, its elements with positive probability). Under this view, because $\varnothing$ is a possible value of Γ with a nonzero probability, in order to make a perfect correspondence between the two frameworks, it is natural to allow a positive value for $m(\varnothing)$. Doing so, the correspondence between the two frameworks [see Remark 7.4(iv)] becomes simpler. Indeed, m can be directly defined as

$$m(A) = P(\Gamma^{-1}(A)) \qquad (A \in 2^X)$$

Now, observe that in (7.2), which serves to establish the lower probability of some event $A \subseteq X$, $\Gamma(\omega) = \varnothing$ is discarded, which corresponds to $B \neq \varnothing$ in the sum computing the belief function in (7.5). Under this view, Bel(A) corresponds to $P(A_*)$, but *not* to $\mu_*(A)$ as given by (7.4), since there is no normalization. Some authors, as Kramosil [216], proposes the two versions: the belief function Bel as defined here, and the normalized belief function which we denote[3] by Bel*:

$$\mathrm{Bel}^*(A) = \frac{\sum_{\varnothing \neq B \subseteq A} m(B)}{\sum_{\varnothing \neq B \subseteq X} m(B)} = \frac{\sum_{\varnothing \neq B \subseteq A} m(B)}{1 - m(\varnothing)}, \tag{7.9}$$

assuming $m(\varnothing) < 1$. Then $\mathrm{Bel}^*(A)$ corresponds to $\mu_*(A)$.

In both frameworks, the withdrawal of $\varnothing$ when computing the belief or lower probability of an event is natural, if not mandatory: the empty set does not bring support nor evidence to an event A. The consequence, however, is that belief functions (respectively, plausibility functions) are, strictly speaking, *not* belief measures (respectively, plausibility measures), as defined in Sect. 2.8.4, because they are not normalized:

$$\mathrm{Bel}(X) = 1 - m(\varnothing), \quad \mathrm{Bel}(\varnothing) = 0$$

$$\mathrm{Pl}(X) = 1 - m(\varnothing), \quad \mathrm{Pl}(\varnothing) = 0.$$

More importantly, m is *not* the Möbius transform of Bel, but instead m^* defined by

$$m^*(A) = \begin{cases} m(A), & \text{if } A \neq \varnothing \\ 0, & \text{if } A = \varnothing \end{cases}$$

is the Möbius transform of Bel. Since it is nonnegative, Bel is a totally monotone and monotone capacity. On the other hand, as it can be checked, Pl remains the conjugate of Bel and therefore is a totally alternating and monotone capacity.

7.2.2 *Kramosil's Probabilistic Approach*

We briefly describe the approach of Kramosil [216], which is based on a compatibility relation between observations and states of a system, and resembles the approach of Dempster, although being more complex.

Let X be the set of states of a system under investigation, whose actual state is unknown. The state of the system cannot be directly observed nor guessed with certainty, however observations, results of experiments, measurements, etc. are available. We denote by E the space of such observations (which could be

[3] In his book, Kramosil takes the opposite convention for the notation.

vectors, etc.). The link between observations and states is done through the *compatibility relation* $\rho : X \times E \to \{0, 1\}$, and $\rho(x, y) = 0$ means that state $x \in X$ cannot be the actual state of the system given that the observation is $y \in E$; i.e., y and x are incompatible. Then $\rho(x, y) = 1$ means that given that the observation is y, one cannot avoid the possibility that the actual state is x (y and x are compatible). We introduce $U_\rho(y) = \{x \in X : \rho(x, y) = 1\}$ the set of states compatible with observation y.

The next step is to consider that observations are pervaded with uncertainty; i.e., y is the value taken by a random variable Y defined on a probability space $(\Omega, \mathcal{B}, P)$, taking its values in a measurable space $(E, \mathcal{E})$. One may ask if given $\omega \in \Omega$ and $A \subseteq X$, the inclusion $U_\rho(Y(\omega)) \subseteq A$ holds or does not hold. If $\{\omega \in \Omega : U_\rho(Y(\omega)) \subseteq A\}$ belongs to $\mathcal{B}$, we may quantify the size of this set by its probability $P(\{\omega \in \Omega : U_\rho(Y(\omega)) \subseteq A\})$. If this is also true for $A = \varnothing$, then the following quantity

$$\mathrm{Bel}_\rho(A) = P(\{\omega \in \Omega : \varnothing \neq U_\rho(Y(\omega)) \subseteq A\})$$

is defined, and is called the non-normalized degree of belief of A. Similarly, the normalized degree of belief is defined by, provided $P(\{\omega \in \Omega : U_\rho(Y(\omega)) = \varnothing\}) < 1$,

$$\mathrm{Bel}^*_\rho(A) = \frac{P(\{\omega \in \Omega : \varnothing \neq U_\rho(Y(\omega)) \subseteq A\})}{1 - P(\{\omega \in \Omega : U_\rho(Y(\omega)) = \varnothing\})}.$$

This defines the belief and the normalized belief functions. Plausibility functions are defined by conjugation.

The correspondence with the framework of Shafer is clear by defining $m : 2^X \to [0, 1]$ by

$$m(A) = P(\{\omega \in \Omega : U_\rho(Y(\omega)) = A\}),$$

letting E, X to be finite and taking $\mathcal{E} = 2^E$.

7.2.3 *Random Sets*

An alternative view of the theories of Dempster and of Shafer is brought by the theory of random sets, as proposed by Matheron [237] and Kendall [206], originally in the field of stochastic geometry. A comprehensive treatment and thorough development of random sets, putting them into perspective with evidence theory as well as possibility theory (see below), and proposing them as a universal means to represent uncertainty, can be found in the monograph of Goodman and Nguyen [160]. We give in this section only a flavor of it, keeping original notation and terminology of [160].

Random sets are just a natural generalization of the notion of random variable, and as will become apparent, are very close to the framework given by Dempster. Let $(\Omega, \mathcal{B}, P)$ be a probability space, and X be a nonempty set. A *random set* is a measurable mapping $S : (\Omega, \mathcal{B}, P) \to (\mathcal{B}_0, \sigma(2^{\mathcal{B}_0}), \nu)$, where $\mathcal{B}_0 \subseteq 2^X$, $\sigma(2^{\mathcal{B}_0})$ a sigma-algebra including $\mathcal{B}_0$, and $\omega \in \Omega \mapsto S(\omega) \subseteq X$. The probability measure ν induced by S on $\mathcal{B}_0$ is $\nu = P \circ S^{-1}$ (we note that S is nothing other than the multivalued-mapping Γ of Dempster).

In words, each realization of a random set yields a subset. One of the oldest and most famous example of a random set is Buffon's needle.

Example 7.7 (Buffon's needle) Consider a needle of length l. The needle falls down on a floor made of parallel strips of wood, each of the same width d. Supposing $d < l$, the needle may fall across one or several lines formed by the strips. Denoting by X the set of lines formed by the strips, we define $S(\omega)$ (or $\Gamma(\omega)$) as the set of lines crossed by the needle in drop ω. Note that one may find $S(\omega) = \varnothing$. ◇

Let us give another example.

Example 7.8 (Couso et al. [62, Sect. 2.1.2]) Let Ω be a finite population of individuals (say, the participants of some international conference). Consider X to be the set of languages (e.g., English, French, Spanish, Polish, German, Japanese, etc.) and the mapping S assigning to every participant ω the set of languages that ω can speak. Endowing Ω with some probability distribution (e.g., a uniform one), S is a random set, enabling the study of the distribution of languages across the population Ω. One can for example compute the probability that a participant is able to speak Polish and Japanese, or that a participant speaks at least three languages, etc. ◇

We introduce the following collections of sets:

(i) The class of *superset coverages* of $B \subseteq X$:

$$\mathcal{C}_B(\mathcal{B}_0) = \{C \in \mathcal{B}_0 \, : \, C \supseteq B\};$$

(ii) The class of *subset coverages* of $B \subseteq X$:

$$\mathcal{D}_B(\mathcal{B}_0) = \{C \in \mathcal{B}_0 \, : \, C \subseteq B\};$$

(iii) The class of *incidences* relative to $B \subseteq X$:

$$\mathcal{E}_B(\mathcal{B}_0) = \{C \in \mathcal{B}_0 \, : \, C \cap B \neq \varnothing\}.$$

Note that $\mathcal{E}_B \supseteq \mathcal{C}_B, \mathcal{D}_B$, and $2^X \setminus \mathcal{D}_B = \mathcal{E}_{X \setminus B}$. Next, we define the following functions, where S is a random set with range $\mathcal{B}_0$

(i) The *subset coverage function* for S:

$$\begin{aligned}\mu_S^{(1)}(B) &= P(\omega \in \Omega \, : \, S(\omega) \in \mathcal{C}_B(\mathcal{B}_0)) = P \circ S^{-1}(\mathcal{C}_B(\mathcal{B}_0)) \\ &= P(B \subseteq S) \qquad (B \subseteq X);\end{aligned}$$

(ii) The *superset coverage function* for S:

$$\mu_S^{(2)}(B) = P(\omega \in \Omega : S(\omega) \in \mathcal{D}_B(\mathcal{B}_0)) = P \circ S^{-1}(\mathcal{D}_B(\mathcal{B}_0))$$
$$= P(B \supseteq S) \qquad (B \subseteq X);$$

(iii) The *incidence function* for S:

$$\mu_S^{(3)}(B) = P(\omega \in \Omega : S(\omega) \in \mathcal{E}_B(\mathcal{B}_0)) = P \circ S^{-1}(\mathcal{E}_B(\mathcal{B}_0))$$
$$= P(B \cap S \neq \varnothing) \qquad (B \subseteq X);$$

(iv) The *complement-incidence function* for S:

$$\mu_S^{(4)}(B) = P \circ S^{-1}(\mathcal{E}_{X\setminus B}(\mathcal{B}_0)) = P(S \nsubseteq B) \qquad (B \subseteq X).$$

It is easy to check the following correspondences ($B \in 2^X$):

$$\mu_S^{(2)}(X \setminus B) = \mu_{X\setminus S}^{(1)}(B) \tag{7.10}$$

$$1 - \mu_S^{(3)}(B) = \mu_{X\setminus S}^{(1)}(B) \tag{7.11}$$

$$1 - \mu_S^{(4)}(X \setminus B) = \mu_{X\setminus S}^{(1)}(B). \tag{7.12}$$

We now show that the above functions are closely related to some families of set functions introduced by Goodman and Nguyen [160] based on Choquet [53, Chap. 3], which can be interpreted as measures of some type of uncertainty. Let $\mathcal{B}_0$ be an algebra on X, ξ a set function defined on $\mathcal{B}_0$, and consider a family of subsets $B, B_1, \ldots, B_k \in \mathcal{B}_0$, $k \in \mathbb{N}_0$. We define the quantity [generalizing difference functions, see Remark 2.19(vi)]:

$$\nabla_k(\xi, \circledast; B, B_1, \ldots, B_k) = \sum_{K\subseteq\{1,\ldots,k\}} (-1)^{|K|}\xi(B \circledast (\circledast_{i\in K} B_i)),$$

where $\circledast$ is either $\cup$ or $\cap$. Then

(i) ξ is a *plausibility measure* or *totally* $\cup$*-alternating capacity* if $\xi(\varnothing) = 0$ and

$$\nabla_k(\xi, \cup; B, B_1, \ldots, B_k) \leqslant 0 \qquad (k = 1, 2, \ldots)(B, B_1, \ldots, B_k \in \mathcal{B}_0)$$

(ii) ξ is a *belief measure* or *totally* $\cap$*-monotone capacity* if $\xi(X) = 1$ and

$$\nabla_k(\xi, \cap; B, B_1, \ldots, B_k) \geqslant 0 \qquad (k = 1, 2, \ldots)(B, B_1, \ldots, B_k \in \mathcal{B}_0)$$

(iii) ξ is a *doubt measure* or *totally* $\cup$*-monotone anti-capacity* if $\xi(\varnothing) = 1$ and

$$\nabla_k(\xi, \cup; B, B_1, \ldots, B_k) \geqslant 0 \qquad (k = 1, 2, \ldots)(B, B_1, \ldots, B_k \in \mathcal{B}_0)$$

(iv) ξ is a *disbelief measure* or *totally* $\cap$*-alternating anti-capacity* if $\xi(X) = 0$ and

$$\nabla_k(\xi, \cap; B, B_1, \ldots, B_k) \leqslant 0 \qquad (k = 1, 2, \ldots)(B, B_1, \ldots, B_k \in \mathcal{B}_0).$$

The readers can check that the two first definitions coincide with what we called monotone and totally alternating set functions, and monotone and totally monotone set functions (up to additional boundary conditions), see Definition 2.18. Observe that by letting $k = 1$ and $B \subseteq B_1$, $\nabla_1(\xi, \cup; B, B_1) \leqslant 0$ is simply monotonicity. The same holds for the second definition, hence they can be considered to be capacities.[4] Doing similarly for the two next definitions, we see that they are anti-monotone, hence the name anti-capacity.

Proceeding as in Theorem 2.20(ii), the following relations are easy to show:

Lemma 7.9

(i) ξ *is a plausibility measure on* $\mathcal{B}_0$ *if and only if* $1 - \xi$ *is a doubt measure on* $\mathcal{B}_0$*;*
(ii) $\xi(\cdot)$ *is a belief measure on* $\mathcal{B}_0$ *if and only if* $\xi(X \setminus \cdot)$ *is a doubt measure on* $\overline{\mathcal{B}_0} = \{X \setminus B : B \in \mathcal{B}_0\}$*;*
(iii) $\xi(\cdot)$ *is a disbelief measure on* $\mathcal{B}_0$ *if and only if* $1 - \xi(X \setminus \cdot)$ *is a doubt measure on* $\overline{\mathcal{B}_0}$*.*

The following theorem relates previous functions of uncertainty to random sets.

Theorem 7.10 *Let S be a random set on* $\mathcal{B}_0$ *corresponding to the probability space* $(\mathcal{B}_0, \sigma(2^{\mathcal{B}_0}), \nu)$*, where* $\mathcal{B}_0$ *is an algebra, and that* $\mathcal{C}_B(\mathcal{B}_0)$*,* $\mathcal{D}_B(\mathcal{B}_0)$*,* $\mathcal{E}_B(\mathcal{B}_0) \in \sigma(2^{\mathcal{B}_0})$ *for all* $B \in \mathcal{B}_0$*. Then*

(i) $\mu_S^{(1)}$ *is a doubt measure;*
(ii) $\mu_S^{(2)}$ *is a belief measure;*
(iii) $\mu_S^{(3)}$ *is a plausibility measure;*
(iv) $\mu_S^{(4)}$ *is a disbelief measure.*

The framework of Shafer can easily be recovered from the above one, up to an important difference however. Indeed, a mass distribution m defines a random set S from $(\Omega, \mathcal{B}, P)$ to $(2^X, 2^{(2^X)}, \nu)$, with:

$$\nu(\{A\}) = P \circ S^{-1}(\{A\}) = m(A) \qquad (A \in 2^X).$$

[4] Up to the notable difference that $\xi(\varnothing) = 0$ is not ensured for the second one. Similarly, $\xi(X) = 1$ is not ensured for the first one. Hence, the above definitions are more general than what we presented in Chap. 2. As Theorem 7.10 will show, these definitions are dictated by the coverage functions of random sets, and the unusual normalization conditions come from the fact that $S = \varnothing$ may have a positive probability to occur.

By definition of m, ν is indeed a probability distribution on 2^X. Moreover, we obtain

$$\mu_S^{(1)}(B) = q(B) \tag{7.13}$$

$$\mu_S^{(2)}(B) = \overset{\varnothing}{\text{Bel}}\,(B) \tag{7.14}$$

$$\mu_S^{(3)}(B) = \text{Pl}(B), \tag{7.15}$$

where $\overset{\varnothing}{\text{Bel}}$ is the modified belief function counting the mass on the empty set:

$$\overset{\varnothing}{\text{Bel}}\,(B) = \sum_{C \subseteq B} m(C).$$

Indeed, for any $B \in 2^X$,

$$\begin{aligned}\overset{\varnothing}{\text{Bel}}\,(B) &= \sum_{C \subseteq B} m(C) = \sum_{C \subseteq B} \nu(\{C\}) \\ &= P \circ S^{-1}(\{C : C \subseteq B\}) = P \circ S^{-1}(\mathcal{D}_B(2^X)) \\ &= \mu_S^{(2)}(B),\end{aligned}$$

and similarly for the other functions. Note that the fourth function $\mu_S^{(4)}$ has no equivalent in the framework of Shafer. Also, the *doubt function* of Shafer, defined as $\text{Dou}(A) = \text{Bel}(X \setminus A)$, is not a doubt measure in the above sense.

Remark 7.11

(i) One can see from (7.13) that the commonality function, which had so far no clear interpretation in Shafer's framework, appears to be the subset coverage function $\mu_S^{(1)}$ of a random set S. This function is quite easy to interpret and useful in the context of random sets. Consider again Example 7.8 and observe that the answer to the question "What is the probability to find a participant speaking both Polish and Japanese?" is given by $\mu_S^{(1)}(\{\text{Polish,Japanese}\})$.

(ii) The readers may find curious that a small discrepancy is left in the equivalence between the two theories, although one could have easily remedied to this by changing the definition of the class of subset coverages as follows:

$$\mathcal{D}'_B(\mathcal{B}_0) = \{C \in \mathcal{B}_0 : \varnothing \neq C \subseteq B\}.$$

But doing so, all the nice symmetries between the four coverage functions are destroyed. Also, allowing the empty set as a possible outcome of an experiment is meaningful in the context of random sets (see the Buffon's needle experiment: it could happen that the needle does not cross any line!). Incidentally, in the works of Goodman, both definitions appear, but sometimes in a mixed way! (see [159, 160]).

◇

7.2.4 *Ontic vs. Epistemic View of Sets*

The different frameworks we have presented so far have been shown to be almost equivalent on the mathematical point of view. This mathematical similarity hides however a deep difference in the usage of these theories, which comes from the two different interpretations of a set, namely the *conjunctive* or *ontic* view, and the *disjunctive* or *epistemic* view (see Couso et al. [62, Sect. 2.1.1] and Couso and Dubois [61]).

In the ontic view, a set is regarded as a collection of elements satisfying some property (it is a conjunctive view in the sense that *all* elements satisfy the property). It is the natural representation of attributes which are by essence set-valued, or, closer to the topic of this chapter, of outcomes of an experiment, answers to a question, which are set-valued. This is the case of Buffon's needle, or of Example 7.8: a needle may intersect several lines, or no line at all, and $S(\omega) = \{\text{German,Spanish}\}$ means that ω can speak German and Spanish, and no other language.

By contrast, in the epistemic view, a set contains all possible (and mutually exclusive, hence the name "disjunctive") values of some variable of interest under a given state of knowledge. Only one of them is the true value, and it is assumed that the true value is contained in the set. Examples 7.1–7.3 are all of this type, since they pertain to observations which are imprecise.

By construction, the frameworks of Dempster, Shafer, and Kramosil deal with sets of the epistemic type, while random sets adopt by essence (at least, in the original view of Kendall and Matheron) the ontic view. This explains why the commonality function has no clear interpretation in the framework of Shafer, while it has in the random set framework [Remark 7.11(i)].

7.3 Dempster's Rule of Combination

In his 1967 paper, Dempster proposed a rule to combine different sources of information pertaining to the same set X; i.e., multivalued mappings $\Gamma_1, \ldots, \Gamma_q$, with $\Gamma_i : \Omega_i \to 2^X$. We present first this rule in the usual form, given by Shafer (recall that X is finite and that $m(\varnothing) = 0$).

7.3.1 *The Rule of Combination in the Framework of Evidence Theory*

Let m_1, m_2 be two mass distributions on X, representing two pieces of evidence on the outcomes in X that are supposed to be independent (i.e., coming from two independent sources). The aim is to build a new mass distribution $m = m_1 \otimes m_2$, combining the information conveyed by m_1 and m_2, and supposing, apart independence, that the two sources are perfectly reliable.

Definition 7.12 (Dempster's rule of combination) Let m_1, m_2 be two mass distributions on X. If

$$\sum_{\substack{B \in \mathbf{supp}(m_1), C \in \mathbf{supp}(m_2) \\ B \cap C = \varnothing}} m_1(B) m_2(C) < 1, \tag{7.16}$$

then $m = m_1 \otimes m_2$ is defined as

$$m(A) = (m_1 \otimes m_2)(A) = \frac{\sum\limits_{\substack{B \in \mathbf{supp}(m_1), C \in \mathbf{supp}(m_2) \\ B \cap C = A}} m_1(B) m_2(C)}{1 - \sum\limits_{\substack{B \in \mathbf{supp}(m_1), C \in \mathbf{supp}(m_2) \\ B \cap C = \varnothing}} m_1(B) m_2(C)} \qquad (A \in 2^X \setminus \{\varnothing\}) \tag{7.17}$$

and $m(\varnothing) = 0$.

It is easy to see that the above defined m is indeed a mass distribution on X (Shafer [296, Theorem 3.1]).

The quantity $\sum_{\substack{B \in \mathbf{supp}(m_1), C \in \mathbf{supp}(m_2) \\ B \cap C = \varnothing}} m_1(B) m_2(C)$ is the *level of conflict* between m_1, m_2. If the level of conflict is equal to 1, the combination is not defined, and the pieces of evidence m_1, m_2 are said to be *contradictory*. Indeed, it means that there is no outcome $x \in X$ on which the two pieces of evidence agree; i.e., such that there exist two focal sets $A \in \mathbf{supp}(m_1), B \in \mathbf{supp}(m_2)$ such that $x \in A \cap B$. The next lemma explains this and still gives other equivalent conditions.

Lemma 7.13 *Let m_1, m_2 be two mass distributions on X with induced belief functions* $\mathrm{Bel}_1, \mathrm{Bel}_2$ *and commonality functions q_1, q_2 respectively. The following propositions are equivalent.*

(i) m_1, m_2 *are contradictory;*
(ii) *If $A \in \mathbf{supp}(m_1)$ and $B \in \mathbf{supp}(m_2)$, then $A \cap B = \varnothing$;*
(iii) *There exists $\varnothing \neq A \subset X$ such that* $\mathrm{Bel}_1(A) = 1$ *and* $\mathrm{Bel}_2(X \setminus A) = 1$*;*
(iv) $q_1(A) q_2(A) = 0$ *for all $A \in 2^X \setminus \{\varnothing\}$.*

Proof (i)⇔(ii) Suppose there exist $A \in \mathbf{supp}(m_1)$, $B \in \mathbf{supp}(m_2)$ such that $A \cap B \neq \varnothing$. Then (7.16) holds because

$$\sum_{\substack{A \in \mathbf{supp}(m_1) \\ B \in \mathbf{supp}(m_2)}} m_1(A)m_2(B) = \sum_{A \in \mathbf{supp}(m_1)} m_1(A) \sum_{B \in \mathbf{supp}(m_2)} m_2(B) = 1. \tag{7.18}$$

Conversely, if (ii) holds then by (7.18) m_1 and m_2 are contradictory.

(ii)⇒(iii) Consider $A = \bigcup_{C \in \mathbf{supp}(m_1)} C$. Then $\mathrm{Bel}_1(A) = 1$ and by (ii), $X \setminus A \supseteq \bigcup_{C \in \mathbf{supp}(m_2)}$, hence $\mathrm{Bel}_2(X \setminus A) = 1$.

(iii)⇒(iv) Consider any $B \in 2^X \setminus \{\varnothing\}$. Note that by definition of A, for any set $C \nsubseteq A$, we have $m_1(C) = 0$, and similarly $m_2(C) = 0$ if $C \cap A \neq \varnothing$. Suppose $B \setminus A \neq \varnothing$. Then

$$q_1(B) = \sum_{C \supseteq B} m_1(C) = 0$$

because each $m_1(C) = 0$. Now, if $B \setminus A = \varnothing$ then $B \cap A \neq \varnothing$, hence $q_2(B) = 0$.

(iv)⇒(ii) Suppose there exist $A \in \mathbf{supp}(m_1)$, $B \in \mathbf{supp}(m_2)$ such that $A \cap B \neq \varnothing$. Then $q_1(A \cap B) \geqslant m_1(A) > 0$ and $q_2(A \cap B) \geqslant m_2(B) > 0$, therefore $q_1(A \cap B)q_2(A \cap B) \neq 0$. □

We give some properties of the Dempster's rule of combination.

Theorem 7.14 *The Dempster's rule of combination, seen as an algebraic operator $\otimes$ on the set of mass distributions on X, satisfies the following properties for every m, m', m'':*

(i) Commutativity: $m \otimes m' = m' \otimes m$;
(ii) Associativity: $m \otimes (m' \otimes m'') = (m \otimes m') \otimes m''$;
(iii) The vacuous mass distribution is a neutral element: $m \otimes m_X = m$;
(iv) Nonidempotence: $m \otimes m \neq m$.

(the easy proof is left to the readers; (iv) is illustrated by Example 7.16).

Remark 7.15 The properties listed in Theorem 7.14 almost make the set of mass distributions on X a group for $\otimes$, because only the existence of an inverse is missing. The question is: For a given mass distribution m, does there exist a distribution m' such that $m \otimes m' = m_X$, the vacuous mass distribution? The answer is no: no such inverse mass distribution exists. Indeed, suppose that $m_1(A) > 0$ for some $A \neq X$. The condition to get $(m_1 \otimes m_2)(A) = 0$ by some adequate m_2 is

$$\sum_{\substack{B \in \mathbf{supp}(m_2) \\ B \supseteq A}} m_1(A)m_2(B) = 0,$$

which is obviously impossible to satisfy because m_2 is nonnegative. This can be interpreted by saying that, once some belief is assigned to some event, it is not possible to erase it completely by combining it with another piece of evidence. To get out of this deadlock, Kramosil [216, Chap. 11] proposes to define belief functions by signed basic probability assignments; i.e., m takes its values in $\mathbb{R}$, so that belief functions take values outside the unit interval. ◇

The fact that $\otimes$ is not idempotent should not be seen as a drawback of the rule: it is rather a consequence of the assumption that the two pieces of evidence should be independent. Then, a combination of m with itself acts as a reinforcement, enhancing the difference between events with high and low belief. This is shown in the next example.

Example 7.16 Consider $m = m_{A,0.3}$ (simple mass distribution). Then

$$\begin{aligned}(m \otimes m)(A) &= m(A)m(A) + m(A)m(X) + m(X)m(A) \\ &= 0.49 + 0.21 + 0.21 = 0.91 \\ (m \otimes m)(X) &= m(X)m(X) = 0.09\,.\end{aligned}$$

Hence we have found that $m_{A,0.3} \otimes m_{A,0.3} = m_{A,0.09}$. ◇

7.3.2 *The Normalized and the Nonnormalized Rules*

To Normalize or Not to Normalize?

An important issue concerns the denominator in (7.17), which performs a normalization in order to get a well-defined mass distribution m (i.e., such that $\sum_{A \subseteq X} m(A) = 1$). Some authors do not perform a normalization, because normalization erases any trace of conflict, which could be a valuable information for further processing. This is illustrated by the following example. Let us denote by $\otimes^*$ the nonnormalized rule.

Example 7.17 (After Zadeh [358]) A person suffers from some illness and calls a doctor. After auscultation, the doctor declares that it is almost surely illness a (with 99 % certainty), although there exists some very small probability that it is illness b (with 1 % certainty).

The issue being serious, a second doctor is called, who after examination comes to the conclusion that it is almost sure to be illness c (with 99 % certainty), although there is very little chance that it is illness b (with 1 % certainty). What to conclude from these very conflicting diagnoses?

Modeling the two pieces of evidence through Dempster-Shafer theory, we set $X = \{a, b, c\}$ and arrive at

$$\begin{aligned}m_1(\{a\}) &= 0.99, \quad m_1(\{b\}) = 0.01 \\ m_2(\{c\}) &= 0.99, \quad m_2(\{b\}) = 0.01\,.\end{aligned}$$

We note that the two mass distributions are not contradictory since $\{b\}$ is common, although highly conflicting because the level of conflict is

$$\sum_{A\cap B=\varnothing} m_1(A)m_2(B) = 1 - m_1(\{b\})m_2(\{b\}) = 0.9999 .$$

Applying the (normalized) rule of combination yields $(m_1 \otimes m_2)(\{b\}) = 1$, and 0 for a and c, while the nonnormalized rule yields $(m_1 \otimes^* m_2)(\{b\}) = 0.0001$, and 0 for a and c.

The result given by the normalized rule is very surprising because the two doctors agreed only on one thing: that the belief assigned to b is very low! ◇

One can conduct the analysis of this example a little bit further and try to reconciliate the two views. In fact, both are meaningful, but in different contexts. The normalized rule is suitable for the *Sherlock Holmes' world* (*closed world hypothesis*). There, the murderer is one of the characters of the novel, and he/she has committed the crime (in our terminology: X contains the true answer and it is unique). Also, every piece of evidence must be taken into account with precision, every detail in the novel is relevant, and the murderer is most often the most improbable person, obtained by successive elimination of the other hypotheses (in our terminology: every piece of evidence is trustable, and even the most improbable element (here b) must be the solution if every other has been eliminated). By contrast, the nonnormalized rule is suitable for the *real everyday world* (*open world hypothesis*). There, it is not sure that X is exhaustive and that the true answer is a single element of X. Also, pieces of evidence are not always fully trustable, so that in case of high conflict as in our example, it is better to call a third doctor! See also Smets [309, Sect. 6] for further comments on this issue.

The $m(\varnothing) > 0$ Issue Again and the Nonnormalized Rule

The normalization issue is also related to the problem of admitting or not that $m(\varnothing)$ could be positive. Relaxing the assumption $m(\varnothing) = 0$ gives in fact a much more clear view of the rule of combination, as we show now.

Take two mass distributions m_1, m_2. We define the *nonnormalized combination* of the two distributions as:

$$(m_1 \otimes^* m_2)(A) = \sum_{\substack{B\in\mathrm{supp}(m_1),C\in\mathrm{supp}(m_2)\\ B\cap C=A}} m_1(B)m_2(C) \qquad (A \subseteq X). \tag{7.19}$$

Note that, unlike the normalized rule, the formula is *extended to* $\varnothing$. Recall that allowing $m(\varnothing) > 0$ amounts to considering an open world, and $m(\emptyset) > 0$ is the mass of belief committed to the fact that the true answer lies outside X. Therefore, $(m_1 \otimes^* m_2)(\varnothing)$ is the amount of belief that, after combination of the evidences, the true answer is outside X. In other words, the level of conflict (a term that somehow suggests a kind of failure or blocking of the procedure) of Shafer is reinterpreted as a belief that the true answer must be sought outside, a conclusion that sounds more optimistic, and opens new horizons(!).

The nonnormalized rule satisfies all properties of Theorem 7.14, and can be computed in a very simple way through commonality functions.

Lemma 7.18 *Given m_1, m_2 two mass distributions on X with induced commonality functions q_1, q_2, the commonality function q induced by $m = m_1 \otimes^* m_2$ is given by*

$$q(A) = q_1(A)q_2(A) \qquad (A \subseteq X). \tag{7.20}$$

Proof For any $A \subseteq X$, we have

$$\begin{aligned} q(A) &= \sum_{B \supseteq A} m(B) = \sum_{B \supseteq A} \sum_{C \cap D = B} m_1(C)m_2(D) \\ &= \sum_{C \cap D \supseteq A} m_1(C)m_2(D) = \sum_{C \supseteq A} \sum_{D \supseteq A} m_1(C)m_2(D) \\ &= q_1(A)q_2(A). \end{aligned}$$

□

It is interesting to note that Dempster in his 1967 paper directly proposed (7.20) as definition of the rule of combination, which tends to indicate that the original Dempster's rule is nonnormalized. It seems however that Dempster implicitly considered normalization when returning to upper and lower probabilities from the commonality functions. Normalization was explicitly considered by Shafer, in order to get a well-defined mass distribution as the result of combination. Also, (7.19) is the formula given in Kramosil [216], and the one used with random sets (see below).

The Combination Rule in the Framework of Random Sets

It is instructive to write the (nonnormalized) Dempster's rule of combination in the framework of random sets. In short, the nonnormalized Dempster's rule corresponds to the intersection of two statistically independent random sets. This opens the door to other rules of combination, based on other set operations, as well as the combination of dependent random sets.

Specifically, let $(2^X, 2^{(2^X)}, \nu_1)$ and $(2^X, 2^{(2^X)}, \nu_2)$ be the two probability spaces associated to two random sets S_1, S_2. Let us define $S = S_1 \cap S_2$ and let us find the corresponding probability measure ν. For any $A \in 2^X$, we have

$$\begin{aligned} S^{-1}(A) &= \{\omega \in \Omega \,:\, S(\omega) = A\} \\ &= \{\omega \in \Omega \,:\, S_1(\omega) \cap S_2(\omega) = A\} \\ &= \bigcup_{C \cap D = A} \{\omega \in \Omega \,:\, S_1(\omega) = C, S_2(\omega) = D\}. \end{aligned}$$

Applying statistical independence between S_1, S_2, we find

$$\begin{aligned}\nu(\{A\}) &= P \circ S^{-1}(A) \\ &= \sum_{C\cap D=A} P(\{\omega \in \Omega : S_1(\omega) = C\})P(\{\omega \in \Omega : S_2(\omega) = D\}) \\ &= \sum_{C\cap D=A} \underbrace{\nu_1(\{C\})}_{m_1(C)}\underbrace{\nu_2(\{D\})}_{m_2(D)}.\end{aligned}$$

7.3.3 *Decomposition of Belief Functions into Simple Belief Functions*

Considering the rule of combination $\otimes^*$ as an algebraic operation on the set of mass distributions, the following question is natural: Does there exist a decomposition of any mass distribution into a combination by $\otimes^*$ of elementary mass distributions? As "elementary" mass distributions we have at disposal the simple mass distributions defined in (7.8). The next result shows that it is indeed possible to make such a decomposition, in most cases.

Theorem 7.19 (Decomposition of a mass distribution into simple mass distributions) *(Smets [311]) Let m be a mass distribution on X such that* $0 < m(X) < 1$. *Then m can be written as*

$$m = \bigotimes^*_{A\in 2^X\setminus\{\varnothing,X\}} m_{A,\alpha_A},$$

with

$$\alpha_A = \prod_{B\supseteq A} q(B)^{(-1)^{|B\setminus A|+1}} \qquad (A \in 2^X \setminus \{\varnothing, X\}).$$

The proof will be given in Sect. 7.8, where the same result is shown in a more general framework. Note that the condition $m(X) < 1$ simply excludes the vacuous mass function, for which obviously no decomposition is needed.

It is important to note that in the above decomposition, it may happen that $\alpha_A > 1$ for some A. If this happens, the corresponding m_{A,α_A} is no longer nonnegative and as a consequence, $\mathrm{Bel}_{A,\alpha}$ is not a belief function any more.

7.4 Compatible Probability Measures

Given a mass distribution m on X and the induced belief and plausibility functions Bel, Pl, a probability measure P on X is said to be *compatible* with m (or with Bel, Pl) if the following bracketting holds:

$$\mathrm{Bel}(A) \leqslant P(A) \leqslant \mathrm{Pl}(A) \qquad (A \in 2^X). \tag{7.21}$$

Suppose that we find some probability measure P such that $P(A) \geqslant \mathrm{Bel}(A)$ for every $A \in 2^X$. Then it follows that

$$\mathrm{Pl}(A) = 1 - \mathrm{Bel}(X \setminus A) \geqslant 1 - P(X \setminus A) = P(A) \qquad (A \in 2^X),$$

from which it follows that if the left inequality holds for every A in (7.21), the right inequality immediately follows, and vice versa. Alternatively, we may say that Bel (or Pl, or m) is compatible with a given P if (7.21) holds, and due to this symmetry, it makes sense to speak of compatible Bel and P.

Since $P(X) = \mathrm{Bel}(X) = 1$, a compatible probability is nothing but a core element of Bel, considered to be a game on X, and conversely (see Chap. 3). Hence, properties of compatible probability measures are straightforwardly obtained from Chap. 3, and we denote the set of probability measures compatible with Bel as **core**(Bel).[5] We summarize them below.

A belief function being totally monotone, it is also supermodular (2-monotone) and therefore the structure of its core is completely known because its vertices are known (Theorem 3.15). In addition, Theorem 3.62(iii) applies and reveals that the set of compatible probability measures is the selectope of the belief function, otherwise said, the set of all sharing values (see Sect. 3.5). These considerations lead to the following.

Theorem 7.20 (The set of compatible probability measures) *Let* Bel *be a belief function and its associated mass distribution m.*

(i) The set **core**(Bel) *of probability measures compatible with* Bel *is never empty, and the set of corresponding probability distributions is a polytope whose extreme points are the marginal vectors* $\varphi^{\sigma,\mathrm{Bel}} \in \mathbb{R}^X$ *defined by*

$$\varphi^{\sigma,\mathrm{Bel}}_{x_{\sigma(i)}} = \mathrm{Bel}(\{x_{\sigma(1)}, \ldots, x_{\sigma(i)}\}) - \mathrm{Bel}(\{x_{\sigma(1)}, \ldots, x_{\sigma(i-1)}\})$$
$$= \sum_{A \subseteq \{x_{\sigma(1)}, \ldots, x_{\sigma(i-1)}\}} m(A \cup x_{\sigma(i)}) \qquad (i = 1, \ldots, n)$$

where σ is any permutation on $\{1, \ldots, n\}$, and $X = \{x_1, \ldots, x_n\}$.

[5] Up to the fact that the core is defined in Chap. 3 as a set of n-dim vectors, not additive games! But the two views are of course equivalent and we use them indifferently in this section.

(ii) **core**(Bel) = **sel**(Bel), *the selectope of* Bel*; i.e., each compatible probability distribution has the form*

$$P(\{x\}) = \sum_{\substack{A \in \mathbf{supp}(m) \\ A \ni x}} \lambda(A, x) m(A) \qquad (x \in X)$$

with $\lambda(A, x) \geqslant 0$ *for all* $\varnothing \neq A \subseteq X$, $x \in A$, *and* $\sum_{x \in A} \lambda(A, x) = 1$, *and each distribution obtained in this way is an element of* **core**(Bel).

Another important feature of the set of compatible probability distributions is that its lower envelope coincides with the belief function, because belief functions are supermodular games, hence exact [Definition 3.43 and Lemma 3.44(iii)]. Formally,

$$P_*(A) = \min_{P \in \mathbf{core}(\mathrm{Bel})} P(A) = \mathrm{Bel}(A) \qquad (A \in 2^X).$$

Hence, belief functions are indeed lower probability measures. Similarly, the corresponding plausibility measure is the upper envelope of **core**(Bel).

In addition, the lower expected value of any random variable f over X is the Choquet integral of f w.r.t Bel:

$$\min_{P \in \mathbf{core}(\mathrm{Bel})} \mathbb{E}[f] = \int f \, \mathrm{dBel}, \tag{7.22}$$

as is immediate from Theorem 4.39.

The notion of compatible probability was already present in Dempster [77, Sect. 2], where almost all of the above properties were shown.

7.5 Conditioning

As conditional probability is a key concept in probability theory, and especially in its application to decision making and statistical inference, an adequate definition of conditional belief and plausibility functions is of primary importance, and for this reason, has already been proposed in the seminal paper of Dempster. Nevertheless, several other definitions have been proposed later (which we call in a generic way *conditioning rules*), with different purposes, and all of them collapse to the ordinary conditional probability when belief functions become additive. We review in this section the most relevant ones and study their properties. For this, we mainly follow Denneberg [79], whose study is in fact more general since valid for capacities and not restricted to belief functions (the readers are referred to [79] for more results). In the whole section, unless otherwise indicated, X is arbitrary (not necessarily finite).

7.5.1 The General Conditioning Rule

It was proposed by Fagin and Halpern [127], Jaffray [202], De Campos et al. [69], and also by Dempster [77]. Following Denneberg [79], we define it for any capacity.

Let μ be a capacity on an algebra $\mathcal{X} \subseteq 2^X$. For any $B \in \mathcal{X} \setminus \{\varnothing\}$, the *(general) conditional capacity* given B is defined by

$$\mu_B(A) = \frac{\mu(A \cap B)}{\mu(A \cap B) + \overline{\mu}(A^c \cap B)} \qquad (A \in \mathcal{X}), \tag{7.23}$$

provided the denominator is nonzero, and where $\overline{\mu}$ is the conjugate capacity. An interesting feature of this conditioning rule is that it commutes with conjugation, as is easy to check:

$$\overline{\mu_B}(A) = \mu_B(X) - \mu_B(A^c) = \overline{\mu}_B(A) \qquad (A \in \mathcal{X}). \tag{7.24}$$

It may happen that the conditional capacity given in (7.23) is not defined for some A, B. We have the following result.

Lemma 7.21 *If μ is subadditive and $\overline{\mu}(B) > 0$, then μ_B is defined on $\mathcal{X}$.*

Proof By subadditivity, we have

$$\mu((A^c \cap B)^c) = \mu(A \cup B^c) \leqslant \mu(A \cap B) + \mu(B^c).$$

Hence

$$\begin{aligned}\mu(A \cap B) + \overline{\mu}(A^c \cap B) &= \mu(A \cap B) + \mu(X) - \mu((A^c \cap B)^c) \geqslant \mu(X) - \mu(B^c)\\ &= \overline{\mu}(B) > 0.\end{aligned}$$

□

Based on the above, we are in a position to define (general) conditional belief and plausibility functions.

Consider a belief function Bel and its conjugate Pl. For any $B \in \mathcal{X}$ such that $\mathrm{Bel}(B) > 0$, the *(general) conditional plausibility function* is defined by

$$\mathrm{Pl}_B(A) = \frac{\mathrm{Pl}(A \cap B)}{\mathrm{Pl}(A \cap B) + \mathrm{Bel}(A^c \cap B)} \qquad (A \in \mathcal{X}), \tag{7.25}$$

while the *conditional belief function* is defined as the conjugate of Pl_B:

$$\mathrm{Bel}_B(A) = 1 - \mathrm{Pl}_B(A^c) = \frac{\mathrm{Bel}(A \cap B)}{\mathrm{Bel}(A \cap B) + \mathrm{Pl}(A^c \cap B)} \qquad (A \in \mathcal{X}), \tag{7.26}$$

by using (7.24). By Lemma 7.21, Pl_B is defined for every B such that $\mathrm{Bel}(B) > 0$, and so is Bel_B by conjugation.

The following lemma provides an interpretation of the conditional capacity.

Lemma 7.22 *Let $A, B \in \mathcal{X}$. If $\mu_B(A)$ is defined, then it is the unique solution of the following equation in $t \in [0,1]$*

$$\int_B (1_A - t \cdot 1_X)\,\mathrm{d}\mu = 0, \tag{7.27}$$

where the integral is the Choquet integral. Conversely, if (7.27) has a unique solution in $[0,1]$, then it is $\mu_B(A)$.

The quantity t represents the amount the decision maker is ready to pay for winning 1 if A occurs, but the transaction is cancelled if B does not occur. Then the lemma shows that $\mu_B(A)$ is the amount t that makes the decision maker indifferent between gambling and not gambling.

Proof The integrand is $(1_A - t1_X)1_B$, which takes the value $1 - t \geqslant 0$ on $A \cap B$, 0 on B^c, and $-t$ on $B \setminus A$. It follows that, using (4.12), Lemma 4.9 and positive homogeneity,

$$\begin{aligned}\int_B (1_A - t \cdot 1_X)\,\mathrm{d}\mu &= \int \Big((1-t)1_{A\cap B} - t1_{A^c\cap B}\Big)\,\mathrm{d}\mu\\ &= (1-t)\mu(A\cap B) - t\overline{\mu}(A^c \cap B).\end{aligned}$$

The integral is zero iff $t\big(\mu(A\cap B) + \overline{\mu}(A^c\cap B)\big) = \mu(A\cap B)$. Since $\mu_B(A) \in [0,1]$ when defined, the assertion follows. □

We give some properties of conditional capacities.

Lemma 7.23 *Let $B \in \mathcal{X} \setminus \{\varnothing\}$ and μ a capacity on X. The following propositions hold.*

(i) $\mu_X = \dfrac{1}{\mu(X)}\mu$; $\mu_B(X) = \mu_B(B) = 1$ *supposing* $\mu(B) > 0$;

(ii) For any $A_1, A_2 \in \mathcal{X}$ such that $\mu_B(A_1), \mu_B(A_2)$ are defined, $A_1 \subseteq A_2$ implies $\mu_B(A_1) \leqslant \mu_B(A_2)$;

(iii) If ν is another capacity on X with $\nu(X) = \mu(X)$ and supposing that $\mu_B(A), \nu_B(A)$ are defined,

$$\nu \leqslant \mu \Rightarrow \mu_B(A) \leqslant \nu_B(A).$$

Proof

(i) is clear from the definition.

(ii) Consider the function $f_{i,t} = (1_{A_i} - t \cdot 1_X)1_B$ and its integral $g_i(t) = \int f_{i,t}\,\mathrm{d}\mu$ for $i = 1, 2$. We know by Lemma 7.22 that the unique solutions of $g_i(t) = 0$ in

[0, 1] are $t_i = \mu_B(A_i)$, $i = 1, 2$. Clearly, $f_{i,s} \geqslant f_{i,t}$ whenever $s \leqslant t$. It follows from Theorem 4.24(vi) that $g_i(s) \geqslant g_i(t)$, $i = 1, 2$. Similarly, we have $g_1(t) \leqslant g_2(t)$. This implies $t_1 \leqslant t_2$, which proves the result.

(iii) Consider the function $f_t = (1_A - t \cdot 1_X)1_B$ and its integrals $g_\mu(t) = \int f_t \, d\mu$ and $g_\nu(t) = \int f_t \, d\nu$. Now $\mu(X) = \nu(X)$ and $\mu \leqslant \nu$ imply $g_\mu(t) \leqslant g_\nu(t)$. Then as above $\mu_B(A) \leqslant \nu_B(A)$.

□

The next result gives a clear interpretation of this conditioning rule.

Theorem 7.24 (The general conditioning rule as upper and lower conditional probabilities) *Let μ be a submodular capacity, and $B \in \mathcal{X} \setminus \{\varnothing\}$ such that $\overline{\mu}(B) > 0$. Then $\mu_B, \overline{\mu}_B$ are defined on $\mathcal{X}$ and*

$$\mu_B(A) = \max_{\phi \in \mathbf{core}(\overline{\mu})} \phi_B(A), \quad \overline{\mu}_B(A) = \min_{\phi \in \mathbf{core}(\overline{\mu})} \phi_B(A).$$

Note that ϕ being additive, ϕ_B is defined as ordinary conditional probabilities; i.e., $\phi_B(A) = \frac{\phi(A \cap B)}{\phi(B)}$.

Proof The fact that μ_B (and therefore $\overline{\mu}_B$) is defined on $\mathcal{X}$ is established by Lemma 7.21. Now, by Lemma 7.22 we have

$$\int (1_A - \mu_B(A) \cdot 1_X)1_B \, d\mu = 0.$$

Since μ is submodular, we can apply the dual version (Remark 4.40) of Theorem 4.39, and we find that there exists some additive measure ϕ in the core of $\overline{\mu}$ such that

$$\int (1_A - \mu_B(A) \cdot 1_X)1_B \, d\mu = \int (1_A - \mu_B(A) \cdot 1_X)1_B \, d\phi.$$

Because $\phi(B) \geqslant \overline{\mu}(B) > 0$, it follows from Lemma 7.21 that ϕ_B is defined on $\mathcal{X}$, and by Lemma 7.22, $\int (1_A - \mu_B(A) \cdot 1_X)1_B \, d\phi = 0$ implies that $\mu_B(A) = \phi_B(A)$. Now, for any other additive $\phi' \in \mathbf{core}(\overline{\mu})$, we deduce from Lemma 7.23(iii) that $\phi'_B(A) \leqslant \mu_B(A)$. □

Remark 7.25

(i) The meaning of conditional belief and plausibility functions becomes clear under the light of Theorem 7.24. Indeed, it shows that the conditional belief function $\mathrm{Bel}_B(A)$ [respectively, the conditional plausibility function $\mathrm{Pl}_B(A)$] can be seen as the most pessimistic (respectively, the most optimistic) estimate of the conditional probability of A given B, among the probability measures compatible with Bel (compare with Remark 7.31).

(ii) Since $\overline{\mu}$ is supermodular and therefore exact [see Definition 3.43 and Lemma 3.44(iii)], it follows that

$$\mu(A) = \max_{\phi \in \mathbf{core}(\overline{\mu})} \phi(A), \quad \overline{\mu}(A) = \min_{\phi \in \mathbf{core}(\overline{\mu})} \phi(A) \qquad (A \in \mathcal{X}),$$

and the theorem can be reformulated as

$$(\max_{\phi \in \mathbf{core}(\overline{\mu})} \phi)_B = \max_{\phi \in \mathbf{core}(\overline{\mu})} \phi_B, \quad (\min_{\phi \in \mathbf{core}(\overline{\mu})} \phi)_B = \min_{\phi \in \mathbf{core}(\overline{\mu})} \phi_B. \tag{7.28}$$

Note also that the theorem implies that $\{\phi_B : \phi \in \mathbf{core}(\overline{\mu})\} \subseteq \mathbf{core}(\overline{\mu_B})$, but as shown by Jaffray [202], equality does not hold in general.

◇

A remarkable property of the general conditioning rule is that it preserves k-monotonicity.

Theorem 7.26 (Preservation of k-monotonicity by the general conditioning rule) *Let μ be a k-monotone capacity on 2^X, X finite, for some $k \geqslant 2$, and $E \in 2^X$ such that $\mu(E) > 0$. Then μ_E is k-monotone too, and $\overline{\mu}_E$ is k-alternating.*

The proof of this result is quite technical (Sundberg and Wagner [324]). A simpler proof was provided by Chateauneuf and Jaffray [50], based on the notion of local Möbius transform, which we give below. First, we introduce some necessary notions and facts.

Consider a set function $\xi : 2^X \to \mathbb{R}$, $k \in \mathbb{N}_0$, and a family $\mathcal{S} = \{A, B_1, \ldots, B_k\}$ in 2^X. We build the set function $\xi_{\mathcal{S}} : 2^{[k]} \to \mathbb{R}$, with $I \mapsto \xi(A \cap \bigcap_{i \in I^c} B_i)$, where $I^c = [k] \setminus I$. Denoting by $m^{\xi_{\mathcal{S}}}$ the Möbius transform of $\xi_{\mathcal{S}}$, the *local Möbius transform of ξ relative to $\mathcal{S}$* is the set function $\lambda^{\xi}_{\mathcal{S}} : 2^{[k]} \to \mathbb{R}$ defined by

$$\lambda^{\xi}_{\mathcal{S}}(I) = m^{\xi_{\mathcal{S}}}(I^c) \qquad (I \subseteq [k]).$$

It follows that, for the above fixed $\mathcal{S}$ and k,

$$\xi(A \cap \bigcap_{i \in I} B_i) = \sum_{J \supseteq I} \lambda^{\xi}_{\mathcal{S}}(J) \qquad (I \subseteq [k]), \tag{7.29}$$

and conversely

$$\lambda^{\xi}_{\mathcal{S}}(I) = \nabla_k \xi(A \cap \bigcap_{i \in I} B_i, \{B_\ell\}_{\ell \in I^c}) \qquad (I \subseteq [k]),$$

with the difference function $\nabla_k \xi$ defined in Remark 2.19(vi). Note that from (7.29) $\xi(A) = \sum_{J \subseteq [k]} \lambda^{\xi}_{\mathcal{S}}(J)$.

The following facts are easy to check ($k, \mathcal{S}$ are arbitrary).

Lemma 7.27

(i) *ξ is k-monotone, monotone and nonnegative if and only if $\lambda^{\xi}_{\mathcal{S}}(I) \geqslant 0$, $I \subseteq [k]$, for every family $\mathcal{S}$ of $k+1$ subsets of X;*

(ii) *For any set functions ξ_1, ξ_2 and $\xi = \xi_1\xi_2$,*

$$\lambda^{\xi}_{\mathcal{S}}(I) = \sum_{I_1 \cap I_2 = I} \lambda^{\xi_1}_{\mathcal{S}}(I_1)\lambda^{\xi_2}_{\mathcal{S}}(I_2) \qquad (I \subseteq [k]);$$

(iii) *Fix $E \in 2^X \setminus \{\varnothing\}$ and consider the set function $\gamma(A) = \nabla_1\xi(A \cup E^c, E) = \xi(A \cup E^c) - \xi(A \cap E)$ for any $A \in 2^X$. Then, if ξ is k-monotone,*

$$\lambda^{\gamma}_{\mathcal{S}}(I) \geqslant 0 \qquad (\varnothing \neq I \subseteq [k]) \tag{7.30}$$

$$\lambda^{\gamma}_{\mathcal{S}}(\varnothing) = \lambda^{\xi}_{\mathcal{S}_{\cup E^c}}(\varnothing) - \lambda^{\xi}_{\mathcal{S}}(\varnothing) \tag{7.31}$$

for any family $\mathcal{S} = \{A, B_1, \ldots, B_k\}$ of $k+1$ subsets of X, and $\mathcal{S}_{\cup E^c} = \{A \cup E^c, B_1 \cup E^c, \ldots, B_k \cup E^c\}$.

Proof (of Theorem 7.26) To simplify and without loss of generality, we consider that μ is normalized.

1. For any $A \in 2^E$, we have

$$\mu(A) = \mu_E(A)(1 - \mu(A \cup E^c) + \mu(A))$$

or equivalently

$$\mu_{|E} = \mu_E - \mu_E \cdot \gamma$$

where $\mu_{|E}$ is the restriction of μ to E, and $\gamma(A) = \mu(A \cup E^c) - \mu(A)$ for any $A \subseteq E$.

Observe that $\mu_{|E}$ and μ have the same local Möbius transform relative to any family $\mathcal{S}$ in 2^E. Moreover, by additivity of the local Möbius transform and Lemma 7.27(ii), we obtain

$$\lambda^{\mu}_{\mathcal{S}}(I) = \lambda^{\mu_E}_{\mathcal{S}}(I) - \sum_{I_1 \cap I_2 = I} \lambda^{\mu_E}_{\mathcal{S}}(I_1)\lambda^{\gamma}_{\mathcal{S}}(I_2) \qquad (I \subseteq [k]). \tag{7.32}$$

By Lemma 7.27(i), we have to prove that $\lambda^{\mu_E}_{\mathcal{S}}(I) \geqslant 0$ for all $I \subseteq [k]$ and all family $\mathcal{S}$ of $k+1$ sets in 2^E.

2. We prove $\lambda^{\mu_E}_{\mathcal{S}}(I) \geqslant 0$ for every nonempty I. Fix $\mathcal{S} = \{A, B_1, \ldots, B_k\}$. We rewrite (7.32) as follows:

$$\lambda^{\mu_E}_{\mathcal{S}}(I) = \lambda^{\mu}_{\mathcal{S}}(I) + \lambda^{\mu_E}_{\mathcal{S}}(I)\sum_{I_2 \supseteq I} \lambda^{\gamma}_{\mathcal{S}}(I_2) + \sum_{I_1 \supset I} \lambda^{\mu_E}_{\mathcal{S}}(I_1) \sum_{I \cup I_1^c \supseteq I_2 \supseteq I} \lambda^{\gamma}_{\mathcal{S}}(I_2)$$

and

$$\lambda_S^{\mu_E}(I)(1-\gamma(C)) = \lambda_S^{\mu}(I) + \sum_{I_1 \supset I} \lambda_S^{\mu_E}(I_1) \sum_{I \cup I_1^c \supseteq I_2 \supseteq I} \lambda_S^{\gamma}(I_2) \tag{7.33}$$

because by (7.29) $\sum_{I_2 \supseteq I} \lambda_S^{\gamma}(I_2) = \gamma(C)$, with $C = A \cap \bigcap_{i \in I} B_i$. Observe that $1-\gamma(C) > 0$ because for any set $K \in 2^E$, $1-\gamma(K) = 1-\mu(K \cup E^c) + \mu(K)$; i.e., the denominator of $\mu_E(K)$, which is positive by assumption on E. Moreover, $\lambda_S^{\mu}(I) \geqslant 0$ by k-monotonicity, and $\lambda_S^{\gamma}(I) \geqslant 0$ for all $I \neq \varnothing$ by Lemma 7.27(iii). It follows from (7.33) that for any $r \geqslant 1$, the assertion

$$\left[\lambda_S^{\mu_E}(i) \geqslant 0 \text{ for } k \geqslant |I| > r\right] \Rightarrow \lambda_S^{\mu_E}(I) \geqslant 0 \text{ for } |I| = r$$

is true. Now, for $I = [k]$, (7.33) reduces to

$$\lambda_S^{\mu_E}([k])(1-\gamma(C)) = \lambda_S^{\mu}([k]),$$

which shows that $\lambda_S^{\mu_E}([k]) \geqslant 0$, and by induction $\lambda_S^{\mu_E}(I) \geqslant 0$ for all $I \neq \varnothing$.

3. It remains to prove that $\lambda_S^{\mu_E}(\varnothing) \geqslant 0$. For $I = \varnothing$, (7.33) becomes

$$\lambda_S^{\mu_E}(\varnothing)(1-\gamma(A)) = \lambda_S^{\mu}(\varnothing) + \sum_{I_1 \neq \varnothing} \lambda_S^{\mu_E}(I_1) \sum_{\varnothing \neq I_2 \subseteq I_1^c} \lambda_S^{\gamma}(I_2) + \Big(\sum_{I_1 \neq \varnothing} \lambda_S^{\mu_E}(I_1)\Big)\lambda_S^{\gamma}(\varnothing).$$

Since $\sum_{I_1 \neq \varnothing} \lambda_S^{\mu_E}(I_1) = \mu_E(A) - \lambda_S^{\mu_E}(\varnothing)$ and by Lemma 7.27(iii) $\lambda_S^{\gamma}(\varnothing) = \lambda_{S \cup E^c}^{\mu}(\varnothing) - \lambda_S^{\mu}(\varnothing)$, we get

$$\begin{aligned}\lambda_S^{\mu_E}(\varnothing)(1-\gamma(A)+\lambda_S^{\gamma}(\varnothing)) &= \lambda_S^{\mu}(\varnothing)\big(1-\mu_E(A)\big) + \lambda_{S \cup E^c}^{\mu}(\varnothing)\mu_E(A) \\ &\quad + \sum_{I_1 \neq \varnothing} \lambda_S^{\mu_E}(I_1) \sum_{\varnothing \neq I_2 \subseteq I_1^c} \lambda_S^{\gamma}(I_2) \\ &\geqslant 0.\end{aligned}$$

Hence $\lambda_S^{\mu_E}(\varnothing) \geqslant 0$ as soon as $1-\gamma(A)+\lambda_S^{\gamma}(\varnothing) > 0$. Suppose *per contra* that $1-\gamma(A)+\lambda_S^{\gamma}(\varnothing) \leqslant 0$. Then

$$(1-\gamma(A))\mu_E(A) + \lambda_S^{\gamma}(\varnothing)\mu_E(A) = \mu(A) + \lambda_S^{\gamma}(\varnothing)\mu_E(A) \leqslant 0,$$

hence by (7.31)

$$\mu(A) \leqslant \big(\lambda_S^{\mu}(\varnothing) - \lambda_{S \cup E^c}^{\mu}(\varnothing)\big)\mu_E(A) \leqslant \lambda_S^{\mu}(\varnothing)\mu_E(A) \leqslant \lambda_S^{\mu}(\varnothing),$$

which implies, because $\mu(A) = \sum_{I\subseteq[k]} \lambda^{\mu}_{S}(I)$ and $\lambda^{\mu}_{S}(I) \geqslant 0$ for all $I \subseteq [k]$, that $\mu(A) = \lambda^{\mu}_{S}(\varnothing)$ and $\lambda^{\mu}_{S}(I) = 0$ for all $I \neq \varnothing$. This implies in turn by (7.32) that $\lambda^{\mu_E}_{S}(I) = 0$ for all $I \neq \varnothing$. Therefore, $\lambda^{\mu_E}_{S}(\varnothing) = \mu_E(A) \geqslant 0$.

□

An immediate consequence of Theorem 7.26 is that conditional belief and plausibility functions are still belief and plausibility functions.

7.5.2 The Bayes' and Dempster-Shafer Conditioning Rules

Let μ be a capacity on an algebra $\mathcal{X} \subseteq 2^X$ and consider $B \in \mathcal{X}$. The *Bayes' conditional capacity given B* is defined exactly as classical conditional probability measures:

$$\mu^{\mathrm{Ba}}_{B}(A) = \frac{\mu(A \cap B)}{\mu(B)} \qquad (A \in \mathcal{X}). \tag{7.34}$$

It is defined if and only if $\mu(B) > 0$. The *Dempster-Shafer conditional capacity given B* is defined by:

$$\mu^{\mathrm{DS}}_{B}(A) = \frac{\mu((A \cap B) \cup B^c) - \mu(B^c)}{\mu(X) - \mu(B^c)} \qquad (A \in \mathcal{X}). \tag{7.35}$$

Note that the denominator is $\overline{\mu}(B)$, hence it is defined if and only if $\overline{\mu}(B) > 0$.

These two rules are in a sense conjugate of each other because, as is easy to check,

$$\overline{\mu^{\mathrm{Ba}}_{B}} = \overline{\mu}^{\mathrm{DS}}_{B}. \tag{7.36}$$

In Dempster [77], the conditional plausibility function $\mathrm{Pl}(A|B)$ is defined by the Bayes' conditioning rule, and the conditional belief function $\mathrm{Bel}(A|B)$ as its conjugate; i.e., using (7.36),

$$\mathrm{Pl}(A|B) = \mathrm{Pl}^{\mathrm{Ba}}_{B}(A), \quad \mathrm{Bel}(A|B) = 1 - \mathrm{Pl}(A^c|B) = \mathrm{Bel}^{\mathrm{DS}}_{B}(A) \qquad (A \in \mathcal{X}).$$

These quantities are defined if and only if $\mathrm{Pl}(B) > 0$.

The analog of Lemma 7.22 is the following.

Lemma 7.28 *Let $A, B \in \mathcal{X}$, and μ a capacity on $\mathcal{X}$. Define the capacity $\mu_{\cap B}$ by $\mu_{\cap B}(A) = \mu(A \cap B)$ for any $A \in \mathcal{X}$. The equation in t*

$$\int (1_A - t \cdot 1_X)\,\mathrm{d}\mu_{\cap B} = 0$$

has a unique solution in $[0, 1]$ *if and only if* $\mu(B) > 0$. *In this case the solution is* $t = \mu_B^{\mathrm{Ba}}(A)$.

Proof Considering $t \geq 0$, the values taken by the integrand are $-t$ and $1 - t$, hence

$$0 = \int (1_A - t \cdot 1_X)\, \mathrm{d}\mu_{\cap B} = -t\mu_{\cap B}(X) + \mu_{\cap B}(A) = \mu(A \cap B) - t\mu(B),$$

which gives the result. □

Since $\int (1_A - t \cdot 1_X)\, \mathrm{d}\mu_{\cap B} = \int (1_A - t \cdot 1_X) 1_B \, \mathrm{d}\mu_{\cap B}$, we can use the same interpretation in terms of gambles as for Lemma 7.22. The only difference is that $\mu_{\cap B}$ is used instead of μ: this means that all what happens outside B is ignored by this rule.

We list some properties of these conditioning rules.

Lemma 7.29 *For any* $B \in \mathcal{X}$ *such that the following conditionals are defined, the following propositions hold:*

(i) $\mu_X^{\mathrm{Ba}} = \mu_X^{\mathrm{DS}} = \frac{1}{\mu(X)}\mu$; $\mu_B^{\mathrm{Ba}}(X) = \mu_B^{\mathrm{Ba}}(B) = \mu_B^{\mathrm{DS}}(X) = \mu_B^{\mathrm{DS}}(B) = 1$;
(ii) *For* $A_1, A_2 \in \mathcal{X}$, $A_1 \subseteq A_2$ *implies* $\mu_B^{\mathrm{Ba}}(A_1) \leqslant \mu_B^{\mathrm{Ba}}(A_2)$, $\mu_B^{\mathrm{DS}}(A_1) \leqslant \mu_B^{\mathrm{DS}}(A_2)$;
(iii) *If* ν *is another capacity on* $\mathcal{X}$ *then*

$$\mu \leqslant \nu, \quad \mu(B) = \nu(B) \text{ implies } \mu_B^{\mathrm{Ba}} \leqslant \nu_B^{\mathrm{Ba}}$$

$$\mu \leqslant \nu, \quad \mu(B^c) = \nu(B^c), \quad \mu(X) = \nu(X) \text{ implies } \mu_B^{\mathrm{DS}} \leqslant \nu_B^{\mathrm{DS}}.$$

(iv) *If* μ *is submodular (respectively, subadditive), then so is* μ_B^{Ba};
(v) *If* μ *is subadditive, then* $\overline{\mu}_B^{\mathrm{DS}} \leqslant \mu_B^{\mathrm{Ba}}$;
(vi) *For* $B_1, B_2 \in \mathcal{X}$,

$$(\mu_{B_1}^{\mathrm{Ba}})_{B_2}^{\mathrm{Ba}} = \mu_{B_1 \cap B_2}^{\mathrm{Ba}}, \quad (\mu_{B_1}^{\mathrm{DS}})_{B_2}^{\mathrm{DS}} = \mu_{B_1 \cap B_2}^{\mathrm{DS}}.$$

(vii) *If* μ *is submodular and* $B \in \mathcal{X}$ *such that* $\overline{\mu}(B) > 0$,

$$\overline{\mu}_B \leqslant \overline{\mu}_B^{\mathrm{DS}} \leqslant \mu_B^{\mathrm{Ba}} \leqslant \mu_B.$$

Proof Proofs for (i) to (iii) are analogous to those for the general conditioning rule and are omitted.

(iv) We have, supposing $\mu(B) > 0$:

$$\begin{aligned}
\mu(B)\big(\mu_B^{\mathrm{Ba}}(A \cup C) + \mu_B^{\mathrm{Ba}}(A \cap C)\big) &= \mu((A \cup C) \cap B) + \mu(A \cap C \cap B) \\
&= \mu((A \cap B) \cup (C \cap B)) + \mu((A \cap B) \cap (C \cap B)) \\
&\leqslant \mu(A \cap B) + \mu(C \cap B) \\
&= \mu(B)\big(\mu_B^{\mathrm{Ba}}(A) + \mu_B^{\mathrm{Ba}}(C)\big).
\end{aligned}$$

(v) Since μ_B^{Ba} is subadditive by (iv), it follows from (7.36) and Theorem 2.20(i) that $\overline{\mu}_B^{\mathrm{DS}} = \overline{\mu_B^{\mathrm{Ba}}} \leqslant \mu_B^{\mathrm{Ba}}$.

(vi)

$$(\mu_{B_1}^{\mathrm{Ba}})_{B_2}^{\mathrm{Ba}}(A) = \frac{\mu_{B_1}^{\mathrm{Ba}}(A \cap B_2)}{\mu_{B_1}^{\mathrm{Ba}}(B_2)} = \frac{\mu(A \cap B_2 \cap B_1)\mu(B_1)}{\mu(B_1)\mu(B_2 \cap B_1)} = \mu_{B_1 \cap B_2}^{\mathrm{Ba}}(A).$$

The assertion for the Dempster-Shafer rule is obtained by conjugation.

(vii) It is enough to prove the last inequality, the first one being obtained through conjugation and the second one being (v). By submodularity,

$$\mu(X) + \mu(A \cap B) = \mu(B \cup (A \cup B^c)) + \mu(B \cap (A \cup B^c)) \leqslant \mu(B) + \mu(A \cup B^c).$$

Then

$$\mu_B^{\mathrm{Ba}}(A) = \frac{\mu(A \cap B)}{\mu(B)} \leqslant \frac{\mu(A \cap B)}{\mu(A \cap B) + \mu(X) - \mu(A \cup B^c)} = \mu_B(A).$$

□

The last result on the comparison of the three rules was already remarked by Dempster [77] for upper and lower probabilities. With our notation:

$$\mathrm{Bel}_B(A) \leqslant \mathrm{Bel}(A|B) \leqslant \mathrm{Pl}(A|B) \leqslant \mathrm{Pl}_B(A) \qquad (A \in \mathcal{X}).$$

The next result, proved by Gilboa and Schmeidler [155], is the analog of Theorem 7.24 and sheds light on the interpretation of these rules.

Theorem 7.30 (The Bayes and DS conditioning rules as upper and lower probabilities on a facet) *Let μ be submodular on $\mathcal{X}$ and $B \in \mathcal{X}$ such that $\overline{\mu}(B) > 0$. Then*

$$\mu_B^{\mathrm{Ba}}(A) = \max_{\substack{\phi \in \mathbf{core}(\overline{\mu}) \\ \phi(B) = \mu(B)}} \phi_B^{\mathrm{Ba}}(A), \quad \overline{\mu}_B^{\mathrm{DS}}(A) = \min_{\substack{\phi \in \mathbf{core}(\overline{\mu}) \\ \phi(B) = \mu(B)}} \phi_B^{\mathrm{DS}}(A) \qquad (A \in \mathcal{X})$$

(notice that for additive ϕ, we have $\phi_B^{\mathrm{Ba}} = \phi_B^{\mathrm{DS}} = \phi_B$), and

$$\mathbf{core}(\overline{\mu_B^{\mathrm{Ba}}}) = \{\phi_B^{\mathrm{Ba}} : \phi \in \mathbf{core}(\overline{\mu}), \phi(B) = \mu(B)\}. \tag{7.37}$$

A short explanation on the title of the theorem: the set of additive measures ϕ in the core of $\overline{\mu}$ satisfying $\phi(B) = \mu(B)$ is a facet of the core (Sect. 1.3.4); i.e., corresponding to a tight inequality among those defining the core, namely $\phi(A) \leqslant \mu(A), A \in \mathcal{X}$.

Proof Let us introduce for ease of notation $M_B = \{\phi \in \mathbf{core}(\overline{\mu}) \,:\, \phi(B) = \mu(B)\}$. Let $A \in \mathcal{X}$. By Lemma 7.28 we have

$$\int (1_A - \mu_B^{\mathrm{Ba}}(A) \cdot 1_X)\,\mathrm{d}\mu_{\cap B} = 0.$$

By submodularity of μ (and therefore of $\mu_{\cap B}$) and Theorem 4.39, there exists some additive measure ϕ_0 on $\mathcal{X}$ in the core of $\overline{\mu_{\cap B}}$ (i.e., $\overline{\mu_{\cap B}} \leqslant \phi_0 \leqslant \mu_{\cap B}$) such that

$$\int (1_A - \mu_B^{\mathrm{Ba}}(A) \cdot 1_X)\,\mathrm{d}\mu_{\cap B} = \int (1_A - \mu_B^{\mathrm{Ba}}(A) \cdot 1_X)\,\mathrm{d}\phi_0.$$

Since $\overline{\mu_{\cap B}}(B) = \mu_{\cap B}(X) - \mu_{\cap B}(B^c) = \mu(B)$, it follows that $\phi_0(B) = \mu(B)$, however $\phi_0(X) = \mu(B) \leqslant \mu(X)$, where the inequality may be strict. We are looking for some additive measure ϕ_1 being nonzero on B^c such that $\phi = \phi_0 + \phi_1 \in M_B$. Because $\overline{\mu}$ is exact, there must exist $\gamma \in \mathbf{core}(\overline{\mu})$ satisfying $\gamma(B^c) = \overline{\mu}(B^c)$. Taking $\phi_1 = \gamma_{\cap B^c}$, the additive measure $\phi = \phi_0 + \phi_1$ satisfies

$$\begin{aligned}
\phi(B) &= \phi_0(B) + \gamma_{\cap B^c}(B) = \mu(B) + \gamma(\varnothing) = \mu(B)\\
\phi(X) &= \phi_0(X) + \gamma_{\cap B^c}(X) = \mu(B) + \overline{\mu}(B^c) = \mu(X).
\end{aligned}$$

Furthermore, for any $C \in \mathcal{X}$,

$$\begin{aligned}
\phi(C) &= \phi_0(C) + \gamma(C \cap B^c) \geqslant \overline{\mu_{\cap B}}(C) + \overline{\mu}(C \cap B^c)\\
&= \mu(B) - \mu(C^c \cap B) + \mu(X) - \mu(C^c \cup B)\\
&\geqslant \mu(X) - \mu(C^c) \text{ (by submodularity of } \mu)\\
&= \overline{\mu}(C).
\end{aligned}$$

It follows that $\phi \in \mathbf{core}(\overline{\mu})$ and $\phi(B) = \mu(B)$.

Now,

$$\int (1_A - \mu_B^{\mathrm{Ba}}(A) \cdot 1_X)\,\mathrm{d}\phi_{\cap B} = \int (1_A - \mu_B^{\mathrm{Ba}}(A) \cdot 1_X)\,\mathrm{d}\mu_{\cap B} = \int (1_A - \mu_B^{\mathrm{Ba}}(A) \cdot 1_X)\,\mathrm{d}\phi_0,$$

hence by Lemma 7.28 we get $\phi_B^{\mathrm{Ba}}(A) = \mu_B^{\mathrm{Ba}}(A)$. Now, for an arbitrary $\phi \in \mathbf{core}(\overline{\mu})$ satisfying $\phi(B) = \mu(B)$, we get by Lemma 7.29(iii) $\phi_B^{\mathrm{Ba}}(A) \leqslant \mu_B^{\mathrm{Ba}}(A)$, so that $\mu_B^{\mathrm{Ba}}(A) = \max_{\substack{\phi \in \mathbf{core}(\overline{\mu})\\ \phi(B) = \mu(B)}} \phi_B^{\mathrm{Ba}}(A)$. The corresponding result for $\overline{\mu}_B^{\mathrm{DS}}$ follows by conjugation.

It remains to prove (7.37). The above considerations show that $\{\alpha_B^{\text{Ba}} : \alpha \in M_B\} \subseteq \mathbf{core}(\overline{\mu_B^{\text{Ba}}})$. To prove the reverse inclusion let $\psi \in \mathbf{core}(\overline{\mu_B^{\text{Ba}}})$. Then

$$\overline{\mu_{\cap B}} \leqslant \mu(B)\psi \leqslant \mu_{\cap B}. \tag{7.38}$$

For $\phi_0 = \mu(B)\psi$ we can find as above an additive measure ϕ_1 being nonzero on B^c such that $\phi = \phi_0 + \phi_1 \in M_B$. For any $A \in \mathcal{X}$, we have

$$\phi_B^{\text{Ba}}(A) = \frac{\phi(A \cap B)}{\phi(B)} = \frac{\mu(B)\psi(A \cap B)}{\mu(B)} = \psi(A \cap B) = \psi(A),$$

where the last equality is explained as follows. First, by (7.38), $\psi(B') = 1$ for any $B' \supseteq B$. Then by additivity, $\psi(A) = \psi(A \cap B) + \psi(A \setminus B)$, and $1 = \psi((A \setminus B) \cup B) = \psi(A \setminus B) + \underbrace{\psi(B)}_{1}$, which implies $\psi(A \setminus B) = 0$. Hence $\psi(A \cap B) = \psi(A)$, and $\psi = \phi_B^{\text{Ba}}$ as desired. □

Remark 7.31 Theorem 7.30 clearly shows the difference between the Bayesian and Dempster rule, and the general rule (see Remark 7.25). The conditional belief and plausibility functions, $\text{Bel}(A|B), \text{Pl}(A|B)$, as defined by Dempster, are respectively the most pessimistic and most optimistic estimates of the conditional probability $P(A \mid B)$, among those compatible probability measures satisfying $P(B) = \text{Pl}(B)$. Recalling that $\text{Pl}(B)$ is interpreted as the upper probability of B, i.e., $\text{Pl}(B) = \max_{P \in \mathbf{core}(\text{Bel})} P(B)$, it turns out that only the probability measures maximizing the probability of B are taken into account. In other words, one considers that the event B has realized, and consequently only those probability measures maximizing $P(B)$ are considered. This explains why this rule is sometimes called the *revision rule*, because the knowledge pertaining to the set of X of possible answers has been modified. ◇

We end this section by relating the Dempster-Shafer rule to combination of evidences. Dempster [77] mentions the following property.

Theorem 7.32 *Let* $B \in \mathcal{X}$ *such that* $\text{Pl}(B) > 0$. *Then*

$$\text{Bel}(\cdot|B) = \text{Bel} \otimes u_B$$

where u_B *is the unanimity game centered on* B; *i.e.,* $\text{Bel}_{B,0}$.

Proof Let us put $\text{Bel}' = \text{Bel} \otimes u_B$, with mass distribution m', and check if we recover for Bel' the conditional belief defined by Dempster. We have, denoting by

m the mass distribution of Bel and recalling that u_B has only B as focal set, for any $A \neq \varnothing$,

$$
\begin{aligned}
m'(A) &= \frac{\sum_{\substack{C \in \mathbf{supp}(m) \\ C \cap B = A}} m(C)}{1 - \sum_{\substack{C \in \mathbf{supp}(m) \\ C \cap B = \varnothing}} m(C)} \\
&= \frac{\sum_{\substack{C \in \mathbf{supp}(m) \\ C \cap B = A}} m(C)}{1 - \sum_{\substack{C \in \mathbf{supp}(m) \\ C \subseteq B^c}} m(C)} = \frac{\sum_{\substack{C \in \mathbf{supp}(m) \\ C \cap B = A}} m(C)}{1 - \mathrm{Bel}(B^c)}.
\end{aligned}
$$

It follows that, for any $A \neq \varnothing$,

$$
\begin{aligned}
\mathrm{Bel}'(A) &= \sum_{C \subseteq A, C \neq \varnothing} m'(C) = \frac{\sum_{C \subseteq A, C \neq \varnothing} \sum_{D \cap B = C} m(D)}{1 - \mathrm{Bel}(B^c)} \\
&= \frac{\sum_{D \cap B \subseteq A, D \cap B \neq \varnothing} m(D)}{1 - \mathrm{Bel}(B^c)} = \frac{\sum_{D \cap B \subseteq A} m(D) - \sum_{D \cap B = \varnothing} m(D)}{1 - \mathrm{Bel}(B^c)} \\
&= \frac{\sum_{D \subseteq A \cup B^c} m(D) - \sum_{D \subseteq B^c} m(D)}{1 - \mathrm{Bel}(B^c)} \\
&= \frac{\mathrm{Bel}(A \cup B^c) + m(\varnothing) - \mathrm{Bel}(B^c) - m(\varnothing)}{1 - \mathrm{Bel}(B^c)} \\
&= \mathrm{Bel}(A|B).
\end{aligned}
$$

□

7.6 The Transferable Belief Model

The transferable belief model is a reworking of the framework of evidence theory given by Shafer, which was proposed by Smets [309, 313]. It is not related to any notion of lower and upper probabilities nor imprecise probability, and likewise Shafer, its primitive ingredient is the mass distribution (called *basic belief assignment*). The model has two levels:

(i) The *credal level*, where the set X is built (set of possible worlds, or states of nature, etc.), as well as the basic belief assignment m quantifying the belief committed to some events. This is the level of knowledge representation;
(ii) The *pignistic level* (from the Latin word *pignus*, bet), where a probability measure is built from the belief function synthesizing the knowledge gathered in the credal level. This is the decision level.

In the credal level, the only notable difference with the framework proposed by Shafer is that it is assumed that $m(\varnothing)$ could be positive (open world assumption), and consistently, the nonnormalized combination rule is used. Its aim is to build eventually a belief function from all the pieces of knowledge that are obtained, after possibly several operations like combination of mass distributions or conditioning. Assuming that at the final stage some decision has to be made, the belief function is mapped to the set of probability measures by what is called the *pignistic transform*. This transformation into a probability measure is done in order to avoid at the decision level a Dutch book argument.[6]

The pignistic transformation is based on the principle of insufficient reason, used in probability theory since Laplace. It says that in the case of a finite set of k outcomes, if there is no information relative to the different outcomes, an equal probability of $\frac{1}{k}$ is assigned to each outcome. This principle can be extended to the case of a mass distribution as follows. Consider a mass distribution m on X, and $A \in \mathbf{supp}(m)$. As explained in Sect. 7.2, the quantity $m(A)$ represents the quantity of belief committed to the event A only, and not to its proper subsets. Since there is no information regarding the elements in A, the principle of insufficient reason applied to A tells that an equal probability should be assigned to each element in A. However, in this case, the initial quantity of "probability" to share is not 1, but $m(A)$. Therefore, to each element of A, a probability equal to $\frac{m(A)}{|A|}$ is assigned. Doing so for each $A \in \mathbf{supp}(m)$ and summing, we come up with the probability distribution BetP^m on X, given by

$$\mathrm{BetP}^m(x) = \sum_{A \ni x} \frac{m(A)}{|A|} \qquad (x \in X). \tag{7.39}$$

The above equation defines what is called the *pignistic transform*, which to each mass distribution m assigns a probability distribution BetP^m (and therefore is not a transform in the sense we used in Chap. 2, Sect. 2.12). With some abuse of notation, we use the same symbol BetP^m for the corresponding probability measure.

Keeping in mind that m is the Möbius transform of the associated belief function Bel, the comparison of (7.39) and (2.41) reveals that $\mathrm{BetP}^m(x)$ is nothing other than $I^{\mathrm{Bel}}(\{x\})$; i.e., the pignistic probability distribution is the interaction transform of Bel restricted to singletons. Recall that this is precisely the Shapley value associated to Bel, denoted by $\phi^{\mathrm{Sh}}(\mathrm{Bel})$ (see Sect. 2.11), hence the pignistic transform coincides with the Shapley value.

Because $\sum_{A \in \mathbf{supp}(m)} m(A) = 1$ and m is nonnegative, it is clear from (7.39) that the pignistic probability distribution is indeed a probability distribution. Moreover, the following property is noteworthy.

[6] A Dutch book is a sequence of bets so that the agent/decision maker is doomed to a sure loss if his subjective belief on events is not representable by a probability measure; see Sect. 5.3.1 for details.

Lemma 7.33 *Let m be a mass distribution on X and* Bel *be the corresponding belief function. Then the pignistic probability measure* BetP^m *is compatible with* Bel*; i.e.,* $\mathrm{BetP}^m(A) \geqslant \mathrm{Bel}(A)$ *for every* $A \subseteq X$.

Proof For any $A \subseteq X$, by nonnegativity of m,

$$\mathrm{BetP}^m(A) = \sum_{x\in A}\sum_{B\ni x}\frac{m(B)}{|B|} \geqslant \sum_{x\in A}\sum_{\substack{B\ni x\\ B\subseteq A}}\frac{m(B)}{|B|} = \sum_{B\subseteq A}\frac{m(B)}{|B|}\sum_{x\in B}1$$

$$= \sum_{B\subseteq A} m(B) = \mathrm{Bel}(A).$$

□

Remark 7.34

(i) The name "pignistic" probability or transform was coined in 1990 by Smets [310], and already proposed in 1982 by Dubois and Prade [102], but none of them noticed the connection with the Shapley value. Smets proposed several axiomatizations of the pignistic transform (see, e.g., [312, 313]), which resembles the one given by Shapley [298].

(ii) The pignistic transform is an example of a mapping assigning a probability distribution to some belief function. Other mappings can be imagined as well (Sect. 7.7.3). In addition, this mapping is obtained by a sharing of the mass distribution: it is therefore a sharing value and hence an element of the selectope of Bel (see Sect. 3.5), more precisely, it is the uniform sharing value (Example 3.61). Lemma 7.33 is then a simple consequence of Theorem 3.62(iii).

◇

7.7 Possibility Theory

Possibility theory is a theory of representation of uncertainty based on possibility measures (Sect. 2.8.3). It was initiated by Zadeh [357], in relation with the theory of fuzzy sets [356], and developed by Dubois and Prade [106].

There are several ways to present possibility theory, each of them having its own interest. The first one is to see it as a (quantitative) theory of uncertainty that is parallel to probability theory, changing addition and multiplication (the basic operations in probability theory corresponding to the union of disjoint events and the intersection of independent events) into maximum and minimum. The second view is to see it as a particular case of Dempster-Shafer theory, which immediately connects it to upper and lower probabilities, as well as to any of the interpretations of Dempster-Shafer theory this chapter has presented. The third view is related to logic and leads to a qualitative view of the representation of uncertainty, as well as a new theory of reasoning with approximate information. The present book being rather far from the field of logic, we essentially stick to the two first views, with a natural emphasis on the second one. The readers interested into the logical view should consult [100, 112]. Comprehensive surveys covering most of these aspects can be found in [109, 110, 113]. Also, a measure-theoretic mathematical analysis of possibility measures was done by de Cooman, see, e.g., [70–72].

7.7.1 The Framework

Throughout this section X is a finite set with $|X| = n$. The basic piece of knowledge is supposed to be a *possibility distribution* $\pi : X \to [0,1]$, where $\pi(x)$ for $x \in X$ quantifies the degree to which it is possible that element x is the true outcome of the experiment (or true answer to a question, true state of nature, etc.), with the following convention:

- $\pi(x) = 0$ means that it is impossible that x is the true outcome/state of nature;
- $\pi(x) = 1$ means that x is quite possibly the true state, more exactly, there is no evidence that x could not be the true state of nature;
- $\pi(x) > \pi(x')$ means that x is more plausible than x'.

It is important to note that in the absence of information (evidence), the default distribution is the constant function 1_X. Hence, a possibility distribution is built by adding *negative* information: any added piece of evidence diminishes the possibility of some outcome/state of nature. In the closed world assumption, the true state of nature lies in X, therefore it must be that $\pi(x) = 1$ for some $x \in X$. It follows that the most informative situation is when the possibility of every $x \in X$ is 0 but one, say x_0, for which $\pi(x_0) = 1$. This motivates the next definition: possibility distribution π is more *specific* (or informative) than π' if $\pi(x) \leqslant \pi'(x)$ for every $x \in X$; see Remark 7.35 for additional material on distributions and their interpretation.

From a given possibility distribution π, one derives a *possibility measure* Π defined by

$$\Pi(A) = \sup_{x \in A} \pi(x) \qquad (A \in 2^X). \tag{7.40}$$

The interpretation of $\Pi(A)$ is similar to the one for belief and plausibility functions: it quantifies the degree to which it is possible that the true state of nature lies in A. Note that $\pi(x) = \Pi(\{x\})$ for every $x \in X$, and Π is a normalized capacity satisfying maxitivity:

$$\Pi(A \cup B) = \Pi(A) \vee \Pi(B) \qquad (A, B \in 2^X). \tag{7.41}$$

Conversely, any set function Π vanishing on the empty set, satisfying $\Pi(X) = 1$ and (7.41) is a possibility measure. Note that the latter property is no longer true when X is infinite: take for example $X = \mathbb{N}$ and define Π by $\Pi(A) = 0$ if A is finite, otherwise $\Pi(A) = 1$. Then Π is a maxitive capacity, but no possibility distribution can generate it. Indeed, supposing such a distribution π exists, there would exist $n \in \mathbb{N}$ such that $\pi(n) = 1$. Then $\Pi(\{n\}) = 1$, contradicting the definition of Π.

It is immediate from the definition that

$$\Pi(A \cap B) \leqslant \Pi(A) \wedge \Pi(B) \text{ for every } A, B \in 2^X. \tag{7.42}$$

Also, for any $A \in 2^X$, either $\Pi(A) = 1$ or $\Pi(A^c) = 1$. The latter property implies that, unlike probability measures, the value $\Pi(A)$ does not give any clue on the value of $\Pi(A^c)$, except that one of them should be equal to 1. Assuming $1 = \Pi(A) \geqslant \Pi(A^c)$, the quantity $\Pi(A) - \Pi(A^c)$ represents in a sense the certainty that the true state of nature lies in A rather than in A^c. We then naturally introduce the quantity $\mathrm{Nec}(A) = 1 - \Pi(A^c)$, representing the certainty or necessity that the true state of nature lies in A, and hence defining what is called a *necessity measure*. From the definition, we immediately obtain that

$$\mathrm{Nec}(A) = 1 - \sup_{x \notin A} \pi(x) = \inf_{x \notin A}(1 - \pi(x)) \qquad (A \in 2^X), \tag{7.43}$$

$$\mathrm{Nec}(A \cap B) = \mathrm{Nec}(A) \wedge \mathrm{Nec}(B) \qquad (A, B \in 2^X). \tag{7.44}$$

Moreover, we have $\mathrm{Nec}(A \cup B) \geqslant \mathrm{Nec}(A) \vee \mathrm{Nec}(B)$ for all $A, B \in 2^X$.

$\mathrm{Nec}(A) = 1$ indicates an event A with full certainty because $\Pi(A^c) = 1 - \mathrm{Nec}(A) = 0$, which means that the complement event is impossible. On the other hand, $\mathrm{Nec}(A) = 0$ means that there is no certainty on A, because the complement event A^c is fully possible ($\Pi(A^c) = 1$). In addition, note that $N(A) > 0$ can happen only if $\Pi(A) = 1$, which is in accordance with the intuition that an event can have some certainty only if it is fully possible. Figure 7.1 depicts the possible values taken by $\Pi(A), \mathrm{Nec}(A)$ for a given event A, with the interpretation of remarkable points. As it will be shown in Sect. 7.7.2, possibility and necessity measures are special cases of plausibility and belief functions, respectively. We show in Fig. 7.1 the locus of the possible pairs $(\mathrm{Pl}(A), \mathrm{Bel}(A))$. We see clearly how possibility and necessity measures, as well as probability measures, are limit cases of plausibility and belief functions. Moreover, these two limit cases have no common point, except the trivial ones.

Still two kinds of measures derived from a possibility distribution are noteworthy. The *guaranteed possibility measure* (Dubois and Prade [107]) is defined by

$$\Delta(A) = \inf_{x \in A} \pi(x) \qquad (A \in 2^X).$$

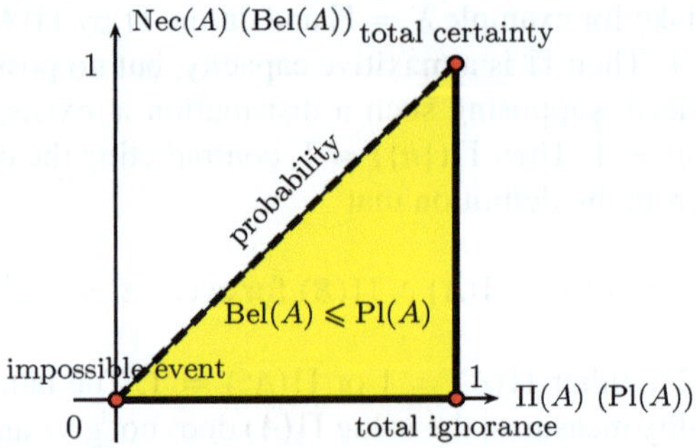

Fig. 7.1 Possibility and necessity of an event A: the *thick line* indicates the locus of the pair $(\Pi(A), \text{Nec}(A))$. The *yellow triangle* is the locus of the pair $(\text{Pl}(A), \text{Bel}(A))$, and the *dashed line* indicates the case where Bel, Pl coincide with a probability measure

This set function is anti-monotone, and satisfies $\Delta(\varnothing) = 1$ (see our convention, Sect. 1.1(xxi)). $\Delta(A)$ is the degree to which all elements in A are possible. The conjugate measure is called the *potential certainty measure*:

$$\nabla(A) = 1 - \Delta(A^c) = \sup_{x \in A^c}(1 - \pi(x)) \qquad (A \in 2^X).$$

It is also anti-monotone and satisfies $\nabla(X) = 0$. The quantity $\nabla(A)$ estimates to what extent there exists at least one element outside A that is impossible (which is a necessary condition that A is somewhat certain). Their characteristic properties are

$$\Delta(A \cup B) = \Delta(A) \wedge \Delta(B) \tag{7.45}$$

$$\nabla(A \cap B) = \nabla(A) \vee \nabla(B) \tag{7.46}$$

for every $A, B \in 2^X$. Comparison of (7.41), (7.44), (7.45) and (7.46) shows that the four measures $\Pi, \text{Nec}, \Delta, \nabla$ exhaust all combinations of transformation between $\cup, \cap$ and $\vee, \wedge$. See Remark 7.35 below for further interpretation in a slightly different framework.

Remark 7.35 There exists in some sense a "bipolar version" of the above framework, proposed by Dubois and Prade (see [111] for a survey). We recall that π encodes a negative information, in the sense that $\pi(x) = 0$ excludes x as a possible state of nature, while $\pi(x) = 1$ does not bring any information on x. One may then introduce a second kind of possibility distribution, called *actual* or *guaranteed possibility* and denoted by δ, which encodes a positive information: $\delta(x)$ expresses to what extent x is supported by evidence. Here $\delta(x) = 1$ means that x is actually fully possible (e.g., because it has been observed) and $\delta(x) = 0$ does not bring any information. By contrast, π is a distribution expressing *potential possibility*, if one may say so. The following example illustrates this.

Example 7.36 (Dubois and Prade [111]) Let X be the set of possible opening hours of a museum in some town. You may know that the museum is opened around 2 p.m. (because when passing in front of it at this time, you have seen it was open). This is a positive piece of information that can be encoded by a distribution δ. On the other hand, you guess that due to usual working time regulations, the museum cannot be open during the night, say, from 10 p.m. to 8 a.m. This is a negative piece of information that can be encoded by a distribution π. ◇

Evidently, one has to impose $\delta(x) \leqslant \pi(x)$ for all $x \in X$. In the case of the bipolar model, the definition of the guaranteed possibility measure has to be modified as follows:

$$\Delta(A) = \inf_{x \in A} \delta(x) \qquad (A \in 2^X).$$

◇

Remark 7.37 Although the term "possibility measure" was coined by Zadeh in 1978 [357], the idea of maxitive measures was introduced before in several domains and with different purposes. Let us cite for example Shackle[7] who introduced the notion of potential surprise in 1961 [294], Shilkret [304] who initiated measure theory with maxitive measures in 1971, and Cohen [55], who introduced "Baconian probabilities" in the context of legal reasoning.

Another noticeable example is given by the works of Litvinov and Maslov, around *idempotent calculus*, related to (max, +)-algebra. Maxitive measure are called by them *idempotent measures* (see [224, 225] and the references therein). ◇

[7]George Lennox Sharman Shackle (Cambridge, 1903 – 1992) is an English economist who developed a whole theory of decision leaving room for unpredictable hypotheses (causing surprise), by opposition to the standard framework of Savage based on probability theory. We quote from [294, p. 68] his definition of potential surprise:

> *It is the degree of surprise to which we expose ourselves, when we examine an imagined happening as to its possibility, in general or in the prevailing circumstances, and assess the obstacles, tensions and difficulties which arise in our minds when we try to imagine it occurring, that provides the indicator of degree of possibility. This is the surprise we should feel, if the given thing did happen; it is potential surprise.*

See the interesting analysis by Fioretti [140] of the theory of potential surprise, compared to evidence theory.

7.7.2 Link with Dempster-Shafer Theory

Theorem 7.38 *Let m be a mass distribution such that $m(\varnothing) = 0$. Then the corresponding belief function is a necessity measure if and only if* **supp**(m) *is a chain in 2^X.*

Moreover in this case, letting **supp**$(m) = \{A_1, \dots, A_q\}$ *with* $A_1 \subset A_2 \subset \cdots \subset A_q$, *the corresponding possibility distribution π is given by*

$$\pi(x) = 1 \textit{ for } x \in A_1$$

$$\pi(x) = 1 - m(A_1) \textit{ for } x \in A_2 \setminus A_1$$

$$\vdots \quad \vdots$$

$$\pi(x) = \sum_{i=k}^{q} m(A_i) \textit{ for } x \in A_k \setminus A_{k-1}$$

$$\vdots \quad \vdots$$

$$\pi(x) = m(A_q) \textit{ for } x \in A_q \setminus A_{q-1}$$

$$\pi(x) = 0 \textit{ otherwise.}$$

Proof $\Rightarrow$) See Theorem 2.36.

$\Leftarrow$) Suppose **supp**(m) is a chain and consider any $A, B \in 2^X$. We have to prove that (7.44) holds for A, B, where Nec is the belief function induced by m. Recall that $\mathrm{Nec}(A) = \sum_{C \subseteq A} m(C)$. We have:

$$\mathrm{Nec}(A) = \sum_{C \subseteq A \cap B} m(C) + \sum_{\substack{D \subseteq A \\ D \setminus (A \cap B) \neq \emptyset}} m(D)$$

$$\mathrm{Nec}(B) = \sum_{C \subseteq A \cap B} m(C) + \sum_{\substack{D \subseteq B \\ D \setminus (A \cap B) \neq \emptyset}} m(D).$$

Observe that by the chain assumption, the second term in the right-hand side cannot be nonzero for both equations, hence equality (7.44) holds.

For the proof of the second statement, see formulas (2.24) and (2.25). □

Remark 7.39 Due to the characteristic property that the focal sets of a necessity measure form a chain, these belief functions are sometimes called *consonant belief functions*, and accordingly one can speak of *consonant mass distribution*. It is interesting to note that, by contrast, a belief function is a probability measure if and only if its focal sets are reduced to singletons (Remark 7.6), in other words, the focal sets form an antichain in 2^X. This clearly shows, in a different way than

Fig. 7.1, how possibility theory and probability theory differ, and that they can never coincide, except on trivial cases. ◇

We now show using Theorem 7.38 that the commonality function associated to a consonant mass distribution is in fact the guaranteed possibility. Indeed, take a mass distribution m with $\mathbf{supp}(m) = \{A_1, \ldots, A_q\}$, $A_1 \subset A_2 \subset \cdots \subset A_q$, and corresponding possibility distribution π. We obtain for any $A \in 2^X$

$$q(A) = \sum_{B \supseteq A} m(B) = \sum_{i \geq \ell_0} m(A_i) = \pi^{\ell_0},$$

with ℓ_0 such that $\pi^{\ell_0} = \min_{x \in A} \pi(x)$. We conclude that $q(A) = \Delta(A)$.

This relation sheds light on the interpretation of the commonality function. Recall also that the commonality function is a doubt measure in the sense of random sets functions (also called totally ∪-monotone anti-capacity; see Sect. 7.2.3). It follows that the potential certainty, conjugate of the guaranteed possibility, is a disbelief measure (totally ∩-alternating anti-capacity).

7.7.3 *Links Between Possibility Measures and Probability Measures*

As necessity measures are special cases of belief functions, all results shown in Sect. 7.4 concerning compatible probability measures still hold:

(i) The set of extreme points of the set of probability measures compatible with Nec (denoted by **core**(Nec)) is the set of marginal vectors associated to Nec;
(ii) **core**(Nec) is the selectope of Nec; i.e., each compatible probability measure is a sharing value and vice-versa. In particular, the pignistic transform (uniform sharing value) yields a particular compatible probability measure;
(iii) The lower envelope of the set of compatible probability measures is Nec.

Yet, there are other noteworthy results in the particular case of necessity measures that we present below. In what follows, we say that a probability distribution is compatible with a possibility distribution if the corresponding measures are.

The following result characterizes imprecise probability measures yielding possibility/necessity measures.

Theorem 7.40 (When imprecise probability measures yield necessity measures) *(Dubois and Prade [108]) Let $\mathcal{P}$ be a set of probability measures of the form*

$$\mathcal{P} = \{P \text{ probability measure on } X : P(A_i) \geqslant \alpha_i, i = 1, \ldots, q\}$$

with $\varnothing \neq A_1 \subset A_2 \subset \cdots \subset A_q$ and $0 < \alpha_1 < \cdots < \alpha_q$. Then the upper and lower envelopes of $\mathcal{P}$ are a pair of conjugate possibility and necessity measures.

Conversely, any possibility measure Π *is the upper envelope of a set of probability measures* $\mathcal{P}$ *of the above type, with*

$$A_i = \{x \in X : \pi(x) \geqslant \pi^i\}, \quad \alpha_i = 1 - \pi^{i+1} \qquad (i = 1, \ldots, q)$$

where π *is the associated possibility distribution whose range (with 0 excluded) is* $\mathbf{ran}\,\pi \setminus \{0\} = \{\pi^1, \ldots, \pi^q\}$*, with* $1 = \pi^1 > \pi^2 > \cdots > \pi^q > 0 = \pi^{q+1}$.

Proof $\Rightarrow$) Let m be a mass distribution defined by $\mathbf{supp}(m) = \{A_1, \ldots, A_q\}$ and $m(A_i) = \alpha_i - \alpha_{i-1}, i = 1, \ldots, q$ with $\alpha_0 = 0$. The corresponding Bel is a necessity measure Nec by Theorem 7.38. Let us prove that the lower envelope P_* coincides with Nec for every $A \in 2^X$.

If $A \not\supseteq A_i$ for every $i = 1, \ldots, q$, there is no constraint on $P(A)$, therefore $P_*(A) = 0 = \sum_{A_i \subseteq A} m(A_i) = \mathrm{Nec}(A)$. Otherwise, take the largest i such that $A \supseteq A_i$. Then $P(A) \geqslant P(A_i) \geqslant \alpha_i = N(A_i)$, hence the minimum is $P_*(A) = N(A_i) = N(A)$.

$\Leftarrow$) Since Nec associated to Π is exact, we have that Nec is the lower envelope of $\mathbf{core}(\mathrm{Nec}) = \{P : P(A) \geqslant N(A), A \in 2^X, P(X) = 1\}$. By Theorem 7.38, we know that the mass distribution m associated to Nec has support $A_1, \ldots, A_q$ with $\varnothing \neq A_1 \subset \cdots \subset A_q$. Let us define $\alpha_1, \ldots, \alpha_q$ as follows:

$$\mathrm{Nec}(A_1) = m(A_1) = \alpha_1$$

$$\vdots \quad \vdots$$

$$\mathrm{Nec}(A_i) = m(A_1) + m(A_2) + \cdots + m(A_i) = \alpha_i$$

$$\vdots \quad \vdots$$

$$\mathrm{Nec}(A_q) = 1 = \alpha_q.$$

Hence, with $\mathcal{P} = \mathbf{core}(\mathrm{Nec})$, we have established that any $P \in \mathcal{P}$ satisfies $P(A_i) \geqslant \alpha_i$ for $i = 1, \ldots, q$, and that $A_1, \ldots, A_q$ and $\alpha_1, \ldots, \alpha_q$ satisfy the requirements. It remains to show that any other inequality $P(B) \geqslant \mathrm{Nec}(B)$, $B \neq A_1, \ldots A_q$, of the core of Nec is satisfied.

If $B \not\supset A_i$ for every i, then $\mathrm{Nec}(B) = 0$, and $P(B) \geqslant 0$ is always satisfied. Otherwise, take the largest i such that $B \supset A_i$. We have $N(B) = \sum_{j \leqslant i} m(A_j)$, and by additivity of P:

$$P(B) = \underbrace{P(A_i)}_{\geqslant \alpha_i} + \underbrace{P(B \setminus A_i)}_{\geqslant 0},$$

hence $P(B) \geqslant \alpha_i = \mathrm{Nec}(B)$, as desired.

□

The next result concerns the pignistic transform for possibility measures (proposed by Dubois and Prade in [102, 103]).

Theorem 7.41 (Pignistic transform for possibility measures) *Let π be a possibility distribution on $X = \{x_1, \dots, x_n\}$, numbering the elements of X such that $\pi(x_1) \geqslant \pi(x_2) \geqslant \cdots \geqslant \pi(x_n)$. The following holds.*

(i) The pignistic probability distribution BetP^{π} *is compatible with π and is given by*

$$\mathrm{BetP}^{\pi}(x_i) = \sum_{j=i}^{n} \frac{\pi(x_j) - \pi(x_{j+1})}{j} \qquad (i = 1, \dots, n), \tag{7.47}$$

with $\pi(x_{n+1}) = 0$;

(ii) Every compatible probability distribution p (and therefore, BetP^{π} *as well) satisfies*

$$\pi(x) \geqslant \pi(x') \Rightarrow p(x) \geqslant p(x') \qquad (x, x' \in X);$$

Proof

(i) BetP^{π} is a probability distribution compatible with π by Lemma 7.33, and its expression is straightforwardly obtained from Theorem 7.38.

(ii) If p is a compatible probability distribution, it is a sharing value of Nec, the associated necessity measure, with some sharing system λ. Letting m be the Möbius transform of Nec and $A_i = \{x_1, \dots, x_i\}$, we have, by nonnegativity of m,

$$p(x_i) = \sum_{j=i}^{n} \lambda(A_j, x_i) m(A_j) \geqslant \sum_{j=i'}^{n} \lambda(A_j, x_{i'}) m(A_j),$$

for any $i' \geqslant i$, which proves the result.

□

Remark 7.42 The pignistic probability distribution is the most even or uniform compatible probability distribution from a *local* point of view because it distributes equally the mass assigned to focal sets. However, this is not true from a global point of view, because the pignistic probability distribution does not maximize the Shannon entropy H among all possible compatible probability distributions, as shown by the next example: Take $X = \{x_1, x_2\}$ and $\pi(x_1) = 1$, $\pi(x_2) = 0.2$. Then

$$\mathrm{BetP}^{\pi}(x_1) = 0.8 + \frac{1}{2}0.2 = 0.9, \quad \mathrm{BetP}^{\pi}(x_2) = \frac{1}{2}0.2 = 0.1,$$

which yields $H(\mathrm{BetP}^{\pi}) = -0.9 \log 0.9 - 0.1 \log 0.1 \approx 0.1412$. Now, modifying the equal distribution of 0.2 on $\{x_1, x_2\}$ as follows:

$$P'(x_1) = 0.8 + \frac{1}{2}0.2 - 0.01 = 0.89, \quad P'(x_2) = \frac{1}{2}0.2 + 0.01 = 0.11,$$

we obtain $H(P') \approx 0.1505$. ◇

The pignistic transform is a canonical way to obtain a probability distribution from a possibility distribution. The converse problem can be considered too: Given a probability distribution, which possibility distribution to choose among those which are compatible? The most specific (i.e., the most informative) one is given in the following theorem.

Theorem 7.43 (From probability to possibility) *(Dubois and Prade [102, 104]) Let p be a probability distribution on $X = \{x_1, \ldots, x_n\}$, supposing $p(x_1) \geqslant \cdots \geqslant p(x_n)$. The most specific possibility distribution π; i.e., minimizing $\sum_{x \in X} \pi(x)$, which is compatible with p is given by*

$$\pi(x_i) = \sum_{j=i}^{n} p(x_j) \qquad (i = 1, \ldots, n).$$

Proof From its definition, we have clearly $\pi(x) \geqslant \pi(x')$ if $p(x) \geqslant p(x')$. Then, for any $A \subseteq X$, letting x_{i_0} be the element in A with smallest index, we have

$$\Pi(A) = \max_{x \in A} \pi(x) = \pi(x_{i_0}) = \sum_{j=i_0}^{n} p(x_j) \geqslant \sum_{x \in A} p(x) = P(A),$$

where Π is the possibility measure associated to π. Hence p is compatible with π. Now, consider π' with $\pi'(x_i) < \pi(x_i)$ for some $x_i \in X$. Then the associated possibility measure Π' would satisfy

$$\Pi'(\{x_i, \ldots, x_n\}) = \pi'(x_i) < \sum_{j=i}^{n} p(x_i) = P(\{x_i, \ldots, x_n\}),$$

and therefore π' would not be compatible with p. □

Remark 7.44 (Relation with Lorenz majorization) The above transformation is closely related to the majorization relation of Lorenz (see Hardy et al. [192], and Marshall and Olkin [236]). Given two vectors $a, b \in \mathbb{R}^n$ and assuming $a_1 \geqslant \cdots \geqslant a_n$ and $b_1 \geqslant \cdots \geqslant b_n$, we say that a (weakly) *majorizes* b if

$$\sum_{i=1}^{j} a_i \geqslant \sum_{i=1}^{j} b_i \qquad (j = 1, \ldots, n).$$

If in addition $\sum_{i=1}^{n} a_i = \sum_{i=1}^{n} b_i$ (like for probability distributions), this is equivalent to $\sum_{i=j}^{n} a_i \leqslant \sum_{i=j}^{n} b_i$ for all j. It follows that given two probability distributions p, p' with corresponding possibility distributions π, π', p majorizes p' if and only if $\pi \leqslant \pi'$ pointwise. In [99], Dubois and Hüllermeier interpret this by saying that p is more *peaked* than p', that is, p is more centered around its peak value than p'. It is known from Hardy et al. that p majorizes p' if and only if for any strictly concave function φ,

$$\sum_{i=1}^{n} \varphi(p_i) \leqslant \sum_{i=1}^{n} \varphi(p'_i),$$

(also proved in [99]), which implies that the entropy of p is smaller than the entropy of p' whenever p majorizes p'. Indeed, the Shannon entropy corresponds to $\varphi(x) = -x \log x$, which is strictly concave. We also mention another remarkable result proved by Hardy et al.: a majorizes b if and only if there exists a bistochastic matrix M such that $b = Ma$. ◇

7.7.4 The Possibilistic Core and Totally Monotone Anticore

By analogy with the notion of core of a game, the *possibilistic core* of a normalized capacity μ is the set of possibility measures dominating μ:

$$\pi\textbf{-core}(\mu) = \{\pi \text{ possibility distribution on } X : \Pi(A) \geqslant \mu(A), \forall A \subset X\}.$$

Obviously, π**-core**(μ) is never empty since the possibility distribution $\pi(x) = 1$ $\forall x \in X$ always belongs to π**-core**(μ) for every normalized capacity μ. The next theorem gathers the main results on the possibilistic core (recall that $\mathfrak{S}(n)$ is the set of permutation on $\{1, \ldots, n\}$).

Theorem 7.45 *Let μ be a normalized capacity on X. The possibilistic core of μ has the following properties:*

(i) π**-core**(μ) *is a nonempty upper semilattice, whose top element is 1_X;*

(ii) For any permutation $\sigma \in \mathfrak{S}(n)$, the possibility distribution $\pi^{\sigma,\mu}$ defined by

$$\pi^{\sigma,\mu}(x_{\sigma(i)}) = \mu(\{x_{\sigma(i)}, \ldots, x_{\sigma(n)}\} \qquad (i \in [n]) \tag{7.48}$$

is an element of the possibilistic core. Moreover, denoting by $\Pi^{\sigma,\mu}$ the corresponding possibility measure,

$$\Pi^{\sigma,\mu}(A_i^\sigma) = \mu(A_i^\sigma) \qquad (i \in [n]) \tag{7.49}$$

with $A_i^\sigma = \{x_{\sigma(i)}, \ldots, x_{\sigma(n)}\}$;

(iii) $\mu = \min_{\pi \in \pi-\mathbf{core}(\mu)} \Pi = \min_{\sigma \in \mathfrak{S}(n)} \Pi^{\sigma,\mu}$;

(iv) $⨍ f \,\mathrm{d}\mu = \min_{\pi \in \pi-\mathbf{core}(\mu)} ⨍ f \,\mathrm{d}\Pi$ *and* $\int f \,\mathrm{d}\mu = \min_{\pi \in \pi-\mathbf{core}(\mu)} \int f \,\mathrm{d}\Pi$;

(v) *If μ is strictly monotone, the minimal elements of $\pi-\mathbf{core}(\mu)$ are the possibility distributions $\pi^{\sigma,\mu}$, for all $\sigma \in \mathfrak{S}(n)$.*

Proof

(i) The assertion on the top element is trivial, and proves nonemptiness. Now, if $\pi, \pi' \in \pi-\mathbf{core}(\mu)$, then obviously the pointwise maximum of π and π' is also a possibility distribution which dominates μ.

(ii) Let us fix σ, μ and prove that $\Pi^{\sigma,\mu}(A) \geqslant \mu(A)$ for every A, with equality if $A = A_i^\sigma$ for some i.

Suppose $A = A_i^\sigma$. Then

$$\begin{aligned}\Pi^{\sigma,\mu}(A_i^\sigma) &= \max_{k=i}^{n} \pi^{\sigma,\mu}(x_{\sigma(k)}) \\ &= \pi^{\sigma,\mu}(x_{\sigma(i)}) \text{ (by monotonicity of } \mu) \\ &= \mu(A_i^\sigma).\end{aligned}$$

Suppose now A is not one of the A_i^σ's. Find the greatest i such that $A_i^\sigma \supset A$, and observe that $x_{\sigma(i)} \in A$. Then

$$\mu(A) \leqslant \mu(A_i^\sigma) = \Pi^{\sigma,\mu}(A_i^\sigma) = \pi^{\sigma,\mu}(x_{\sigma(i)}) = \Pi^{\sigma,\mu}(A).$$

(iii) Consider $A \subseteq X$, and a permutation σ such that $A = A_i^\sigma$ for some i. Then $\mu(A) = \Pi^{\sigma,\mu}(A)$, which proves the result.

(iv) By monotonicity of the Sugeno integral [Theorem 4.43(vii)], we have

$$⨍ f \,\mathrm{d}\mu \leqslant ⨍ f \,\mathrm{d}\Pi$$

for any $\pi \in \pi-\mathbf{core}(\mu)$. It suffices then to find a core element such that equality occurs. Take $\sigma \in \mathfrak{S}(n)$ such that $f_{\sigma(1)} \leqslant \cdots \leqslant f_{\sigma(n)}$. By definition of the Sugeno integral and using (ii)

$$\begin{aligned}⨍ f \,\mathrm{d}\mu &= \bigvee_{i=1}^{n} \Big(f_{\sigma(i)} \wedge \mu(A_i^\sigma)\Big) \\ &= \bigvee_{i=1}^{n} \Big(f_{\sigma(i)} \wedge \Pi^{\sigma,\mu}(A_i^\sigma)\Big),\end{aligned}$$

hence $\pi^{\sigma,\mu}$ is the desired core element. The proof is much the same with the Choquet integral.

(v) We prove that $\pi^{\sigma,\mu}$ is minimal, for every $\sigma \in \mathfrak{S}(n)$. Observe that by strict monotonicity of μ,

$$1 = \pi^{\sigma,\mu}(x_{\sigma(1)}) > \cdots > \pi^{\sigma,\mu}(x_{\sigma(n)}).$$

Take $i \in N$ and define a possibility distribution π' by $\pi'(x_{\sigma(i)}) = \pi^{\sigma,\mu}(x_{\sigma(i)}) - \epsilon$ for some $\epsilon > 0$, and $\pi', \pi^{\sigma,\mu}$ identical otherwise. Then, by the above observation,

$$\mu(A_i^\sigma) = \Pi^{\sigma,\mu}(A_i^\sigma) = \pi^{\sigma,\mu}(x_{\sigma(i)}) > \pi'(x_{\sigma(i)}) \vee \cdots \vee \pi'(x_{\sigma(n)}) = \Pi'(A_i^\sigma),$$

hence π' is not a core element. Now, by (iii), any minimal element must be of the form $\pi^{\sigma,\mu}$.

□

Remark 7.46

(i) The possibilistic core was proposed and studied by Dubois et al. [114]. However, Denneberg [81] already in 2000 proposed to consider the set of belief functions dominated by some capacity, which he called the *totally monotone (anti)core*, which amounts to considering the core of plausibility functions, a set including the possibilistic core. He showed that any capacity can be written as a minimum over the core of plausibility functions, and identified $\Pi^{\sigma,\mu}$ as special plausibility functions coinciding with μ on the maximal chain induced by σ.

(ii) It is interesting to compare the classical core (see Chap. 3) with the possibilistic core. In the latter, everything is very simple compared to the classical core: every capacity is balanced (its core is nonempty) and exact (it coincides with the lower envelope of the core). Moreover, the possibilistic core is not a convex polyhedron but is an upper semilattice. The particular possibility distributions $\pi^{\sigma,\mu}$ play the rôle of the marginal vectors in the classical core: just as the latter are vertices of the core for convex games, the possibility distributions $\pi^{\sigma,\mu}$ are minimal elements of the possibilistic core when the capacity is strictly monotone.

◇

Finding the minimal elements of the possibilistic core when the capacity is not strictly monotone is more difficult, and this problem was solved by Dubois et al. [114]. Its solution is based on selectors, as in the definition of the selectope (Sect. 3.5). We recall that a selector is a function $\alpha : 2^X \setminus \{\varnothing\} \to X$ selecting an element in each nonempty set. We denote the set of selectors on X by $\mathcal{S}(X)$. A selector is consistent if for any nonempty sets S, T such that $S \subset T$ and $\alpha(T) \in S$, then $\alpha(S) = \alpha(T)$. For a given selector α and a normalized capacity μ, define the following possibility distribution:

$$\pi^{\alpha,\mu}(x) = \max_{S \subseteq X : \alpha(S) = x} m_*^\mu(S) \qquad (x \in X),$$

where m_*^μ is the ordinal Möbius transform of μ [see (4.64)]. The corresponding possibility measure is denoted by $\Pi^{\alpha,\mu}$. We summarize the main properties of these distributions below.

Theorem 7.47 *Let μ be a normalized capacity on X. The following holds.*

*(i) For any selector α, $\pi^{\alpha,\mu} \in \pi-$**core**$(\mu)$;*
(ii) $\mu = \min_{\alpha \in \mathcal{S}(X)} \Pi^{\alpha,\mu}$;

Proof

(i) For any set $A \subseteq X$, we have

$$\Pi^{\alpha,\mu}(A) = \max_{x \in A} \max_{T : \alpha(T) = x} m_*^\mu(T) \geqslant \max_{T \subseteq A} m_*^\mu(T) = \mu(A),$$

the last equality coming from the definition of the ordinal Möbius transform. Now, if $A = X$, equality holds throughout, so that $\pi^{\alpha,\mu}$ is a possibility distribution.

(ii) Let us consider $A \subseteq X$. From (i), it suffices to build a selector α such that $\Pi^{\alpha,\mu}(A) = \mu(A)$. Take $S \subseteq A$ such that $m_*^\mu(S) = \mu(A)$ and put $\alpha(S) = x$ for some $x \in S$. Now, for any set T such that $m_*^\mu(T) > \mu(A)$, put $\alpha(T) = y$ for some $y \notin A$. This is possible because by monotonicity of μ, $T \nsubseteq A$. It follows that $\Pi^{\alpha,\mu}(A) = \pi^{\alpha,\mu}(x) = \mu(A)$.

□

It should be noted that, contrarily to the classical selectope where marginal vectors correspond to selector values where the selector is consistent, possibility distributions generated by a permutation cannot always be generated by a selector: take $X = \{x_1, x_2\}$ and $\mu(X) = \mu(\{x_2\}) = 1$, $\mu(\{x_1\}) = 0.5$ and consider the identity permutation. Then the possibility distribution generated by this permutation is $(1, 1)$. Because $m_*^\mu(X) = 0$ and $m_*^\mu(\{x_1\}) = 0.5$, there is no way to recover this distribution with a selector.

Dubois et al. give in [114] an algorithm to find the minimal elements of the possibilistic core, which are possibility distributions induced by selectors. The corresponding selectors are found as follows:

Step 0. Put $\mathcal{F} = \{S \in 2^X : m_*^\mu(S) > 0\}$ and order the sets in decreasing value of the ordinal Möbius transform;
Step 1. Pick the first set S in $\mathcal{F}$; select an element $x \in S$; delete S from $\mathcal{F}$; for any set T in $\mathcal{F}$ such that $T \ni x$, put $\alpha(T) = x$ and delete T from $\mathcal{F}$;
Step 2. Repeat Step 1 till $\mathcal{F} = \varnothing$.

Observe that the selector given by the algorithm is consistent. The proof of minimality and completeness is given in [114].

7.8 Belief Functions and Possibility Measures on Lattices and Infinite Spaces

We give a brief account of results when belief functions are not defined on the power set of some finite set X, but on some more general structure, especially a lattice.

7.8.1 *Finite Lattices*

This case was considered by the author in [167], along the line of the work of Barthélemy [18], and completed by Zhou [359]. Some of these results have been presented in Chap. 2, Sect. 2.19.2, where the lattice is a lattice of subsets of X. We do not impose this restriction here, which is in fact unnecessary for all the results of Sect. 2.19.2. The readers should consult Sect. 1.3.2 for all definitions and useful results on lattices.

Belief Functions

Let $(L, \leqslant)$ be a finite lattice, with top and bottom elements denoted by $\top, \bot$ respectively, hereafter always abbreviated by L. By a *game* on L, we mean a mapping $v : L \to \mathbb{R}$ such that $v(\bot) = 0$. A *capacity* μ on L is an isotone game. It is *normalized* if in addition $\mu(\top) = 1$. A function $\mathrm{Bel} : L \to [0, 1]$ is a *belief function on* L if $\mathrm{Bel}(\bot) = 0$, $\mathrm{Bel}(\top) = 1$, and its Möbius transform m is nonnegative. Recall that the Möbius transform is defined for functions on lattices in Remark 2.32(ii), and we have

$$\mathrm{Bel}(x) = \sum_{y \leqslant x} m(y) \qquad (x \in L). \tag{7.50}$$

By contrast to the classical case $L = 2^X$, it is not possible to give a general inverse formula for (7.50), because it depends on the structure of L. Note that by nonnegativity of m, Bel is an isotone function on L, hence it is a normalized capacity.

The commonality function is defined naturally as the co-Möbius transform of the belief function; i.e.,

$$q(x) = \sum_{y \geqslant x} m(y) \qquad (x \in L). \tag{7.51}$$

Plausibility Functions

Plausibility functions can be defined as in the classical case as being the conjugate of a belief function, provided a suitable definition of conjugation is taken. To this end, we suppose that L is autodual. We define a $\vee$-*negation* as a bijective mapping $n : L \to L$ satisfying $n(x \vee y) = n(x) \wedge n(y)$, for all $x, y \in L$, and $n(\top) = \bot$. The following is easy to show.

Lemma 7.48 *(Grabisch [167]) Let L be an autodual lattice. Then, there exists a $\vee$-negation n on L. Moreover, any such negation satisfies the following properties:*

(i) $n(\bot) = \top$;
(ii) $n^{-1}(x \wedge y) = n^{-1}(x) \vee n^{-1}(y)$;
(iii) *x is join-irreducible (respectively, meet-irreducible) if and only if $n(x)$ is meet-irreducible (respectively, join-irreducible).*

Observe that there is in general no unique $\vee$-negation on an autodual lattice L. Indeed, it suffices to take for $n(x)$ the element in the Hasse diagram of L^∂ that takes the position of x in L. But there are in general several ways to draw Hasse diagrams, each of them inducing a different $\vee$-negation. Another typical example of $\vee$-negation is $n(x) = x'$, the complement of x, provided that L is complemented. Also, $n^{-1} \neq n$ in general, so that n has not the property of *involution* ($n \circ n = \mathrm{Id}$) in general.

Based on these considerations, the *conjugate* of a capacity μ on L, relative to some $\vee$-negation n, is a capacity defined as

$$\overline{\mu}(x) = \mu(\top) - \mu(n(x)) \qquad (x \in L).$$

Now, the conjugate of a belief function Bel is called a *plausibility function*: $\mathrm{Pl} = \overline{\mathrm{Bel}}$. Recall that its definition depends on the chosen $\vee$-negation.

Remark 7.49 Observe that $\overline{\overline{\mu}} = \mu$ if and only if n is involutive, implying that the conjugate relative to n of a plausibility function $\mathrm{Pl} = \overline{\mathrm{Bel}}$, where the latter conjugate is taken relative to n, gives back Bel if and only if n is involutive. If n is not involutive, one should take the conjugate relative to n^{-1} (the property to be a $\vee$-negation is irrelevant in the definition of the conjugation), denoted by $\overline{(\cdot)}^{-1}$. Then $\mathrm{Bel} = \overline{\mathrm{Pl}}^{-1} = \overline{\overline{\mathrm{Bel}}}^{-1}$. ◇

k-Monotone and Totally Monotone Functions

Here the whole subsection on k-monotone and totally monotone games of Sect. 2.19.2 can be copied and pasted here, substituting lattices of sets by lattices. For the sake of convenience, we recall the main elements of it.

A game v on L is *k-monotone* for some $k \geqslant 2$ if for any family of elements $x_1, \ldots, x_k$ of L,

$$v\Big(\bigvee_{i=1}^{k} x_i\Big) \leqslant \sum_{\substack{I\subseteq\{1,\ldots,k\}\\ I\neq\varnothing}} (-1)^{|I|+1} v\Big(\bigwedge_{i\in I} x_i\Big). \tag{7.52}$$

Now, v is *totally monotone* if it is k-monotone for every $k \geqslant 2$. The finiteness of L implies the following result.

Lemma 7.50 *(Barthélemy [18]) Let v be a game on a lattice L. Then v is totally monotone if and only if v is* $(|L| - 2)$*-monotone.*

The fact that L is modular or distributive induces some properties on the existence of *k-valuations*, that is, functions satisfying (7.52) with equality, and ∞-valuations (functions that are k-valuations for any $k \geqslant 2$). The following is well known in lattice theory (see, e.g., Birkhoff [30, Chap. X]).

Lemma 7.51 *Let L be a lattice. The following holds.*

(i) L is modular if and only if it admits a strictly monotone 2-valuation;
(ii) L is distributive if and only if it admits a strictly monotone 3-valuation;
(iii) L is distributive if and only if it is modular and every strictly monotone 2-valuation is a 3-valuation;
(iv) L is distributive if and only if it is modular and every strictly monotone 2-valuation is an ∞-valuation.

Interestingly, no such restriction holds for isotone and totally monotone functions.

Lemma 7.52 (Barthélemy [18]) *Any lattice L admits an isotone and totally monotone game on L.*

The next result is fundamental because it establishes the link between belief functions on L and totally monotone capacities (which are called *belief measures* in Sect. 2.8.4).

Theorem 7.53 (Barthélemy [18] for the "if" part, Zhou [359] for the "only if") *Let μ be a capacity on a lattice L. Then μ has a nonnegative Möbius transform if and only if μ is totally monotone.*

Properties of Belief Functions

We give some properties of belief functions related to the nonnormalized combination rule $\otimes^*$. It is easily checked that Lemma 7.18 is still valid: if q_1, q_2 are the commonality functions associated to belief functions $\mathrm{Bel}_1, \mathrm{Bel}_2$, then the commonality function q associated to $\mathrm{Bel} = \mathrm{Bel}_1 \otimes^* \mathrm{Bel}_2$ is given by

$$q(x) = q_1(x)q_2(x) \qquad (x \in L). \tag{7.53}$$

More importantly, Theorem 7.19 on the decomposition of a belief function into simple belief functions is still valid, and we give a proof of it in this general case. We restate it in its general form, as well as the definition of a simple belief function.

A *simple belief function* is a belief function whose Möbius transform has the following form:

$$m_{y,\alpha}(x) = \begin{cases} 1-\alpha, & \text{if } x = y \\ \alpha, & \text{if } x = \top \\ 0, & \text{otherwise} \end{cases}, \tag{7.54}$$

for some $y \in L, y \neq \top, \bot$, and $\alpha \in [0, 1[$.

Theorem 7.54 (Decomposition of a belief function into simple belief functions) *(Grabisch [167]) Let* Bel *be a belief function on L such that its Möbius transform m satisfies* $0 < m(\top) < 1$. *Then m can be decomposed as*

$$m = \bigotimes^*_{y \in L \setminus \{\bot,\top\}} m_{y,\alpha_y},$$

with

$$\alpha_y = \prod_{x \geqslant y} q(x)^{-\mu(x,y)} \qquad (y \in L \setminus \{\bot, \top\}),$$

where $\mu(x, y)$ *is the Möbius function of L [see (2.19)].*

Proof We try to find quantities α_y such that

$$m(x) = \bigotimes^*_{y \in L \setminus \{\bot,\top\}} m_{y,\alpha_y}(x) \qquad (x \in L).$$

This expression can be written in terms of the commonality functions using (7.53):

$$q(x) = \prod_{y \in L \setminus \{\bot,\top\}} q_{y,\alpha_y}(x) \qquad (x \in L), \tag{7.55}$$

where q_{y,α_y} is the commonality function induced by m_{y,α_y}:

$$q_{y,\alpha_y}(x) = \begin{cases} 1, & \text{if } x \leqslant y \\ \alpha_y, & \text{otherwise.} \end{cases}$$

From (7.55), we obtain:

$$\log q(x) = \sum_{y \in L \setminus \{\bot, \top\}} \log q_{y,\alpha_y}(x) = \sum_{y \not\geqslant x, y \neq \top} \log \alpha_y$$
$$= \sum_{y \in L \setminus \{\bot, \top\}} \log \alpha_y - \sum_{y \geqslant x} \log \alpha_y,$$

letting $\alpha_\top = \alpha_\bot = 1$. On the other hand,

$$q(\top) = \prod_{y \in L \setminus \{\bot, \top\}} q_{y,\alpha_y}(\top) = \prod_{y \in L \setminus \{\bot, \top\}} \alpha_y.$$

We have supposed that $q(\top) = m(\top) \neq 0$, hence:

$$\log q(x) = \log q(\top) - \sum_{y \geqslant x} \log \alpha_y \qquad (x \in L).$$

We set $Q(x) = \log q(x)$ and $A(y) = \log \alpha_y$ for every $y \in L$. Then the last equality becomes:

$$Q(x) = Q(\top) - \sum_{y \geqslant x} A(y) \qquad (x \in L).$$

If we define $Q'(x) = Q(\top) - Q(x)$, we finally obtain:

$$Q'(x) = \sum_{y \geqslant x} A(y).$$

We recognize here the equation defining the Möbius transform of Q', up to an inversion of the order (dual order) [see (2.17)]. Hence, using (2.18):

$$A(y) = \sum_{x \geqslant y} \mu(x, y) Q'(x)$$

with μ defined by (2.19). Rewriting this with the original notation, we obtain:

$$\log \alpha_y = \sum_{x \geqslant y} \mu(x, y) \big(\log q(\top) - \log q(x) \big).$$

Remarking that $\sum_{x \geqslant y} \mu(x, y) \log q(\top)$ is zero for every $y \neq \bot$, because it corresponds to the Möbius transform of a constant function [Remark 2.72(ii)], we finally get:

$$\alpha_y = \prod_{x \geqslant y} q(x)^{-\mu(x,y)} \qquad (y \in L \setminus \{\bot, \top\}).$$

□

Probability Measures on Distributive Lattices

Recall from Remark 2.19(v) that additive measures are characterized by the fact that they satisfy the total monotonicity property with equality, in other words, they are ∞-valuations. Therefore, it is natural to define probability measures on L by those ∞-valuations P that satisfy $P(\bot) = 0$, $P(\top) = 1$.

If L is distributive (hence modular), by Lemma 7.51(i) and (iv), we see that there always exists an ∞-valuation. The following theorem shows that probability measures are those belief functions on L such that the support of their Möbius transform lives on $\mathcal{J}(L)$, the set of join-irreducible elements. This is a generalization of the classical case $L = 2^X$, where the focal sets of a probability measure are singletons (see Remark 7.6).

Theorem 7.55 (When belief functions are probability measures) *A belief function on a distributive lattice L is a probability measure if and only if the support of its Möbius transform is included in $\mathcal{J}(L)$.*

Proof ⇐) Denoting by Bel the belief function, we have to prove that

$$\mathrm{Bel}\Big(\bigvee_{i=1}^{k} x_i\Big) = \sum_{\substack{I \subseteq \{1,\dots,k\} \\ I \neq \varnothing}} (-1)^{|I|+1} \mathrm{Bel}\Big(\bigwedge_{i \in I} x_i\Big),$$

holds for every family $x_1, \dots, x_k \in L$, $k \geqslant 2$. Using the mapping η on L (Sect. 1.3.2) defined by $\eta(x) = \{j \in \mathcal{J}(L) : j \leqslant x\}$, we have by assumption on m, the Möbius transform of Bel:

$$\begin{aligned}
\mathrm{Bel}\Big(\bigvee_{i=1}^{k} x_i\Big) &= \sum_{j \in \eta\left(\bigvee_{i=1}^{k} x_i\right)} m(j) \\
&= \sum_{j \in \bigcup_{i=1}^{k} \eta(x_i)} m(j)
\end{aligned}$$

$$
\begin{aligned}
&= M\Big(\bigcup_{i=1}^{k} \eta(x_i)\Big) \\
&= \sum_{\substack{I\subseteq\{1,\dots,k\}\\ I\neq\varnothing}} (-1)^{|I|+1} M\Big(\bigcap_{i\in I} \eta(x_i)\Big) \\
&= \sum_{\substack{I\subseteq\{1,\dots,k\}\\ I\neq\varnothing}} (-1)^{|I|+1} \sum_{j\in\bigcap_{i\in I}\eta(x_i)} m(j) \\
&= \sum_{\substack{I\subseteq\{1,\dots,k\}\\ I\neq\varnothing}} (-1)^{|I|+1} \sum_{j\in\eta\left(\bigwedge_{i\in I} x_i\right)} m(j) \\
&= \sum_{\substack{I\subseteq\{1,\dots,k\}\\ I\neq\varnothing}} (-1)^{|I|+1} \mathrm{Bel}\Big(\bigwedge_{i\in I} x_i\Big),
\end{aligned}
$$

where M is an additive measure on 2^L whose distribution is m. The second and sixth equality comes from the distributivity of L, and the fourth one from the fact that M is a ∞-valuation on 2^L.

$\Rightarrow$) Suppose there exists $z \notin \mathcal{J}(L)$ such that $m(z) > 0$, and take a smallest such element. Define Bel^0 on $\downarrow z$ by $\mathrm{Bel}^0(x) = \sum_{j\in\eta(x)} m(j)$ for all $x \leqslant z$. By $\Leftarrow$), and because $\downarrow z$ is a distributive lattice, it follows that Bel^0 is a probability measure on $\downarrow z$. By assumption, letting $\eta(z) = \{j_1,\dots,j_k\}$,

$$
\mathrm{Bel}(z) = \mathrm{Bel}\Big(\bigvee_{i=1}^{k} j_i\Big) = \sum_{\substack{I\subseteq\{1,\dots,k\}\\ I\neq\varnothing}} (-1)^{|I|+1} \mathrm{Bel}\Big(\bigwedge_{i\in I} j_i\Big). \tag{7.56}
$$

On the other hand,

$$
\begin{aligned}
\mathrm{Bel}(z) &= \sum_{x\leqslant z} m(x) = m(z) + \mathrm{Bel}^0(z) \\
&= m(z) + \mathrm{Bel}^0\Big(\bigvee_{i=1}^{k} j_i\Big) \\
&= m(z) + \sum_{\substack{I\subseteq\{1,\dots,k\}\\ I\neq\varnothing}} (-1)^{|I|+1} \mathrm{Bel}^0\Big(\bigwedge_{i\in I} j_i\Big).
\end{aligned} \tag{7.57}
$$

Since $\mathrm{Bel}^0 = \mathrm{Bel}$ on $\downarrow z \setminus \{z\}$, the comparison of (7.56) and (7.57) yields $m(z) = 0$, a contradiction. □

A similar result was proved by Zhou [359], where probability measures are defined as 2-valuations.

Possibility and Necessity Measures

A *necessity measure* on L is a mapping $\text{Nec} : L \to [0, 1]$ satisfying $\text{Nec}(x \wedge y) = \min(\text{Nec}(x), \text{Nec}(y))$ for all $x, y \in L$, and $\text{Nec}(\bot) = 0$, $\text{Nec}(\top) = 1$. Obviously a necessity measure is a normalized capacity.

Theorem 7.38 can be generalized in the following way.

Theorem 7.56 *(Barthélemy [18]) Let μ be a function on L satisfying $\mu(\bot) = 0$ and $\mu(\top) = 1$. Then μ is a necessity measure if and only if μ is a belief function such that* **supp**(m)*, the support of its Möbius transform, is a chain in L.*

The proof given for Theorem 7.38 can be used *mutatis mutandis* to this more general case (the readers can also consult [18], which provides for the "if" part a different proof).

We introduce possibility measures and possibility distributions, which will permit to complement the results of Theorem 7.56. *Possibility measures* on L are defined as conjugate (via a $\vee$-negation) of necessity measures on L, where L is autodual, exactly as plausibility functions are conjugates of belief functions (see also Remark 7.49). The following is easy to show.

Lemma 7.57 *(Grabisch [167]) Let L be an autodual lattice and n be a $\vee$-negation. The following holds.*

(i) For any necessity measure N, its conjugate relative to n, $\Pi = \overline{\text{Nec}}$, satisfies

$$\Pi(x \vee y) = \max(\Pi(x), \Pi(y)) \qquad (x, y \in L). \tag{7.58}$$

(ii) Suppose Π is a mapping on L satisfying $\Pi(\bot) = 0$, $\Pi(\top) = 1$ and (7.58). Then Π is a possibility measure; i.e., the conjugate (relative to n) to a necessity measure Nec *given by* $\text{Nec} = \overline{\Pi}^{-1}$ *(conjugate relative to n^{-1}).*

Due to the above Lemma, one can equivalently define a possibility measure as a mapping on L satisfying $\Pi(\bot) = 0$, $\Pi(\top) = 1$ and (7.58).

Now, distributions arise naturally as the value taken by the measure on the generating elements of the lattice.

Definition 7.58 Let L be an autodual distributive lattice, with $\mathcal{J}(L)$ and $\mathcal{M}(L)$ the set of its join-irreducible and meet-irreducible elements, respectively. The *possibility distribution* of a possibility measure Π on L is the mapping $\pi : \mathcal{J}(L) \to [0, 1]$ defined by $\pi(j) = \Pi(j)$ for all $j \in \mathcal{J}(L)$, and the *necessity distribution* of a necessity measure Nec is the mapping $\nu : \mathcal{M}(L) \to [0, 1]$ defined by $\nu(i) = \text{Nec}(i)$ for all $i \in \mathcal{M}(L)$.

Using the mappings η, λ (Sect. 1.3.2) on L defined by $\eta(x) = \{j \in \mathcal{J}(L) : j \leqslant x\}$ and $\lambda(x) = \{i \in \mathcal{M}(L) : i \geqslant x\}$, we get

$$\Pi(x) = \max(\pi(j) : j \in \eta(x)), \quad \mathrm{Nec}(x) = \min(\nu(i) : i \in \lambda(x)) \qquad (x \in L).$$

Note that by definition of the necessity and possibility measures, we have necessarily $\pi(j) = 1$ for some j and $\nu(i) = 0$ for some m. If Π and Nec are related through conjugation relative to a $\vee$-negation n, π and ν are in bijection and related as follows:

$$\pi(j) = 1 - \mathrm{Nec}(n(j)) = 1 - \nu(i_j) \tag{7.59}$$
$$\nu(i) = 1 - \Pi(n^{-1}(i)) = 1 - \pi(j_i),$$

where $i_j = n(j)$, and $j_i = n^{-1}(i)$. Given a necessity measure Nec, Eq. (7.59) gives immediately the corresponding possibility distribution π. However, unlike the classical case $L = 2^X$, given a possibility distribution π, it is not immediate to find the corresponding chain in L that is the support of the Möbius transform of Nec, the necessity measure induced by π. The following theorem explains how to obtain it.

Theorem 7.59 *(Grabisch [167]) Let L be an autodual and distributive lattice, and n be a $\vee$-negation on L. Let π be a possibility distribution, and assume that the join-irreducible elements of L are numbered such that $\pi(j_1) < \cdots < \pi(j_r) = 1$. Denote by* Nec *the corresponding necessity measure (relative to n). Then the Möbius transform m of* Nec *is given by the following procedure:*

(i) Going from j_r to j_1, at each step $k = r, r-1, \ldots, 2$, select the smallest join-irreducible element ι_k in $\eta(n(j_{k-1})) \setminus \eta(n(j_k))$;
(ii) Then **supp**$(m) = \{\iota_r, \iota_r \vee \iota_{r-1}, \ldots, \iota_r \vee \cdots \vee \iota_2, \top\}$, *and*

$$m(\iota_r \vee \iota_{r-1} \vee \cdots \vee \iota_k) = \pi(j_k) - \pi(j_{k-1}), \quad k = 1, \ldots, r, \tag{7.60}$$

with $\pi(j_0) = 0$.

The proof, which is rather technical, is omitted (see [167]). We put an example from the same reference, to illustrate the whole procedure.

Example 7.60 Let us consider the distributive autodual lattice given on Fig. 7.2. Join-irreducible elements are a, b, c, d, e, f (in red), while meet-irreducible ones

x	$n(x)$	x	$n(x)$
a	α	d	ϵ
b	f	e	δ
c	γ	f	b

are $\alpha, b, \gamma, \delta, \epsilon, f$. We take the following $\vee$-negation: Let us consider a possibility

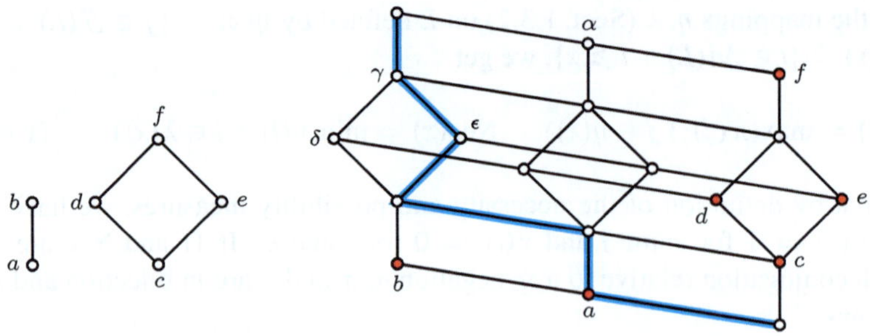

Fig. 7.2 Example of an autodual distributive lattice L (*right*), with $\mathcal{J}(L)$ (*left*)

distribution satisfying

$$\pi(c) < \pi(d) < \pi(e) < \pi(a) < \pi(f) < \pi(b) = 1.$$

Let us apply the procedure of Theorem 7.59. For b, we have $n(b) = f = c \vee d \vee e \vee f$, and for f, we have $n(f) = b = a \vee b$. Hence the first join-irreducible element of the sequence, ι_6, is a. Table 7.1 summarizes all the steps.

Table 7.1 Computation of **supp**(m)

Step k	x	$n(x)$	$\eta(n(x))$	ι_k	Chain
6	b	f	c, d, e, f	a	a
5	f	b	a, b	c	$a \vee c$
4	a	α	a, c, d, e, f	b	$a \vee c \vee b$
3	e	δ	a, b, c, d	e	$a \vee c \vee b \vee e$
2	d	ϵ	a, b, c, e	d	$a \vee c \vee b \vee e \vee d$
1	c	γ	a, b, c, d, e	f	$\top$

The support of m is in blue on Fig. 7.2. We deduce that:

$$\pi(b) = 1$$
$$\pi(f) = 1 - m(a)$$
$$\pi(a) = 1 - m(a) - m(a \vee c)$$
$$\pi(e) = 1 - m(a) - m(a \vee c) - m(a \vee c \vee b)$$
$$\pi(d) = 1 - m(a) - m(a \vee c) - m(a \vee c \vee b) - m(a \vee c \vee b \vee e)$$
$$\pi(c) = 1 - m(a) - m(a \vee c) - m(a \vee c \vee b) - m(a \vee c \vee b \vee e) - m(a \vee c \vee b \vee e \vee d)$$

from which we deduce

$$m(a) = \pi(b) - \pi(f)$$
$$m(a \vee c) = \pi(f) - \pi(a)$$
$$m(a \vee c \vee b) = \pi(a) - \pi(e)$$
$$m(a \vee c \vee b \vee e) = \pi(e) - \pi(d)$$
$$m(a \vee c \vee b \vee e \vee d) = \pi(d) - \pi(c)$$

and $m(\top) = 1 - m(a) - m(a \vee c) - m(a \vee c \vee b) - m(a \vee c \vee b \vee e) - m(a \vee c \vee b \vee e \vee d) = \pi(c)$.

7.8.2 *Infinite Spaces*

The literature on belief functions defined on infinite spaces being rather limited and the topic becoming rapidly complex, we limit ourselves to indicating some literature.

A first important observation is that even if X becomes infinite, nothing changes as far as the support of the mass distribution is finite. Otherwise, the way of proceeding is very close to the usual framework of probability theory; i.e., one endows X with an algebra or a σ-algebra $\mathcal{X}$, which is the set of measurable events. The belief of nonmeasurable events can be computed using the notion of inner and outer measure of classical measure theory (see Kramosil [216, Chap. 9]). Note also that the general theory of random sets is a possible framework.

It is to be noted that Shafer himself provided a framework for belief functions on infinite spaces already in his thesis [295], and also in [297].

An important contribution was done by Rébillé [273], concerning belief functions on compact topological spaces. He generalized in this context the representation of the Choquet integral w.r.t. a belief function as the minimum of means over the core (see (7.22) and Theorem 4.39) and as a mean of minima [Eq. (4.55)]. We mention also the paper of Brüning and Denneberg [40] studying the extreme points of the set of belief functions.

As we have already indicated, a measure-theoretic mathematical analysis of possibility measures was done by de Cooman [70–72].

Lastly, although the topic is different, we again mention Kramosil [216, Chaps. 11 and 14] who studied belief functions taking values in some Boolean algebra, or with nonstandard values.

Appendix A
Tables

A.1 Bases and Transforms of Set Functions

	Transform	Basis
Möbius	$m^{\xi}(S) = \sum_{T \subseteq S} (-1)^{\lvert S \setminus T \rvert} \xi(T)$	$u_T(S) = \begin{cases} 1, & \text{if } S \supseteq T \\ 0, & \text{otherwise} \end{cases}$
Co-Möbius (commonality)	$\check{m}^{\xi}(S) = \sum_{T \supseteq X \setminus S} (-1)^{n-\lvert T \rvert} \xi(T)$	$\check{u}_T(S) = \begin{cases} (-1)^{\lvert T \rvert}, & \text{if } S \cap T = \varnothing \\ 0, & \text{otherwise} \end{cases}$
Conjugate unanimity games	$\overline{U}^{\xi}(S) = (-1)^{\lvert S \rvert + 1} \sum_{T \supseteq X \setminus S} (-1)^{n-\lvert T \rvert} \xi(T)$	$\overline{u_T}(S) = \begin{cases} 1, & \text{if } S \cap T \neq \varnothing \\ 0, & \text{otherwise.} \end{cases}$
Shapley interaction	$I^{\xi}(S) = \sum_{K \subseteq X} \frac{\lvert X \setminus (S \cup K) \rvert ! \lvert K \setminus S \rvert !}{(n-s+1)!} (-1)^{\lvert S \setminus K \rvert} \xi(K)$	$b^I_T(S) = \beta^{\lvert T \rvert}_{\lvert T \cap S \rvert}$
Banzhaf interaction	$I^{\xi}_{\mathrm{B}}(S) = \left(\frac{1}{2}\right)^{n-s} \sum_{K \subseteq X} (-1)^{\lvert S \setminus K \rvert} \xi(K)$	$b^{I_B}_T(S) = \left(\frac{1}{2}\right)^{\lvert T \rvert} (-1)^{\lvert T \setminus S \rvert}$
Fourier	$F^{\xi}(S) = \widehat{\xi}(S) = \frac{1}{2^n} \sum_{K \subseteq X} (-1)^{\lvert S \cap K \rvert} \xi(K)$	$b^F_T(S) = \chi_T(S) = (-1)^{\lvert S \cap T \rvert}$
Walsh	$W^{\xi}(S) = \frac{1}{2^n} \sum_{K \subseteq X} (-1)^{\lvert S \setminus K \rvert} \xi(K)$	$w_T(S) = (-1)^{\lvert T \setminus S \rvert}$
Yokote ($T \neq \varnothing$)	$Y^v(S) = \sum_{L \subseteq X} (-1)^{\lvert S \cap L \rvert + 1} \frac{(n-s-l)!(s+l-1)!}{n!} v(L)$	$\upsilon_T(S) = \begin{cases} 1, & \text{if } \lvert S \cap T \rvert = 1 \\ 0, & \text{otherwise} \end{cases}$

Table A.1 Linear invertible transforms and their associated bases

M. Grabisch, *Set Functions, Games and Capacities in Decision Making*, Theory and Decision Library C 46, DOI 10.1007/978-3-319-30690-2

A.2 Conversion Formulae Between Transforms

We summarize all formulas established in Chap. 2 for passing from one representation of a set function ξ to another. The readers can also consult [178], where all conversion formulae between the Möbius, co-Möbius, interaction and Banzhaf interaction transforms are proved (Tables A.2, A.3, A.4, and A.5).

The superscript $\cdot^{\xi}$ is omitted in $m, \check{m}$, etc. Cardinality of sets are indicated in corresponding small letters. We recall that $|X| = n$.

	ξ	m	$\check{m}$
$\xi(A) =$	$\xi(A)$	$\sum_{B\subseteq A} m(B)$	$\sum_{B\subseteq X\setminus A} (-1)^b \check{m}(B)$
$m(A) =$	$\sum_{B\subseteq A}(-1)^{a-b}\xi(B)$	$m(A)$	$\sum_{B\supseteq A}(-1)^{b-a}\check{m}(B)$
$\check{m}(A) =$	$\sum_{B\supseteq X\setminus A} (-1)^{n-b}\xi(B)$	$\sum_{B\supseteq A} m(B)$	$\check{m}(A)$
$I(A) =$	$\sum_{B\subseteq X} \frac{(-1)^{\lvert A\setminus B\rvert}}{(n-a+1)\binom{n-a}{\lvert B\setminus A\rvert}}\xi(B)$	$\sum_{B\supseteq A} \frac{1}{b-a+1} m(B)$	$\sum_{B\supseteq A} \frac{(-1)^{b-a}}{b-a+1}\check{m}(B)$
$I_{\mathrm{B}}(A) =$	$\left(\frac{1}{2}\right)^{n-a} \sum_{B\subseteq X}(-1)^{\lvert A\setminus B\rvert}\xi(B)$	$\sum_{B\supseteq A}\left(\frac{1}{2}\right)^{b-a} m(B)$	$\sum_{B\supseteq A}\left(-\frac{1}{2}\right)^{b-a}\check{m}(B)$

Table A.2 Conversion formulae between $\xi, m, \check{m}, I$, and I_{B}

	I	I_{B}
$\xi(A) =$	$\sum_{D\subseteq X} \beta^d_{\lvert A\cap D\rvert} I(D)$	$\sum_{B\subseteq X}\left(\frac{1}{2}\right)^b (-1)^{\lvert B\setminus A\rvert} I_{\mathrm{B}}(B)$
$m(A) =$	$\sum_{B\supseteq A} B_{a-b} I(B)$	$\sum_{B\supseteq A}\left(-\frac{1}{2}\right)^{b-a} I_{\mathrm{B}}(B)$
$\check{m}(A) =$	$\sum_{B\supseteq A}(-1)^{b-a}B_{b-a}I(B)$	$\sum_{B\supseteq A}\left(\frac{1}{2}\right)^{b-a} I_{\mathrm{B}}(B)$
$I(A) =$	$I(A)$	$\sum_{B\supseteq A} \frac{1+(-1)^{b-a}}{(b-a+1)2^{b-a+1}} I_{\mathrm{B}}(B)$
$I_{\mathrm{B}}(A) =$	$\sum_{B\supseteq A}\left(\frac{1}{2^{b-a-1}}-1\right)B_{b-a}I(B)$	$I_{\mathrm{B}}(A)$

Table A.3 Conversion formulae between $\xi, m, \check{m}, I$, and I_{B} (ctd')

Relations between the Fourier, Banzhaf interaction and Walsh transforms:

$$\widehat{\xi}(A) = \left(\frac{-1}{2}\right)^a I^{\xi}_{\mathrm{B}}(A) = (-1)^a W(A)$$

	ξ	m	$\widehat{\xi}$
$\xi(A) =$	$\xi(A)$	$\sum_{B\subseteq A} m(B)$	$\sum_{B\subseteq X}(-1)^{\lvert A\cap B\rvert}\widehat{\xi}(B)$
$m(A) =$	$\sum_{B\subseteq A}(-1)^{a-b}\xi(B)$	$m(A)$	$(-2)^a \sum_{B\supseteq A}\widehat{\xi}(B)$
$\widehat{\xi}(A) =$	$\frac{1}{2^n}\sum_{B\subseteq X}(-1)^{\lvert A\cap B\rvert}\xi(B)$	$(-1)^a \sum_{B\supseteq A}\frac{1}{2^b}m(B)$	$\widehat{\xi}(A)$

Table A.4 Conversion formulae between ξ, m and $\widehat{\xi}$

	v	m	Y
$v(A) =$	$v(A)$	$\sum_{B\subseteq A} m(B)$	$\sum_{B:\lvert B\cap A\rvert=1} Y(B)$
$m(A) =$	$\sum_{B\subseteq A}(-1)^{a-b}v(B)$	$m(A)$	$a(-1)^{a+1}\sum_{B\supseteq A} Y(B)$
$Y(A) =$	$\sum_{B\subseteq X}(-1)^{\lvert A\cap B\rvert+1}\frac{(n-a-b)!(a+b-1)!}{n!}v(B)$	$(-1)^{a+1}\sum_{B\supseteq A}\frac{1}{b}m(B)$	$Y(A)$

Table A.5 Conversion formulae between v, m and Y. These formulae are valid for games only

List of Symbols

$\sqsubseteq$	Order relation for coverings	126
$\sqcap$	Intersection of coverings	128
$\sqcup$	Union of coverings	128
$\wedge$	Lattice operation : infimum/minimum	3
$\vee$	Lattice operation : supremum/maximum	3
$\varovee$	Symmetric maximum	201
$\varowedge$	Symmetric minimum	207
$\bigwedge Q$	Infimum of the poset Q	8
$\bigvee Q$	Supremum of the poset Q	8
$\bigcap Q$	Infimum of the poset Q when $\wedge = \cap$	8
$\bigcup Q$	Supremum of the poset Q when $\vee = \cup$	8
$\oplus$	Binary addition	96
$\oplus$	Pseudo-addition	271
$\ominus$	Pseudo-difference	271
$\otimes$	Pseudo-multiplication	270
$\otimes$	Dempster's rule of combination	392
$\otimes^*$	Nonnormalized rule of combination	394
$*$	Convolution product	97
$*$	Concatenation	328
$\succ_{\text{lex}}$	Lexicographic ordering	331
$\succcurlyeq^*$	Quaternary relation	334
$\succcurlyeq_{\text{lmax}}$	Leximax order	373
$\succcurlyeq_{\text{lmin}}$	Leximin order	373
$\prec\!\cdot\,, \succ\!\cdot$	Covering relation	9
$\subset\!\cdot\,, \supset\!\cdot$	Covering relation for sets	9
$\geqslant^{\mu}_{\text{SD}}$	Stochastic dominance w.r.t. μ	211
$\top$	Transposition of matrices and vectors	2

M. Grabisch, *Set Functions, Games and Capacities in Decision Making*,
Theory and Decision Library C 46, DOI 10.1007/978-3-319-30690-2

References

1. M. Abramowitz and I. Stegun, editors. *Handbook of Mathematical Functions*, volume 55 of *Applied Mathematics Series*. National Bureau of Standards, 1972.
2. M. Aigner. *Combinatorial Theory*. Springer Verlag, 1979.
3. E. Algaba. *Extensión de juegos definidos en sistemas de conjuntos*. PhD thesis, Univ. of Seville, Spain, 1998.
4. E. Algaba, J. M. Bilbao, P. Borm, and J. J. López. The position value for union stable systems. *Math. Meth. Oper. Res.*, 52:221–236, 2000.
5. E. Algaba, J. M. Bilbao, P. Borm, and J. J. López. The Myerson value for union stable structures. *Math. Meth. Oper. Res.*, 54:359–371, 2001.
6. E. Algaba, J. M. Bilbao, R. van den Brink, and A. Jiménez-Losada. Cooperative games on antimatroids. *Discrete Mathematics*, 282:1–15, 2004.
7. C. Aliprantis and K. Border. *Infinite Dimensional Analysis*. Springer, 3d edition, 2006.
8. M. Allais. Le comportement de l'homme rationnel devant le risque, critique des postulats et axiomes de l'école américaine. *Econometrica*, 21:503–546, 1953.
9. F. Anscombe and R. Aumann. A definition of subjective probability. *The Annals of Mathematical Statistics*, 34:199–205, 1963.
10. R. J. Aumann and J. H. Drèze. Cooperative games with coalition structures. *Int. J. of Game Theory*, 3:217–237, 1974.
11. R. J. Aumann and L. S. Shapley. *Values of Non-Atomic Games*. Princeton University Press, 1974.
12. Y. Azrieli and E. Lehrer. Extendable cooperative games. *J. Public Economic Theory*, 9:1069–1078, 2007.
13. C. A. Bana e Costa, J.-M. D. Corte, and J.-C. Vansnick. MACBETH. *Int. J. of Information Technology & Decision Making*, 11:359–387, 2012.
14. C. A. Bana e Costa and J.-C. Vansnick. A theoretical framework for Measuring Attractiveness by a Categorical Based Evaluation TecHnique (MACBETH). In *Proc. XIth Int. Conf. on MultiCriteria Decision Making*, pages 15–24, Coimbra, Portugal, August 1994.
15. C. A. Bana e Costa and J.-C. Vansnick. Applications of the MACBETH approach in the framework of an additive aggregation model. *J. of Multicriteria Decision Analysis*, 6:107–114, 1997.
16. C. A. Bana e Costa and J.-C. Vansnick. The MACBETH approach: basic ideas, software and an application. In N. Meskens and M. Roubens, editors, *Advances in Decision Analysis*, pages 131–157. Kluwer Academic Publishers, 1999.
17. J. Banzhaf. Weighted voting doesn't work: A mathematical analysis. *Rutgers Law Review*, 19:317–343, 1965.

M. Grabisch, *Set Functions, Games and Capacities in Decision Making*,
Theory and Decision Library C 46, DOI 10.1007/978-3-319-30690-2

18. J.-P. Barthélemy. Monotone functions on finite lattices: an ordinal approach to capacities, belief and necessity functions. In J. Fodor, B. De Baets, and P. Perny, editors, *Preferences and Decisions under Incomplete Knowledge*, pages 195–208. Physica Verlag, 2000.
19. P. Benvenuti, R. Mesiar, and D. Vivona. Monotone set functions-based integrals. In E. Pap, editor, *Handbook of Measure Theory*, pages 1329–1379. Elsevier Science, 2002.
20. C. Berge. *Principles of Combinatorics*. Academic Press, 1971.
21. J. M. Bilbao. Axioms for the Shapley value on convex geometries. *European Journal of Operational Research*, 110:368–376, 1998.
22. J. M. Bilbao. *Cooperative games on combinatorial structures*. Kluwer Academic Publishers, Boston, 2000.
23. J. M. Bilbao. Cooperative games under augmenting systems. *SIAM J. Discrete Math.*, 17:122–133, 2003.
24. J. M. Bilbao, J. R. Fernandez, A. Jiménez Losada, and E. Lebrón. Bicooperative games. In J. M. Bilbao, editor, *Cooperative games on combinatorial structures*. Kluwer Acad. Publ., 2000.
25. J. M. Bilbao, E. Lebrón, and N. Jiménez. Probabilistic values on convex geometries. *Annals of Operations Research*, 84:79–95, 1998.
26. J. M. Bilbao, E. Lebrón, and N. Jiménez. The core of games on convex geometries. *European Journal of Operational Research*, 119:365–372, 1999.
27. J. M. Bilbao and M. Ordóñez. Axiomatizations of the Shapley value for games on augmenting systems. *Eur. J. of Operational Research*, 196:1008–1014, 2009.
28. A. Billot and J.-F. Thisse. How to share when context matters: The Möbius value as a generalized solution for cooperative games. *J. of Math. Economics*, 41:1007–1029, 2005.
29. G. Birkhoff. On the combination of subalgebras. *Proc. Camb. Phil. Soc.*, 29:441–464, 1933.
30. G. Birkhoff. *Lattice Theory*. American Mathematical Society, 3d edition, 1967.
31. A. K. Biswas, T. Parthasarathy, J. A. M. Potters, and M. Voorneveld. Large cores and exactness. *Games and Economic Behavior*, 28:1–12, 1999.
32. A. K. Biswas, T. Parthasarathy, and G. Ravindran. Stability and largeness of the core. *Games and Economic Behavior*, 34:227–237, 2001.
33. A. Björklund, T. Husfeldt, P. Kaski, and M. Koivisto. Fourier meets Möbius: fast subset convolution. In *Proc. of the 39th ACM Symposium on Theory of Computing*, pages 67–74, New York, NY, USA, 2007.
34. O. Bondareva. Theory of the core in the n-person game. *Vestnik Leningradskii Universitet*, pages 141–142, 1962. in Russian.
35. O. Bondareva. Some applications of linear programming to the theory of cooperative games. *Problemy Kibernetiki*, 10:119–139, 1963. in Russian.
36. E. Boros and P. Hammer. Pseudo-boolean optimization. *Discrete Applied Mathematics*, 2002.
37. D. Bouyssou, T. Marchant, and M. Pirlot. A conjoint measurement approach to the discrete Sugeno integral. In S. J. Brams, W. V. Gehrlein, and F. S. Roberts, editors, *The Mathematics of Preference, Choice and Order: Essays in Honor of Peter C. Fishburn*, pages 85–109. Springer-Verlag, 2009.
38. D. Bouyssou and M. Pirlot. Preferences for multi-attributed alternatives: traces, dominance, and numerical representations. *J. of Mathematical Psychology*, 48:167–185, 2004.
39. D. Bouyssou and M. Pirlot. Conjoint measurement tools for MCDM. In J. Figueira, S. Greco, and M. Ehrgott, editors, *Multiple Criteria Decision Analysis*, pages 73–130. Kluwer Academic Publishers, 2005.
40. M. Brüning and D. Denneberg. The extreme points of the set of belief measures. *Int. J. of Approximate Reasoning*, 48:670–675, 2008.
41. J. T. Cacioppo, W. L. Gardner, and G. G. Berntson. Beyond bipolar conceptualizations and measures: the case of attitudes and evaluative space. *Personality and Social Psychology Review*, 1(1):3–25, 1997.
42. G. Cantor. Beiträge zur Begründung der transfiniten Mengenlehre. *Mathematische Annalen*, 46:481–512, 1895.

43. M. Cardin, M. Couceiro, S. Giove, and J.-L. Marichal. Axiomatizations of signed discrete Choquet integrals. *Int. J. of Uncertainty, Fuzziness and Knowledge-Based Systems*, 19:193–199, 2011.
44. N. Caspard, B. Leclerc, and B. Monjardet. *Finite ordered sets — Concepts, results and uses*. Number 144 in Encyclopedia of Mathematics and Its Applications. Cambridge University Press, 2012.
45. A. Charnes, B. Golany, M. Keane, and J. Rousseau. Extremal principle solutions of games in characteristic function form: core, Chebychev and Shapley value generalizations. In J. Sengupta and G. Kadekodi, editors, *Econometrics of Planning and Efficiency*, pages 123–133. Kluwer Academic Publisher, 1988.
46. A. Chateauneuf. Modeling attitudes towards uncertainty and risk through the use of Choquet integral. *Annals of Operations Research*, 52:3–20, 1994.
47. A. Chateauneuf and M. Cohen. Risk-seeking with diminishing marginal utility in a non-expected utility model. *J. of Risk and Uncertainty*, 9:77–91, 1994.
48. A. Chateauneuf and M. Cohen. Choquet expected utility model: a new approach to individual behavior under uncertainty and to social welfare. In M. Grabisch, T. Murofushi, and M. Sugeno, editors, *Fuzzy Measures and Integrals — Theory and Applications*, pages 289–313. Physica Verlag, 2000.
49. A. Chateauneuf and J.-Y. Jaffray. Some characterizations of lower probabilities and other monotone capacities through the use of Möbius inversion. *Mathematical Social Sciences*, 17:263–283, 1989.
50. A. Chateauneuf and J.-Y. Jaffray. Local Möbius transforms of monotone capacities. In C. Froidevaux and J. Kohlas, editors, *Symbolic and Quantitative Approaches to Reasoning and Uncertainty*, pages 115–124. Springer Verlag, 1995.
51. S. Chew, E. Karni, and Z. Safra. Risk aversion in the theory of expected utility with rank dependent preferences. *J. of Economic Theory*, 42:370–381, 1987.
52. S. H. Chew and P. Wakker. The comonotonic sure-thing principle. *J. of Risk and Uncertainty*, 12:5–27, 1996.
53. G. Choquet. Theory of capacities. *Annales de l'Institut Fourier*, 5:131–295, 1953.
54. V. Chvátal. *Linear Programming*. Freeman and Company, 1983.
55. L. J. Cohen. *The Probable and the Provable*. Clarendon Press, Oxford, 1977.
56. E. Combarro and P. Miranda. On the polytope of non-additive fuzzy measures. *Fuzzy Sets and Systems*, 159:2145–2162, 2008.
57. E. Combarro and P. Miranda. On the structure of the *k*-additive fuzzy measures. *Fuzzy Sets and Systems*, 161:2314–2327, 2010.
58. M. Couceiro and J.-L. Marichal. Axiomatizations of Lovász extensions of pseudo-boolean functions. *Fuzzy Sets and Systems*, 181:28–38, 2011.
59. M. Couceiro and J.-L. Marichal. Axiomatizations of quasi-Lovász extensions of pseudo-boolean functions. *Aequat. Math.*, 82:213–231, 2011.
60. M. Couceiro and J.-L. Marichal. Discrete integrals based on comonotonic modularity. *Axioms*, 3:390–403, 2013.
61. I. Couso and D. Dubois. Statistical reasoning with set-valued information: ontic vs. epistemic views. *Int. J. of Approximate Reasoning*, 55:1502–1518, 2014.
62. I. Couso, D. Dubois, and L. Sánchez. *Random sets and random fuzzy sets as ill-perceived random variables*. Springer, 2014.
63. Y. Crama and P.L. Hammer. *Boolean functions: Theory, Algorithms, and Applications*. Number 142 in Encyclopedia of Mathematics and Its Applications. Cambridge University Press, 2011.
64. Y. Crama, P. Hammer, and R. Holzman. A characterization of a cone of pseudo-Boolean functions via supermodularity-type inequalities. In P. Kall, J. Kohlas, W. Popp, and C. A. Zehnder, editors, *Quantitative Methoden in den Wirtschaftswissenschaften*, pages 53–55. Springer-Verlag, Berlin-Heidelberg, 1989.
65. A. Damasio. *Descartes' Error: Emotion, Reason, and the Human Brain*. G.P. Putman's Sons, New York, 1994.

66. V. I. Danilov and G. A. Koshevoy. Cores of cooperative games, superdifferentials of functions, and the Minkovski difference of sets. *J. of Math. Analysis and Applications*, 247:1–14, 2000.
67. B. A. Davey and H. A. Priestley. *Introduction to Lattices and Orders*. Cambridge University Press, 1990.
68. L. de Campos and M. J. Bolaños. Characterization and comparison of Sugeno and Choquet integrals. *Fuzzy Sets & Systems*, 52:61–67, 1992.
69. L. de Campos, M. T. Lamata, and S. Moral. The concept of conditional fuzzy measure. *Int. J. of Intelligent Systems*, 5:237–246, 1990.
70. G. de Cooman. Possibility theory I: the measure- and integral-theoretic groundwork. *Int. J. of General Systems*, 25:291–323, 1997.
71. G. de Cooman. Possibility theory II: conditional possibility. *Int. J. of General Systems*, 25:325–351, 1997.
72. G. de Cooman. Possibility theory III: possibilistic independence. *Int. J. of General Systems*, 25:353–371, 1997.
73. R. de Wolf. A brief introduction to Fourier analysis on the Boolean cube. *Theory of Computing Library Graduate Surveys*, 1:1–20, 2008.
74. G. Debreu. Representation of a preference ordering by a numerical function. In R. Thrall, C. Coombs, and R. Davis, editors, *Decision Processes*, pages 159–165. J. Wiley, 1954.
75. R. Dedekind. Über Zerlegungen von Zahlen durch ihre größten gemeinsamen Teiler. *Gesammelte Werke*, 2:103–148, 1897.
76. C. Dellacherie. Quelques commentaires sur le prolongement de capacités. In *Séminaire de Probabilités V*, volume 191 of *Lecture Notes in Math.* Springer Verlag, 1971.
77. A. P. Dempster. Upper and lower probabilities induced by a multivalued mapping. *Ann. Math. Statist.*, 38:325–339, 1967.
78. A. P. Dempster. The Dempster-Shafer calculus for statisticians. *Int. J. of Approximate Reasoning*, 48:365–377, 2008.
79. D. Denneberg. Conditioning (updating) non-additive measures. *Annals of Operations Research*, 52:21–42, 1994.
80. D. Denneberg. *Non-Additive Measure and Integral*. Kluwer Academic, 1994.
81. D. Denneberg. Totally monotone core and products of monotone measures. *Int. J. of Approximate Reasoning*, 24:273–281, 2000.
82. D. Denneberg and M. Grabisch. Interaction transform of set functions over a finite set. *Information Sciences*, 121:149–170, 1999.
83. J. Derks. A short proof of the inclusion of the core in the Weber set. *Int. J. of Game Theory*, 21:149–150, 1992.
84. J. Derks and R. Gilles. Hierarchical organization structures and constraints on coalition formation. *Int. J. of Game Theory*, 24:147–163, 1995.
85. J. Derks, H. Haller, and H. Peters. The selectope for cooperative games. *Int. J. of Game Theory*, 29:23–38, 2000.
86. J. Derks and H. Peters. Orderings, excess functions, and the nucleolus. *Mathematical Social Sciences*, 36:175–182, 1998.
87. J. Derks and H. Reijnierse. On the core of a collection of coalitions. *Int. J. of Game Theory*, 27:451–459, 1998.
88. R. Deschamps and L. Gevers. Leximin and utilitarian rules: A joint characterization. *J. of Economic Theory*, 17:143–163, 1978.
89. R. P. Dilworth. Lattices with unique irreducible representations. *Annals of Mathematics*, 41:771–777, 1940.
90. G. Ding, R. Lax, J. Chen, and P. Chen. Formulas for approximating pseudo-boolean random variables. *Discrete Applied Mathematics*, 156:1581–1597, 2008.
91. G. Ding, R. Lax, J. Chen, P. Chen, and B. Marx. Transforms of pseudo-boolean random variables. *Discrete Applied Mathematics*, 158:13–24, 2010.
92. J.-P. Doignon. Threshold representations of multiple semiorders. *SIAM J. on Algebraic and Discrete Methods*, 8:77–84, 1987.

93. I. Dragan. The potential basis and the weighted Shapley value. *Libertas Mathematica*, 11:139–150, 1991.
94. I. Dragan. On the inverse problem for semivalues of cooperative TU games. *International Journal of Pure and Applied Mathematics*, 22(4):539–555, 2005.
95. I. Dragan. The least square values and the Shapley value for cooperative TU games. *TOP*, 14:61–73, 2006.
96. T. Driessen. *Cooperative Games, Solutions and Applications*. Kluwer Academic Publishers, 1988.
97. P. Dubey, A. Neyman, and R. J. Weber. Value theory without efficiency. *Mathematics of Operations Research*, 6:122–128, 1981.
98. D. Dubois and H. Fargier. Making discrete Sugeno integrals more discriminant. *Int. J. of Approximate Reasoning*, 50:880–898, 2009.
99. D. Dubois and E. Hüllermeier. Comparing probability measures using possibility theory: A notion of relative peakedness. *Int. J. of Approximate Reasoning*, 45:364–385, 2007.
100. D. Dubois, J. Lang, and H. Prade. Possibilistic logic. In D. Gabbay, C. Hogger, and J. Robinson, editors, *Handbook of logic in artificial intelligence and logic programming*, volume 3, pages 439–513. Clarendon Press, 1994.
101. D. Dubois and H. Prade. A class of fuzzy measures based on triangular norms. *Int. J. General Systems*, 8:43–61, 1982.
102. D. Dubois and H. Prade. On several representations of an uncertain body of evidence. In M. Gupta and E. Sanchez, editors, *Fuzzy information and decision processes*, pages 167–181. North Holland, 1982.
103. D. Dubois and H. Prade. Unfair coins and necessity measures: towards a possibilistic interpretation of histograms. *Fuzzy Sets & Systems*, 10:15–20, 1983.
104. D. Dubois and H. Prade. A set-theoretic view of belief functions. logical operations and approximations by fuzzy sets. *Int. J. of General Systems*, 12:193–226, 1986.
105. D. Dubois and H. Prade. Weighted minimum and maximum operations in fuzzy set theory. *Information Sciences*, 39:205–210, 1986.
106. D. Dubois and H. Prade. *Possibility Theory*. Plenum Press, 1988.
107. D. Dubois and H. Prade. Fuzzy rules in knowledge-based systems — Modelling gradedness, uncertainty and preference. In R. R. Yager and L. A. Zadeh, editors, *An Introduction to Fuzzy Logic Applications in Intelligent Systems*, pages 45–68. Kluwer Academic Publ., 1992.
108. D. Dubois and H. Prade. When upper probabilities are possibility measures. *Fuzzy Sets and Systems*, 49:65–74, 1992.
109. D. Dubois and H. Prade. Possibility theory: qualitative and quantitative aspects. In D. M. Gabbay and P. Smets, editors, *Handbook of Defeasible Reasoning and Uncertainty Management Systems*, pages 169–226. Kluwer Academic Publishers, 1998.
110. D. Dubois and H. Prade. Possibility theory, probability and fuzzy sets. In D. Dubois and H. Prade, editors, *Fundamentals of Fuzzy Sets*, The Handbooks of Fuzzy Sets Series, pages 343–438. Kluwer Academic, 2000.
111. D. Dubois and H. Prade. An overview of the asymmetric bipolar representation of positive and negative information in possibility theory. *Fuzzy Sets and Systems*, 160:1355–1366, 2009.
112. D. Dubois and H. Prade. Possibilistic logic — an overview. In D. Gabbay, J. H. Siekmann, and J. Woods, editors, *Handbook of the History of Logic*, volume 9, pages 283–342. North Holland, 2014.
113. D. Dubois and H. Prade. Possibility theory and its applications: Where do we stand? In J. Kacprzyk and W. Pedrycz, editors, *Springer Handbook of Computational Intelligence*, pages 31–60. Springer, 2015.
114. D. Dubois, H. Prade, and A. Rico. Representing qualitative capacities as families of possibility measures. *Int. J. of Approximate Reasoning*, 58:3–24, 2015.
115. D. Dubois, H. Prade, and R. Sabbadin. Qualitative decision theory with Sugeno integrals. In M. Grabisch, T. Murofushi, and M. Sugeno, editors, *Fuzzy Measures and Integrals — Theory and Applications*, pages 314–332. Physica Verlag, Heidelberg, 2000.

116. D. Dubois, H. Prade, and R. Sabbadin. Decision-theoretic foundations of qualitative possibility theory. *Eur. J. of Operational Research*, 128:459–478, 2001.
117. D. Dubois, H. Prade, and C. Testemale. Weighted fuzzy pattern matching. *Fuzzy Sets & Systems*, 28:313–331, 1988.
118. F. Durante and C. Sempi. Semicopulæ. *Kybernetika (Prague)*, 41:315–328, 2005.
119. J. S. Dyer. MAUT – Multiattribute utility theory. In J. Figueira, S. Greco, and M. Ehrgott, editors, *Multiple Criteria Decision Analysis*, pages 265–295. Kluwer Academic Publishers, 2005.
120. J. S. Dyer and R. Sarin. Measurable multiattribute value functions. *Operations Research*, 27:810–822, 1979.
121. P. H. Edelman and R. Jamison. The theory of convex geometries. *Geometriae Dedicata*, 19(3):247–270, 1985.
122. J. Edmonds. Submodular functions, matroids, and certain polyhedra. In R. Guy, H. Hanani, N. Sauer, and J. Schönheim, editors, *Proc. of the Calgary Int. Conf. on Combinatorial Structures and Their Applications*, pages 66–87, 1970.
123. W. Edwards. The prediction of decisions among bets. *J. of Experimental Psychology*, 50:201–214, 1955.
124. D. Ellsberg. Risk, ambiguity, and the Savage axioms. *Quart. J. Econom.*, 75:643–669, 1961.
125. A. Estévez-Fernández. New characterizations for largeness of the core. *Games and Economic Behavior*, 76:160–180, 2012.
126. Y. Even and E. Lehrer. Decomposition-integral: Unifying Choquet and the concave integrals. *Economic Theory*, 56:33–58, 2014.
127. R. Fagin and J. Y. Halpern. A new approach to updating beliefs. In *Proc. 6th Conf. on Uncertainty in AI*, 1990. Also in: Uncertainty in Art. Int. 6 (1991) 347–374.
128. U. Faigle. Cores of games with restricted cooperation. *ZOR – Methods and Models of Operations Research*, 33:405–422, 1989.
129. U. Faigle and M. Grabisch. Least square approximations and linear values of cooperative games. arXiv:1601.02831.
130. U. Faigle and M. Grabisch. A discrete Choquet integral for ordered systems. *Fuzzy Sets and Systems*, 168:3–17, 2011. DOI 10.1016/j.fss.2010.10.003.
131. U. Faigle and M. Grabisch. Bases and linear transforms of TU-games and cooperation systems. *Int. J. of Game Theory*, to appear.
132. U. Faigle, M. Grabisch, and M. Heyne. Monge extensions of cooperation and communication structures. *European Journal of Operational Research*, 206:104–110, 2010. 10.1016/j.ejor.2010.01.043.
133. U. Faigle, M. Grabisch, A. Jiménez-Losada, and M. Ordóñez. Games on concepts lattices: Shapley value and the core. *Discrete Applied Mathematics*, 198:29–47, 2016.
134. U. Faigle and W. Kern. The Shapley value for cooperative games under precedence constraints. *Int. J. of Game Theory*, 21:249–266, 1992.
135. U. Faigle and W. Kern. On the core of ordered submodular cost games. *Mathematical Programming*, 87:483–499, 2000.
136. U. Faigle, W. Kern, and G. Still. *Algorithmic Principles of Mathematical Programming*. Springer, Dordrecht, 2002.
137. K. Fan. Entfernung zweier zufälliger Größen und die Konvergenz nach Wahrscheinlichkeit. *Mathematische Zeitschrift*, 49:681–683, 1944.
138. D. Felsenthal and M. Machover. Ternary voting games. *Int. J. of Game Theory*, 26:335–351, 1997.
139. J. Figueira, S. Greco, and M. Ehrgott, editors. *Multiple Criteria Decision Analysis: State of the Art Surveys*. Kluwer Acad. Publ., to appear.
140. G. Fioretti. A mathematical theory of evidence for G. L. S. Shackle. *Mind and Society*, 2:77–98, 2001.
141. P. Fishburn. *Utility Theory for Decision Making*. J. Wiley & Sons, 1970.
142. R. A. Fisher and W. A. Mackenzie. The manurial response of different potato varieties. *J. of Agricultural Science*, 13:311–320, 1923.

143. J. Fodor, J.-L. Marichal, and M. Roubens. Characterization of the ordered weighted averaging operators. *IEEE Tr. on Fuzzy Systems*, 3(2):236–240, 1995.
144. S. Foldes and P. Hammer. Submodularity, supermodularity and higher order monotonicities of pseudo-Boolean functions. *Mathematics of Operations Research*, 30:453–461, 2005.
145. V. Fragnelli, N. Llorca, J. Sánchez-Soriano, S. Tijs, and R. Branzei. Convex games with an infinite number of players and sequencing situations. *J. of Mathematical Analysis and Applications*, 362:200–209, 2010.
146. K. Fujimoto, I. Kojadinovic, and J.-L. Marichal. Axiomatic characterizations of probabilistic and cardinal-probabilistic interaction indices. *Games and Economic Behavior*, 55:72–99, 2006.
147. K. Fujimoto and T. Murofushi. Some characterizations of the systems represented by Choquet and multi-linear functionals through the use of Möbius inversion. *Int. J. of Uncertainty, Fuzziness and Knowledge-Based Systems*, 5:547–561, 1997.
148. K. Fujimoto and T. Murofushi. Some relations among values, interactions, and decomposable structures. *Int. J. of Uncertainty, Fuzziness and Knowledge-Based Systems*, 15:175–191, 2007.
149. S. Fujishige. *Submodular functions and optimization*, volume 58 of *Annals of Discrete Mathematics*. Elsevier, Amsterdam, 2nd edition, 2005.
150. S. Fujishige and N. Tomizawa. A note on submodular functions on distributive lattices. *J. of the Operations Research Society of Japan*, 26:309–318, 1983.
151. M. Ghossoub. Equimeasurable rearrangements with capacities. *Math. of Operations Research*, 40:429–445, 2014.
152. I. Gilboa. Expected utility with purely subjective non-additive probabilities. *J. of Mathematical Economics*, 16(1):65–88, 1987.
153. I. Gilboa. *Theory of Decision under Uncertainty*. Cambridge University Press, New York, NY, 2009.
154. I. Gilboa and D. Schmeidler. Maxmin expected utility with a non-unique prior. *J. of Mathematical Economics*, 18:141–153, 1989.
155. I. Gilboa and D. Schmeidler. Updating ambiguous beliefs. *J. of Economic Theory*, 59:33–49, 1993.
156. I. Gilboa and D. Schmeidler. Additive representations of non-additive measures and the Choquet integral. *Annals of Operations Research*, 52:43–65, 1994.
157. I. Gilboa and D. Schmeidler. Canonical representation of set functions. *Mathematics of Operations Research*, 20:197–212, 1995.
158. D. Gillies. *Some theorems on n-person games*. PhD thesis, Princeton, New Jersey, 1953.
159. I. R. Goodman. Fuzzy sets as equivalence classes of random sets. In R. R. Yager, editor, *Recent Developments in Fuzzy Sets and Possibility Theory*, pages 327–343. Pergamon Press, 1982.
160. I. R. Goodman and H. T. Nguyen. *Uncertainty Models for Knowledge-Based Systems*. North-Holland, 1985.
161. M. Grabisch. The application of fuzzy integrals in multicriteria decision making. *European J. of Operational Research*, 89:445–456, 1996.
162. M. Grabisch. Alternative representations of discrete fuzzy measures for decision making. *Int. J. of Uncertainty, Fuzziness, and Knowledge Based Systems*, 5:587–607, 1997.
163. M. Grabisch. *k*-order additive discrete fuzzy measures and their representation. *Fuzzy Sets and Systems*, 92:167–189, 1997.
164. M. Grabisch. Symmetric and asymmetric fuzzy integrals: the ordinal case. In *6th Int. Conf. on Soft Computing (Iizuka'2000)*, Iizuka, Japan, October 2000.
165. M. Grabisch. The symmetric Sugeno integral. *Fuzzy Sets and Systems*, 139:473–490, 2003.
166. M. Grabisch. The Möbius function on symmetric ordered structures and its application to capacities on finite sets. *Discrete Mathematics*, 287(1-3):17–34, 2004.
167. M. Grabisch. Belief functions on lattices. *Int. J. of Intelligent Systems*, 24:76–95, 2009.
168. M. Grabisch. The lattice of embedded subsets. *Discrete Applied Mathematics*, 158:479–488, 2010. doi: 10.1016/j.dam.2009.10.015.

169. M. Grabisch. Ensuring the boundedness of the core of games with restricted cooperation. *Annals of Operations Research*, 191:137–154, 2011.
170. M. Grabisch. The core of games on ordered structures and graphs. *Annals of Operations Research*, 204:33–64, 2013. doi: 10.1007/s10479-012-1265-4.
171. M. Grabisch and C. Labreuche. Bi-capacities. Part I: definition, Möbius transform and interaction. *Fuzzy Sets and Systems*, 151:211–236, 2005.
172. M. Grabisch and C. Labreuche. Bi-capacities. Part II: the Choquet integral. *Fuzzy Sets and Systems*, 151:237–259, 2005.
173. M. Grabisch and C. Labreuche. Fuzzy measures and integrals in MCDA. In J. Figueira, S. Greco, and M. Ehrgott, editors, *Multiple Criteria Decision Analysis*, pages 563–608. Kluwer Academic Publishers, 2005.
174. M. Grabisch and C. Labreuche. Bipolarization of posets and natural interpolation. *J. of Mathematical Analysis and Applications*, 343:1080–1097, 2008. doi: 10.1016/j.jmaa.2008.02.008.
175. M. Grabisch and C. Labreuche. A decade of application of the Choquet and Sugeno integrals in multi-criteria decision aid. *Annals of Operations Research*, 175:247–286, 2010. doi: 10.1007/s10479-009-0655-8.
176. M. Grabisch and C. Labreuche. A note on the Sobol' indices and interactive criteria. arXiv:1601.02127, 2016.
177. M. Grabisch, J.-L. Marichal, R. Mesiar, and E. Pap. *Aggregation functions*. Number 127 in Encyclopedia of Mathematics and its Applications. Cambridge University Press, 2009.
178. M. Grabisch, J.-L. Marichal, and M. Roubens. Equivalent representations of set functions. *Mathematics of Operations Research*, 25(2):157–178, 2000.
179. M. Grabisch and P. Miranda. Exact bounds of the Möbius inverse of monotone set functions. *Discrete Applied Mathematics*, 186:7–12, 2015.
180. M. Grabisch and M. Roubens. An axiomatic approach to the concept of interaction among players in cooperative games. *Int. Journal of Game Theory*, 28:547–565, 1999.
181. M. Grabisch and P. Sudhölter. The bounded core for games with precedence constraints. *Annals of Operations Research*, 201:251–264, 2012. doi: 10.1007/s10479-012-1228-9.
182. M. Grabisch and P. Sudhölter. On the restricted cores and the bounded core of games on distributive lattices. *Eur. J. of Operational Research*, 235:709–717, 2014. doi: 10.1016/j.ejor.2013.10.027.
183. I. S. Gradshteyn and I. M. Ryzhik. *Table of integrals, series and products*. Academic Press, 7th edition, 2007.
184. S. Graf. A Radon-Nikodym theorem for capacities. *J. für die reine und angew. Math.*, 1980:192–214, 1980.
185. G. Grätzer. *General Lattice Theory*. Birkhäuser, Basel, 2nd edition, 1998.
186. S. Greco, B. Matarazzo, and R. Słowiński. Bipolar Sugeno and Choquet integrals. In *EUROFUSE Workshop on Informations Systems*, pages 191–196, Varenna, Italy, September 2002.
187. S. Greco, B. Matarazzo, and R. Słowiński. Axiomatic characterization of a general utility function and its particular cases in terms of conjoint measurement and rough-set decision rules. *Eur. J. of Operational Research*, 158(2):271–292, 2004.
188. P. R. Halmos. *Measure Theory*. Springer Verlag, 1950.
189. P. Hammer, U. Peled, and S. Sorensen. Pseudo-Boolean functions and game theory. I: core elements and Shapley value. *Cahiers du CERO*, 19:159–176, 1977.
190. P. L. Hammer and R. Holzman. On approximations of pseudo-Boolean functions. *ZOR - Methods and Models of Operations Research*, 36:3–21, 1992.
191. P. L. Hammer and S. Rudeanu. *Boolean Methods in Operations Research and Related Areas*. Springer, 1968.
192. G. H. Hardy, J. E. Littlewood, and G. Pólya. *Inequalities*. Cambridge Univ. Press, Cambridge, 1952.
193. J. C. Harsanyi. A simplified bargaining model for the *n*-person cooperative game. *International Economic Review*, 4:194–220, 1963.

194. G. Herden and G. Mehta. The continuous analogue and generalization of the classical Birkhoff-Milgram theorem. *Math. Social Sciences*, 28:59–66, 1994.
195. J. E. Hirsch. An index to quantify an individual's scientific research output. *Proceedings of the National Academy of Sciences*, 102(46):16569–16572, 2005.
196. W. Hoeffding. A class of statistics with asymptotically normal distribution. *Annals of Mathematical Statistics*, 19:293–325, 1948.
197. A. Honda and M. Grabisch. An axiomatization of entropy of capacities on set systems. *Eur. J. of Operational Research*, 190:526–538, 2008.
198. C. K. Hsee. The evaluability hypothesis: An explanation for preference reversals between joint and separate evaluations of alternatives. *Organizational Behavior and Human Decision Processes*, 67:242–257, 1996.
199. S. Hurst, D. Miller, and J. Muzio. *Spectral techniques in digital logic*. Academic Press, London, 1985.
200. L. Hurwicz. Some specification problems and applications to econometric models. *Econometrica*, 19:343–344, 1951.
201. T. Ichiishi. Super-modularity: applications to convex games and to the greedy algorithm for LP. *J. Econom. Theory*, 25:283–286, 1981.
202. J.-Y. Jaffray. Bayesian updating and belief functions. *IEEE Tr. Syst., Man and Cybern.*, 22:1144–1152, 1992.
203. D. Kahneman and A. Tversky. Prospect theory: an analysis of decision under risk. *Econometrica*, 47:263–291, 1979.
204. A. Kandel and W. Byatt. Fuzzy sets, fuzzy algebra, and fuzzy statistics. *Proc. of the IEEE*, 66:1619–1639, 1978.
205. R. L. Keeney and H. Raiffa. *Decision with Multiple Objectives*. Wiley, New York, 1976.
206. D. G. Kendall. Foundations of a theory of random sets. In E. F. Harding and D. G. Kendall, editors, *Stochastic Geometry*, pages 322–376. J. Wiley, New York, 1974.
207. J. M. Keynes. The general theory of employment. *The Quarterly Journal of Economics*, 51:209–223, 1937.
208. K. Kikuta and L. Shapley. Core stability in n-person games. Unpublished paper, 1986.
209. N. L. Kleinberg and J. H. Weiss. Equivalent n-person games and the null space of the Shapley value. *Mathematics of Operations Research*, 10(2):233–243, 1985.
210. E. P. Klement, R. Mesiar, and E. Pap. *Triangular Norms*. Kluwer Academic Publishers, Dordrecht, 2000.
211. E. P. Klement, R. Mesiar, and E. Pap. A universal integral as common frame for Choquet and Sugeno integral. *IEEE Tr. on Fuzzy Systems*, 18:178–187, 2010.
212. E. P. Klement, R. Mesiar, F. Spizzichino, and A. Stupňanová. Universal integrals based on copulas. *Fuzzy Optimization and Decision Making*, 13:273–286, 2014.
213. F. H. Knight. *Risk, Uncertainty, and Profit*. Houghton Mifflin, Boston and New York, 1921.
214. V. Köbberling and P. Wakker. Preference foundations for nonexpected utility: a generalized and simplified technique. *Mathematics of Operations Research*, 28:395–423, 2003.
215. H. König. *Measure and Integration*. Springer Verlag, 1997.
216. I. Kramosil. *Probabilistic analysis of belief functions*. Kluwer Academic/Plenum Publishers, New York, 2001.
217. D. H. Krantz, R. D. Luce, P. Suppes, and A. Tversky. *Foundations of measurement*, volume 1: Additive and Polynomial Representations. Academic Press, 1971.
218. R. Kruse, E. Schwecke, and J. Heinsohn. *Uncertainty and Vagueness in Knowledge Based Systems*. Springer-Verlag, 1991.
219. C. Labreuche. An axiomatization of the Choquet integral and its utility functions withtout any commensurateness assumption. In *Int. Conf. on Information Processing and Management of Uncertainty in Knowledge-Based Systems (IPMU)*, Catania, Italy, July 2012.
220. C. Labreuche and M. Grabisch. Generalized Choquet-like aggregation functions for handling bipolar scales. *Eur. J. of Operational Research*, 172:931–955, 2006.
221. F. Lange and M. Grabisch. Values on regular games under Kirchhoff's laws. *Mathematical Social Sciences*, 58:322–340, 2009. DOI: 10.1016/j.mathsocsci.2009.07.003.

222. E. Lehrer. A new integral for capacities. *Economic Theory*, 39:157–176, 2009.
223. E. Lehrer and R. Teper. The concave integral over large spaces. *Fuzzy Sets and Systems*, 159:2130–2144, 2008.
224. G. L. Litvinov and V. P. Maslov. Correspondence principle for idempotent calculus and some computer applications. In J. Gunawardena, editor, *Idempotency*, pages 420–443. Cambridge University Press, Cambridge, 1998.
225. G. L. Litvinov, V. P. Maslov, and G. B. Shpiz. Idempotent functional analysis: an algebraic approach. *Mathematical Notes*, 69:696–729, 2001.
226. L. Lovász. Submodular functions and convexity. In A. Bachem, M. Grötschel, and B. Korte, editors, *Mathematical programming. The state of the art*, pages 235–257. Springer Verlag, 1983.
227. J.-L. Marichal. *Aggregation operators for multicriteria decision aid.* PhD thesis, University of Liège, 1998.
228. J.-L. Marichal. On Sugeno integral as an aggregation function. *Fuzzy Sets and Systems*, 114:347–365, 2000.
229. J.-L. Marichal. An axiomatic approach of the discrete Sugeno integral as a tool to aggregate interacting criteria in a qualitative framework. *IEEE Tr. on Fuzzy Systems*, 9(1):164–172, 2001.
230. J.-L. Marichal. Aggregation of interacting criteria by means of the discrete Choquet integral. In T. Calvo, G. Mayor, and R. Mesiar, editors, *Aggregation operators: new trends and applications*, volume 97 of *Studies in Fuzziness and Soft Computing*, pages 224–244. Physica Verlag, 2002.
231. J.-L. Marichal. Weighted lattice polynomials. *Discrete Mathematics*, 309:814–820, 2009.
232. J.-L. Marichal and P. Mathonet. Weighted Banzhaf power and interaction indexes through weighted approximations of games. *Eur. J. of Operations Research*, 211:352–358, 2011.
233. J.-L. Marichal and P. Mathonet. Computing system signatures through reliability functions. *Statistics and Probability Letters*, 83:710–717, 2013.
234. J.-L. Marichal, P. Mathonet, and E. Tousset. Mesures floues définies sur une échelle ordinale. working paper, 1996.
235. M. Marinacci and L. Montrucchio. Introduction to the mathematics of ambiguity. In I. Gilboa, editor, *Uncertainty in economic theory: a collection of essays in honor of David Schmeidler's 65th birthday*. Routledge, New York, 2004.
236. A. Marshall and I. Olkin. *Inequalities: A Theory of Majorization and its Applications*. Academic Press, New York, 1970.
237. G. Matheron. *Random sets and integral geometry*. J. Wiley, New York, 1975.
238. J. Matoušek and B. Gärtner. *Understanding and using linear programming*. Springer, 2007.
239. R. Mesiar. Choquet-like integrals. *J. of Mathematical Analysis and Application*, 194:477–488, 1995.
240. R. Mesiar. k-order pan-additive discrete fuzzy measures. In *7th IFSA World Congress*, pages 488–490, Prague, Czech Republic, June 1997.
241. R. Mesiar and M. Komornikova. Aggregation operators. In *Proc. PRIM'96, 11th Conference on Applied Mathematics*, pages 193–211, 1996.
242. A. N. Milgram. Partially ordered sets, separating systems and inductiveness. In K. Menger, editor, *Reports of a Mathematical Colloquium, Second Series No. 1*. University of Notre-Dame, 1939.
243. P. Miranda, E. Combarro, and P. Gil. Extreme points of some families of non-additive measures. *Eur. J. of Operational Research*, 174:1865–1884, 2006.
244. P. Miranda and M. Grabisch. Optimization issues for fuzzy measures. *Int. J. of Uncertainty, Fuzziness, and Knowledge-Based Systems*, 7(6):545–560, 1999.
245. P. Miranda and M. Grabisch. p-symmetric bi-capacities. *Kybernetika*, 40(4):421–440, 2004.
246. P. Miranda, M. Grabisch, and P. Gil. p-symmetric fuzzy measures. *Int. J. of Uncertainty, Fuzziness, and Knowledge-Based Systems*, 10 (Suppl.):105–123, 2002.
247. S. Moretti and F. Patrone. Transversality of the Shapley value. *TOP*, 16:1–41, 2008.
248. H. Moulin. *Axioms of cooperative decision making*. Wiley, 1988.

249. T. Murofushi. Lexicographic use of Sugeno integrals and monotonicity conditions. *IEEE Tr. on Fuzzy Systems*, 9(6):783–794, 2001.
250. T. Murofushi and M. Sugeno. An interpretation of fuzzy measure and the Choquet integral as an integral with respect to a fuzzy measure. *Fuzzy Sets & Systems*, 29:201–227, 1989.
251. T. Murofushi and M. Sugeno. Fuzzy t-conorm integrals with respect to fuzzy measures : generalization of Sugeno integral and Choquet integral. *Fuzzy Sets & Systems*, 42:57–71, 1991.
252. T. Murofushi and M. Sugeno. A theory of fuzzy measures. Representation, the Choquet integral and null sets. *J. Math. Anal. Appl.*, 159(2):532–549, 1991.
253. T. Murofushi and M. Sugeno. Some quantities represented by the Choquet integral. *Fuzzy Sets & Systems*, 56:229–235, 1993.
254. T. Murofushi and M. Sugeno. Fuzzy measures and fuzzy integrals. In M. Grabisch, T. Murofushi, and M. Sugeno, editors, *Fuzzy Measures and Integrals — Theory and Applications*, pages 3–41. Physica Verlag, 2000.
255. T. Murofushi, M. Sugeno, and M. Machida. Non-monotonic fuzzy measures and the Choquet integral. *Fuzzy Sets and Systems*, 64:73–86, 1994.
256. R. B. Myerson. Graphs and cooperation in games. *Mathematics of Operations Research*, 2:225–229, 1977.
257. Y. Narukawa, V. Torra, and M. Sugeno. Choquet integral with respect to a symmetric fuzzy measure of a function on the real line. *Annals of Operations Research*, 2012. Published on line, Vol. 13.
258. R. O'Donnell. Some topics in analysis of Boolean functions. Technical Report TR08-055, Electronic Colloquium on Computational Complexity, 2008. `http://eccc.hpi-web.de/`.
259. R. O'Donnell. *Analysis of Boolean functions*. Cambridge University Press, 2014.
260. C. Osgood, G. Suci, and P. Tannenbaum. *The measurement of meaning*. University of Illinois Press, Urbana, IL, 1957.
261. G. Owen. Multilinear extensions of games. *Management Sci.*, 18:64–79, 1972.
262. G. Owen. Values of games with a priori unions. In R. Henn and O. Moeschlin, editors, *Essays in Mathematical Economics and Game Theory*, pages 76–88. Springer Verlag, 1977.
263. G. Owen. *Game Theory*. Academic Press, 3d edition, 1995.
264. E. Pap. *Null-Additive Set Functions*. Kluwer Academic, 1995.
265. E. Pap, editor. *Handbook of measure theory*. Elsevier, 2002.
266. B. Peleg. An inductive method for constructing minimal balanced collections of finite sets. *Naval Research Logistics Quarterly*, 12, 1965.
267. B. Peleg and P. Sudhölter. *Introduction to the theory of cooperative games*. Kluwer Academic Publisher, 2003.
268. H. Peters. *Game Theory: A Multilevel Approach*. Springer, 2008.
269. J.-C. Pomerol and S. Barba-Romero. *Multicriterion decision in management: principles and practice*. Kluwer Academic Publishers, 2000.
270. J. Quiggin. *Generalized Expected Utility Theory: the rank-dependent model*. Kluwer Academic, 1993.
271. R. Radner. Satisficing. *J. of Math. Economics*, 2:253–262, 1975.
272. D. Radojevic. The logical representation of the discrete Choquet integral. *Belgian Journal of Operations Research, Statistics and Computer Science*, 38:67–89, 1998.
273. Y. Rébillé. Integral representation of belief measures on compact spaces. *Int. J. of Approximate Reasoning*, 60:37–56, 2015.
274. A. Rico, M. Grabisch, C. Labreuche, and A. Chateauneuf. Preference modelling on totally ordered sets by the Sugeno integral. *Discrete Applied Mathematics*, 147:113–124, 2005.
275. F. S. Roberts. *Measurement Theory*. Addison-Wesley, 1979.
276. J. Rosenmüller and H. G. Weidner. Extreme convex set functions with finite carrier: general theory. *Discrete Mathematics*, 10:343–382, 1974.
277. G. Rota. On the foundations of combinatorial theory I. Theory of Möbius functions. *Zeitschrift für Wahrscheinlichkeitstheorie und Verwandte Gebiete*, 2:340–368, 1964.

278. M. Roubens. Interaction between criteria and definition of weights in MCDA problems. In *44th Meeting of the European Working Group "Multicriteria Aid for Decisions"*, Brussels, Belgium, October 1996.
279. L. M. Ruiz, F. Valenciano, and J. M. Zarzuelo. The least square prenucleolus and the least square nucleolus. two values for TU games based on the excess vector. *Int. J. of Game Theory*, 25:113–134, 1996.
280. L. M. Ruiz, F. Valenciano, and J. M. Zarzuelo. The family of least-square values for transferable utility games. *Games and Economic Behavior*, 24:109–130, 1998.
281. W. Sander and J. Siedekum. Multiplication, distributivity and fuzzy integral I. *Kybernetika*, 41:397–422, 2005.
282. W. Sander and J. Siedekum. Multiplication, distributivity and fuzzy integral II. *Kybernetika*, 41:469–496, 2005.
283. W. Sander and J. Siedekum. Multiplication, distributivity and fuzzy integral III. *Kybernetika*, 41:497–518, 2005.
284. L. J. Savage. *The Foundations of Statistics*. Dover, 2nd edition, 1972.
285. D. Schmeidler. Cores of exact games I. *J. Math. Analysis and Appl.*, 40:214–225, 1972.
286. D. Schmeidler. Integral representation without additivity. *Proc. of the Amer. Math. Soc.*, 97(2):255–261, 1986.
287. D. Schmeidler. Subjective probability and expected utility without additivity. *Econometrica*, 57(3):571–587, 1989.
288. R. Schneider. *Convex bodies: the Brunn-Minkovski theory*. Number 151 in Encyclopedia of Mathematics and its Applications. Cambridge University Press, 2nd edition, 2014.
289. A. Schrijver. A course on combinatorial optimization. http://homepages.cwi.nl/ lex/files/dict.pdf.
290. A. Schrijver. *Theory of linear and integer programming*. J. Wiley & Sons, Chichester, 1986.
291. A. Schrijver. *Combinatorial Optimization*. Springer, 2003.
292. B. Schweizer and A. Sklar. Statistical metric spaces. *Pacific J. Math.*, 10:313–334, 1960.
293. B. Schweizer and A. Sklar. *Probabilistic metric spaces*. Dover Publications, New York, 2005.
294. G. L. Shackle. *Decision, Order and Time in Human Affairs*. Cambridge Univ. Press, 1961.
295. G. Shafer. *Allocations of probability: a theory of partial belief*. PhD thesis, Princeton University, 1973.
296. G. Shafer. *A Mathematical Theory of Evidence*. Princeton Univ. Press, 1976.
297. G. Shafer. Allocations of probability. *The Annals of Probability*, 7:827–839, 1979.
298. L. S. Shapley. A value for n-person games. In H. W. Kuhn and A. W. Tucker, editors, *Contributions to the Theory of Games, Vol. II*, number 28 in Annals of Mathematics Studies, pages 307–317. Princeton University Press, 1953.
299. L. S. Shapley. On balanced sets and cores. *Naval research Logistics Quarterly*, 14:453–460, 1967.
300. L. S. Shapley. On market games. *Journal of Economic Theory*, 1:9–25, 1969.
301. L. S. Shapley. Cores of convex games. *Int. J. Game Theory*, 1:11–26, 1971.
302. W. Sharkey. Cooperative games with large cores. *Int. J. of Game Theory*, 11:175–182, 1982.
303. E. Shellshear and P. Sudhölter. On core stability, vital coalitions, and extendability. *Games and Economic Behavior*, 67:633–644, 2009.
304. N. Shilkret. Maxitive measure and integration. *Nederl. Akad. Wetensch. Proc. Ser. A*, 74:109–116, 1971.
305. H. Simon. Rational choice and the structure of the environment. *Psychological Review*, 63(2):129–138, 1956.
306. H. A. Simon. Theories of bounded rationality. In P. Earl, editor, *The Legacy of Herbert Simon in Economic Analysis*, volume 1. Edward Elgar Publishing Ltd, 2001.
307. I. Singer. Extensions of functions of 0-1 variables and applications to combinatorial optimization. *Numerical Functional Analysis and Optimization*, 7(1):23–62, 1984.
308. P. Slovic, M. Finucane, E. Peters, and D. G. MacGregor. The affect heuristic. In T. Gilovitch, D. Griffin, and D. Kahneman, editors, *Heuristics and biases: the psychology of intuitive judgment*, pages 397–420. Cambridge University Press, 2002.

309. P. Smets. The combination of evidence in the transferable belief model. *IEEE Tr. On Pattern Analysis and Machine Intelligence*, 12(5):447–458, 1990.
310. P. Smets. Constructing the pignistic probability function in a context of uncertainty. In *M. Henrion and R. D. Shachter and K. N. Kanal and J. F. Lemmer*, pages 29–40. Uncertainty in artificial intelligence, 1990.
311. P. Smets. The canonical decomposition of a weighted belief. In *Proc. of the 14th Int. Joint Conf. on Artificial Intelligence (IJCAI'95)*, pages 1896–1901, Montreal, August 1995.
312. P. Smets. Decision making in a context where uncertainty is represented by belief functions. In R. Srivastava and T. J. Mock, editors, *Belief Functions in Business Decisions*, pages 17–61. Physica-Verlag, Heidelberg, Germany, 2002.
313. P. Smets and R. Kennes. The transferable belief model. *Artificial Intelligence*, 66:191–234, 1994.
314. I. M. Sobol'. On sensitivity estimation for nonlinear mathematical models. *Matematicheskoe Modelirovanie*, 2:112–118, 1990. (in Russian).
315. I. M. Sobol'. Sensitivity estimates for nonlinear mathematical models. *Mathematical Modeling and Computational Experiment*, 1:407–414, 1993.
316. R. Stanley. Two poset polytopes. *Discrete and Computational Geometry*, 1:9–23, 1986.
317. M. Studený and T. Kroupa. Core-based criterion for extreme supermodular functions. arXiv:1410.8395v1, 2014.
318. M. Sugeno. Fuzzy measures and fuzzy integrals. *Trans. S.I.C.E. (Keisoku Jidōseigyō Gakkai)*, 8(2):218–226, 1972. In Japanese.
319. M. Sugeno. *Theory of fuzzy integrals and its applications*. PhD thesis, Tokyo Institute of Technology, 1974.
320. M. Sugeno. Fuzzy measures and fuzzy integrals — a survey. In Gupta, Saridis, and Gaines, editors, *Fuzzy Automata and Decision Processes*, pages 89–102. North Holland, 1977.
321. M. Sugeno. A note on derivatives of functions with respect to fuzzy measures. *Fuzzy Sets and Systems*, 222:1–17, 2013.
322. M. Sugeno. A way to Choquet calculus. *IEEE Tr. on Fuzzy Systems*, 23:1439–1457, 2015.
323. M. Sugeno, K. Fujimoto, and T. Murofushi. A hierarchical decomposition of Choquet integral model. *Int. J. of Uncertainty, Fuzziness and Knowledge-Based Systems*, 3(1):1–15, 1995.
324. C. Sundberg and C. Wagner. Generalized finite differences and Bayesian conditioning of Choquet capacities. *Adv. in Applied Math.*, 13:262–272, 1992.
325. K. Takemura. *Behavioral decision theory*. Springer Japan, 2014.
326. N. Tomizawa. Theory of hyperspace (XVI)—on the structure of hedrons. Papers of the Technical Group on Circuits and Systems CAS82-172, Inst. of Electronics and Communications Engineers of Japan, 1983. In Japanese.
327. D. Topkis. Minimizing a submodular function on a lattice. *Operations Research*, 26:305–321, 1978.
328. V. Torra and Y. Narukawa. The *h*-index and the number of citations: two fuzzy integrals. *IEEE Tr. on Fuzzy Systems*, 16:795–797, 2008.
329. A. Tversky and D. Kahneman. Advances in prospect theory: cumulative representation of uncertainty. *J. of Risk and Uncertainty*, 5:297–323, 1992.
330. J. van Gellekom, J. Potters, and J. Reijnierse. Prosperity properties of tu-games. *Int. J. of Game Theory*, 28:211–227, 1999.
331. V. Vassil'ev. Polynomial cores of cooperative games. *Optimizacia*, 21:5–29, 1978. in Russian.
332. G. Vitali. Sulla definizione di integrale delle funzioni di una variabile. *Ann. Pura ed Appl.*, 1925.
333. J. von Neumann and O. Morgenstern. *Theory of Games and Economic Behavior*. Princeton University Press, 2nd edition, 1947.
334. J. Šipoš. Integral with respect to a pre-measure. *Math. Slovaca*, 29:141–155, 1979.
335. P. Wakker. *Additive Representations of Preferences*. Kluwer Academic Publishers, 1989.
336. P. Wakker. Transforming probabilities without violating stochastic dominance. In E. Roskam, editor, *Mathematical Psychology in Progress*, pages 29–47. Springer Verlag, Berlin, 1989.

337. P. Wakker. A behavioral foundation for fuzzy measures. *Fuzzy Sets & Systems*, 37:327–350, 1990.
338. P. Wakker. Unbounded utility for Savage's "foundations of statistics," and other models. *Math. of Operations Research*, 18:446–485, 1993.
339. P. Wakker. *Prospect theory for risk and ambiguity*. Cambridge University Press, New York, NY, 2010.
340. P. Walley. Coherent lower (and upper) probabilities. Technical Report 22, University of Warvick, Coventry, 1981.
341. P. Walley. *Statistical Reasoning with Imprecise Probabilities*. Chapman and Hall, London, 1991.
342. J. Walsh. A closed set of normal orthogonal functions. *American Journal of Mathematics*, 45:5–24, 1923.
343. Z. Wang and G. J. Klir. *Generalized measure theory*. Springer, 2009.
344. R. J. Weber. Probabilistic values for games. In A. E. Roth, editor, *The Shapley Value. Essays in Honor of Lloyd S. Shapley*, pages 101–119. Cambridge University Press, 1988.
345. S. Weber. $\perp$-decomposable measures and integrals for Archimedean t-conorms $\perp$. *J. Math. Anal. Appl.*, 101:114–138, 1984.
346. S. Weber. Two integrals and some modified versions: critical remarks. *Fuzzy Sets and Systems*, 20:97–105, 1986.
347. H. Weisberg. The distribution of linear combinations of order statistics from the uniform distributions. *Ann. Math. Statist.*, 42:704–709, 1971.
348. T. Whalen. Decision making under uncertainty with various assumptions about available information. *IEEE Tr. on Systems, Man and Cybernetics*, 14:888–900, 1984.
349. L. J. Xie and M. Grabisch. The core of games on *k*-regular set systems. *Technical Report, Centre d'Economie de la Sorbonne, 09055*, 2009.
350. M. Yaari. The dual theory of choice under risk. *Econometrica*, 55:95–105, 1987.
351. R. R. Yager. Possibilistic decision making. *IEEE Tr. on Systems, Man and Cybernetics*, 9:388–392, 1979.
352. R. R. Yager. On ordered weighted averaging aggregation operators in multicriteria decision making. *IEEE Trans. Systems, Man & Cybern.*, 18:183–190, 1988.
353. R. R. Yager and L. Liu, editors. *Classic works of the Dempster-Shafer theory of belief functions*. Springer, 2008.
354. K. Yokote. Weak addition invariance and axiomatization of the weighted Shapley value. *Int. J. of Game Theory*, 44:275–293, 2015.
355. K. Yokote, Y. Funaki, and Y. Kamijo. Linear basis to the Shapley value. Technical report, Waseda Economic Working Paper Series, 2013.
356. L. A. Zadeh. Fuzzy sets. *Information and Control*, 8:338–353, 1965.
357. L. A. Zadeh. Fuzzy sets as a basis for a theory of possibility. *Fuzzy Sets & Systems*, 1:3–28, 1978.
358. L. A. Zadeh. On the validity of Dempster's rule of combination of evidence. Technical Report UCB.ERL M79/24, University of California, Berkeley, CA, 1979.
359. C. Zhou. Belief functions on distributive lattices. *Artificial Intelligence*, 201:1–31, 2013.
360. G. Ziegler. *Lectures on polytopes*. Springer Verlag, 1995.
361. S. M. Zumsteg. *Non-cooperative aspects of cooperative game theory and related computational problems*. PhD thesis, Eidgenössische Technische Hochschule Zürich, 1995.

Index

M. Grabisch, *Set Functions, Games and Capacities in Decision Making*,
Theory and Decision Library C 46, DOI 10.1007/978-3-319-30690-2

MIX
Papier aus verantwortungsvollen Quellen
Paper from responsible sources
FSC® C105338

If you have any concerns about our products,
you can contact us on
ProductSafety@springernature.com

In case Publisher is established outside the EU,
the EU authorized representative is:
Springer Nature Customer Service Center GmbH
Europaplatz 3, 69115 Heidelberg, Germany

Printed by Libri Plureos GmbH
in Hamburg, Germany